MW00845651

ENGINEERING APPLICATIONS OF DYNAMICS

ENGINEERING APPLICATIONS OF DYNAMICS

DEAN C. KARNOPP
Department of Mechanical and Aeronautical Engineering
University of California, Davis
Davis, California

DONALD L. MARGOLIS
Department of Mechanical and Aeronautical Engineering
University of California, Davis
Davis, California

JOHN WILEY & SONS, INC.

This book is printed on acid-free paper. ♾

Copyright © 2008 by John Wiley & Sons, Inc. All rights reserved

Published by John Wiley & Sons, Inc., Hoboken, New Jersey
Published simultaneously in Canada

Wiley Bicentennial Logo: Richard J. Pacifico.

No part of this publication may be reproduced, stored in a retrieval system, or transmitted in any form or by any means, electronic, mechanical, photocopying, recording, scanning, or otherwise, except as permitted under Section 107 or 108 of the 1976 United States Copyright Act, without either the prior written permission of the Publisher, or authorization through payment of the appropriate per-copy fee to the Copyright Clearance Center, 222 Rosewood Drive, Danvers, MA 01923, (978) 750-8400, fax (978) 646-8600, or on the web at www.copyright.com. Requests to the Publisher for permission should be addressed to the Permissions Department, John Wiley & Sons, Inc., 111 River Street, Hoboken, NJ 07030, (201) 748-6011, fax (201) 748-6008, or online at www.wiley.com/go/permissions.

Limit of Liability/Disclaimer of Warranty: While the publisher and the author have used their best efforts in preparing this book, they make no representations or warranties with respect to the accuracy or completeness of the contents of this book and specifically disclaim any implied warranties of merchantability or fitness for a particular purpose. No warranty may be created or extended by sales representatives or written sales materials. The advice and strategies contained herein may not be suitable for your situation. You should consult with a professional where appropriate. Neither the publisher nor the author shall be liable for any loss of profit or any other commercial damages, including but not limited to special, incidental, consequential, or other damages.

For general information about our other products and services, please contact our Customer Care Department within the United States at (800) 762-2974, outside the United States at (317) 572-3993 or fax (317) 572-4002.

Wiley also publishes its books in a variety of electronic formats. Some content that appears in print may not be available in electronic books. For more information about Wiley products, visit our web site at www.wiley.com.

Library of Congress Cataloging-in-Publication Data:
Karnopp, Dean.
Engineering applications of dynamics/Dean C. Karnopp and Donald L. Margolis.
p. cm.
Includes index.
ISBN 978-0-470-11266-3 (cloth)
1. Structural dynamics. 2. Structural engineering. I. Margolis, Donald L.
II. Title.
TA654. K26 2008
620.1'04–dc22 2007017040

Printed in the United States of America

10 9 8 7 6 5 4 3 2 1

CONTENTS

PREFACE

Mark Twain is alleged to have said "We all grumble about the weather, but nothing is done about it." [1] This quote describes the way a number of instructors feel about the state of undergraduate and even graduate engineering education in the classical field of dynamics. The grumbling of dynamics instructors has to do with the perceived failure of introductory dynamics courses to produce students capable of applying dynamic principles successfully in subsequent courses and in practice. This book is our attempt to do something about the problem.

Dynamics is a subject of obvious importance for mechanical and aeronautical engineering students and it is important for students in many other fields that have to do with things that move. Most popular introductory textbooks on dynamics are all very similar—lengthy and with three-colored illustrations—and covering a standard set of topics. Students who have taken a one-quarter or one-semester course using one of these texts have been exposed to a number of rather difficult topics dealing mainly with plane motion including kinematics and kinetics of particles and rigid bodies. Sometimes a short introduction to vibration theory is included.

In many curricula, the study of dynamics is a required course for all students and thus there will always be some who do not like dynamics or who have no particular talent for it, but the problem is that even interested students have difficulty imagining just what use dynamics has in engineering practice. There are many excellent books on intermediate or advanced dynamics that concentrate on advanced theoretical treatments of dynamics. But they too do not necessarily inspire engineering students who might be interested in the *usefulness* of dynamics.

[1] www.twainquotes.com

Another point is that many books on dynamics do not pay much attention to the fact that computers are now capable of solving nonlinear differential equations rapidly and with little effort or cost. This means that the formulation of the equations of motion for dynamic systems is more important than it was in the past since computer simulation can significantly aid in the design and optimization of any number of devices. At one time dynamics texts focused on simplified dynamic systems that had analytical solutions or on linearized equations but now nonlinear equations of motion can be solved readily.

The present book is our attempt to address these issues. We assume that the users of this book have a basic understanding of the principles of dynamics, but have not had the opportunity of applying these principles to any systems related to actual engineering practice. We emphasize the derivation of equations of motion, the analysis of the system response when this is possible, and the manipulation of the equations into a form for use with any suitable computer simulation program.

Our main goal is to introduce engineering students who are potentially interested in dynamics to a wide range of interesting applications that exist in practice. In doing so we hope to inspire at least some of them to regard dynamics as a fascinating field of interest rather than just a hurdle to be surmounted on the way to a degree.

It may be worth emphasizing some things this book does not attempt to do. One thing we do not do in this book is give long derivations of principles which should have been discussed in an introductory dynamics course, although theoretical results are restated and reviewed.

At the risk of our academic reputation we have even resisted the impulse to give a derivation of Lagrange's equations even though we do not expect that the students will necessarily have seen such a derivation previously. It is our feeling that students can learn to use the equations effectively without studying the rather tedious derivations. Many excellent books provide the derivations if the student wishes to study them. Furthermore, we emphasize the interpretation of the terms in Lagrange equations of motion using the principles of Newtonian mechanics. In our experience, students who have two independent ways to analyze a dynamic system are much more likely to find their mistakes than those who have studied only one formulation method.

We are aware that we may be criticized for promoting a "cookbook" approach to a sophisticated intellectual topic. Our experience has been that most students are better off learning how to get results first. Later, if their interest has been piqued, they may wish to deepen their understanding by further study of the underlying theory.

We also have resisted the idea of using programs that automate the simulation of mechanical systems. Our feeling is that it is better to formulate equations of motion, transform them into a form suitable for computer solution and then to experiment with input and parameter variation. We realize that this means that each instructor will have to make up problems and recommend equation solvers if the students are to have a simulation laboratory experience.

It is our feeling that the automated simulation programs for dynamics are not as helpful for promoting understanding of problems in dynamics as the experience of formulating equations of motion by hand and then turning to an equation solver to study the system response. Furthermore, the most automated dynamics programs continually evolve and any particular program may well prove to be out of date in just a few years time.

We have provided a comprehensive solutions manual for the problems for use by instructors. Several problems extend the chapter discussions significantly and provide the possibility of customizing the course to suit the individual instructor's interests.

It is our fervent hope that this book will help to revitalize interest in one of the most intellectually satisfying areas of engineering. Although we are Americans and feel at home with style of the opening quote from the quintessential American Mark Twain, we close with a flowery quote from a Hungarian-German-American-Irishman, Cornelius Lanczos.[2] He was describing analytical mechanics, a field that leads to Lagrange's equations for dynamics, but it shows the intellectual enthusiasm that a subject like engineering dynamics can engender even among those who are fundamentally interested in applications.

> The variational principles of mechanics are firmly rooted in that great century of Liberalism which starts with Descartes and ends with the French Revolution and which has witnessed the lives of Leibniz, Spinoza, Goethe and Johann Sebastian Bach. It is the only period of cosmic thinking in the entire history of Europe since the time of the Greeks. If the author has succeeded in conveying an inkling of that cosmic spirit, his efforts will be amply rewarded.

Our aims are perhaps more modest than those of Lanczos, but we do aim to inspire students to appreciate the beauty as well as the usefulness of the subject of dynamics.

DEAN C. KARNOPP
DONALD L. MARGOLIS

Davis, California
June, 2007

[2]Lanczos, C. *The Variational Principles of Mechanics,* The University of Toronto Press, Toronto, Canada, 1949.

1

NEWTON'S LAWS FOR PARTICLES AND RIGID BODIES

In the basic course in dynamics you learned about various coordinate frames and how to express displacement, velocity, and acceleration with respect to these different frames of reference (see, e.g., [1]). You also covered Newton's laws and learned the basics of relating forces to acceleration when the system motion was prescribed and calculations were made for particular instantaneous system configurations.

In your first exposure to engineering dynamics it was appropriate to separate the concepts of kinematics and kinetics, the former dealing with how physical elements fit and move together and the latter dealing with the forces required for the prescribed motion to occur. A further distinction was to treat the motion of single particles as separate from that of rigid bodies. Here we treat kinematics and kinetics as simply all part of the dynamic system analysis, and particle and rigid, body mechanics are treated together. You will see that there really are only a few principles that govern the dynamics of all systems, and that, dynamically, all systems are pretty much the same.

The emphasis in this chapter is on choosing appropriate coordinates for deriving complete equations of motion such that if the resulting equations are solved, the entire motion–time history of the system can be predicted, including any special configurations of interest. The resulting equations of motion for most engineering systems will be nonlinear, and thus solutions will be obtained using simulation. This topic is covered in later sections.

We start by reviewing the most common coordinate frames and learn how to express position, velocity, and acceleration with these coordinates. This is followed by a discussion of free-body diagrams, where internal forces are exposed acting on the inertial components of the system, and how Newton's laws are used to derive the equations of motion. We then show the increased complexity of working with rigid bodies rather than particles. In a later chapter you will see how to solve equations of motion using time-step simulation.

1.1 NEWTON'S SECOND LAW

The fundamental principle upon which dynamics is based is *Newton's second law,* which states that force is equal to mass times acceleration:

$$\mathbf{F} = m\mathbf{a} \tag{1.1}$$

where the boldface notation indicates vector or matrix quantities. Equation (1.1) simply states that the force acting in some direction on a mass particle causes an acceleration of that particle in the same direction as the force. It is implied that the acceleration is measured with respect to an inertial reference frame, and for most engineering problems the inertial reference frame is attached to the Earth. The proportionality constant is the particle mass expressed in kilograms (kg). We use the SI system of units in this book, where force is measured in newtons (N), length is measured in meters (m), and time is measured in seconds (s).

You should remember that for a particle currently at position $\mathbf{r}$ moving at velocity $\mathbf{v}$ and having acceleration $\mathbf{a}$, the position, velocity, and acceleration are related by

$$\begin{aligned} \mathbf{v} &= \frac{d}{dt}\mathbf{r} = \dot{\mathbf{r}} \\ \mathbf{a} &= \frac{d}{dt}\mathbf{v} = \dot{\mathbf{v}} = \ddot{\mathbf{r}} \end{aligned} \tag{1.2}$$

or

$$\begin{aligned} \mathbf{v} &= \int \mathbf{a}\, dt \\ \mathbf{r} &= \int \mathbf{v}\, dt \end{aligned} \tag{1.3}$$

Note in Eq. (1.2) that we introduce the "dot" notation for derivatives. This is used interchangeably with $\frac{d}{dt}$ throughout the book.

Although Eq. (1.1) is often seen emblazoned on posters and T-shirts, it is not very applicable in the form shown. To derive equations of motion that lead to predictions of system response, we must be able to express acceleration in terms of position and velocity coordinates. Next we describe the most common coordinates to use for this purpose.

1.2 COORDINATE FRAMES AND VELOCITY AND ACCELERATION DIAGRAMS

In this section the most common coordinate frames for expressing position, velocity, and acceleration are shown.

Rectangular Coordinates

Figure 1.1*a* shows an inertial reference frame and a particle traveling along some path. At the instant shown, the particle has a position vector **r**, velocity vector **v**, and acceleration vector **a**. The velocity vector is aligned tangent to the path of the particle, but the acceleration vector could be in any direction. In Figure 1.1*b* the general vector representation has been removed and the perpendicular components of the vectors are

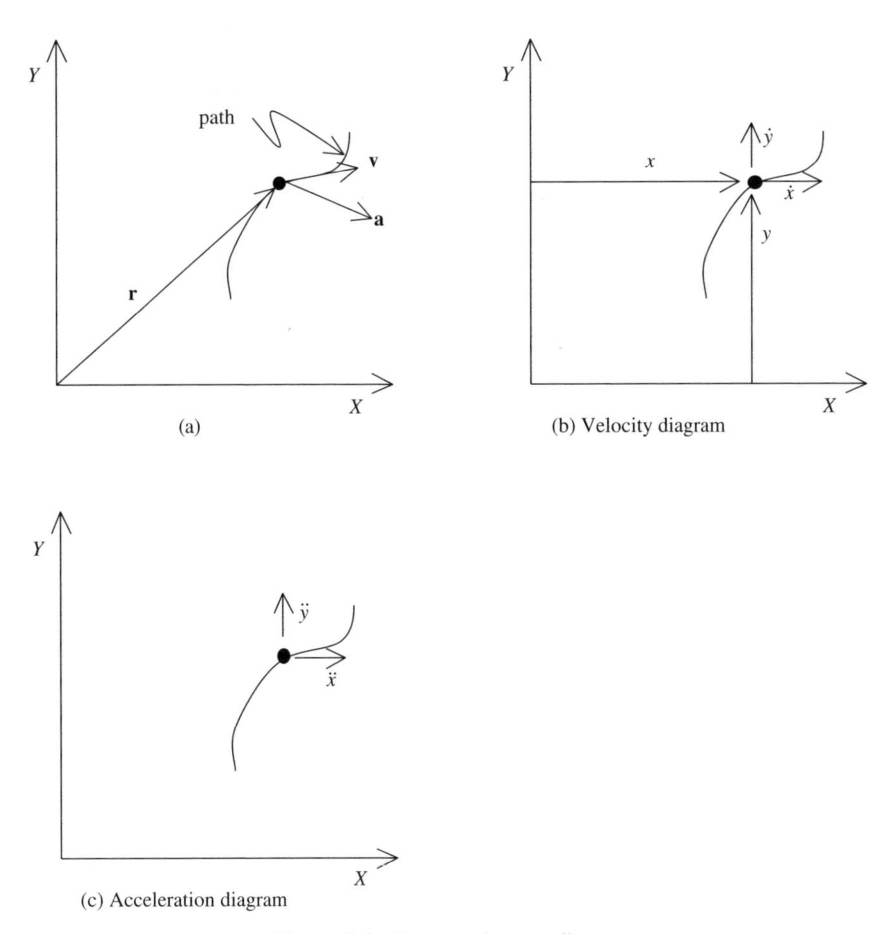

Figure 1.1 Rectangular coordinates.

shown aligned in the inertial X and Y directions. At this point we do not know whether the particle is moving right or left or up or down. We do not know in which direction the acceleration vector is actually pointing. So we choose the directions that we define as positive and point the components in their assumed positive directions. When we ultimately derive the equations of motion for a particular system, the solution will tell us in which direction the system parts are actually moving. In Figure 1.1*b* the velocity components are pointing in the positive X and Y directions. These are the rectangular components of the velocity vector with respect to the inertial X,Y frame and they are displayed on a *velocity diagram*. Figure 1.1*c* shows the components of the acceleration vector displayed on an *acceleration diagram*.

Polar Coordinates

With polar coordinates we start with the position vector shown in Figure 1.1*a* and use the unit vectors $\boldsymbol{\varepsilon}_r$ and $\boldsymbol{\varepsilon}_\theta$ as indicated in Figure 1.2*a*. The unit vector $\boldsymbol{\varepsilon}_r$ is pointing in the positive r-direction and the unit vector $\boldsymbol{\varepsilon}_\theta$ is perpendicular to the r-direction. Polar coordinates are less obvious than rectangular coordinates because when we take the time rate of change of the position vector to obtain the velocity, it is necessary to account for the change in direction of the position vector. In rectangular coordinates this is not necessary. The x-direction component of velocity is always pointing in the x-direction. Its length can change, but its direction never does.

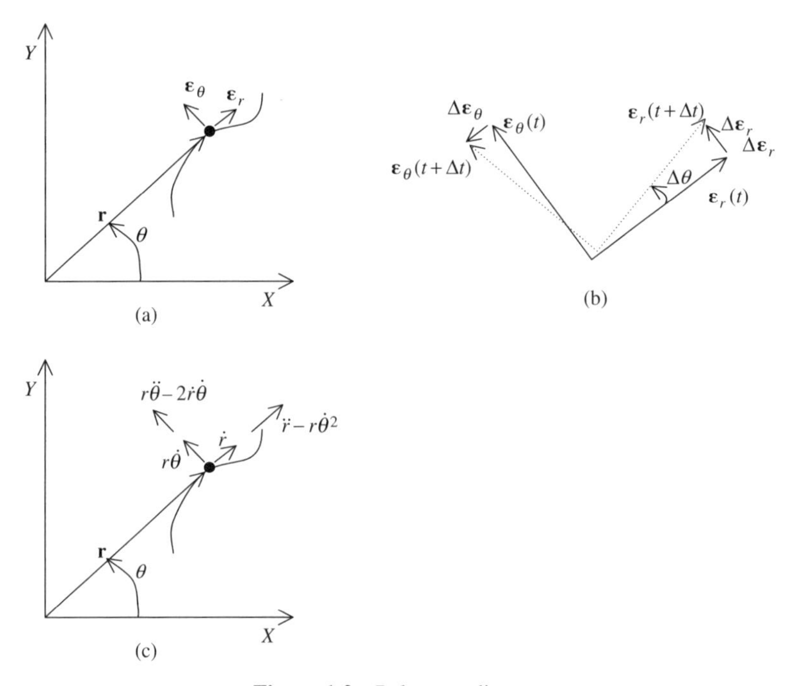

Figure 1.2 Polar coordinates.

Figure 1.2*b* shows an expanded view of the unit vectors at time t and their orientation Δt seconds later. To determine the change of the unit vectors over this time period, we need

$$\begin{aligned} \Delta\boldsymbol{\varepsilon}_r &= \frac{\boldsymbol{\varepsilon}_r(t+\Delta t) - \boldsymbol{\varepsilon}_r(t)}{\Delta t} \\ \Delta\boldsymbol{\varepsilon}_\theta &= \frac{\boldsymbol{\varepsilon}_\theta(t+\Delta t) - \boldsymbol{\varepsilon}_\theta(t)}{\Delta t} \end{aligned} \tag{1.4}$$

and these vectors are shown in Figure 1.2*b*. The reader should remember that the difference between two vectors, say $\mathbf{A} - \mathbf{B}$, is the vector drawn between their tips, from $\mathbf{B}$ pointing toward $\mathbf{A}$. From Figure 1.2*b* it is seen that $\Delta\boldsymbol{\varepsilon}_r$ is pointing in the $\boldsymbol{\varepsilon}_\theta$-direction and that $\Delta\boldsymbol{\varepsilon}_\theta$ is pointing in the negative $\boldsymbol{\varepsilon}_r$-direction. Since $\Delta\theta$ is very small, the magnitudes of $\Delta\boldsymbol{\varepsilon}_r$ and $\Delta\boldsymbol{\varepsilon}_\theta$ are very nearly equal to the magnitude of $\boldsymbol{\varepsilon}_r$ or $\boldsymbol{\varepsilon}_\theta$ (which is unity by definition), times the angle that has been swept through, $\Delta\theta$. Thus,

$$\begin{aligned} \frac{\Delta\boldsymbol{\varepsilon}_r}{\Delta t} &\rightarrow 1\,\frac{\Delta\theta}{\Delta t}\,\boldsymbol{\varepsilon}_\theta \\ \frac{\Delta\boldsymbol{\varepsilon}_\theta}{\Delta t} &\rightarrow -1\,\frac{\Delta\theta}{\Delta t}\,\boldsymbol{\varepsilon}_r \end{aligned} \tag{1.5}$$

and the final result is

$$\begin{aligned} \dot{\boldsymbol{\varepsilon}}_r &= \dot{\theta}\boldsymbol{\varepsilon}_\theta \\ \dot{\boldsymbol{\varepsilon}}_\theta &= \dot{\theta}\boldsymbol{\varepsilon}_r \end{aligned} \tag{1.6}$$

Later we will see that the change in these unit vectors can be expressed as a cross-product,

$$\begin{aligned} \dot{\boldsymbol{\varepsilon}}_r &= \boldsymbol{\omega} \times \boldsymbol{\varepsilon}_r \\ \dot{\boldsymbol{\varepsilon}}_\theta &= \boldsymbol{\omega} \times \boldsymbol{\varepsilon}_\theta \end{aligned} \tag{1.7}$$

where the angular velocity vector $\boldsymbol{\omega}$ has length $\dot{\theta}$ and direction given by the right-hand rule, where the fingers of your right hand are oriented in the $\mathbf{r}$-direction and swept in the direction of defined positive angular motion. The direction of $\boldsymbol{\omega}$ is the direction your thumb is pointing after sweeping your fingers appropriately. For the case of Figure 1.2, $\boldsymbol{\omega}$ is pointing toward the reader, out of the page. We would call this the $\boldsymbol{\varepsilon}_z$-direction.

Starting with the position vector

$$\mathbf{r} = r\boldsymbol{\varepsilon}_r \tag{1.8}$$

the velocity vector is determined from

$$\dot{\mathbf{r}} = \dot{r}\boldsymbol{\varepsilon}_r + r\dot{\boldsymbol{\varepsilon}}_r = \dot{r}\boldsymbol{\varepsilon}_r + r\dot{\theta}\boldsymbol{\varepsilon}_\theta \tag{1.9}$$

and the acceleration vector is determined by differentiating again:

$$\ddot{\mathbf{r}} = \ddot{r}\boldsymbol{\varepsilon}_r + \dot{r}\dot{\boldsymbol{\varepsilon}}_r + \dot{r}\dot{\theta}\boldsymbol{\varepsilon}_\theta + r\ddot{\theta}\boldsymbol{\varepsilon}_\theta + r\dot{\theta}\dot{\boldsymbol{\varepsilon}}_\theta \tag{1.10}$$

which becomes the final result when Eqs. (1.6) are substituted; thus,

$$\ddot{r} = \ddot{r}\boldsymbol{\varepsilon}_r + \dot{r}\dot{\theta}\boldsymbol{\varepsilon}_\theta + \dot{r}\dot{\theta}\boldsymbol{\varepsilon}_\theta + r\ddot{\theta}\boldsymbol{\varepsilon}_\theta - r\dot{\theta}^2\boldsymbol{\varepsilon}_r \quad \text{or} \quad \ddot{\mathbf{r}} = (\ddot{r} - r\dot{\theta}^2)\boldsymbol{\varepsilon}_r + (r\ddot{\theta} + 2\dot{r}\dot{\theta})\boldsymbol{\varepsilon}_\theta \tag{1.11}$$

The components of velocity and acceleration in polar coordinates are shown in Figure 1.2*c*.

The reader may wonder why one would ever use polar coordinates over rectangular coordinates since polar components seem so much more complicated. We demonstrate through examples that there are cases when polar coordinates are far more convenient than rectangular coordinates for setting up and deriving equations of motion. For now it is worthwhile to remember that velocity components come from Eq. (1.9) and acceleration components from Eq. (1.11) when using polar coordinates.

As an example of the use of rectangular and polar coordinates, consider the system in Figure 1.3*a*. It consists of a spring and a mass pendulum in a gravity field. It is desired to derive the equations of motion that, if solved, would predict the motion–time history of this system from any set of initial conditions. We first need to show acceleration and force components so that Newton's laws can be used to derive equations. The position components in both coordinate frames are shown in part *a* of the figure. The velocity components and acceleration components in rectangular coordinates are shown in Figure 1.3*b*, and the velocity and acceleration components in polar coordinates are shown in part (*c*). As introduced above and indicated in Figure 1.3, it is convenient to show velocity components using arrows and symbols on a velocity diagram. Similarly, we show the acceleration components on an acceleration diagram. When these diagrams are coupled with a force diagram we will discover that it is very straightforward to derive the equations of motion directly using Newton's law.

It should be noted that both coordinate representations are describing exactly the same motion. At this instant in time, the mass particle has some absolute velocity vector and absolute acceleration vector measured with respect to the inertial frame. Rectangular coordinates represent these vectors with mutually perpendicular components in the X and Y directions, and polar coordinates use mutually perpendicular components in the r and θ directions. If the respective components from each description were added vectorially, the result would be the instantaneous absolute velocity and acceleration of the particle.

Another example of choosing coordinates is the system of Figure 1.4*a*. In this system the spring and mass pendulum are connected to a cart of mass m_c and the cart is attached to ground through a spring k_c. An external force $F(t)$ is prescribed on the cart. Our ultimate desire is to derive equations of motion which, if solved, would predict the motion–time history of this system. For now we just want to choose

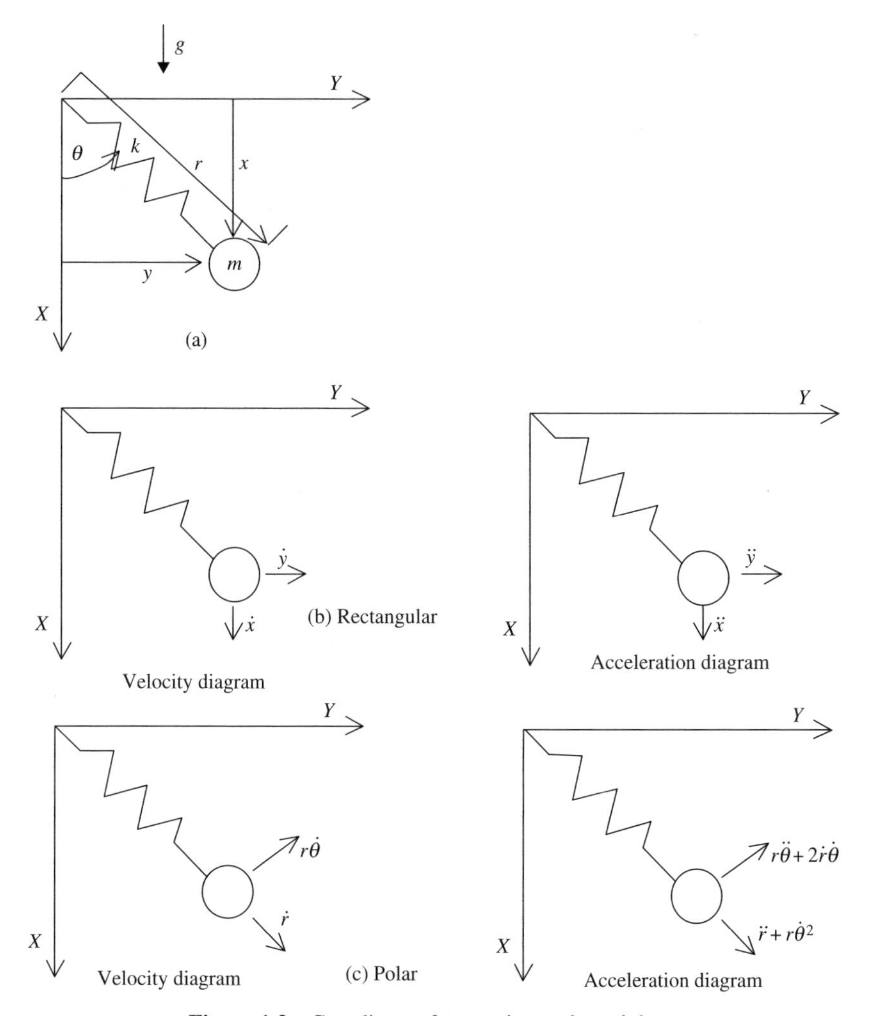

Figure 1.3 Coordinates for a spring and pendulum.

reasonable coordinates with which we can conveniently express the velocity and acceleration of the inertial components.

In Figure 1.4*b* rectangular coordinates are used for both mass elements. We only need one component for the cart, as it is constrained to move only horizontally. We need both the x and y components for the pendulum mass since it can move in the entire plane. In part c of the figure, rectangular coordinates are used for the cart and polar coordinates are used for the pendulum mass. The velocity and acceleration components are indicated. In addition to the polar coordinate components, the velocity and acceleration components of the cart have been transferred to the pendulum mass. This is required because the polar coordinates are defined with respect to a frame

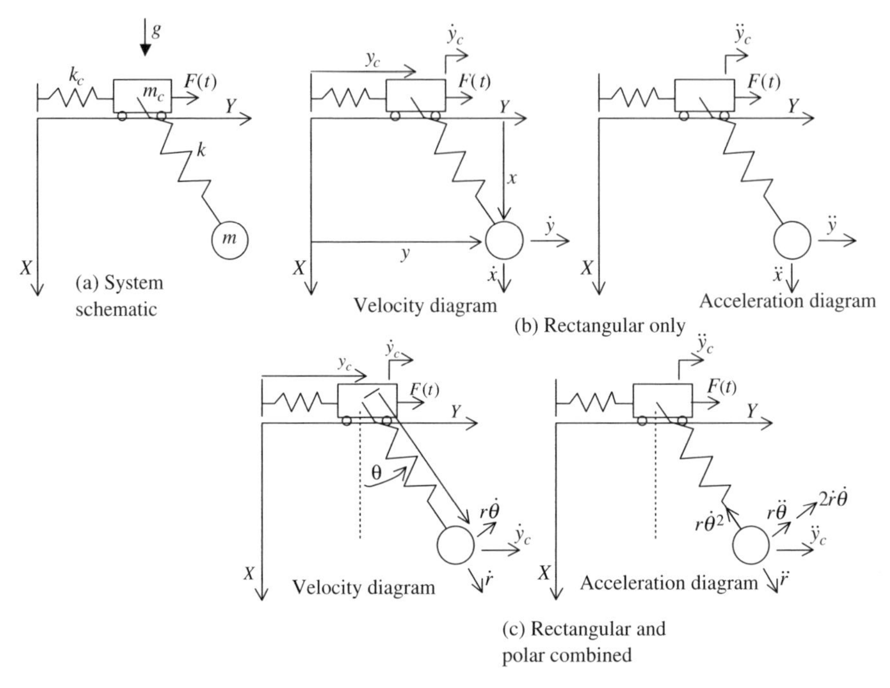

Figure 1.4 Coordinates for spring and pendulum attached to a cart.

that is attached to the cart, and we must account for the fact that the pendulum mass experiences the motion of the cart plus additional motions described by the polar coordinates. In other words, if r and θ were fixed and not permitted to change, it is pretty obvious that the pendulum mass would move identically to the cart. Later we formalize this idea of transferring velocity and acceleration components from one location to another. For now it is simply stated that when the equations of motion are derived for the system of Figure 1.4, the reader will discover that it is much more convenient to use combined coordinates rather than rectangular components only.

Coordinate Choice and Degrees-of-Freedom

A major emphasis of this book is the derivation of equations of motion that allow prediction of the response of a system. These equations will typically be nonlinear differential equations in terms of the coordinates chosen. Choosing the most convenient coordinates for large, complex problems is not an easy task. Sometimes there are multiple choices for appropriate coordinates, and some choices are better than others (meaning easier to work with), but prior to working on the problem, it is difficult to know which choice is better. For smaller systems, as demonstrated primarily in this book, choosing coordinates is not so hard.

We typically need a position coordinate for each direction in which a mass particle can move, which can include an angular position coordinate, and we need a

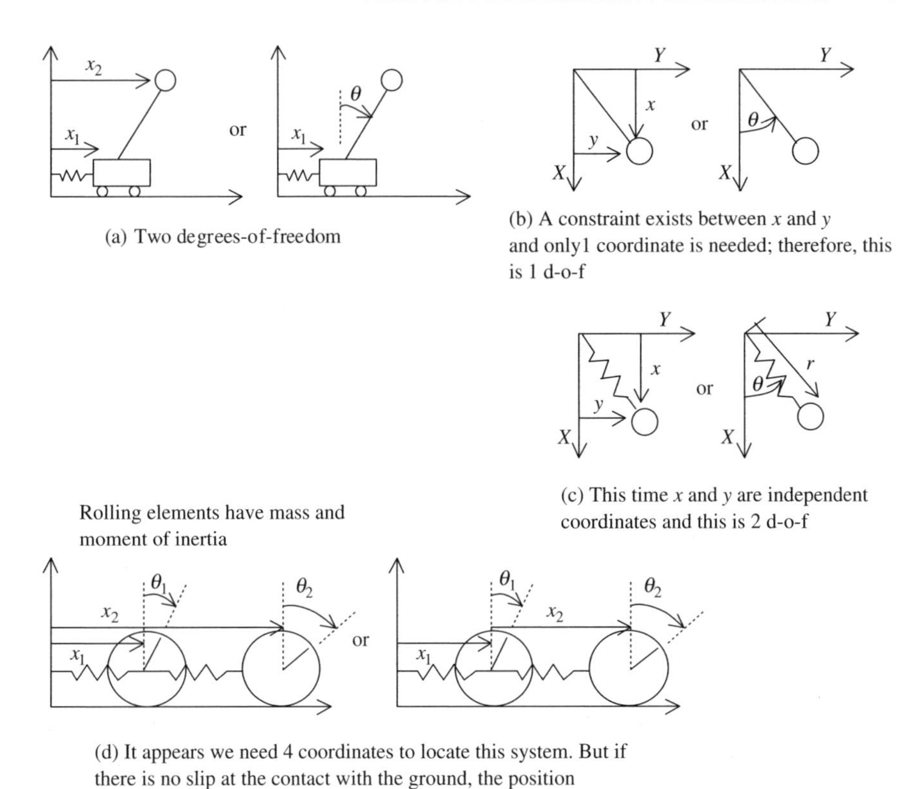

Figure 1.5 Coordinates and degrees of freedom for several systems.

position coordinate for each direction in which the center of gravity of rigid bodies can move. We also need the angular position of each rigid body if that body can rotate. Sometimes there exist constraints among the coordinates, and sometimes we can use the constraint relationships to reduce the number of coordinates needed. With some restrictions, the minimum number of geometric specifications needed to describe any possible position that a system can attain is called its ***degrees-of-freedom***. Thus, a single degrees-of-freedom system requires only one coordinate to describe its motion, a two degree-of-freedom system requires two coordinates, and so on. In Figure 1.5 we show several systems, some possible coordinate choices, and identify the degrees-of-freedom. Equations of motion are derived in terms of these coordinates and their derivatives.

1.3 FREE-BODY DIAGRAMS AND FORCE DIAGRAMS

Newton's second law states that forces in a given direction cause acceleration in that direction. Force is a vector quantity just as position, velocity, and acceleration are vector quantities. There are a few forces that act mysteriously from afar. Electromagnetic and gravity forces are examples of these. All other forces and torques are a direct result of physical contact between system components, and these forces and torques must be identified before equations of motion can be derived. It is convenient to show all forces and torques acting on inertial components of a system using arrows and symbols in a separate diagram called a *force diagram*. With the exception of some prescribed input forces, and gravity forces, we do not know the magnitude or actual direction of any of the system force vectors. So we choose a positive direction and use arrows and symbols to indicate the force components. For plane motion we need to show two mutually perpendicular components for each force. There are some elements that can sustain forces only along their axis. For such elements we need only show one force component.

Consider the system shown in Figure 1.6. It consists of two rigid bodies pinned together at the right end of body 1 and the left end of body 2. Body 1 is pinned to ground at its left end and has an applied torque $\tau_1(t)$. Body 2 is constrained to move only horizontally at its right end, and there is a spring attached at the right end of body 2. In Figure 1.6*b* the bodies are separated and the forces at their connection points are exposed. These force vectors are indicated by their respective components. At the left end of body 1 we have the force components F_{x1} and F_{y1}. At the left end of body 2 we have the force components F_{x2} and F_{y2}. According to Newton's laws, we must show these force components as equal and opposite in direction on the right end of body 1. We do not know if at this instant of time component F_{x2} is acting to the right on body 2 as indicated, but if it is, a force of the same magnitude must be acting to the left on body 1. This is a requirement of Newton's laws which must not be forgotten when making a diagram such as Figure 1.6*b*. At the right end of body 2 the spring force is indicated as F_s and the normal force required to constrain body 2 to move horizontally at its right end is indicated as F_N. This normal force has an equal but opposite effect on the constraining slot. There are equal and opposite forces acting on the ground at the left end of body 1, but these are not shown. The spring force passes

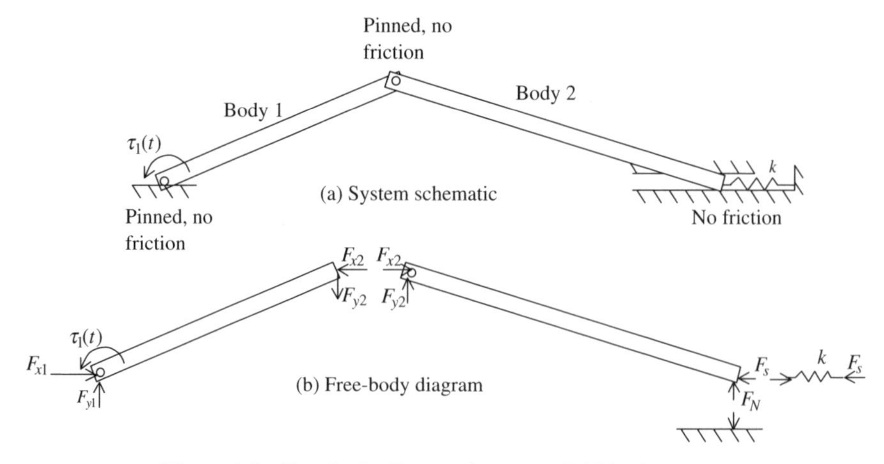

Figure 1.6 Free-body diagram for a two-rigid-body system.

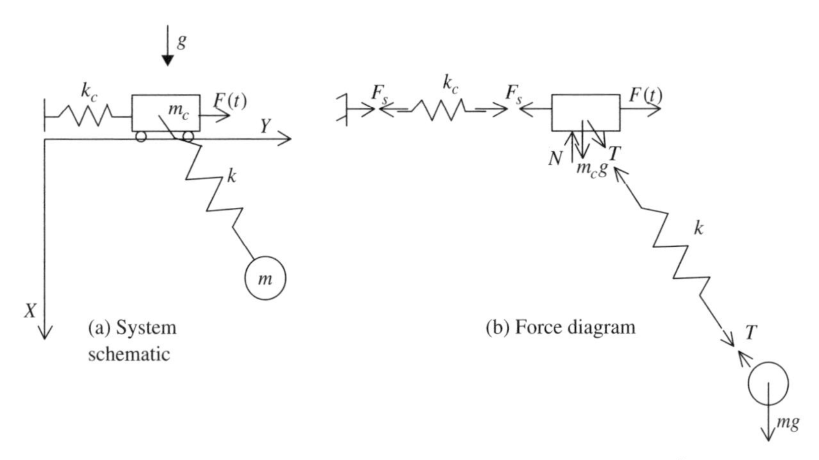

Figure 1.7 Force diagram for the system of Figure 1.4.

through the spring as shown. A diagram in which forces are exposed and arrows and symbols are used to indicate force components is known as a *free-body diagram* or *force diagram*. We will be sketching a lot of these.

Figure 1.7 repeats the system from Figure 1.4. In part *b* the system elements have been separated and the internal forces have been exposed. The forces have been given symbolic names and are pointing in arbitrarily chosen positive directions. It makes absolutely no difference whether or not any of these force components ever actually point in the directions chosen. When a solution is obtained for the dynamic motion of this system and we plot the respective force components, if they are sometimes positive, they are pointing in the directions shown in the figure. If they plot negative, they are pointing in the opposite direction. We ultimately know in what direction the forces are pointing, but we have to wait until we solve the equations of motion. Obviously, we first need to derive the equations of motion.

In Figure 1.7 several different symbols have been used to represent force components. The pendulum and spring can support a force only along the axis; thus, only one force component is shown. It is called T and acts equal and oppositely on the two mass elements connected to the spring. The gravity forces on both masses are shown acting downward. The force between the cart and the ground is called a *normal force* and is given the symbol N in this example. The horizontal spring force is called F_S. Both spring forces have been defined arbitrarily to be positive in tension. With the acceleration diagram of Figure 1.4 and the force diagram of Figure 1.7, the equations of motion are derived straightforwardly, which we do in Chapter 2.

1.4 TRANSFERRING VELOCITY AND ACCELERATION COMPONENTS

The concept of relative motion is introduced in any first course in dynamics [1]. The idea is to be able to express absolute velocity and absolute acceleration when

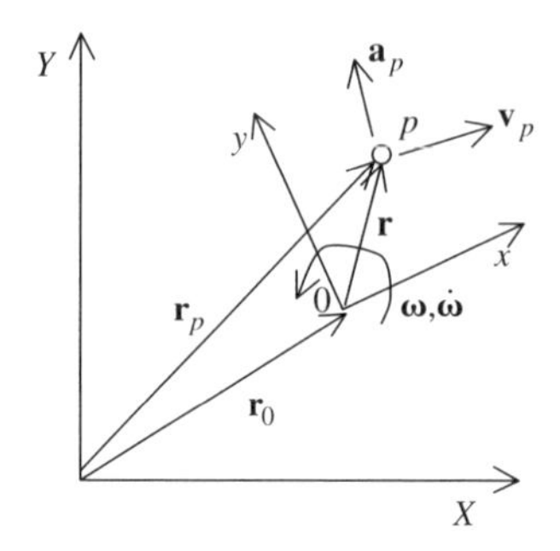

Figure 1.8 Absolute motion of a particle with respect to a moving frame.

coordinate descriptions are stated with respect to an intermediate coordinate frame that accelerates and rotates. Figure 1.8 shows such a situation. The particle p has instantaneous absolute velocity $\mathbf{v}_p$ and absolute acceleration $\mathbf{a}_p$. The particle has absolute position vector $\mathbf{r}_p$, and this vector is the vector sum of the vector that locates the base of the moving frame $\mathbf{r}_0$ and the vector that locates the particle relative to the moving frame $\mathbf{r}$. The frame itself has angular velocity and acceleration $\boldsymbol{\omega}$ and $\dot{\boldsymbol{\omega}}$, with vector orientation defined by the right hand rule, and the base of the moving frame has absolute velocity $\mathbf{v}_0$ and absolute acceleration $\mathbf{a}_0$. To keep the figure less busy, these vectors are not shown in Figure 1.8. It is straightforward to write

$$\begin{aligned}\mathbf{r}_p &= \mathbf{r}_0 + \mathbf{r} \\ \mathbf{v}_p &= \dot{\mathbf{r}}_0 + \dot{\mathbf{r}} = \mathbf{v}_0 + \dot{\mathbf{r}} \\ \mathbf{a}_p &= \ddot{\mathbf{r}}_0 + \ddot{\mathbf{r}} = \mathbf{a}_0 + \ddot{\mathbf{r}}\end{aligned} \tag{1.12}$$

The motion of the base of the frame can be described using either of the coordinate frames already discussed, but the rate of change of the relative vector $\mathbf{r}$ must take into account the rotational motion of the relative frame. It turns out [1] that for any vector $\mathbf{A}$ expressed with respect to a frame that is rotating, its absolute rate of change is given by

$$\frac{d\mathbf{A}}{dt} = \left.\frac{\partial \mathbf{A}}{\partial t}\right|_{\text{rel}} + \boldsymbol{\omega} \times \mathbf{A} \tag{1.13}$$

where the first term on the right-hand side symbolizes the change in the vector as observed from the moving frame and accounts for the change in length of the vector. The second term in (1.13) accounts for the change in direction of the vector. We used the second term in (1.13) above in Eq. (1.7) when determining the rate of change of the unit vectors in polar coordinates. Unit vectors do not change their length, so only the second term in Eq. (1.13) is needed.

When the vector in Eq. (1.13) is a position vector like **r**,

$$\dot{\mathbf{r}} = \mathbf{v}_{\text{rel}} + \boldsymbol{\omega} \times \mathbf{r} \tag{1.14}$$

and

$$\ddot{\mathbf{r}} = \dot{\mathbf{v}}_{\text{rel}} + \dot{\boldsymbol{\omega}} \times \mathbf{r} + \boldsymbol{\omega} \times \dot{\mathbf{r}} \quad \text{or} \quad \ddot{\mathbf{r}} = \mathbf{a}_{\text{rel}} + \boldsymbol{\omega} \times \mathbf{v}_{\text{rel}} + \dot{\boldsymbol{\omega}} \times \mathbf{r} + \boldsymbol{\omega} \times (\mathbf{v}_{\text{rel}} + \boldsymbol{\omega} \times \mathbf{r})$$

where Eq. (1.13) has been used for $\dot{\mathbf{v}}_{\text{rel}}$, to yield

$$\dot{\mathbf{v}}_{\text{rel}} = \mathbf{a}_{\text{rel}} + \boldsymbol{\omega} \times \mathbf{v}_{\text{rel}} \tag{1.15}$$

and Eq. (1.14) has been substituted for $\dot{\mathbf{r}}$. The final result for (1.15) is

$$\ddot{\mathbf{r}} = \dot{\boldsymbol{\omega}} \times \mathbf{r} + \boldsymbol{\omega} \times (\boldsymbol{\omega} \times \mathbf{r}) + 2\boldsymbol{\omega} \times \mathbf{v}_{\text{rel}} + \mathbf{a}_{\text{rel}} \tag{1.16}$$

Combining (1.12), (1.14), and (1.15) yields the final result,

$$\begin{aligned} \mathbf{v}_p &= \mathbf{v}_0 + \boldsymbol{\omega} \times \mathbf{r} + \mathbf{v}_{\text{rel}} \\ \mathbf{a}_p &= \mathbf{a}_0 + \dot{\boldsymbol{\omega}} \times \mathbf{r} + \boldsymbol{\omega} \times (\boldsymbol{\omega} \times \mathbf{r}) + 2\boldsymbol{\omega} \times \mathbf{v}_{\text{rel}} + \mathbf{a}_{\text{rel}} \end{aligned} \tag{1.17}$$

Equation (1.17) expresses symbolically the absolute velocity and acceleration of a point when the position of that point is defined with respect to a frame that is moving and rotating. In (1.16), $\boldsymbol{\omega}$, $\dot{\boldsymbol{\omega}}$ is the angular motion of the moving frame and **r** is the vector drawn from the base of the moving frame to point p.

It is important to be able to apply Eqs. (1.17) and to transfer velocity and acceleration components from one part of a system to another. Consider the rotating disk and slot of Figure 1.9. The disk is pinned at its center and has angular velocity $\dot{\theta} = \omega$ and angular acceleration $\ddot{\theta} = \dot{\omega}$. The angular motion vectors have the magnitudes $\dot{\theta}$ and $\ddot{\theta}$, respectively, and the direction given by the right-hand rule, in this example pointing out of the page toward the reader. Within the radial slot is a mass particle m that can move within the slot. Inertial axes are indicted, as are axes attached to the disk and rotating with the disk. The rotating coordinate frame executes the same motion as the disk.

The velocity and acceleration diagrams are shown in Figure 1.9*b* and *c*. It should be noted that the particle motion could be represented simply using polar coordinates for this system, but this is a good starting example to demonstrate use of the relative motion relationships of Eqs. (1.17). On velocity and acceleration diagrams the components are identified along with the term from Eq. (1.17) from which each comes. For the velocity diagram the $\boldsymbol{\omega} \times \mathbf{r}$ term comes from pointing the fingers of your right hand in the $\boldsymbol{\omega}$-direction (out of the page in this instance) and sweeping your fingers toward the vector **r** (of length x in this instance), resulting in your thumb pointing in the component direction indicated in the figure with a magnitude of $x\dot{\theta}$.

The $\mathbf{v}_{\text{rel}}$-component comes from imagining that you are located on the moving frame at the location of the particle and asking if the particle has any additional

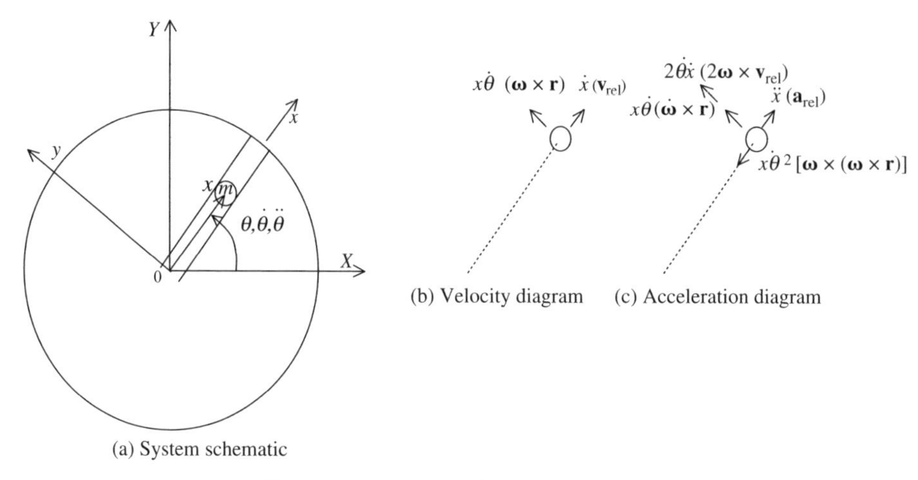

Figure 1.9 Rotating disk with a radial slot.

velocity that you are not experiencing. In this case the particle can move radially within the slot and has an additional velocity component $\dot{x}$ as indicated. Another test for the existence of $\mathbf{v}_{\text{rel}}$ is to imagine that the moving frame, as you have defined it, is stopped. Then assess if the particle has additional velocity. In this case the particle could continue moving radially.

The acceleration diagram is shown in Figure 1.9*c*. The cross product for $\dot{\boldsymbol{\omega}} \times \mathbf{r}$ is carried out exactly as for the $\boldsymbol{\omega} \times \mathbf{r}$ term in the velocity diagram. The $\boldsymbol{\omega} \times (\boldsymbol{\omega} \times \mathbf{r})$ term is made easier to assess due to the presence of the velocity diagram. Notice that the $\boldsymbol{\omega} \times \mathbf{r}$ term is already exposed on the velocity diagram. To cross $\boldsymbol{\omega}$ into $\boldsymbol{\omega} \times \mathbf{r}$ we take the $\boldsymbol{\omega}$ vector and place it at the base of the $\boldsymbol{\omega} \times \mathbf{r}$ vector and take the cross product using our right hand. Point the fingers of the right hand in the direction of $\boldsymbol{\omega}$ (out of the page for this example) and sweep the fingers toward the $\boldsymbol{\omega} \times \mathbf{r}$ vector. The thumb will be pointing in the resultant direction, in this case toward the center of the disk. Thus, the resulting component has magnitude equal to $|\mathbf{r}| \times |\boldsymbol{\omega}|^2$, or $\times \omega^2$, as shown. Similarly, the cross product for $2\boldsymbol{\omega} \times \mathbf{v}_{\text{rel}}$ is obtained by putting the $\boldsymbol{\omega}$ vector at the base of the $\mathbf{v}_{\text{rel}}$ vector and carrying out the cross product with the right hand. The resultant vector is oriented as shown in the figure. The relative acceleration $\mathbf{a}_{\text{rel}}$ is determined in a manner similar to that described for $\mathbf{v}_{\text{rel}}$. If one were either located on the defined moving frame at the location of the particle or if the moving frame were stopped temporarily, we would determine that the particle could still have an acceleration component $\ddot{x}$ directed along the slot. When creating velocity and acceleration diagrams using the relative motion, Eqs. (1.17), it is quite convenient to identify the components as is done in Figure 1.9.

As a second example, consider the rotating pinned disk shown in Figure 1.10. This time a vertical tube is located a distance a from the center and a particle can move within the tube. Inertial and rotating frames are indicated. To use the relative motion, Eqs. (1.17), the $\mathbf{r}$ vector is officially the vector from the base of the moving frame to the particle m. To always take cross products of vectors at right angles, it is convenient to think of the $\mathbf{r}$ vector as the sum of vector components of length a along

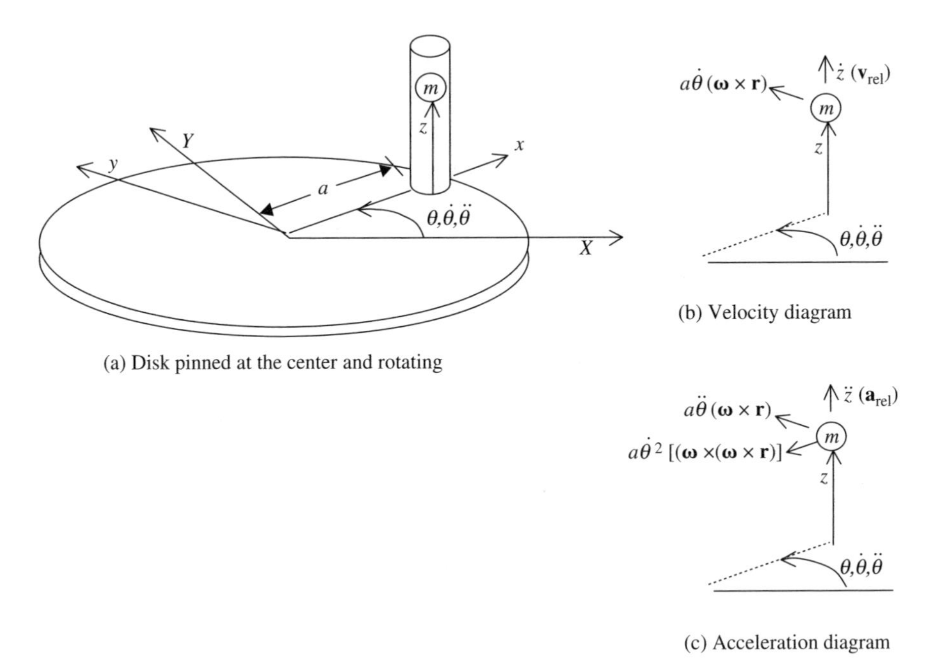

Figure 1.10 Rotating disk with a vertical slot.

the disk plane and of length z vertically upward. When cross products are needed, say $\boldsymbol{\omega} \times \mathbf{r}$, we simply take the cross product of $\boldsymbol{\omega}$ with each component of $\mathbf{r}$. In the velocity diagram, Figure 1.10*b* the $\boldsymbol{\omega} \times \mathbf{r}$ term was generated by first taking the cross product of $\boldsymbol{\omega}$ with the a-component of $\mathbf{r}$, resulting in the component $a\dot{\theta}$ in the direction indicated. The cross product of $\boldsymbol{\omega}$ with the z-component of $\mathbf{r}$ results in zero, since the two vector components are parallel. The relative velocity component $\dot{z}$ comes from the thought experiment described above. If the moving frame were stopped temporarily, the particle could still have a velocity component along the tube.

The acceleration diagram, Figure 1.10 *c*, is generated term by term according to Eq. (1.17). The $\dot{\boldsymbol{\omega}} \times \mathbf{r}$ term is just like the $\boldsymbol{\omega} \times \mathbf{r}$ term, but in terms of angular acceleration instead of angular velocity. The $\boldsymbol{\omega} \times (\boldsymbol{\omega} \times \mathbf{r})$ term is generated conveniently, due to the presence of the $\boldsymbol{\omega} \times \mathbf{r}$ term in the velocity diagram. Placing the $\boldsymbol{\omega}$ vector at the base of $\boldsymbol{\omega} \times \mathbf{r}$ and carrying out the cross product with the right hand results in the appropriate component in the acceleration diagram. The $2\boldsymbol{\omega} \times \mathbf{v}_{\text{rel}}$ term is zero because the $\boldsymbol{\omega}$ vector is parallel to the $\mathbf{v}_{\text{rel}}$ vector. Finally, there is the $\mathbf{a}_{\text{rel}}$-component, which comes from stopping the moving frame and assessing any additional acceleration.

As a third example, consider the pinned rotating disk in Figure 1.11, with a slot creating a chord out a distance a from the center. In the slot is a mass particle located with the relative position vector of length y. The inertial frame and a rotating frame are shown in the figure. The position vector is officially the vector drawn from the base of the moving frame out to the mass particle. It is more convenient to consider the position vector as composed of vector components of length a oriented radially and of length y oriented along the slot. The velocity diagram is shown in Figure 1.11*b*.

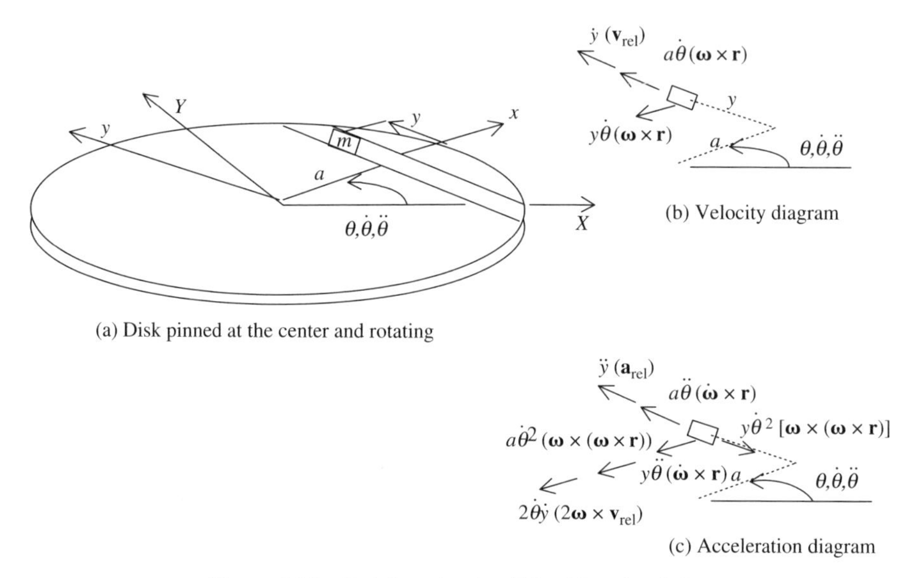

(a) Disk pinned at the center and rotating

(b) Velocity diagram

(c) Acceleration diagram

Figure 1.11 Rotating circular disk with a chord slot.

The $\boldsymbol{\omega} \times \mathbf{r}$ term has two components. This comes from first placing the $\boldsymbol{\omega}$ vector at the base of the a component of the position vector and carrying out the cross product with the right hand. This produces $a\dot{\theta}$, as shown. We next place the $\boldsymbol{\omega}$ vector at the base of the y-component of the position vector and carry out the cross product, producing the $y\dot{\theta}$-component. The $\mathbf{v}_{rel}$ vector comes from the familiar assessment determining any additional velocity of the particle if the moving frame were stopped.

There are quite a few components in the acceleration diagram shown in Figure 1.11c, but each was generated term by term from Eqs. (1.17). The base of the moving frame has no acceleration; thus, $\mathbf{a}_0 = 0$. The $\dot{\boldsymbol{\omega}} \times \mathbf{r}$ term is identical to $\boldsymbol{\omega} \times \mathbf{r}$ but using angular acceleration rather than angular velocity. Then $\boldsymbol{\omega} \times (\boldsymbol{\omega} \times \mathbf{r})$ vector benefits from having the $\boldsymbol{\omega} \times \mathbf{r}$ vector exposed on the velocity diagram. The $\boldsymbol{\omega}$ vector is placed at the base of each component of $\boldsymbol{\omega} \times \mathbf{r}$ and the cross product is carried out using the right hand. This produces the two terms shown for components of $\boldsymbol{\omega} \times (\boldsymbol{\omega} \times \mathbf{r})$. Having $\mathbf{v}_{rel}$ exposed on the velocity diagram facilitates generation of $2\boldsymbol{\omega} \times \mathbf{v}_{rel}$. We simply place the base of the $\boldsymbol{\omega}$ vector at the base of the $\mathbf{v}_{rel}$ vector and carry out the cross product.

1.5 TRANSFERRING MOTION COMPONENTS OF RIGID BODIES AND GENERATING KINEMATIC CONSTRAINTS

Equations (1.17) are very useful when dealing with rigid bodies. It is very typical to need the velocity and acceleration of the center of mass [or center of gravity (c. g.)] of rigid bodies, and it is also very typical to need the velocity and acceleration at attachment points of a rigid body to other rigid bodies or system components. We will find that when rigid bodies share motion at an attachment point, a kinematic constraint is generated. It turns out that using Eqs. (1.17) is actually easier when dealing with rigid bodies than in the particle motion examples done above. The particles that comprise a rigid body by definition do not execute any relative motion. Points on a rigid body located some distance apart remain at that distance regardless of the forces acting. Thus, in Eqs. (1.17) the terms involving relative velocity and relative acceleration are all zero when working with rigid bodies. For rigid bodies,

$$\begin{aligned} \mathbf{v}_p &= \mathbf{v}_0 + \boldsymbol{\omega} \times \mathbf{r} \\ \mathbf{a}_p &= \mathbf{a}_0 + \dot{\boldsymbol{\omega}} \times \mathbf{r} + \boldsymbol{\omega} \times (\boldsymbol{\omega} \times \mathbf{r}) \end{aligned} \tag{1.18}$$

We say that the velocity of some point p on a rigid body is equal to the velocity some other point 0 on the rigid body plus $\boldsymbol{\omega}$ of the rigid body crossed into the vector drawn from point 0 to point p. Similarly, we can relate the acceleration of some point on a rigid body to the acceleration of any other point on the rigid body. This is demonstrated through example.

The rigid bodies from Figure 1.6 are shown again in Figure 1.12. The velocity components of the center of mass of each body are shown in inertial directions. For this example, body 1 uses inertial coordinates with the X-axis pointing to the right and positive rotation counterclockwise, and body 2 uses inertial coordinates with the X-axis pointing to the left and positive rotation clockwise. Using the right-hand rule for rotational vectors, the positive direction for the $\boldsymbol{\omega}$ vector of body 1 is pointing out of the page and the positive direction for the $\boldsymbol{\omega}$ vector of body 2 is pointing into the page. The first of Eqs. (1.18) is used to transfer the velocity components to the ends of the two bodies. This is accomplished by treating the center of gravity of the body as point 0 and the end to which velocity is transferred as point p. For example, to transfer

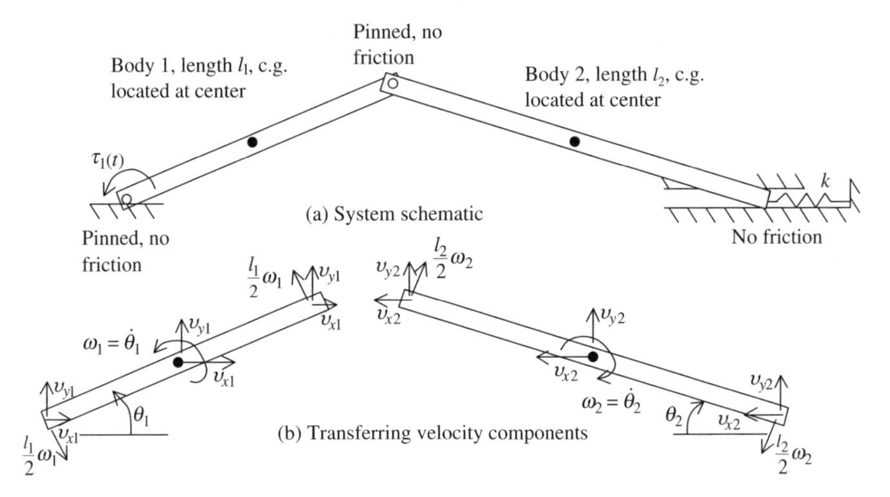

Figure 1.12 Rigid bodies with kinematic constraints.

components from the center of gravity of body 1 to its right end, we first transfer the c.g. velocity components to accomplish the first requirement of Eqs. (1.18) followed by using our right hands to perform the cross-product requirement. For body 1 we point the fingers of the right hand out of the page (positive $\boldsymbol{\omega}$ direction) and sweep the fingers toward the **r** vector (length $l_{1/2}$ and from the center of gravity to the right end of body 1). The thumb is pointing as indicated in Figure 1.12. We similarly transfer velocity components to the left end of body 1, except that this time the **r** vector is directed from the center of gravity toward the left end of body 1. The cross-product term points in the direction shown. Identical steps are carried out for body 2. The only difference is that the positive angular velocity vector points into the page. The cross-product terms are carried out the same way as for body 1 using the right-hand rule.

Kinematic Constraints

The bodies in Figure 1.11 do not move independently, and one of the reasons for transferring velocity components to the attachment points of rigid bodies is to allow specification of kinematic constraints. At the left end of body 1, the body is pinned to inertial ground and the absolute velocity in the X and Y directions must be zero at the left end. From Figure 1.12 we can write down the horizontal and vertical components of velocity as

$$
\begin{aligned}
v_{H1_L} &= v_{x1} + \frac{l_1}{2}\,\omega_1 \sin\theta_1 = 0 \\
v_{V1_L} &= v_{y1} - \frac{l_1}{2}\,\omega_1 \cos\theta_1 = 0
\end{aligned}
\tag{1.19}
$$

These components must each equal zero to enforce the kinematic constraint at the left end of body 1. For the right end of body 2 there can be no vertical velocity, as motion is constrained to be horizontal only. Thus, we have the kinematic constraint

$$
v_{y2} - \frac{l_2}{2}\omega_2 \cos\theta_2 = 0 \tag{1.20}
$$

For a perfect pin joint at the center of the two bodies it is simplest to write down the horizontal and vertical components at the ends of each body and then enforce that the respective components must be equal. Thus, in the horizontal direction,

$$
v_{x2} - \frac{l_2}{2}\omega_2 \sin\theta_2 = -\left(v_{x1} - \frac{l_1}{2}\omega_1 \sin\theta_1\right) \tag{1.21}
$$

and in the vertical direction,

$$
v_{y2} + \frac{l_2}{2}\omega_2 \cos\theta_2 = v_{y1} + \frac{l_1}{2}\omega_1 \cos\theta_1 \tag{1.22}
$$

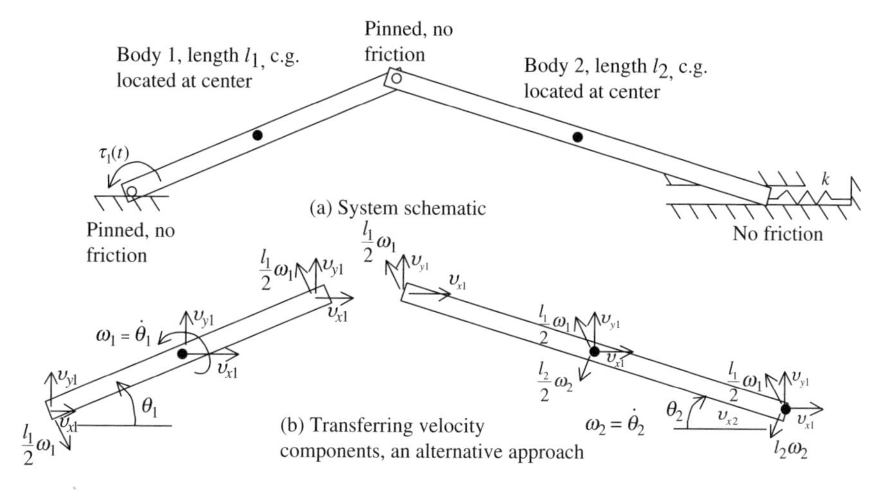

Figure 1.13 Rigid bodies with an alternative approach to kinematic constraints.

These constraint equations can be used to eliminate some variables from the formulation of system equations while retaining others. This is demonstrated in subsequent chapters.

An alternative way to enforce kinematic constraints for the system of Figure 1.12 is shown in Figure 1.13. The velocity components of the center of gravity of body 1 are transferred to the left and right ends exactly is in Figure 1.12. The velocity components at the right end of body 1 are shared by the left end of body 2. Then the velocity components are transferred to the center of gravity of body 2 and to the right end of body 2 using the first of Eqs. (1.18). We never introduce independent specification of the velocity of body 2. The kinematic constraint at the pin joint of bodies 1 and 2 is enforced automatically, and the constraint of zero vertical velocity at the right end of body 2 becomes

$$v_{y1} + \frac{l_1}{2}\omega_1 \cos\theta_1 - l_2\omega_2 \cos\theta_2 = 0 \tag{1.23}$$

As a final example of transferring velocity components and enforcing kinematic constraints, consider the slider–crank mechanism of Figure 1.14. This mechanism comprises the fundamental geometry of virtually every internal combustion engine. The astute reader will notice that the slider–crank device is really identical to the system of Figure 1.12. It is shown here to introduce the concept of body-fixed coordinates. In Figure 1.14 the absolute velocity vector of the center of gravity of the connecting rod is shown as two mutually perpendicular components, one aligned along the rod and the other oriented perpendicular to the rod. The axes shown in Figure 1.14*a* are called *body fixed* in that they are attached to the body and execute all the motions of the body. The velocity components referred to this coordinate frame, called *body-fixed velocity components*, differ from inertial direction components in that they change direction

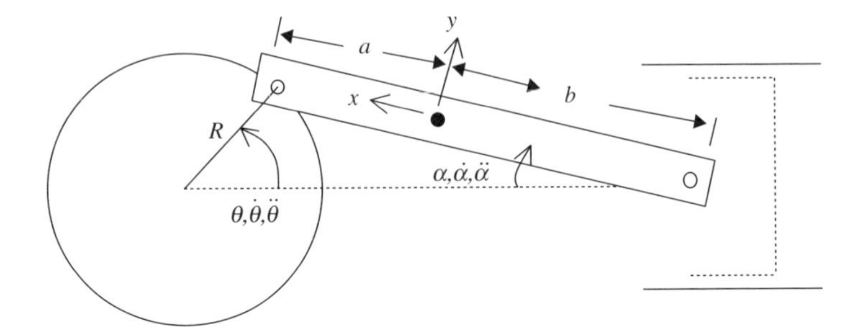

(a) Slider-crank mechanism with body-fixed coordinates

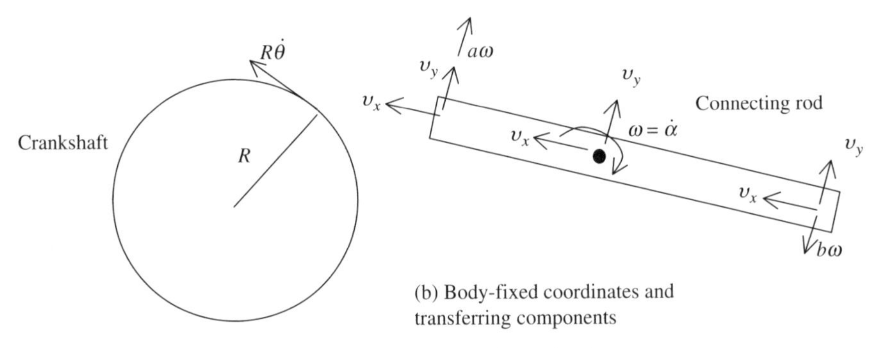

(b) Body-fixed coordinates and transferring components

Figure 1.14 Slider–crank mechanism.

as the body moves. More will be said of body-fixed coordinates in later chapters. For the problem here we simply want to transfer the components to the ends of the rod in order to generate kinematic constraints. The angular motion of the connecting rod is indicated by the angle α and its derivatives. The angular velocity vector is pointing into the page by the right-hand rule.

Applying Eqs. (1.18) allows us to transfer the body-fixed velocity components at the center of gravity to the body-fixed components at the left and right ends of the rod. The crankshaft rotational motion is indicated by the angle θ and its derivatives. For the crankshaft we use polar coordinates and indicate the velocity at the connection with the rod as shown. At the right end of the rod only horizontal motion is permitted, so the constraint that the vertical velocity at the right end equals zero becomes

$$v_x \sin\alpha + (v_y - b\omega)\cos\alpha = 0 \tag{1.24}$$

At the left end of the rod, one way to enforce the kinematic constraint is to equate the horizontal and vertical velocity components from the crankshaft and connecting rod; thus,

$$R\dot{\theta}\,\sin\theta = v_x \cos\alpha - (v_y + a\omega)\sin\alpha$$

for the horizontal direction and

$$R\dot{\theta}\cos\theta = v_x \sin\alpha + (v_y + a\omega)\cos\alpha \tag{1.25}$$

for the vertical direction. Another way to enforce the constraint at the left end of the rod is to equate components in the direction tangent to the crankshaft and set to zero the component perpendicular to the crankshaft. Thus,

$$\begin{aligned} v_x \sin(\theta + \alpha) + (v_y + a\omega)\cos(\theta + \alpha) &= R\dot{\theta} \\ v_x \cos(\theta + \alpha) - (v_y + a\omega)\sin(\theta + \alpha) &= 0 \end{aligned} \tag{1.26}$$

It is left as an exercise to show that Eqs. (1.25) and (1.26) are the same.

1.6 REVIEW OF CENTER OF MASS, LINEAR MOMENTUM, AND ANGULAR MOMENTUM FOR RIGID BODIES

In elementary courses in dynamics using texts such as ref. [1] or very similar texts, the concepts of center of mass, linear momentum, and angular momentum are presented. For plane motion the concept is as shown in Figure 1.15. The rigid body is composed of many small particles glued together such that no single particle executes relative motion with respect to any other; each particle obeys Newton's laws. One such particle is shown in Figure 1.15. Axes are attached at an arbitrary point 0, and a special point called the *center of mass* or center of gravity of the rigid body is located by the position vector $\mathbf{r}_{c/0}$. The ith particle of mass dm_i, which is part of the rigid body, is located by the position vector $\mathbf{r}_i$ and has the absolute velocity $\mathbf{v}_i$. Acting on the particle is the internal force vector $\mathbf{f}_i$, which is a result of contact with other particles surrounding this one, and the external force $\mathbf{F}_i$, which is due to contact at that point with some external element not shown.

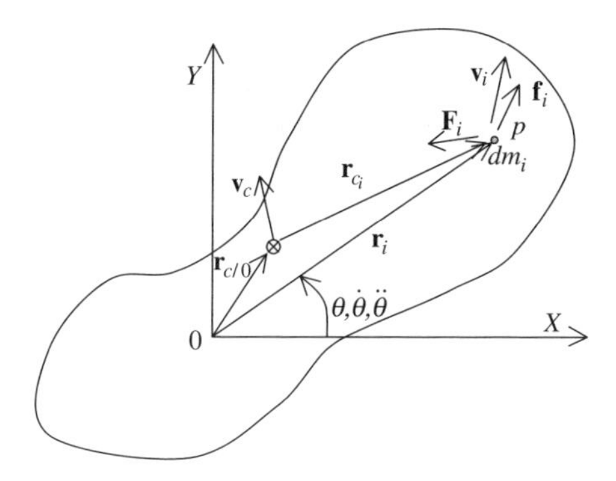

Figure 1.15 Rigid body that demonstrates angular momentum.

The linear momentum of a particle is a vector quantity equal to the product of mass and velocity. Thus, for the mass particle that is part of the rigid body of Figure 1.15, the momentum contribution to the to the total momentum of the body is

$$d\mathbf{p}_i = dm_i\, \mathbf{v}_i \tag{1.27}$$

The total momentum of the body comes from summing all the momentum contributions from all the particles:

$$\mathbf{p} = \sum \mathbf{v}_i\, dm_i = \sum (\mathbf{v}_c + \boldsymbol{\omega} \times \mathbf{r}_{c_i})\, dm_i \tag{1.28}$$

where the transfer relationship has been used to express the velocity of the ith particle in terms of the velocity of the center of mass plus the cross-product term, which accounts for the rotation of the position vector $\mathbf{r}_{c_i}$. As the number of particles approaches infinity, Eq. (1.28) can be expressed as

$$\mathbf{p} = \int (\mathbf{v}_c + \boldsymbol{\omega} \times \mathbf{r}_c)\, dm = m\mathbf{v}_c + \boldsymbol{\omega} \times \int \mathbf{r}_c\, dm \tag{1.29}$$

where integration takes place over the continuous mass distribution. Since $\mathbf{v}_c$ is independent of any *dm* particle, it can be taken outside the integral to produce the first term on the right-hand side of (1.29). Also, $\boldsymbol{\omega}$ is the angular velocity of the frame and is independent of any *dm* particle or its location; thus, $\boldsymbol{\omega}$ can be taken outside the integral, resulting in the second term of (1.29).

The final integral on the right side of (1.29) is zero, due to the definition of the center of mass. Consider the axes through zero in Fig. 1.15. The center of mass is defined as

$$m\mathbf{r}_{c/0} = \int \mathbf{r}\, dm \tag{1.30}$$

To actually find the center of mass of a rigid body or collection of mass particles, we would cast Eq. (1.30) into component form and carry out the operations. In this way we could locate the x, y, z location of the center of mass relative to the original axes. A nice physical way to think about the center of gravity is to imagine pinning the body to a wall with a pin through the center of gravity. You could rotate the body to any orientation, release it, and it would remain in that location and not move under the influence of gravity.

For our purposes here we need to recognize that if we are working with axes attached at the center of gravity as in Eq. (1.29), $\mathbf{r}_{c/0}$ is zero and the integral on the right side of (1.29) is zero when measuring from the center of gravity [1]. Thus, the linear momentum of a rigid body is equal to the mass of the body times the velocity of the center of mass,

$$\mathbf{p}_c = m\mathbf{v}_c \tag{1.31}$$

where the subscript c has been appended to the momentum to enforce that it is the momentum of the center of mass of the rigid body that characterizes the total momentum.

Angular momentum is defined as the moment of the momentum vector with respect to some axes. The moment of a force is probably familiar to the reader. In Figure 1.15, if we desired the moment of the force $\mathbf{F}_i$ about the axes through zero, we would need to resolve the force into components perpendicular to $\mathbf{r}_i$ and parallel to $\mathbf{r}_i$. The moment of the component parallel to $\mathbf{r}_i$ would be zero, the moment of the component perpendicular to $\mathbf{r}_i$ would be the magnitude of the perpendicular component times the length of $\mathbf{r}_i$, and the direction of the moment vector would be given by the right-hand rule. The vector notation that accounts for the total moment $\mathbf{M}_i$ of the force vector $\mathbf{F}_i$ is given by

$$\mathbf{M}_i = \mathbf{r}_i \times \mathbf{F}_i \tag{1.32}$$

This relationship is easy to demonstrate from Figure 1.16. The force vector $\mathbf{F}$ is resolved into two components and the position vector $\mathbf{r}$ is resolved into two components. If we carry out the operation $\mathbf{r} \times \mathbf{F}$, we need to cross each component of $\mathbf{r}$ into each component of $\mathbf{F}$. We first cross the x-component of $\mathbf{r}$ into the F_x-component and then into the F_y-component of $\mathbf{F}$. Since x and F_x are parallel, their cross product is zero. As shown in Figure 1.16, by placing the x-component at the base of the F_y component and using our right hands to take the cross product, we align out fingers along the x-component and sweep them toward the F_y-component, with the result indicated in Figure 1.16*b*. We do a similar operation to cross the y-component of position into the F_x-component of force, with the result indicated in the figure. The final result is

$$\mathbf{r} \times \mathbf{F} = xF_y - yF_x \quad \text{with a vector direction out of the page} \tag{1.33}$$

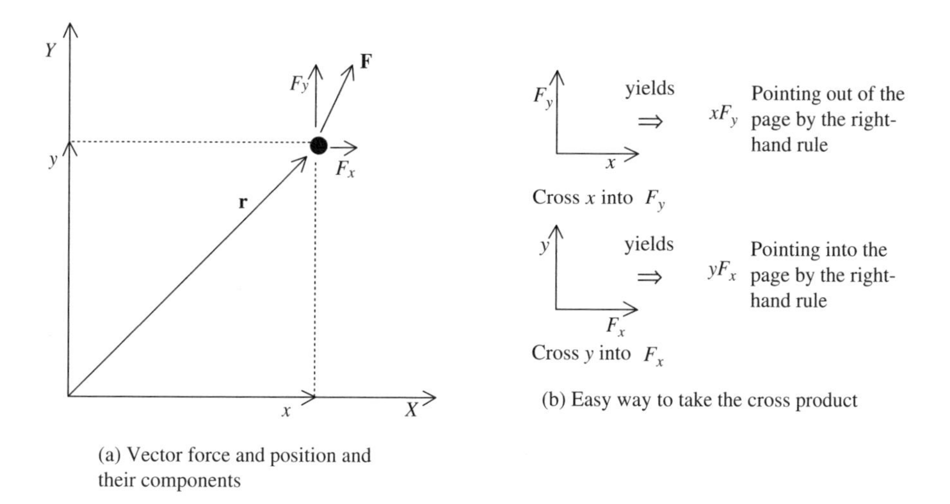

Figure 1.16 Demonstration of the vector moment equation.

We see that carrying out the operations indicated by Eq. (1.32) results in properly taking the moment of a force with respect to some axes.

Let's return to angular momentum, which is the moment of the momentum vector. With respect to the axes through zero in Figure 1.15, the angular momentum contribution of the ith particle is

$$d\mathbf{h}_i = \mathbf{r}_i \times \mathbf{v}_i \, dm_i = \mathbf{r}_i \times (\mathbf{v}_0 + \boldsymbol{\omega} \times \mathbf{r}_i) \, dm_i \tag{1.34}$$

using the transfer of velocity relationship. The total angular momentum of all the particles comes from summing the contribution from each:

$$\mathbf{H}_0 = \sum \mathbf{r}_i \, dm_i \times \mathbf{v}_0 + \sum \mathbf{r}_i \times (\boldsymbol{\omega} \times \mathbf{r}_i) \, dm_i \tag{1.35}$$

As the number of particles becomes infinite, the summations become integrations over the continuous mass distribution, and the angular momentum of a rigid body with respect to axes through the general point zero is

$$\mathbf{H}_0 = \int \mathbf{r} \, dm \times \mathbf{v}_0 + \int \mathbf{r} \times (\boldsymbol{\omega} \times \mathbf{r}) \, dm \tag{1.36}$$

We now make a simplification that is used throughout the book. The first term on the right side of (1.36) adds unnecessary complexity. If the rigid body happens to have a fixed point (i.e., a point that is literally pinned to inertial ground) and the base of the reference frame is placed there, then for a fixed point $\mathbf{v}_0 = 0$ and the first term of (1.36) is zero. If the base of the reference frame happens to be on the center of gravity, then $\int \mathbf{r} \, dm = 0$. This comes from the definition of the center of mass as described in Eq. (1.30). Thus, if the base of the reference frame is placed on a fixed point or the center of gravity, the angular momentum of the body is given by

$$\begin{aligned} \mathbf{H}_0 &= \int \mathbf{r}_0 \times (\boldsymbol{\omega} \times \mathbf{r}_0) \, dm \\ \mathbf{H}_c &= \int \mathbf{r}_c \times (\boldsymbol{\omega} \times \mathbf{r}_c) \, dm \end{aligned} \tag{1.37}$$

where sub 0 and sub c are used to indicate measurements from a fixed point or the center of gravity, respectively. For all the applications discussed in this book we will always work with axes attached at a fixed point if one exists or at the center of gravity. If you want to remember only one thing, it would be always to attach axes to rigid bodies at the center of gravity. Then the angular momentum is always the second of Eqs. (1.37). We next have to interpret the operational meaning of Eqs. (1.37).

These expressions for angular momentum are applicable in three dimensions, and such use is described in later chapters. It is much easier to get a physical understanding of Eqs. (1.37) by first considering plane motion. Angular momentum is a vector quantity and the vector direction is given by the cross-product terms in the integrand of (1.37) using the right-hand rule. Figure 1.17 shows a planar rigid body with axes

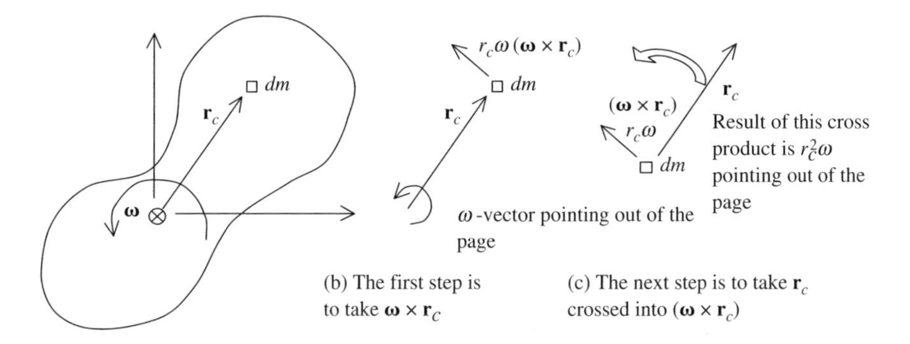

Figure 1.17 Interpretation of angular momentum.

attached at the center of gravity. The angular velocity vector is defined positive counterclockwise and is pointing out of the page by the right-hand rule. In part *b* of the figure, the cross-product $\boldsymbol{\omega} \times \mathbf{r}_c$ has been done using the right hand. It is pointing orthogonally to the position vector and has length $r_c\omega$. If the vector $\mathbf{r}_c$ is placed at the base of $\boldsymbol{\omega} \times \mathbf{r}_c$ and the required cross product is taken, the result will be a vector pointing out of the page with magnitude $r_c^2\omega$. Thus, for plane motion, Eq. (1.37) becomes

$$\mathbf{H}_c = \omega \int r_c^2\, dm = I_c\omega \quad \text{pointing out of the page} \tag{1.38}$$

The quantity $I_c = \int r_c^2\, dm$, called the *moment of inertia*, is a property of the body. Because we are working with axes attached at the center of gravity, I_c is called the *centroidal moment of inertia*. For simple geometric shapes such as disks and rods, the moment of inertia can be calculated directly from its integral definition. For complex shapes such as engine case castings, the moment of inertia might be determined by testing or by finite-element analyses. The thing that changes in Eq. (1.37) if a fixed point is used rather than the center of gravity is the moment of inertia, since measurements are made from a different location for the two axes choices. The reader may recall the parallel axis theorem, which allows transfer of inertial properties from a set of axes through the center of gravity to another set of parallel axes at some other location on a rigid body. We discuss this further in another section.

1.7 NEWTON'S LAW APPLIED TO RIGID BODIES

Returning to Figure 1.15, Newton's second law applied to the particle shown has the result

$$\mathbf{F}_i + \mathbf{f}_i = dm_i \frac{d\mathbf{v}_i}{dt} = \frac{d}{dt}\mathbf{p}_i \tag{1.39}$$

where $\mathbf{p}_i$ is the momentum of the ith particle. Summing over all the particles yields

$$\sum \mathbf{F}_i + \sum \mathbf{f}_i = \frac{d}{dt} \sum \mathbf{v}_i \, dm_i = \frac{d}{dt} \sum \mathbf{p}_i \tag{1.39a}$$

The first summation on the left side of (1.39a) is the sum of all the external forces acting on the collection of particles that make up the rigid body. The second summation on the left is zero since the internal forces are a result of physical contact among particles, and each internal force will have an equal but opposite counterpart. As the summation is carried out, each positive internal force will be canceled by its negative counterpart, resulting in zero net force. The summation on the right side of (1.39a) is the momentum of the rigid body as developed above, resulting in Eq. (1.31). Thus, the final result becomes

$$\mathbf{F} = m \frac{d\mathbf{v}_c}{dt} = \frac{d\mathbf{p}_c}{dt} \tag{1.39b}$$

where $\mathbf{F}$ is the vector sum of all the external forces acting on the body, and these forces cause an acceleration of the center of mass or center of gravity of the rigid body.

Taking the moment of the forces about the center of gravity from Figure 1.15 results in

$$\mathbf{r}_{c_i} \times \mathbf{F}_i + \mathbf{r}_{c_i} \times \mathbf{f}_i = \mathbf{r}_{c_i} \times \frac{d\mathbf{v}_i}{dt} \, dm_i \tag{1.40}$$

Using the velocity transfer relationship, we get

$$\mathbf{r}_{c_i} \times \mathbf{F}_i + \mathbf{r}_{c_i} \times \mathbf{f}_i = \mathbf{r}_{c_i} \times \frac{d(\mathbf{v}_c + \boldsymbol{\omega} \times \mathbf{r}_{c_i})}{dt} \, dm_i \tag{1.41}$$

We next sum over all the particles, yielding

$$\sum \mathbf{r}_{c_i} \times \mathbf{F}_i + \sum \mathbf{r}_{c_i} \times \mathbf{f}_i = \sum \mathbf{r}_{c_i} \, dm_i \times \frac{d\mathbf{v}_c}{dt} + \frac{d}{dt} \sum \mathbf{r}_{c_i} \times (\boldsymbol{\omega} \times \mathbf{r}_{c_i}) \, dm_i \tag{1.42}$$

The first term on the left side is the moment of all the external forces about the center of gravity. The second term on the left is zero because these are the moments of the internal forces. This means that for every positive contribution there will be an equal and opposite contribution. In other words, if particle a pushes on particle b with an internal force, and this force has a moment about the center of gravity, then particle b is pushing on particle a with equal but opposite internal force and the moment contribution is canceled. The first term on the right side of (1.42) would become an integral over the mass distribution as the number of particles goes toward infinity. The first summation or integral will be zero from the definition of the center of mass since we are measuring from the center of gravity. The second term on the right side of (1.42) as it becomes an integral is the time rate of change of the angular momentum as it was defined in Eq. (1.37). Thus, we have the rotational equation of motion,

$$\mathbf{M}_c = \frac{d}{dt}\mathbf{H}_c \tag{1.43}$$

where $\mathbf{H}_c$ comes from Eq. (1.38).

In Chapter 2 we make use of velocity, acceleration, and force diagrams to derive equations of motion which, if solved, would predict the motion–time history of the physical system represented.

REFERENCE

[1] F. P. Beer, E. R. Johnston, Jr., and W. E. Clausen, *Vector Mechanics for Engineers: Dynamics*, 8th ed., McGraw-Hill, New York, 2004.

PROBLEMS

1.1 For the systems shown in Figure P1.1, state the number of degrees-of-freedom and choose and name coordinates that could be used to describe the motion of the systems.

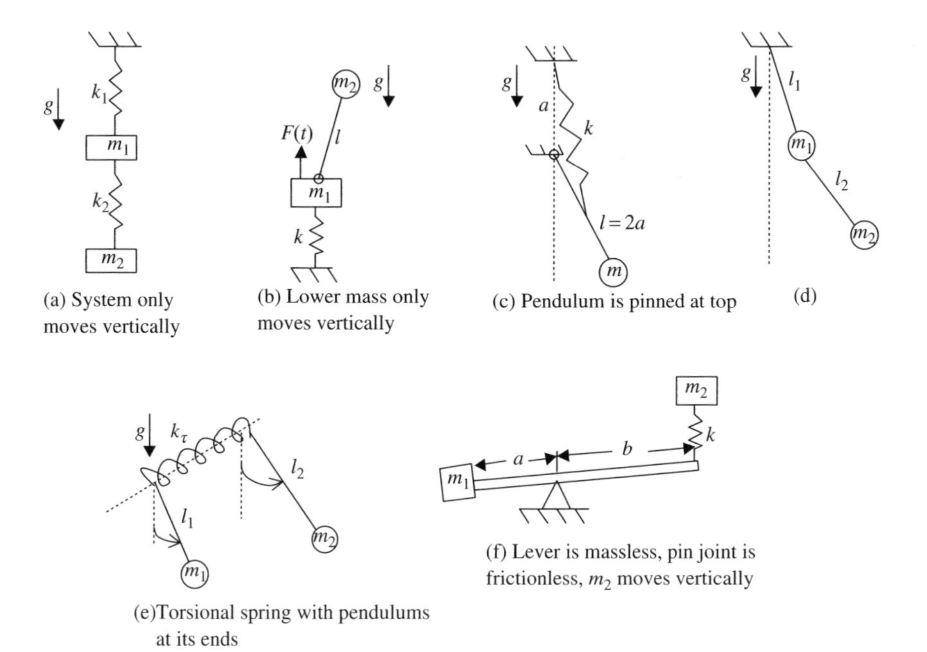

Figure P1.1

1.2 The system shown in Figure P1.2 is sometimes called the *half-car model*. It can be thought of as the side view of a car, with the right side being the front and the left side being the rear. It can also be thought of as the front view of a car where the right side of the figure is the left/front of the car and the left side of the figure is the right/front of the car. The rigid body has mass and c.g. moment of inertia. The suspension elements have a spring and a damper at each end. The unsprung mass is represented by the point masses m_r and m_l, and the tire stiffnesses are represented by the springs k_t. The ground inputs are indicated at the bottom of the right and left sides of the figure. Identify the number of degrees of freedom and show several different choices for coordinates that would fully describe the motion of the system. Make it clear from which reference point each coordinate is measured.

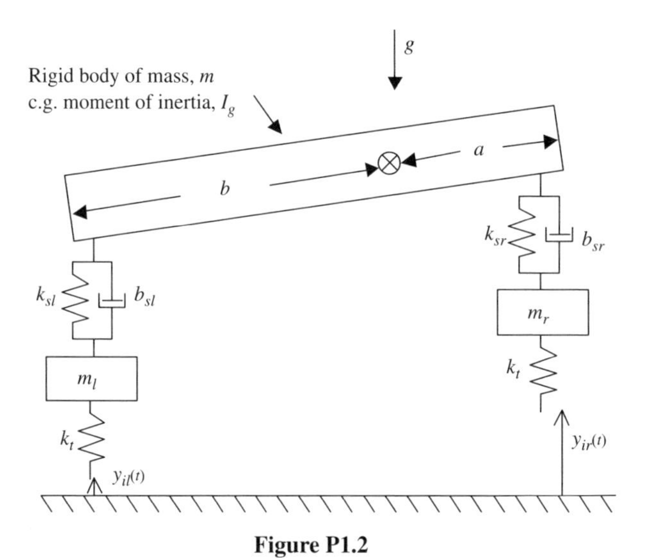

Figure P1.2

1.3 For the systems of Problem 1.1, use your chosen coordinates and show velocity, acceleration, and force diagrams. An example is shown in Figure P1.3 from Problem 1.1 (b). The coordinate y is measured from equilibrium in a gravity field such that the spring force, positive in tension, is $F_s = k(y - y_{\text{eq}})$ where $y_{\text{eq}} = [(m_1 + m_2)g]/k$. Note that you must make use of the velocity transfer formula from Eq. (1.17).

1.4 For the half-car model of Problem 1.2, using the coordinates indicated in the Figure P1.4, show velocity, acceleration, and force diagrams. You can define spring free lengths if you find it necessary. The coordinates are defined such that when all are zero, the system is at equilibrium under the influence of gravity.

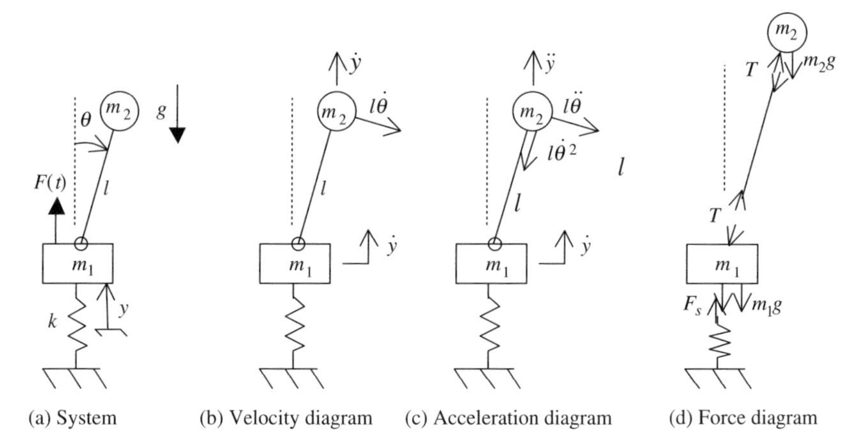

Figure P1.3

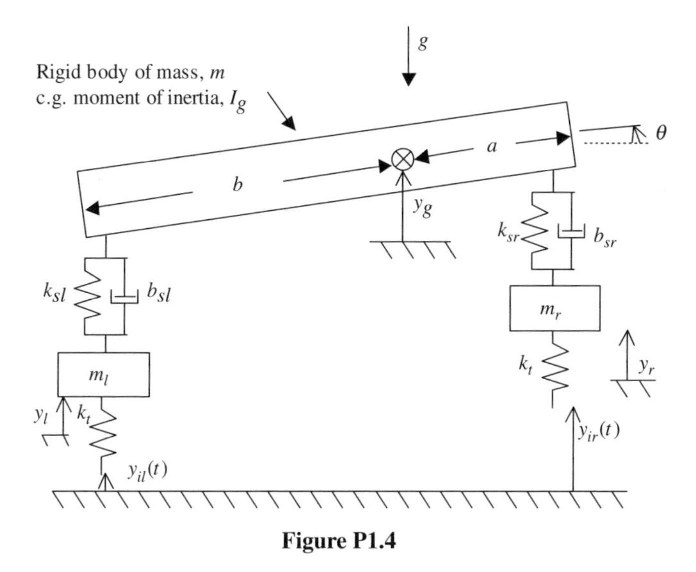

Figure P1.4

1.5 Figure P1.5 shows the front view of what might be an engine block in a heavy truck. The engine is represented as a rigid body and is suspended by soft mounts at the right and left at the bottom. What would probably be a single self-contained isolator in application is shown here as separate horizontal and vertical stiffness elements. The horizontal springs are constrained to remain horizontal and the vertical springs are constrained to remain vertical. The coordinates that describe the possible motions of the engine block are the inertial location of the center of gravity x_g, y_g, and the angular rotation θ.

(a) Show a velocity diagram for the center of gravity and transfer these components to the bottom right and left corners so that the spring displacements

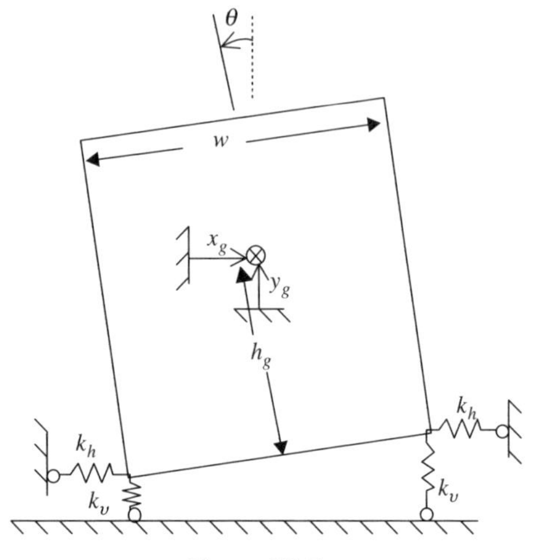

Figure P1.5

could ultimately be determined. You will be making use of the velocity transfer Eq. (1.18), $\mathbf{v}_p = \mathbf{v}_0 + \boldsymbol{\omega} \times \mathbf{r}$.

(b) Show acceleration and force diagrams. You will need to assign appropriate names for the spring forces.

(c) Try to express the spring forces in terms of the coordinates.

1.6 The system shown in the Figure P1.6 is a cylinder that rolls without slip on a horizontal ground surface and has an offset center of mass. If this device were given some initial velocity to the left, it would roll in a not-very-smooth motion. We ultimately desire the equations of motion for this system, but first we need the velocity, acceleration, and force diagrams. The constraint of rolling without slip makes this a single degree-of-freedom system, and either the location of

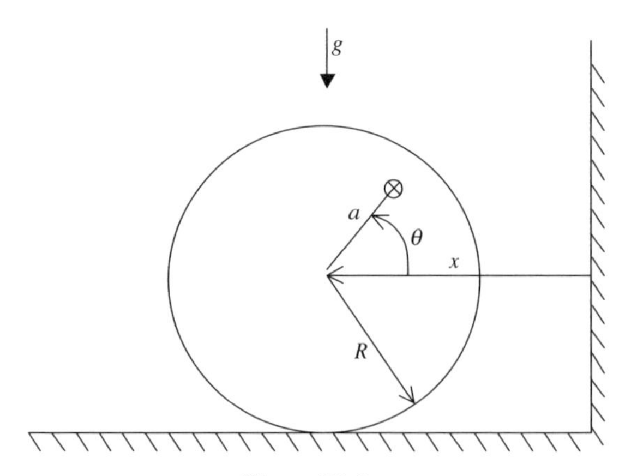

Figure P1.6

the center of the cylinder x or the angular position of the center of gravity, θ can be used as the coordinate for the problem. Assuming that $\theta = 0$ when $x = 0$, the constraint is $x = R\theta$.

The mass of the cylinder is m and the c.g. moment of inertia is I_g. Construct velocity, acceleration, and force diagrams for this rigid body in terms of coordinates x and in terms of the coordinate θ.

1.7 A cylinder of mass m_c and c.g. moment of inertia I_c rolls without slip on a horizontal surface shown in Figure P1.7 . On the inner surface of the cylinder, a mass particle m can slide without friction. This is a two degree-of-freedom system and two coordinates are needed to describe the system motion. Since the cylinder rolls without slip, either X or θ can be used to locate the cylinder. The angle ϕ is used to locate the mass particle. Using arrows and symbols, show velocity, acceleration, and force diagrams for the cylinder and the mass particle. Imagine that a set of axes is attached to the cylinder at the center of gravity and is rotating with the cylinder. Use the transfer of velocity and acceleration formulas, Eqs. (1.17), to identify all the motion components of the mass particle. Be methodical—there are a lot of individual components.

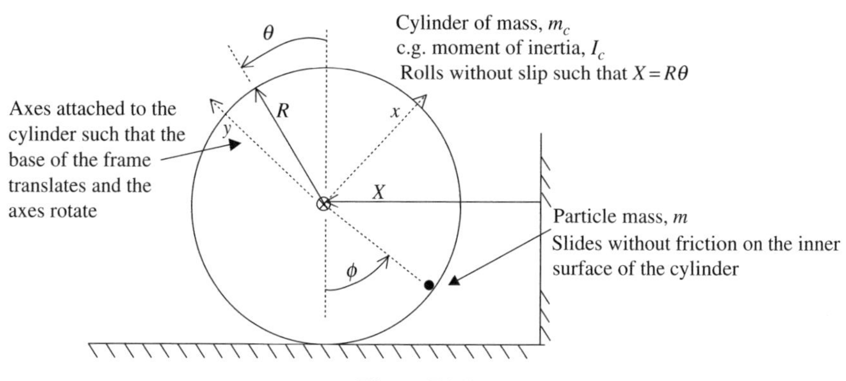

Figure P1.7

1.8 This is basically the same problem as Problem 1.7 except that this time a small cylinder rolls without slip on the inner surface of a big cylinder (Figure P1.8). This remains a two degree-of-freedom system, and either X or θ can be used to locate the big cylinder, with ϕ locating the small cylinder. Using arrows and symbols for the coordinates, construct velocity, acceleration, and force diagrams for this system. When you show the forces, be sure to include a rolling friction force between the bottom of the small cylinder and the inner surface of the big cylinder.

1.9 A cylinder of radius R rolls without slip on a horizontal surface shown in Figure P1.9. The construction is such that there is an inner radius r wrapped with string that extends out horizontally as shown. This is somewhat like a spool of thread. A force F is applied to the end of the string and the cylinder will execute some

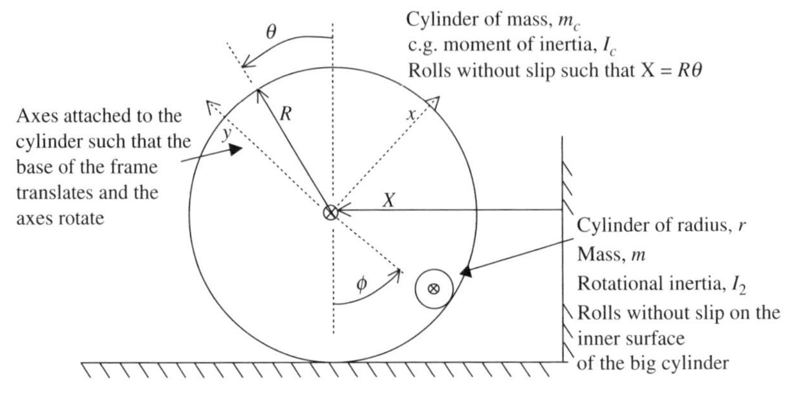

Figure P1.8

motion. Either X or θ can be used as the coordinate for this single degree-of-freedom system. Using arrows and symbols, construct velocity, acceleration, and force diagrams for this system.

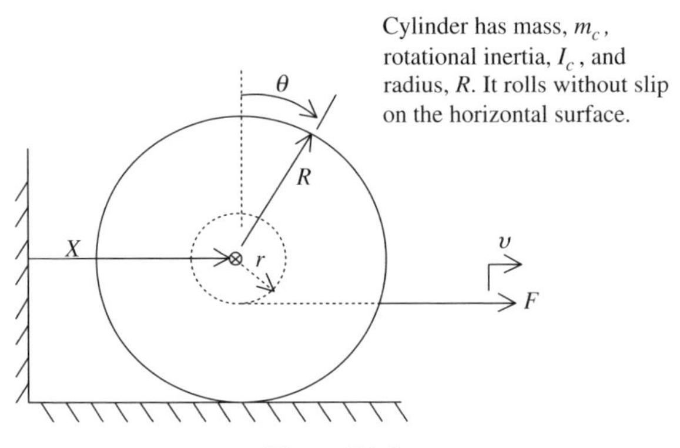

Figure P1.9

1.10 This problem is similar to Problem 1.6, but this time a spring is attached at a height R on the left side wall and extends to the edge of the cylinder diametrally across from the offset center of gravity (Figure P1.10). The cylinder still rolls without slip on the horizontal surface such that either x or θ can be used as the coordinate for the single degree-of-freedom system. The system is set up such that when $\theta = 0$, $x = 0$ and the spring is horizontal and relaxed with no force. In this position the distance from the left wall to the center of the cylinder is D. The mass of the cylinder is m and the c.g. moment of inertia is I_g. Starting from the velocity diagram for the center of gravity of the cylinder, transfer the velocity components to the attachment point of the spring on the cylinder and determine the velocity component in the instantaneous longitudinal spring direction. You might want to use the angle α to make evaluation easier, but as a last step, try to express α as a function of θ.

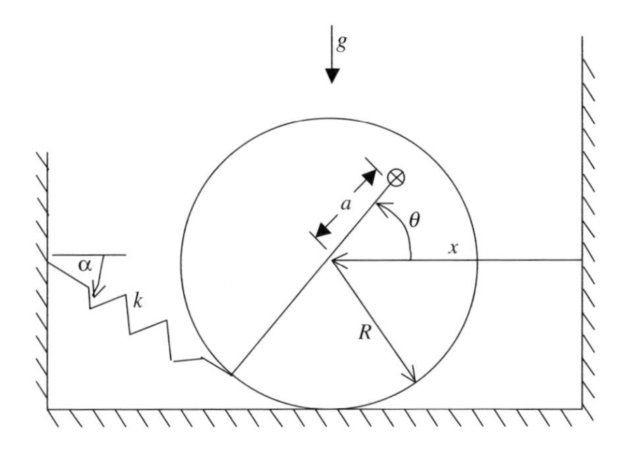

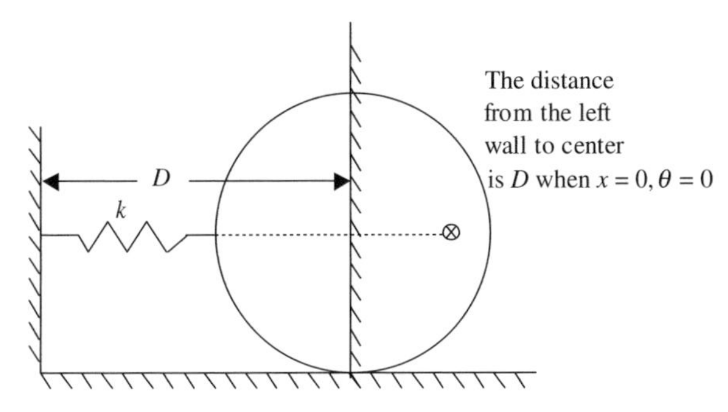

Figure P1.10

1.11 Figure P1.11 shows the top view of a vehicle that has mass m and c.g. moment of inertia about the axis out of the page, I_g. The center of gravity is located a distance a from the front axle and a distance b from the rear axle. The half-width of the vehicle is $w/2$. The front wheels can be steered, indicated by the steer angle δ. A body-fixed coordinate frame is attached to the vehicle at its center of gravity and aligned as shown. The body-fixed velocity components of the center of gravity and the yaw angular velocity are indicated.

(a) Using arrows and symbols, transfer the c.g. velocity to body-fixed directions at the four wheels.

(b) If each wheel is constrained to have no velocity perpendicular to the plane of the wheel, state the kinematic constraints for each wheel.

1.12 This system is identical to that of Problem 1.11, but this time inertial coordinates are used to locate the center of gravity of the vehicle, and the angle θ indicates the angular orientation (Figure P1.12).

(a) Using arrows and symbols, transfer the c.g. velocity to body-fixed directions at the four wheels.

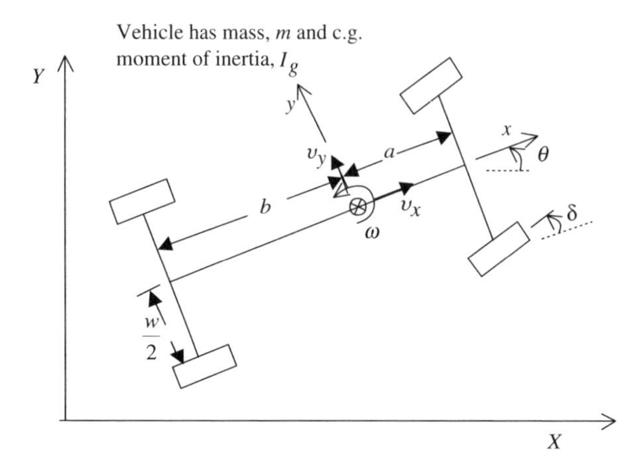

Figure P1.11

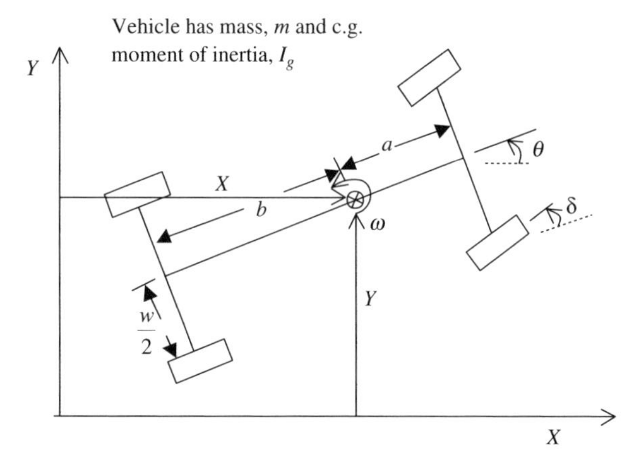

Figure P1.12

(b) If each wheel is constrained to have no velocity perpendicular to the plane of the wheel, state the kinematic constraints for each wheel.

1.13 This is the same problem as Problem 1.5, with an engine block of mass m and c.g. moment of inertia I suspended at the base by some horizontal and vertical springs (Figure P1.13). In this problem the inertial coordinates have been replaced with body-fixed coordinates and the body-fixed velocity components of the center of gravity are indicated.

(a) Using arrows and symbols, show a velocity diagram where the body-fixed c.g. components are transferred to body-fixed directions at the attachment points of the springs.

(b) Resolve the body-fixed components at the attachment points into inertial components and write an expression for the velocity at the end of each spring. For each spring state whether the spring velocities are compressing or extending the spring.

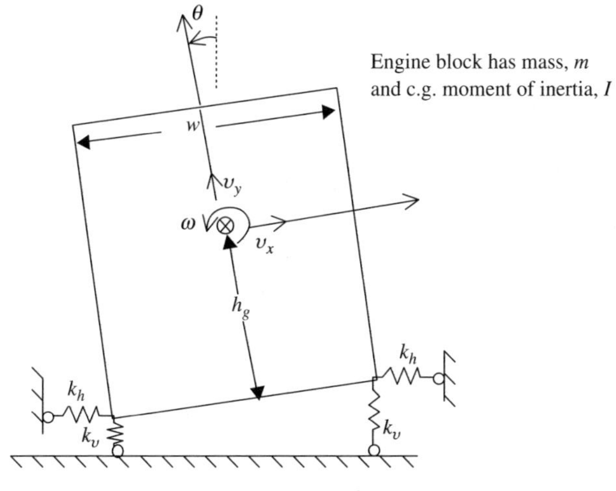

Figure P1.13

1.14 This is the same vehicle as Problem 1.12, but this time the vehicle is towing a trailer of mass m_t and c.g. moment of inertia I_t. The trailer is pinned to the towing vehicle as shown in Figure P1.14 and the center of gravity of the trailer is located c distance behind the hitch point. The rear axle of the trailer is located d from the trailer center of gravity. Body-fixed coordinates are used for the vehicle, and the body-fixed c.g. velocity components v_x, and v_y are shown.

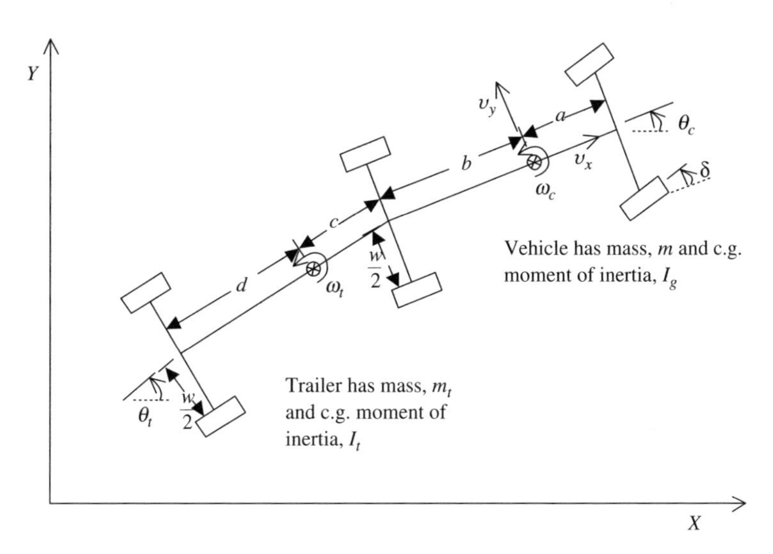

Figure P1.14

(a) Using arrows and symbols, transfer the velocity components of the vehicle to body-fixed directions at the wheels and at the trailer hitch point.

(b) Transfer the velocity components at the hitch point to the center of gravity of the trailer and to its wheels. Show this on a velocity diagram.

(c) If the wheels of the vehicle and the trailer are constrained to have no sideways velocity, state the kinematic constraints that enforce this for all the wheels of the system.

1.15 On fixed tabletop sliding without friction is a mass particle attached to a string (Figure P1.15). The string runs through a hole in the center of the table and another mass particle is attached hanging vertically in a gravity field. The string has length L and this is a two degree-of-freedom system in that position r and angular position θ fully locate the entire system. The mass particle and radial string on the table have been given some initial angular velocity $\dot{\theta}_{ini}$ and some initial radius r_{ini} and released. Ultimately, we want the equations of motion that would predict the motion–time history of this system, but right now we want the velocity, acceleration, and force diagrams in terms of the two coordinates and their derivatives.

(a) Using arrows and symbols, construct velocity, acceleration, and force diagrams for this system. The string can only support a tensile force, and you will have to give this force a symbolic name.

(b) You may have noticed that there is no force perpendicular to the r-direction. This means that the acceleration in that direction must be zero. Using your acceleration diagram, write an expression which states that the acceleration perpendicular to the r-direction is zero.

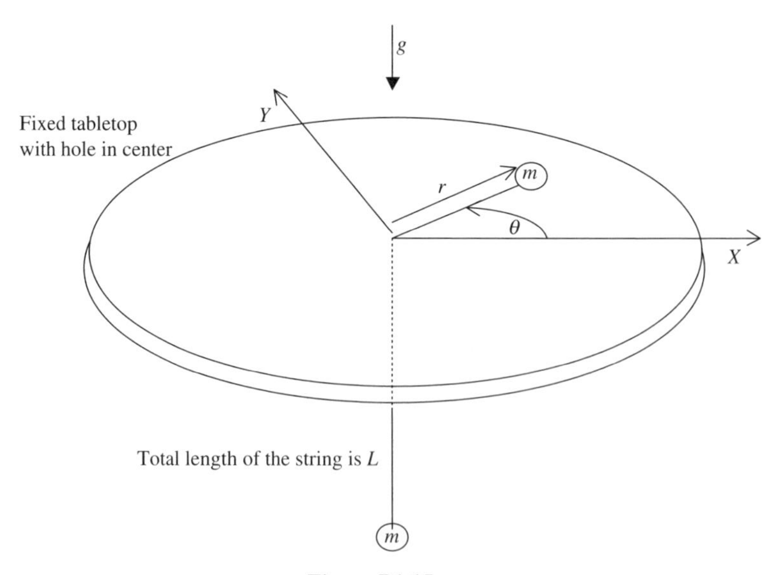

Figure P1.15

(c) The angular momentum of the mass on the table about the axes through the center is the moment of the momentum vector. From your velocity diagram, confirm that the angular momentum is $h = mr^2\dot{\theta}$. Since the tensile force in the string has no moment about these axes, the angular momentum must be constant; thus, $mr^2\dot{\theta} = \text{const}$. Differentiate this expression with respect to time and see if it states the result from part (b).

1.16 A rigid body of mass m, c.g. moment of inertia I_g, and length L rotates without slip on a fixed rigid cylinder (Figure P1.16). Because the rigid body rotates without slip, this is really only a single degree-of-freedom system and only the angle θ is needed to locate the rotated body. Several coordinate systems are shown. The angle θ tracks the contact point on the cylinder, the inertial coordinates X, and Y locate the center of gravity, and body-fixed velocity components are shown in the body-fixed x, and y directions. Springs attached to the rigid body at its ends are constrained to remain vertical regardless of the rotational motion of the rigid body.

(a) For each of the coordinate systems indicated, show velocity diagrams for the attachment points of the springs. Derive the velocity components in the vertical spring directions.

(b) For the θ and X,Y coordinates, show acceleration diagrams for the center of gravity of the rigid body.

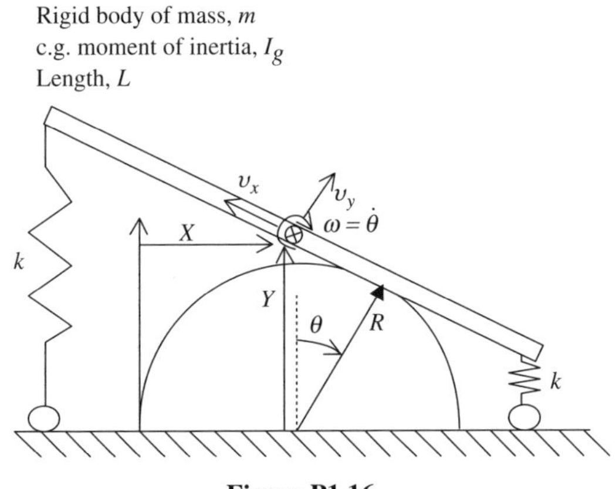

Figure P1.16

1.17 Angular momentum is the moment of the momentum vector about some specified axes. In this chapter, expressions for the angular momentum were derived for rigid bodies with axes attached at a fixed point and for axes attached to the center of gravity. In this problem a rigid body is executing plane motion and has c.g. velocity $\mathbf{v}_g$ and angular velocity ω with respect to inertial X,Y axes. The rigid body is constructed from an infinite number of particles of mass dm, and the ith particle is shown in Figure P1.17. From the inertial X,Y axes, the center of gravity is located by the position vector $\mathbf{r}_g$, the ith particle is located by the position vector $\mathbf{r}_0$, and the ith particle is located with respect to the center of gravity by the position vector $\mathbf{r}$. Using procedures discussed in the chapter text, derive an expression for the angular momentum with respect to the inertial axes in terms of the c.g. velocity and the angular velocity. You will need to recognize that $\mathbf{r}_0 = \mathbf{r}_g + \mathbf{r}$, and you will need to make use of the definition of the center of gravity, where $\int \mathbf{r}\, dm = 0$. The result of your work should be $\mathbf{H}_0 = \mathbf{r}_g \times m\mathbf{v}_g + \int (\mathbf{r} \times \boldsymbol{\omega} \times \mathbf{r})\, dm$, and the interpretation of this result is that the angular momentum of a rigid body with respect to inertial axes is the angular momentum of the center of gravity plus angular momentum due to rotation about the center of gravity. Keep in mind that angular momentum is a vector quantity with direction given by the right-hand rule.

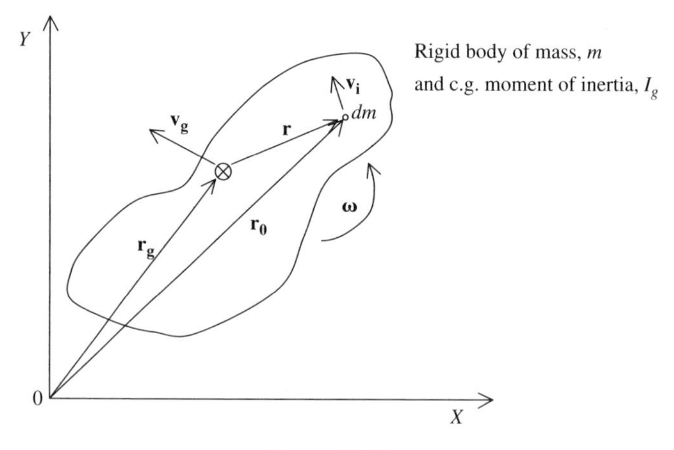

Figure P1.17

1.18 A rigid body with the inertial properties indicated in the Figure P1.18 is traveling at constant initial velocity v_{g_i} toward the left. It is about to encounter an impediment that stops the lower left corner instantaneously. When this happens the body attains a rotational speed ω and the center of gravity attains a velocity v_{g_f}.

(a) Before impact with the impediment, what is the angular momentum of the body with respect to axes attached to the impediment?

(b) Using the results from Problem 1.17, derive an expression for the angular momentum just after impact in terms of v_{g_f} and ω.

(c) Relate v_{g_f} to ω and derive an expression for the angular momentum after impact in terms of ω only. Using your knowledge of the angular momentum of a rigid body about a fixed point and your knowledge of the parallel axis theorem, does your result support what must be the angular momentum of the body just after impact?

(d) It turns out that angular momentum is conserved during the short duration of the impact. Set the initial and final angular momentum equal to each other and derive an expression for the angular velocity just after impact.

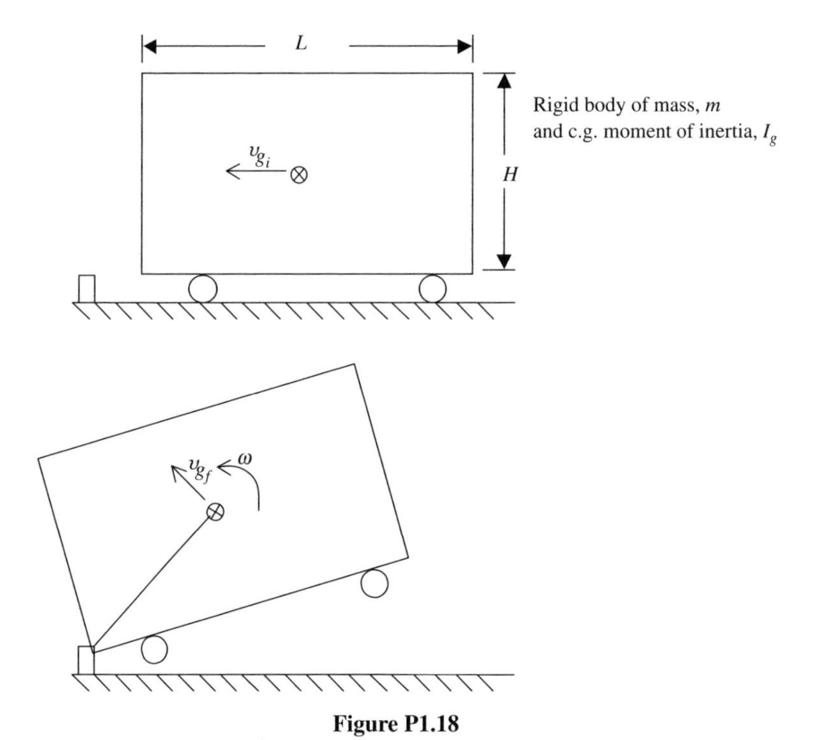

Figure P1.18

1.19 This is a pretty interesting system (Figure P1.19). A pinned cylinder is rotating at an initial angular velocity ω_i. A string is wrapped around the cylinder and a mass particle m is attached to the end of the string and is somehow attached to the surface of the cylinder. At some instant the mass is released and begins to swing out from the cylinder until it attains the position shown in Figure P1.19 *b*. It is released at this point and the mass flies away at some final velocity vf

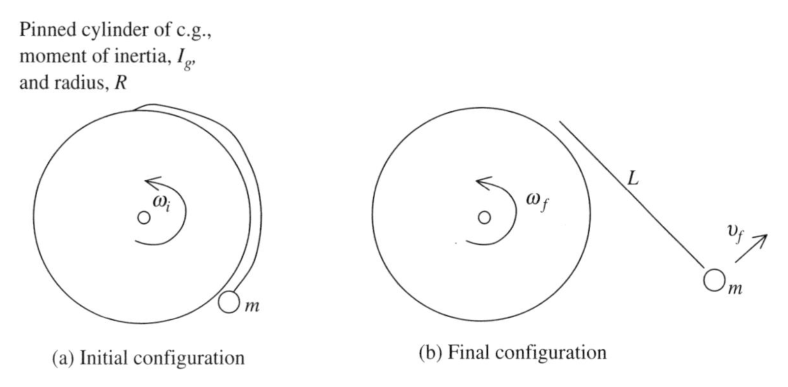

(a) Initial configuration

(b) Final configuration

Figure P1.19

and the cylinder has some final angular velocity ω_f. What makes this system interesting is that it is possible to choose the proper mass and proper string length such that the final angular velocity is absolutely zero.

(a) Derive expressions for the initial angular momentum and final angular momentum of the system.

We have not yet covered energy of rotating and translating bodies, but from a first course in dynamics the kinetic energy of the cylinder in configuration a is $T_{c_i} = \frac{1}{2} I_g \omega_i^2$ and the kinetic energy of the mass particle is $T_{m_i} = \frac{1}{2} m (R\omega_i)^2$. The final kinetic energy in configuration b is $T_{c_f} = \frac{1}{2} I_g \omega_f^2$ for the cylinder and $T_{m_f} = \frac{1}{2} m v_f^2$ for the mass particle.

(b) It turns out that both angular momentum and energy are conserved during the motion from configuration a to b. Equate initial and final angular momentum and initial and final total system energy, postulate that $\omega_f = 0$, and determine a length L when the mass is released that will make the postulate true.

1.20 In this system a rigid body can move without friction on a horizontal surface, and it can rotate with positive direction clockwise (Figure P1.20). This is a two degree-of-freedom system and it is decided to use coordinate x to locate the bottom of the body and coordinate θ to locate the angular orientation.

(a) Using arrows and symbols, construct velocity, acceleration, and force diagrams for this system. You will have to assign names to the forces.

(b) Write an expression for the angular momentum of the body with respect to the X,Y axes.

(c) Write an expression for the linear momentum of the body in the inertial directions.

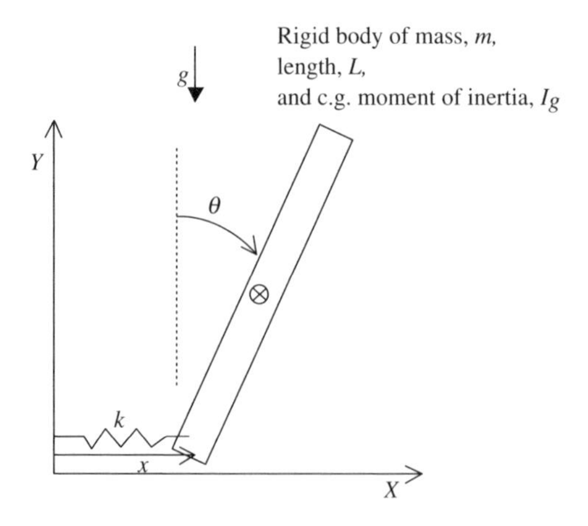

Figure P1.20

2

EQUATIONS OF MOTION IN SECOND- AND FIRST-ORDER FORM

In Chapter 1, coordinate frames were reviewed and the concept of using arrows and symbols to represent velocity, acceleration, and force components was introduced. Our goal is to derive equations of motion which if solved would predict the motion–time history of the physical system represented. To accomplish this goal we must apply Newton's laws and ensure that sufficient equations exist for the variables being used. Generally, real physical systems produce nonlinear differential equations that govern their motion, and computation is the only practical way of solving the governing equations. We will see that if the equations are organized in a specific format, computer solution is quite straightforward; however, simulation of systems of equations is considered in Chapter 3. In this chapter we focus on deriving equations of motion and ensuring that the result is a computable set of equations.

2.1 DERIVING EQUATIONS OF MOTION FOR SYSTEMS OF PARTICLES

Newton's second law states that forces acting on a particle of mass cause an acceleration of the mass in the same direction as the force. By starting with velocity, acceleration, and force diagrams it is straightforward to align force components and acceleration components in mutually perpendicular directions and derive the equations of motion.

Example 2.1: Spring–Pendulum Consider the system shown in Figure 2.1 (introduced in Figure 1.3). The system consists of a mass m attached to inertial ground through a linear spring of spring constant k. Gravity is acting vertically downward. The idea would be to derive and solve the equations of motion and predict the complex

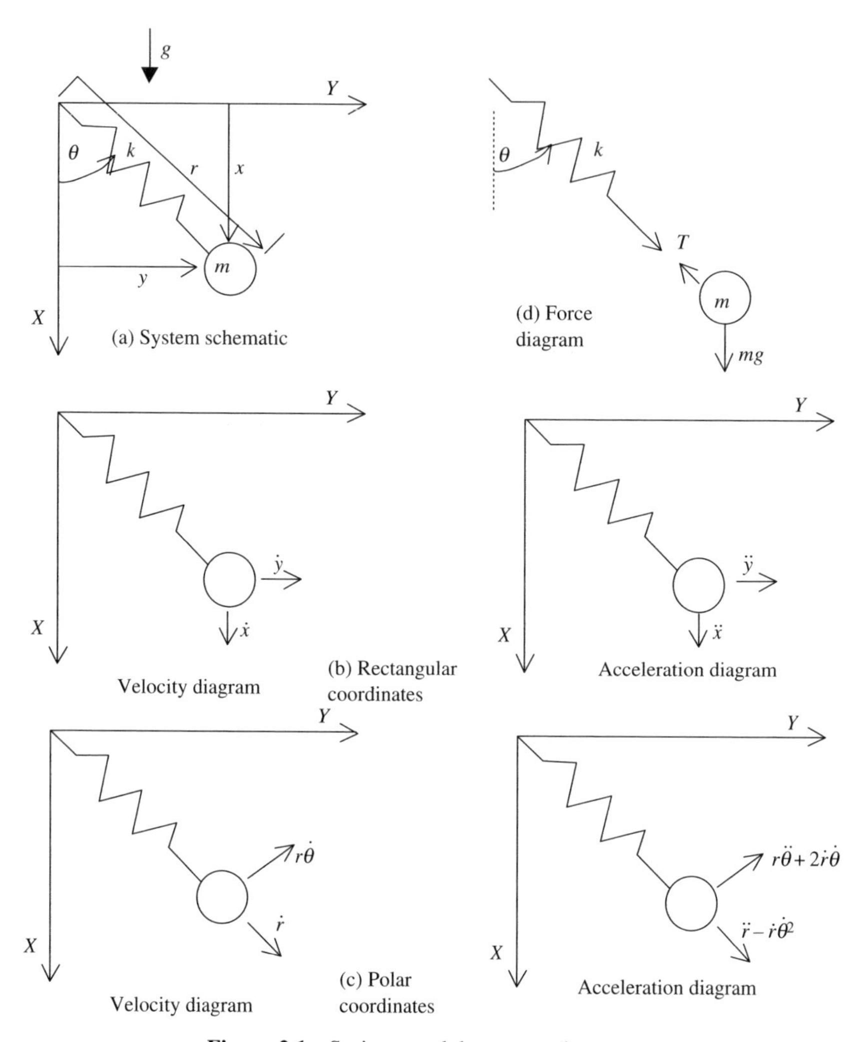

Figure 2.1 Spring–pendulum example system.

motion of the spring–pendulum from any specified initial conditions. Figure 2.1 shows velocity and acceleration diagrams for both rectangular and polar coordinates. A force diagram is included as Figure 2.1*d*. The spring can only support forces along its length; and thus just one force is shown acting equal and opposite between the mass and spring. We will derive the equations of motion using both coordinate frames. Regardless of the coordinates used, this system has two degrees of freedom.

Newton's second law must be applied in mutually perpendicular or orthogonal directions. It really does not matter which orthogonal component directions are used, but for plane motion the force and acceleration vectors can and must be resolved into orthogonal components. Using rectangular components first, we derive equations in the horizontal and vertical directions. We always add the forces acting in the direction of positive acceleration and subtract the forces opposing the positive acceleration. For this example the X-direction acceleration is positive downward; thus, the force components would add as

$$mg - T\cos\theta = m\ddot{x} \tag{2.1}$$

In the positive Y-direction the equation of motion becomes

$$-T\sin\theta = m\ddot{y} \tag{2.2}$$

We have made use of the force and acceleration diagrams and have applied Newton's law, $\mathbf{F} = m\mathbf{a}$, but the result from Eqs. (2.1) and (2.2) is not a computable set of equations. There are too many unknowns for the number of equations. As a reminder, an *unknown* is any variable that changes with time. For this example the unknowns include T, θ, x, and y. The derivatives of variables do not contribute additional unknowns since if, for example, x was a known function of time, we would also know the derivatives of x. Also, m and g are not unknowns, they are parameters. We certainly do not know them at the moment, but they would have to be specified prior to obtaining a solution. It appears that more steps are needed to obtain a computable set of equations than simply applying Newton's law.

The spring force T depends on how much it is stretched from its free length l_0. For a linear spring characterized by a spring constant k, the force would be

$$T = k(l - l_0) \tag{2.3}$$

where l is the current length of the spring. The spring constant is dimensionally force per unit length, or N/m. The problem with (2.3) is that a new unknown has been introduced. We need to recognize that

$$l = \sqrt{x^2 + y^2} \tag{2.4}$$

Equations (2.3) and (2.4) can be used in Eqs. (2.1) and (2.2), thus eliminating T and introducing the desired x,y variables.

Next, we must recognize that θ can be related to the desired variables by

$$\begin{aligned} \sin\theta &= \frac{y}{\sqrt{x^2 + y^2}} \\ \cos\theta &= \frac{x}{\sqrt{x^2 + y^2}} \end{aligned} \tag{2.5}$$

Combining Eqs. (2.1) through (2.5) yields two second-order nonlinear differential equations,

$$\begin{aligned}\ddot{x} &= g - \frac{k}{m}\left(\sqrt{x^2+y^2} - l_0\right)\frac{x}{\sqrt{x^2+y^2}} \\ \ddot{y} &= -\frac{k}{m}\left(\sqrt{x^2+y^2} - l_0\right)\frac{y}{\sqrt{x^2+y^2}}\end{aligned} \tag{2.6}$$

Equations (2.6) can be solved computationally once parameters are specified. We would start from some set of initial conditions and work out the solution one time step at a time and predict the motion–time history of the system.

In Figure 2.1*c* polar coordinates are used in the acceleration diagram. This time we derive the equations of motion in the r-direction and orthogonal to the r-direction. Using the force diagram, we can write

$$mg\cos\theta - T = m(\ddot{r} - r\dot{\theta}^2) \tag{2.7}$$

and perpendicular to this direction,

$$-mg\sin\theta = m(r\ddot{\theta} + 2\dot{r}\dot{\theta}) \tag{2.8}$$

Equations (2.7) and (2.8) result from using Newton's law. This time we desire to retain the variables r and θ, so we need to determine the spring force in terms of these variables. This is pretty easy in this case, since r is the current length of the spring; thus,

$$T = k(r - l_0) \tag{2.9}$$

The final result is

$$\begin{aligned}\ddot{r} &= r\dot{\theta}^2 + g\cos\theta - \frac{k}{m}(r - l_0) \\ \ddot{\theta} &= -\frac{g}{r}\sin\theta - \frac{2\dot{r}\dot{\theta}}{r}\end{aligned} \tag{2.10}$$

Equations (2.10) are two nonlinear second-order differential equations which, if solved, would predict the motion–time history of the system. These equations look nothing like Eqs. (2.6), yet they represent the same system. Only the coordinates being used are different. Polar coordinates appeared to be much more difficult than rectangular coordinates when they were introduced in Chapter 1, but the final result for this example was obtained much more easily using polar coordinates.

The reader might have noticed that the final results for both coordinate frames were presented with the highest derivative on the left-hand side of the equations and all lower derivative terms on the right-hand side. This is a very typical presentation for starting the next step toward a solution.

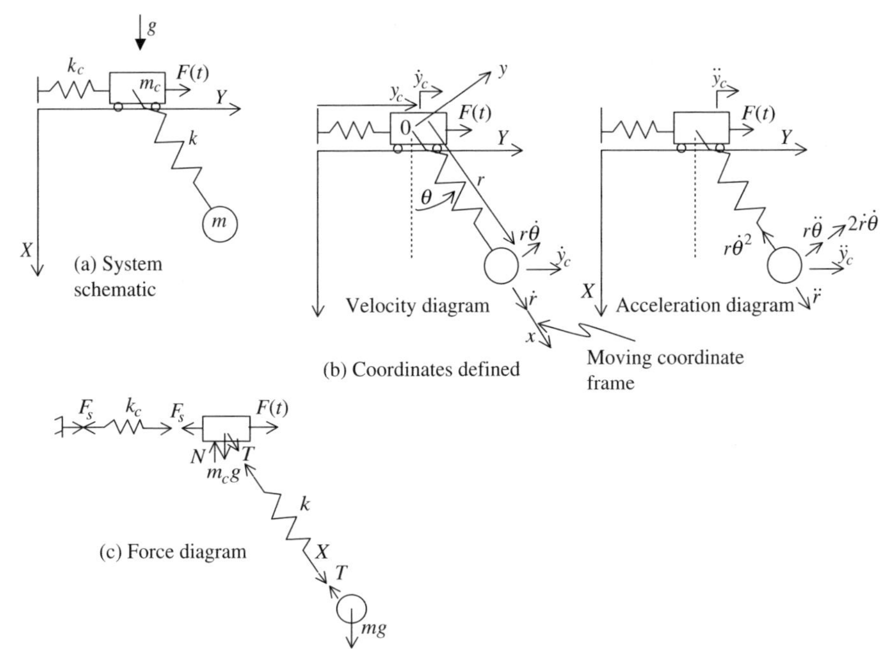

Figure 2.2 Equations of motion for the spring–pendulum attached to a cart.

Example 2.2: Spring Pendulum Attached to a Cart In Figure 2.2 the system if Figure 1.4 is repeated along with the force diagram from Figure 1.6, where the spring–pendulum is attached to a cart of mass m_c. The coordinates for this example are the rectangular coordinate y_c for the cart, and polar coordinates r and θ locate the pendulum mass. The velocity and acceleration diagrams made use of the relative motion relationships of Eqs. (1.17). For the velocity and acceleration diagrams of the pendulum mass, consider the moving coordinate frame shown in Figure 2.2*b*. The base of the frame is the point 0, and the x-axis of the frame is aligned along the spring. The angular velocity vector of the frame has length $\dot{\theta}$ and is pointing out of the page by the right-hand rule. The vector from the base of the frame to the mass element has length r and is oriented as shown. Using Eqs. (1.17), the velocity and acceleration components shown in the figure are obtained.

For the cart we derive the equation of motion in the horizontal direction as

$$F(t) + T\sin\theta - F_s = m_c\ddot{y}_c \tag{2.11}$$

For the pendulum mass one choice is to use the orthogonal directions along the spring and perpendicular to the spring: thus,

$$mg\cos\theta - T = m(\ddot{r} - r\dot{\theta}^2 + \ddot{y}_c\sin\theta)$$

along the spring and

$$-mg\sin\theta = m(r\ddot{\theta} + 2\dot{r}\dot{\theta} + \ddot{y}_c\cos\theta) \tag{2.12}$$

perpendicular to the spring. Equations (2.11) and (2.12) complete the use of Newton's law for this system. Once again it is pretty clear that we are not done. We have three equations but more than three unknowns.

The pendulum spring generates a force proportional to the amount it is stretched beyond its free length, l_{0p}:

$$T = k(r - l_{0\,p}) \tag{2.13}$$

and the cart spring generates a force proportional to its stretch from its free length l_{0c}:

$$F_s = k_c(y_c - l_{0c}) \tag{2.14}$$

Equations (2.11) through (2.14) can be combined and cast into the final form

$$\begin{aligned} \ddot{y}_c &= \frac{F(t)}{m_c} + \frac{k}{m_c}(r - l_{0\,p})\sin\theta - \frac{k_c}{m_c}(y_c - l_{0c}) \\ \ddot{r} + \ddot{y}_c \sin\theta &= g\cos\theta - \frac{k}{m}(r - l_{0\,p}) + r\dot{\theta}^2 \\ r\ddot{\theta} + \ddot{y}_c \cos\theta &= -g\sin\theta - 2\dot{r}\dot{\theta} \end{aligned} \tag{2.15}$$

Notice that once again the highest-order derivatives are placed on the left-hand side of the equations. In this example, however, more than one second-order term appears in two of the equations; thus, both second-order terms appear on the left-hand side. It turns out that simulating sets of equations when multiple second-order terms appear in several of the equations adds increased complexity to obtaining the solution. The excited reader will have to wait until Chapter 3 to see this.

2.2 DERIVING EQUATIONS OF MOTION WHEN RIGID BODIES ARE PART OF THE SYSTEM

When rigid bodies are part of a system then we must apply Newton's law to account for acceleration of the center of gravity, as described in Chapter 1 [Eq. (1.39b)] and account for rotation of the body according to Eq. (1.43). Thus, we isolate the mass elements and construct velocity, acceleration, and force diagrams, and we isolate the rigid bodies and construct acceleration diagrams for the centers of mass as well as indicate all the forces and moments acting on the body.

Example 2.3: Pinned Disk with a Slot and Mass Particle This is demonstrated in the example shown in Figure 2.3, which is repeated from Figure 1.10 The disk has rotational inertia I_d and is pinned at its center. Within the slot is a mass element m attached to the disk through a spring k. A prescribed input torque or moment $\tau_i(t)$ is acting on the disk. Friction is neglected at this time. The idea is to derive the

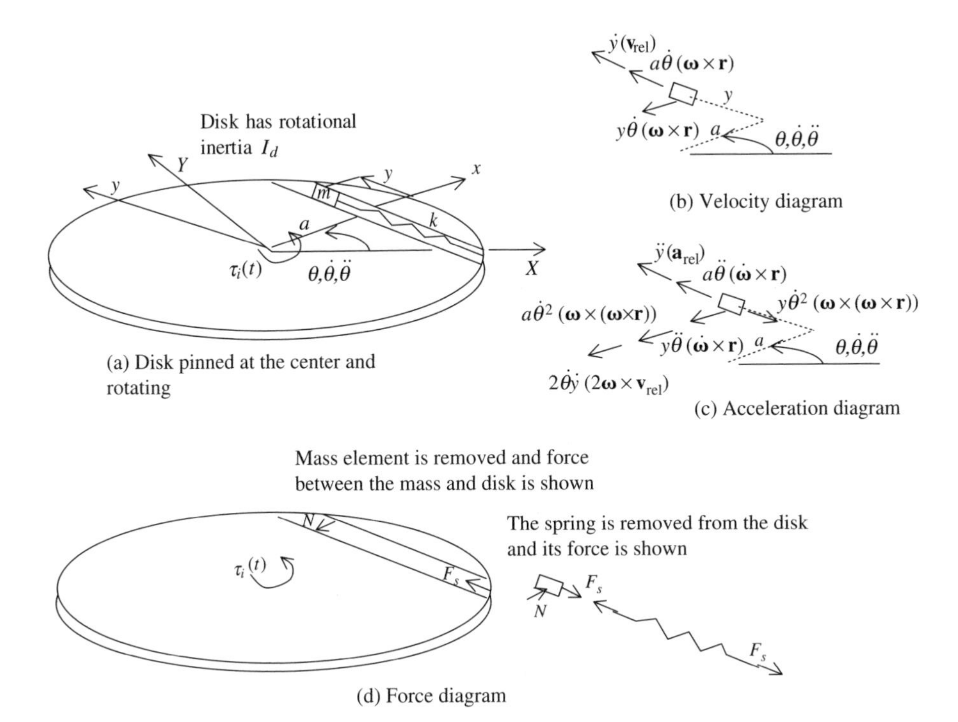

Figure 2.3 Rotation if circular disk with mass particle and spring.

equations of motion that would predict the motion–time history of this system for any prescribed $\tau_i(t)$. Figure 2.3*b* and *c* show the velocity and acceleration diagrams developed previously in Figure 1.10. In Figure 2.3*d* the mass element and spring are removed from the slot and the forces acting on the mass element and slot are shown using symbols and arrows. The normal force N is the force between the slot and the mass. It is shown in one direction on the mass and as equal in magnitude but opposite in direction on the slot. We really do not know in which direction the force is acting, so we just choose one and use it consistently while deriving equations. The spring force F_s is assumed positive in tension, as shown in the figure. It pulls one way on the mass and oppositely where the spring attaches to the disk. With accelerations and forces exposed, we are ready to write down the equations of motion.

For the disk, since it is pinned at its center of gravity, we do not have to use Newton's law for acceleration of the center of gravity. We need only write the moment equation, which becomes

$$\tau_i(t) + Ny + F_s a = \frac{d\mathbf{H}_c}{dt} = \frac{dI_d\boldsymbol{\omega}}{dt} = I_d\dot{\omega} \quad \text{direction out of the page}$$

or

$$\tau_i(t) + Ny + F_s a = I_d\ddot{\theta} \tag{2.16}$$

With N oriented as shown in Figure 2.3d, its moment arm about axes through the center of the disk is the length y. Its contribution to the total moment is Ny directed out of the page, which is the positive direction for angular rotation, as shown in the figure. The same is true for the spring force F_s. Its moment arm is the length a and it also contributes a positive moment $F_s a$. The sum of the moments equals the rate of change of angular momentum, also with positive orientation out of the page. Notice that we could have defined the normal force positive in the direction opposite that shown in the figure. This would change the sign of the moment term in Eq. 2.16. It would also have required a change in the normal force direction on the mass element. These two changes would result in a consistent use of the normal force, and the final equations would be the same regardless of positive force direction.

For the mass element we must apply Newton's law in any two orthogonal directions. The logical choice is to use the direction along the slot and perpendicular to it. Along the slot in the positive y-direction, we get

$$-F_s = m(\ddot{y} + a\ddot{\theta} - y\dot{\theta}^2) \tag{2.17}$$

and in the positive x-direction we get

$$N = m(-y\ddot{\theta} - a\dot{\theta}^2 - 2\dot{\theta}\dot{y}) \tag{2.18}$$

Having used Newton's laws, we see that there are more unknowns than equations, so our work is not quite done. We need to recognize that the spring force is related to the stretch in the spring from its free length. Let's assume here that the free length of the spring is equal to the distance from the attachment point on the disk to the y-axis, where the coordinate y equals zero. Thus, the force in the spring becomes

$$F_s = ky \tag{2.19}$$

The unknowns are now N, F_s, θ, and y, and there are four equations, (2.16)–(2.19).To present the final equations in a nicer form we can use (2.18) in (2.16) to get rid of N and use (2.19) in (2.16) and (2.17) to get rid of F_s. The final result is

$$\begin{aligned} (I_d + my^2)\ddot{\theta} &= -may\dot{\theta}^2 - 2my\dot{\theta}\dot{y} + kay + \tau_i(t) \\ \ddot{y} + a\ddot{\theta} &= y\dot{\theta}^2 - \frac{k}{m}y \end{aligned} \tag{2.20}$$

These are coupled, nonlinear differential equations that the reader probably thinks were too difficult to obtain and impossible to solve. But please notice that there is a certain repetitive consistency to our approach. We start with a schematic diagram of the system, we choose appropriate coordinates, we make velocity, acceleration, and force diagrams, and we then use Newton's laws to derive component equations in orthogonal directions for each mass particle and each rigid body. The rest is manipulation of the result into a nice presentation of the equations of motion. The reader will just have to wait until Chapter 3 to see how straightforward it is to numerically work out the solution to complicated sets of equations such as (2.20).

Before leaving this example it should be noted that the system of Figure 2.3 is actually an engineering device where the rotational speed of the disk is related to the frequency of vibration of the mass and spring. This device can be used as a tachometer to measure the rpm rate of the disk.

Example 2.4: Cart with Rigid-Body Pendulum Another example of equation derivation is shown in Figure 2.4*a*. This system consists of a cart of mass m_c with attached spring of stiffness k. The cart moves only horizontally. Attached to the cart is the rigid-body pendulum of mass m and centroidal moment of inertia I_c. A torsional spring is located between the cart and the pendulum at the pinned attachment point. A torsional spring ideally generates a torque opposing the angular twist of the spring. For a torsional spring stiffness k_τ the torque generated would be

$$\tau = k_\tau \theta \tag{2.21}$$

The units of the torsional spring constant are N.m/rad. An input force $F_i(t)$ acts on the cart. We desire equations of motion that would predict the motion–time history

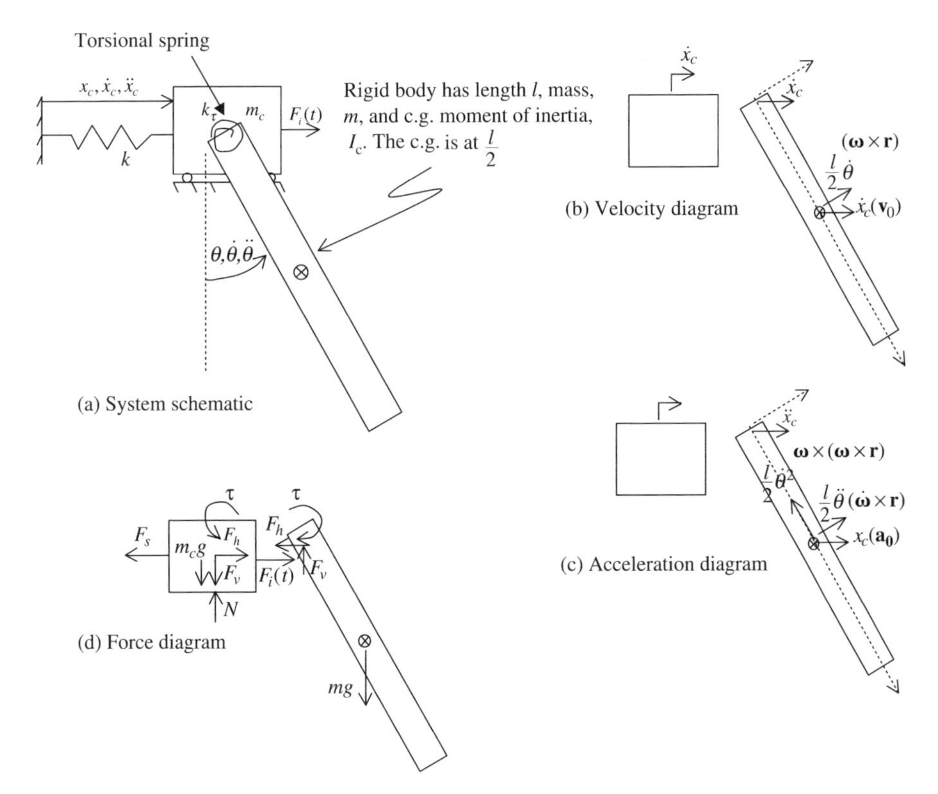

Figure 2.4 Cart with rigid-body pendulum.

of this system for any prescribed input force. The coordinates to be used are the position x_c and its derivatives and the angular position θ and its derivatives. This is a two-degree-of-freedom system.

This example system is actually part of a real engineering system called a *gantry robot*, a large manufacturing machine where an overhead track supports a cart that can move in plane motion. Extending from the cart vertically downward is a beam-like structure that supports a cutting tool at its end. The vertical structure can be extended and retracted. Hydraulic and electric motors drive the various components. In some applications the load being cut causes flexing and vibration of the vertical arm, and the surface finish is compromised. A good problem is to figure out how to control the drive motors so as to prevent vibration at the cutting tool end. To even consider doing such a problem, we need a system model; and a good starting point for such a model is shown in Figure 2.4.

Figure 2.4*b* shows the velocity diagram for the system components. The cart velocity $\dot{x}_c$ is transferred to the end of the rigid body. A moving frame is attached to the rigid body to help us use the transfer relationships derived in Chapter 1 and repeated here:

$$\begin{aligned} \mathbf{v}_p &= \mathbf{v}_0 + \boldsymbol{\omega} \times \mathbf{r} \\ \mathbf{a}_p &= \mathbf{a}_0 + \dot{\boldsymbol{\omega}} \times \mathbf{r} + \boldsymbol{\omega} \times (\boldsymbol{\omega} \times \mathbf{r}) \end{aligned} \tag{2.22}$$

The c.g. velocity components come from using the first of (2.22) as indicated in the figure and using our right hand to take the cross-product term. The acceleration diagram is shown in Figure 2.4*c*. Again, the acceleration is transferred from the end of the rigid body to the center of gravity. We need the acceleration of the center of gravity in order to use Newton's law properly for translational motion of a rigid body.

Figure 2.4*d* shows the forces and moments acting on the cart and rigid body. We know that some force vector exists at the attachment point between cart and rigid body, so we show orthogonal components and arbitrarily assign a positive direction. The forces are equal and opposite on the cart and rigid body. The gravity force is shown acting vertically downward on the cart and rigid-body center of gravity. A normal force is shown between the cart and the ground. The torque from the torsional spring is shown acting equal and opposite on the rigid body and the cart. In this model we are treating the cart as a mass particle or point mass. Strictly speaking, it has no rotational degree of freedom. Nevertheless, the spring torque is acting on the cart, and in a real physical system the cart would need to support this torque. In this example the torque will have no dynamic effect on the cart.

We are now ready to derive the equations of motion. For the cart in the positive x-direction, using the force and acceleration diagrams,

$$F_i(t) + F_h - F_s = m_c \ddot{x}_c \tag{2.23}$$

For the rigid body we first derive the translational equations in the horizontal and vertical directions. For the horizontal direction,

$$-F_h = m\left(\ddot{x}_c + \frac{l}{2}\ddot{\theta}\cos\theta - \frac{l}{2}\dot{\theta}^2\sin\theta\right) \tag{2.24}$$

and for the vertical direction,

$$F_v - mg = m\left(\frac{l}{2}\ddot{\theta}\sin\theta + \frac{l}{2}\dot{\theta}^2\cos\theta\right) \tag{2.25}$$

We next derive the rotational equation for the rigid body using the center of gravity as the point about which moments are taken. Recall that the rotational principle is simplest when using either a fixed point or the center of gravity. For this problem there is no fixed point, so the best choice is to use the center of gravity. Thus, in the positive rotational direction,

$$F_h\frac{l}{2}\cos\theta - F_v\frac{l}{2}\sin\theta - \tau = I_c\ddot{\theta} \tag{2.26}$$

One could write the vertical equation for the cart, but it contributes nothing to the dynamics of the system. Since no vertical cart motion is permitted in this model, we would simply determine that the cart is in force equilibrium in the vertical direction. The spring force F_s in Eq. (2.23) is related to the cart displacement by

$$F_s = k(x_c - x_{fl}) \tag{2.27}$$

Equations (2.23) through (2.26) are the result of applying Newton's laws to our system. After substituting Eq. (2.21) for the spring torque τ and Eq. (2.27) for the spring force F_s, there are equations for the four unknowns F_h, F_v, x_c, and θ. Equations (2.23) through (2.26) are not in a good form for computer solution, but there are sufficient equations for the number of unknowns, so a solution is possible. A nicer presentation would result from using (2.24) and (2.25) in (2.26) and get rid of F_h and F_v, and then use (2.24) in (2.23) to get rid of F_h in that equation. After doing some trigonometric simplifications, the final result is

$$\begin{aligned}(m_c + m)\ddot{x}_c + m\frac{l}{2}\ddot{\theta}\cos\theta &= m\frac{l}{2}\dot{\theta}^2\sin\theta - k(x_c - x_{fl}) + F_i(t)\\ m\frac{l}{2}\ddot{x}_c\cos\theta + \left[I_c + m\left(\frac{l}{2}\right)^2\right]\ddot{\theta} &= -k_\tau\theta - mg\frac{l}{2}\sin\theta\end{aligned} \tag{2.28}$$

All higher-order derivatives are on the left side and lower-order derivatives are on the right of (2.28).

Example 2.5 Slider–Crank System In this example, a kinematic constraint affects the derivation of the equations of motion. The system is the slider–crank mechanism shown in Figure 1.14, but this time there is a spring k and mass m attached at the right end. This end is constrained to move only horizontally, and the attached mass is also constrained to move only horizontally. The crank of radius R has a prescribed input angular velocity; that is, $\dot{\theta} = \omega_i(t)$, so $\theta(t)$ is known and θ does not contribute a degree-of-freedom to the system. The coordinates chosen to describe this system are the x, y location of the center of gravity, the angular position α of the rod measured positive in the clockwise direction, and the relative position of the mass particle, x_c. We will see that a kinematic constraint reduces the number of coordinates needed for this system. But for now it is clear that if one knew x, y, α, and x_c, the system could be placed in any of its possible configurations.

Figure 2.5*b* is a velocity diagram for this system. The c.g. components of velocity of the rod have been transferred to each end using the transfer velocity relationship from Eq. (2.22). When carrying out the cross-product terms, note that the angular velocity vector is positive *into* the page for this example and convince yourself that the component directions are correct for this sign convention. The velocity of the mass particle shares the velocity at the right end of the rod plus the additional relative velocity component $\dot{x}_c$. Noted on Figure 2.5*b* is the kinematic constraint at the right end of the rod. This end can have no vertical velocity, and this component is set to zero in the constraint equation.

Figure 2.5*c* is the acceleration diagram for this system. Again the transfer relationship of Eq. (2.22) was used to assign a direction and magnitude to each component. The point mass has the acceleration of the rod end plus the additional relative acceleration $\ddot{x}_c$. Figure 2.5*d* shows the forces acting in equal but opposite directions at each attachment point on the bodies. Note that the components are given names and shown in orthogonal directions, as is required when we know that the force is a vector but we do not know its magnitude or direction.

We can now derive the equations of motion. In the x-direction for the rod,

$$-F_h - F_s = m_r \ddot{x} \tag{2.29}$$

and for the y-direction,

$$F_v + N = m_r \ddot{y} \tag{2.30}$$

The rotational equation for the rod becomes

$$F_v a \cos\alpha - F_h a \sin\alpha + F_s b \sin\alpha - N b \cos\alpha = I_r \ddot{\alpha} \tag{2.31}$$

Readers should make certain of their understanding that the moment component directions are consistent with the positive force directions and positive angular rotation direction chosen.

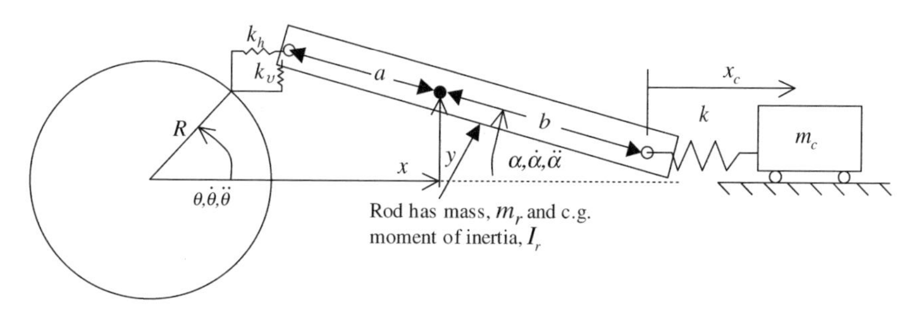

(a) System schematic

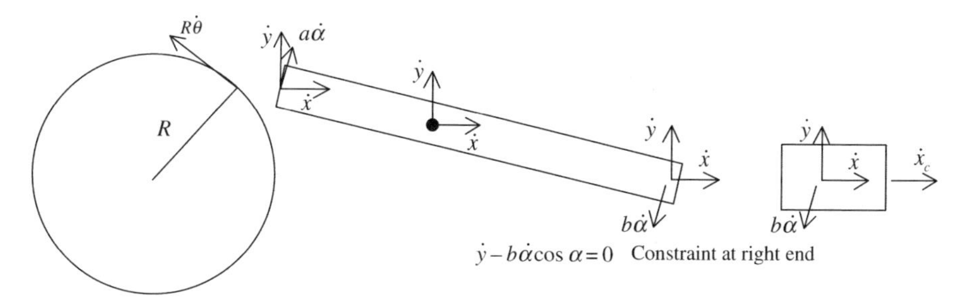

(b) Velocity diagram

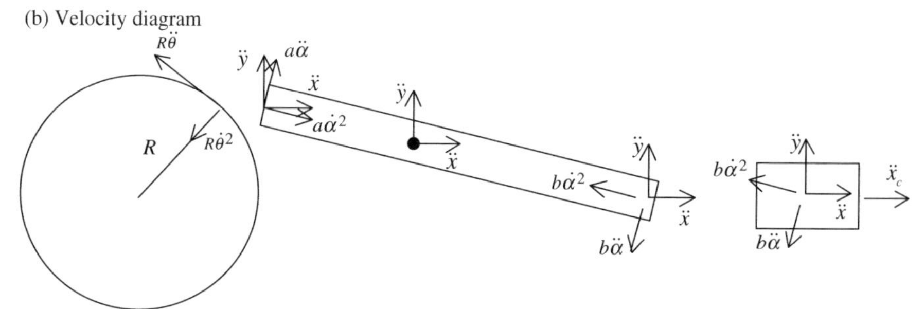

(c) Acceleration diagram

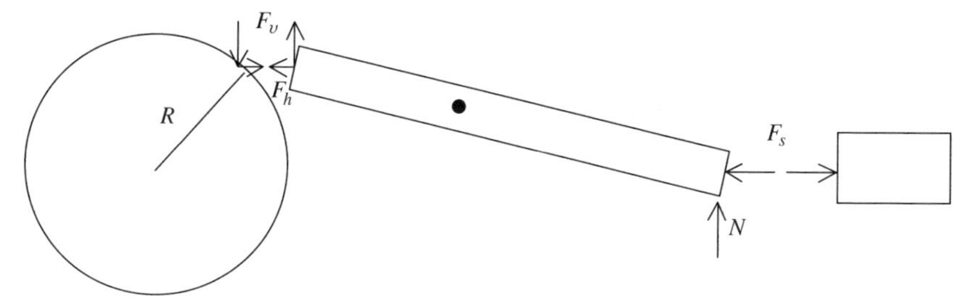

(d) Force diagram

Figure 2.5 Slider–crank mechanism with attached spring and mass.

The equation for the mass particle is

$$F_s = m_c(\ddot{x} + \ddot{x}_c - b\ddot{\alpha}\sin\alpha - b\dot{\alpha}^2\cos\alpha) \tag{2.32}$$

In addition to the equations of motion from Newton's laws, there is the constraint equation for the vertical motion at the right end of the rod,

$$\dot{y} - b\dot{\alpha}\cos\alpha = 0 \tag{2.33}$$

Also at the right end of the rod, the spring force comes from

$$F_s = k(x_c - x_{\text{fl}}) \tag{2.34}$$

where x_{fl} is the spring free length.

At the left end of the rod the horizontal spring is positive in tension and the relative velocity across the spring is

$$v_{\text{rel}_h} = \dot{x} + a\dot{\alpha}\sin\alpha + R\dot{\theta}\sin\theta \tag{2.35}$$

The vertical spring is positive in compression and the relative velocity across this spring is

$$v_{\text{rel}_v} = R\dot{\theta}\cos\theta - (\dot{y} + a\dot{\alpha}\cos\alpha) \tag{2.36}$$

Usually, we express the forces in springs by their dependence on the displacement of the spring from its free length. The time derivative of the displacement of a spring is the relative velocity across the spring. So an alternative expression for the forces in the springs at the left end of the rod is

$$\begin{aligned} \dot{F}_h &= k_h v_{\text{rel}_h} \\ \dot{F}_v &= k_v v_{\text{rel}_v} \end{aligned} \tag{2.37}$$

Since ultimately, we will use a computer to march out a solution to the equations of motion, we can use this form for the spring forces, and a simulation algorithm will integrate Eqs. (2.37) along with the other equations of motion. This possibility is discussed further in Chapter 3.

If we do not use the approach from Eqs. (2.37) to assess the spring forces, we must express the spring displacements using the coordinates of the problem. For the horizontal spring,

$$F_h = k_h[(x - a\cos\alpha) - R\cos\theta] \tag{2.38}$$

and for the vertical spring,

$$F_v = k_v[R\sin\theta - (y + a\sin\alpha)] \tag{2.39}$$

Notice that the time derivative of Eqs. (2.38) and (2.39) reproduces Eq. (2.37) with (2.35) and (2.36). Also, in (2.38), and (2.39) the spring free lengths have been assumed to be zero.

From the equations of motion, Eqs.(2.29)–(2.32), subject to the constraint equation (2.33) along with the spring force Equations (2.34), (2.38), and (2.39), we have eight equations in terms of the eight time-varying variables x, y, α, x_c, F_h, F_v, N, and F_s (remember that θ is not an unknown variable since we are prescribing the angular motion of the crank as an input to the system). Thus, in principle, we have sufficient equations to pursue a solution. At this point we could try to simplify the governing equations by substituting for some variables in terms of others. For example, (2.33) can be used to substitute for y and its derivatives in terms of α and its derivatives. Thus, the variable y would no longer appear in the final equations. The spring forces F_s, F_h, and F_v from Eqs. (2.34), (2.38) and (2.39) can be used in Eqs. (2.29)–(2.32) and these forces will no longer appear in the final equations. Finally, Eq. (2.30) can be solved for the normal force N and used in (2.31) so that N no longer appears. Ultimately, we would end up with three coupled second-order nonlinear differential equations in terms of the coordinate variables x, x_c, and α.

It is certainly a challenging exercise in algebraic manipulation to reduce eight equations in terms of eight unknowns to three equations in terms of three unknowns, but are we any closer to a solution? Chapter 3 is devoted to computer solution of sets of nonlinear and linear differential equations. It will be shown that the simplest form of the equations is not required if a computational solution is the ultimate goal. Many of the manipulations demonstrated in earlier examples and described in this final example are not necessary for computational solution. What is absolutely essential is that appropriate coordinates be chosen, that velocity and acceleration of appropriate points in the system be expressed in terms of these coordinates, that equal and opposite force components be labeled and shown acting orthogonally at appropriate points of the system, and that Newton's laws be applied appropriately in orthogonal directions for point masses and rigid bodies. Kinematic constraints need to be recognized and expressed in terms of the coordinates. If these steps are carried out correctly, computer solution is quite straightforward.

2.3 FORMS OF EQUATIONS AND THEIR COMPUTATIONAL SOLUTION

In Chapter 1 and the sections above we developed fundamental dynamic principles and showed how to apply them to mass particles and rigid bodies. The result of constructing velocity, acceleration, and force diagrams and using Newton's laws is always sets of second-order differential equations. For most engineering systems these equations are nonlinear, and the nonlinearities typically result from geometric considerations of system motion and from nonlinear spring force and damper force relationships. Sometimes we can justify assuming that the system is linear, which means that the governing equations of motion are sets of linear second-order differential equations. When the equations of motion are linear, sometimes an analytical solution can be pursued. For most systems a computational solution is required whether governed by linear or by nonlinear equations of motion. In this section, the most suitable forms of equations for computational solution are developed. The concept of time-step simulation is described. Simulation of resulting sets of equations is demonstrated in Chapter 3.

First-Order State Equations

When equations of motion are derived using direct application of Newton's laws, sets of second-order differential equations result. It turns out that this form of equation, regardless of simplifications carried out to reduce numbers of variables, is not particularly amenable to computer solution. Some work needs to be done to get the governing equations into a form suitable for computation. The following describes various forms of reduced-order equations. This is followed by a description of how sets of second-order equations can be reduced to first-order form.

Explicit Form

The simplest form for equations intended for computer solution is explicit first-order state equations, which have the general appearance

$$\dot{\mathbf{x}} = \mathbf{f}(\mathbf{x}, \mathbf{u}) \tag{2.40}$$

where $\mathbf{x}$ is a vector of state variables and $\mathbf{u}$ is a vector of inputs such as prescribed forces or moments. The vector function $\mathbf{f}(\cdot)$ is shorthand notation indicating that the right-hand side of (2.40) is composed of nonlinear functions of the variables and inputs. A more functional form of (2.40) is

$$\begin{aligned}
\dot{x}_1 &= f_1(x_1, x_2, \ldots, x_n, u_1, u_2, \ldots, u_r) \\
\dot{x}_2 &= f_2(x_1, x_2, \ldots, x_n, u_1, u_2, \ldots, u_r) \\
\dot{x}_3 &= f_3(x_1, x_2, \ldots, x_n, u_1, u_2, \ldots, u_r) \\
&\vdots \\
\dot{x}_n &= f_n(x_1, x_2, \ldots, x_n, u_1, u_2, \ldots, u_r)
\end{aligned} \tag{2.41}$$

where $x_1, x_2, \ldots, x_n$ are the state variables and will be composed mostly of the position and angular position coordinates and their derivatives for the system being modeled. This form of equation is absolutely trivial for a computer to solve. There are many commercially available programs that will march out solutions to equations of the form (2.41). References [1–4] are just a sampling of programs available. The challenge is to get the sets of second-order differential equations that come from application of Newton's laws into this first-order form. When this is accomplished, developing a solution is pretty simple.

The Fundamentals of Computer-Developed Time-Step Simulation

To solve equations of the form (2.41), they are first written as difference equations,

$$\begin{aligned}
\Delta x_1 &= f_1(x_1, x_2, \ldots, x_n, u_1, u_2, \ldots, u_r)\Delta t \\
\Delta x_2 &= f_2(x_1, x_2, \ldots, x_n, u_1, u_2, \ldots, u_r)\Delta t \\
&\vdots \\
\Delta x_n &= f_n(x_1, x_2, \ldots, x_n, u_1, u_2, \ldots, u_r)\Delta t
\end{aligned} \tag{2.42}$$

Then, starting from the initial conditions for each state variable and an appropriately chosen time step, Δt, the functions are evaluated at $t = 0$ and the change in each state variable is computed by multiplying by Δt. The change in each state variable is then added to its previous value, and we have the new values of the state variables at the new time, $t + \Delta t$. The new values of state variables are reinserted into the functions, and new changes are computed, and the solution continues to be developed. What has been described here is called *Euler integration* [2.5], the simplest of all the integration algorithms. Commercial equation solvers typically offer many choices for the integration algorithm and typically have a default algorithm that works for most problems. The computational procedure can be quite involved for the higher-order algorithms, but basically they all do time-step simulation, as just described.

When the physical components comprising an engineering system interact, there can be high- and low-frequency oscillations; some oscillations; decay rapidly and others decay slowly. For computers to provide solutions to systems of equations; the computational step size Δt must be significantly shorter than the period of highest-frequency oscillation or fastest decay rate. For most commercial solvers the user must input an initial time step Δt, but the algorithm determines if the users' choice is a good one and automatically adjusts the time step to be appropriate for the simulation. It is not too difficult to make some estimates as to expected frequencies and speed of response, so choosing a reasonable starting time step is really not that difficult.

Implicit Form

It sometimes happens that when reducing sets of second-order equations to first-order form, the derivative of some variables appears on the right-hand side as shown in the general form

$$\dot{\mathbf{x}} = \mathbf{f}(\mathbf{x}, \mathbf{u}, \dot{\mathbf{x}}) \tag{2.43}$$

This form of the equations is called *implicit first-order form*. Very few numerical algorithms handle implicit equations of this form. They work by "guessing" the derivatives and then using these guesses in the right-hand side of the equations to reproduce the same derivatives on the left-hand side. Much algebra is needed for the guesses to be modified until the right-hand side reproduces the left-hand side to within some level of precision, at which time an integration step takes place in conventional manner.

When this form occurs, some algebra needs to be done to get the derivatives on the right side of the equations over to the left-side. The resulting left-hand side can be put into matrix form and then inverted as part of the computational procedure. In some cases the inversion needs to be done at each time step of the simulation.

Differential Algebraic Form

When reducing sets of second-order equations to first-order form, it sometimes happens that unwanted variables enter the formulation that are dependent on the variables desired as well as dependent on themselves. This typically occurs when using kinematic constraints to reduce the number of coordinate variables. The resulting equations are called *differential algebraic equations* (DAEs). In general form, these look like

$$\begin{aligned}\dot{\mathbf{x}} &= \mathbf{f}(\mathbf{x}, \mathbf{u}, \mathbf{x}) \\ \mathbf{z} &= \mathbf{g}(\mathbf{x}, \mathbf{u}, \mathbf{z})\end{aligned} \tag{2.44}$$

where the vector $\mathbf{z}$ contains unwanted variables and $\mathbf{g}$ is a vector of algebraic functions. The computational procedure is to manipulate the auxiliary variables until the algebraic equations match the right- and left-hand sides to within some prescribed level of precision, at which time a conventional integration step is taken.

There are numerical algorithms that will handle DAEs, but there are far more that will easily produce solutions to many explicit first-order equations. Since models of physical systems are built by the modeler, it is always possible to make modeling decisions that result in explicit first-order state equations.

2.4 REDUCING SETS OF SECOND-ORDER DIFFERENTIAL EQUATIONS TO FIRST-ORDER FORM

Example 2.6: Spring Pendulum Since explicit first-order state equations are the most straightforward to solve computationally, it is a good idea to try and put systems of second-order equations into the explicit form of Eqs. (2.40) or (2.41). As a first example of reducing governing equations to first-order form, we return to the spring pendulum of Figure 2.1 with the final equations of motion repeated here from Eqs. (2.10):

$$\begin{aligned}\ddot{r} &= r\dot{\theta}^2 + g\cos\theta - \frac{k}{m}(r - l_0) \\ \ddot{\theta} &= -\frac{g}{r}\sin\theta - \frac{2\dot{r}\dot{\theta}}{r}\end{aligned} \tag{2.45}$$

As discussed above, it is typical to write the final equations with the highest-order derivative of variables on the left side and all the other terms on the right side. When a single second-order derivative appears in each equation, as is the case in (2.45), it is particularly simple to reduce to first-order form. The procedure is to introduce the first-order derivative of each coordinate variable by defining it with a new symbol;

thus,

$$\begin{aligned} \frac{dr}{dt} &= v_r \\ \frac{d\theta}{dt} &= \omega \end{aligned} \tag{2.46}$$

The symbol v_r was chosen since the rate of change of r is a velocity in the r-direction. The symbol ω was chosen since the rate of change of an angle is angular velocity, and ω is a very typical symbol to use for angular velocity.

The next step is to recognize that the rates of change of v_r and ω are the second derivatives of the coordinates in Eq. (2.45) and are equal to the right-hand side of (2.45):

$$\begin{aligned} \frac{dv_r}{dt} &= r\omega^2 + g\cos\theta - \frac{k}{m}(r - l_0) \\ \frac{d\omega}{dt} &= -\frac{g}{r}\sin\theta - \frac{2v_r\omega}{r} \end{aligned} \tag{2.47}$$

In (2.47), substitutions have been made on the right-hand side for the symbols $\dot{r}$ and $\dot{\theta}$ in terms of their respective definitions from (2.46). The reader should verify that Eqs. (2.46) and (2.47) are a set of explicit first-order equations in terms of the state variables r, θ, v_r, and ω. The first derivative of each state variable appears on the left side of the equations, and only the variables themselves appear on the right. The two second-order equations from (2.45) have been reduced to four first-order equations in explicit form. Once the parameters are specified, an initial condition needs to be assigned to each state variable, and then simulation of these very nonlinear equations is quite simple. Simulation is demonstrated in Chapter 3.

Example 2.7: Spring Pendulum Attached to a Cart Figure 2.2 shows the spring pendulum from the first example attached to a horizontally moving cart with an applied force and a spring. The equations of motion were derived in Eqs. (2.15) and are repeated here for convenience as Eqs. (2.48). As noted above, the second derivatives of the coordinate variables appear on the left-hand side of the equations, and all the remaining terms appear on the right-hand side. The input force $F(t)$ must be prescribed by the user along with the parameters, including masses, spring stiffnesses, and spring free lengths l_{0p} and l_{0c}.

$$\begin{aligned} \ddot{y}_c &= \frac{F(t)}{m_c} + \frac{k}{m_c}(r - l_{0p})\sin\theta - \frac{k_c}{m_c}(y_c - l_{0c}) \\ \ddot{r} + \ddot{y}_c\sin\theta &= g\cos\theta - \frac{k}{m}(r - l_{0p}) + r\dot{\theta}^2 \\ r\ddot{\theta} + \ddot{y}_c\cos\theta &= -g\sin\theta - 2\dot{r}\dot{\theta} \end{aligned} \tag{2.48}$$

Also noted above, there is increased complication in that multiple second-order terms appear in two of the equations. For this example the explicit first-order form can be obtained by using the first equation of (2.48) in the second and third equations of (2.48). If this substitution is done, the result will be one equation with $\ddot{y}_c$ only on the left side, one equation with $\ddot{r}$ only on the left side, and one equation with $\ddot{\theta}$ only on the left. We then would define

$$\begin{aligned}\frac{dy_c}{dt} &= v_c \\ \frac{dr}{dt} &= v_r \\ \frac{d\theta}{dt} &= \omega\end{aligned} \tag{2.49}$$

and then

$$\begin{aligned}\frac{dv_c}{dt} &= \text{right side of the first of (2.48)} \\ \frac{dv_r}{dt} &= \text{right side of the second of (2.48)} \\ \frac{d\omega}{dt} &= \text{right side of the third of (2.48)}\end{aligned} \tag{2.50}$$

all after the substitution just described. The result would be six explicit first-order equations in terms of the six state variables y_c, r, θ, v_c, v_r, and ω.

Since we are after a computational result, not an analytical one, it is not necessary to reduce the equations to their final, most attractive form. For this example we can write the first of Eqs. (2.48) as

$$\begin{aligned}\frac{dy_c}{dt} &= v_c \\ \frac{dv_c}{dt} &= \frac{F(t)}{m_c} + \frac{k}{m_c}(r - l_{0\,p})\sin\theta - \frac{k_c}{m_c}(y_c - l_{0c})\end{aligned} \tag{2.51}$$

and then the second two equations as

$$\begin{aligned}\frac{dr}{dt} &= v_r \\ \frac{d\theta}{dt} &= \omega \\ \frac{dv_r}{dt} &= -\frac{dv_c}{dt}\sin\theta + g\cos\theta - \frac{k}{m}(r - l_{0\,p}) + r\omega^2 \\ \frac{d\omega}{dt} &= -\frac{dv_c}{dt}\frac{\cos\theta}{r} - \frac{g}{r}\sin\theta - \frac{2v_r\omega}{r}\end{aligned} \tag{2.52}$$

Equations (2.51) and (2.52) are not officially in explicit first-order form, but since dv_c/dt is evaluated in terms of the state variables in (2.51), its value is known for use in Eqs. (2.52). Computationally, this form is just fine and we need not reduce the equations further.

Example 2.8: Pinned Disk with a Slot and Mass Particle Figure 2.3 shows the system used above to demonstrate equation derivation. The resulting second-order equations are Eqs. (2.20) and are repeated here:

$$\begin{aligned} (I_d + my^2)\ddot{\theta} &= -may\dot{\theta}^2 - 2my\dot{\theta}\dot{y} + kay + \tau_i(t) \\ \ddot{y} + a\ddot{\theta} &= y\dot{\theta}^2 - \frac{k}{m}y \end{aligned} \tag{2.53}$$

Reducing these equations to first-order form is very similar to that done in Example 2.7. The first equation of (2.53) is cast into first-order form by

$$\begin{aligned} \dot{\theta} &= \omega \\ \dot{y} &= v_y \\ \dot{\omega} &= -\frac{may\omega^2}{I_d + my^2} - \frac{2my\omega v_y}{I_d + my^2} + \frac{kay}{I_d + my^2} + \frac{\tau_i(t)}{I_d + my^2} \end{aligned} \tag{2.54}$$

and the second equation of (2.53) can be written as

$$\dot{v}_y = -a\dot{\omega} + y\omega^2 - \frac{k}{m}y \tag{2.55}$$

Since $\dot{\omega}$ is known from (2.54) in terms of the state variables θ, y, ω, and v_y, we can compute $\dot{\omega}$ from (2.54) starting with initial conditions for the state variables and then use its value in (2.55). No further manipulation of the equations is required.

It is interesting to note that there is no specific dependence on the angular position of the disk, θ. Thus, the first of (2.54) is really not required to simulate the motion of the system. When a state equation exists for a state variable and no other equations depend on that variable, that equation is called a *free integrator*. This means that the computational algorithm will simply integrate that equation along with the other equations and make that variable (θ in this example) available for observation. It may be desirable to plot the angular position of the disk, and the first of (2.54) allows this. However, the dynamics of the system do not depend on θ.

Example 2.9: Inverted Rigid-Body Pendulum on a Cart A device very typically used as a laboratory demonstration in automatic control is the inverted pendulum shown in Figure 2.6. The pendulum is a rigid body of mass m and centroidal moment of inertia I_g. The pendulum is pinned to a cart of mass m_c. The cart is constrained to move only horizontally and has an attached spring of stiffness k. An external force $F(t)$ acts on the cart. In a controls laboratory the idea would be to measure the angle

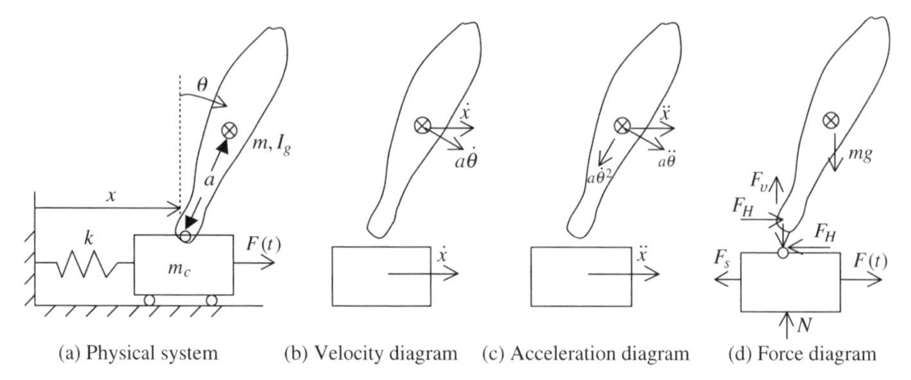

Figure 2.6 Inverted pendulum.

θ and perhaps the cart position x and design an automatic controller for the input force that would balance the pendulum and keep it oriented vertically upward. For the example here we derive the equations of motion and cast them into first-order form.

Figure 2.6 shows velocity, acceleration, and force diagrams for the system. The force vector at the pin joint is shown as two orthogonal components aligned horizontally and vertically. Note that they are shown equal and opposite acting on the pendulum and the cart. The spring force is given by

$$F_s = k(x - x_{\text{fl}}) \tag{2.56}$$

where x_{fl} is the free length of the spring. Newton's law for the cart yields

$$F - k(x - x_{fl}) - F_H = m_c \ddot{x} \tag{2.57}$$

For the center of gravity of the rigid body, the equations of motion become

$$\begin{aligned} F_H &= m(\ddot{x} + a\ddot{\theta}\cos\theta - a\dot{\theta}^2 \sin\theta) \\ F_V - mg &= m(-a\ddot{\theta}\sin\theta - a\dot{\theta}^2 \cos\theta) \end{aligned} \tag{2.58}$$

and for rotation of the rigid body the moment equation becomes

$$F_V a \sin\theta - F_H a \cos\theta = I_g \ddot{\theta} \tag{2.59}$$

One way to reduce Eqs. (2.57) through (2.59) to first-order form is to substitute for the horizontal and vertical forces by substituting for F_H from (2.58) in (2.57) and (2.59) and substituting for F_V from (2.58) in (2.59). The result will be two equations in the two unknowns x and θ and their derivatives. The final result is

$$\begin{aligned} (m_c + m)\ddot{x} + ma\ddot{\theta}\cos\theta &= F - k(x - x_{fl}) + ma\dot{\theta}^2 \sin\theta \\ ma\ddot{x}\cos\theta + (I_g + ma^2)\ddot{\theta} &= mga\sin\theta \end{aligned} \tag{2.60}$$

We see that the second derivative of each of the variables enters into both equations, which further complicates reduction to first-order form. In this example we could solve either of Eqs. (2.60) for, say, $\ddot{\theta}$ and substitute into the other equation. This would result in one equation with an $\ddot{x}$ term on both the left- and right-hand sides. We then could solve for $\ddot{x}$ and proceed toward reduction to first-order form.

Another approach is to write Eqs. (2.60) in matrix form

$$\begin{bmatrix} m_c + m & ma\cos\theta \\ ma\cos\theta & I_g + ma^2 \end{bmatrix} \begin{bmatrix} \ddot{x} \\ \ddot{\theta} \end{bmatrix} = \begin{bmatrix} F - k(x - x_{fl}) + ma\dot{\theta}^2 \sin\theta \\ mga\sin\theta \end{bmatrix} \tag{2.61}$$

and then multiply both sides by the inverse of the first matrix on the left side. For this 2×2 matrix, the inverse can be generated analytically by interchanging the diagonal terms, changing the sign of the off-diagonal terms, and dividing by the determinant of the matrix. For this example, if the first matrix is $\mathbf{M}$, then

$$\mathbf{M}^{-1} = \frac{\begin{bmatrix} I_g + ma^2 & -ma\cos\theta \\ -ma\cos\theta & m_c + m \end{bmatrix}}{(m_c + m)I_g + m^2a^2\left[(m_c/m) + \sin^2\theta\right]} \tag{2.62}$$

Carrying out the matrix multiplication with the vector on the right-hand side of (2.61) yields

$$\begin{aligned} \ddot{x} &= \frac{(I_g+ma^2)\left[F-k(x-x_{fl})+ma\dot{\theta}^2\sin\theta\right]}{(m_c+m)I_g+m^2a^2\left[(m_c/m)+\sin^2\theta\right]} - \frac{ma\cos\theta(mga\sin\theta)}{(m_c+m)I_g+m^2a^2\left[(m_c/m)+\sin^2\theta\right]} \\ \ddot{\theta} &= -\frac{ma\cos\theta\left[F-k(x-x_{fl})+ma\dot{\theta}^2\sin\theta\right]}{(m_c+m)I_g+m^2a^2\left[(m_c/m)+\sin^2\theta\right]} + \frac{(m_c+m)(mga\sin\theta)}{(m_c+m)I_g+m^2a^2\left[(m_c/m)+\sin^2\theta\right]} \end{aligned} \tag{2.63}$$

We would next define the left side of (2.63) as $\dot{v}_x$ and $\dot{\omega}$, respectively, and add the state equations,

$$\begin{aligned} \dot{x} &= v_x \\ \dot{\theta} &= \omega \end{aligned} \tag{2.64}$$

and we finally have explicit first-order state equations ready for simulation.

The reader may find this manipulation procedure rather forbidding and the resulting equations rather complicated. The fact is that you can rarely carry out these steps for real engineering systems because the number of variables is too large, and carrying out the algebraic manipulations by hand is virtually impossible. Fortunately, matrix inversion is quite easy for a computer, and most commercial equation solvers allow the user to define a matrix such as the one on the left side of (2.61) and then compute $\ddot{x}$ and $\ddot{\theta}$ (or $\dot{v}_x$ and $\dot{\omega}$) at each time step of the simulation. In this way the first-order derivatives are computed at each time step and an integration step can be taken.

In fact, there is an even easier stopping point than Eq. (2.61) if computation is the ultimate goal. Returning to Eqs. (2.57), (2.58), and (2.59), we could define a matrix equation at this point as

$$\begin{bmatrix} m_c & 0 & 1 & 0 \\ m & ma\cos\theta & -1 & 0 \\ 0 & ma\sin\theta & 0 & 1 \\ 0 & I_g & a\cos\theta & -a\sin\theta \end{bmatrix} \begin{bmatrix} \ddot{x} \\ \ddot{\theta} \\ F_H \\ F_V \end{bmatrix} = \begin{bmatrix} F - k(x - x_{fl}) \\ ma\dot{\theta}^2 \sin\theta \\ mg - ma\dot{\theta}^2 \cos\theta \\ 0 \end{bmatrix} \tag{2.65}$$

Then $\ddot{x}$ and $\ddot{\theta}$ (or $\dot{v}_x$ and $\dot{\omega}$) would be available through matrix inversion and multiplication. We might as well let the computer do all the work, and we do as little algebraic manipulation as possible. Also note that F_H and F_V are automatically available as outputs when using this approach.

2.5 MATRIX FORMS FOR LINEARIZED EQUATIONS

When physical systems execute small motions about some steady-state operating point, it is justified to linearize the equations about the operating point. Often, we are only interested in small motions of a physical system and we represent the system as linear from the beginning. Whether applying Newton's laws to a nonlinear system and then linearizing the result or starting with an assumed linear system, the resulting equations are sets of linear second-order differential equations. With nonlinear systems we are pretty much constrained to use simulation. However, much more analytical manipulation can be done with linear systems. Some of the analytical tools are discussed in subsequent chapters. For now, we are interested in the forms of these equations.

Quarter-Car Model for Vibration Analysis

A model frequently used for suspension studies in ground vehicles is the quarter-car model shown in Figure 2.7. It consists of a "sprung" mass m_s, representing the body of the vehicle, an "unsprung" mass m_{us}, representing the tires, wheels, and parts of the suspension; suspension elements k_s, and b_s; representing the suspension spring and damping; and the tire stiffness k_t, representing the compliance of the tire. An actual vehicle rolls horizontally over roadway unevenness, and this is represented in the model by the vertical roadway displacement input $x_i(t)$. The two variables needed for this two-degree-of-freedom system is the positions of the two masses x_{us} for the unsprung mass and x_s for the sprung mass, each measured with respect to an inertial reference frame.

Figure 2.7 also shows the velocity, acceleration, and force diagrams for the system. The acceleration diagram is particularly simple since the system is constrained to move only vertically and we are using displacement coordinates measured with respect to an inertial reference. The tire and spring forces, F_t and F_s, are related to their relative displacements from their free length, meaning that when x_i and x_{us} are

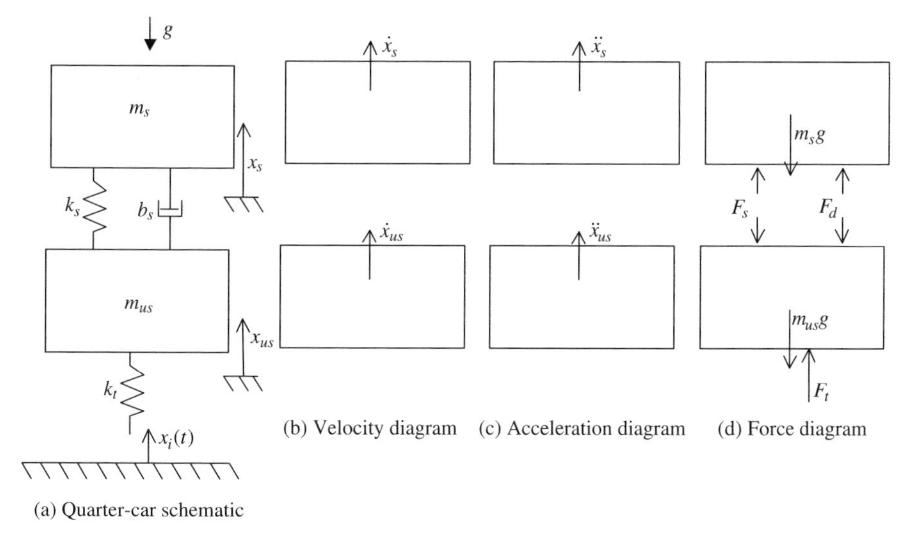

Figure 2.7 Quarter-car model.

both zero, the tire spring is at its free length, generating no force, and when x_{us} and x_s are both zero, the suspension spring is at its free length, generating no force. With this assumption and considering both springs positive in compression, the spring forces are

$$\begin{aligned} F_t &= k_t(x_i - x_{us}) \\ F_s &= k_s(x_{us} - x_s) \end{aligned} \tag{2.66}$$

Thus, for the tire force, if x_i is larger than x_{us}, the spring is in compression and the force is positive, as indicated in the figure. A similar argument can be made for the suspension spring force.

A linear damper generates a force proportional to the relative velocity across it. For the damper defined positive in compression as indicated in the figure by the direction of the damper force, the result is

$$F_d = b_s(\dot{x}_{us} - \dot{x}_s) \tag{2.67}$$

If $\dot{x}_{us}$ is larger than $\dot{x}_s$, the damper is in compression and the force is positive from Eq. (2.67) and has the direction indicated in the figure.

The equations of motion come from application of Newton's laws such that

$$\begin{aligned} k_s(x_{us} - x_s) + b_s(\dot{x}_{us} - \dot{x}_s) - m_s g &= m_s \ddot{x}_s \\ k_t(x_i - x_{us}) - k_s(x_{us} - x_s) - b_s(\dot{x}_{us} - \dot{x}_s) - m_{us} g &= m_{us} \ddot{x}_{us} \end{aligned} \tag{2.68}$$

Since the mass positions are measured from the free length of the springs, there will be a compression in the springs if the system is at rest under the influence of gravity. This steady-state operating point can be found by letting all time-varying terms equal zero in Eqs. (2.68). If there is no acceleration or velocity, then

$$\ddot{x}_s = \dot{x}_s = \ddot{x}_{us} = \dot{x}_{us} = 0 \tag{2.69}$$

and the equations of motion reduce to

$$\begin{aligned} k_s(x_{us} - x_s) - m_s g &= 0 \\ k_t(0 - x_{us}) - k_s(x_{us} - x_s) - m_{us} g &= 0 \end{aligned} \tag{2.70}$$

where the input x_i has also been set to zero. These algebraic equations have the solution

$$\begin{aligned} x_{us0} &= \frac{-(m_s + m_{us})g}{k_t} \\ x_{s0} &= \frac{-(m_s + m_{us})g}{k_t} - \frac{m_s g}{k_s} \end{aligned} \tag{2.71}$$

where the sub 0 is used to indicate the steady-state position. Both steady-state positions are negative because the position coordinates are positive upward and gravity is going to move both masses in the downward direction. If it is desired to study the dynamics of the quarter-car model by starting the analysis or simulation when the system is at rest in a gravity field, initializing the mass positions to the steady-state values of Eqs. (2.71) will ensure this.

Another way to start analysis with a system at rest in a gravity field is to move the frame of reference for the positions of the masses so that the zero position corresponds to the at-rest position for each mass. In Figure 2.7 imagine that the system is at rest in a gravity field and the inertial references for x_s and x_{us} are at these respective locations. In this position there would be a force in each spring equal to their steady-state value,

$$\begin{aligned} F_s &= k_s(x_{us0} - x_{s0}) = m_s g \\ F_t &= k_t(0 - x_{us0}) = (m_s + m_{us})g \end{aligned} \tag{2.72}$$

where the suspension spring must balance the weight of the sprung mass and the tire spring must balance the weight of the both masses. The spring forces for any arbitrary displacements of these inertial references would then be

$$\begin{aligned} F_s &= k_s((x_{us} + x_{us0}) - (x_s + x_{s0})) \\ F_t &= k_t(x_i - (x_{us} + x_{us0})) \end{aligned} \tag{2.73}$$

such that if x_s and x_{us} were both zero, the spring forces would attain their respective steady-state values. Using this frame of reference, the equations of motion become

$$\begin{aligned} k_s((x_{us}+x_{us_0})-(x_s+x_{s_0}))+b_s(\dot{x}_{us}-\dot{x}_s)-m_s g &= m_s\ddot{x}_s \\ k_t(x_i-(x_{us}+x_{us_0})-k_s((x_{us}+x_{us_0})-(x_s+x_{s_0}))-b_s(\dot{x}_{us}-\dot{x}_s)-m_{us}g &= m_{us}\ddot{x}_{us} \end{aligned} \tag{2.74}$$

But all the terms involving the steady-state displacements add up in each equation to exactly balance the weight term in that equation, thus, the equations can be simplified to

$$\begin{aligned} k_s(x_{us}-x_s)+b_s(\dot{x}_{us}-\dot{x}_s) &= m_s\ddot{x}_s \\ k_t(x_i-x_{us})-k_s(x_{us}-x_s)-b_s(\dot{x}_{us}-\dot{x}_s) &= m_{us}\ddot{x}_{us} \end{aligned} \tag{2.75}$$

where the gravity terms and the steady-state spring force terms no longer appear. Thus, if we measure mass positions from their respective steady-state positions in a gravity field, we need not include the gravity force in the equations. As a side note it should be realized that if this same system from Figure 2.7 were on its side and not in a gravity field, the steady-state positions for the masses would obviously correspond to locations where the springs were relaxed with no force. For systems not in a gravity field it makes sense to choose the inertial reference such that the springs are relaxed when the coordinates are zero.

Equations (2.75) are a set of linear second-order differential equations. They lend themselves to the matrix formulation

$$\begin{bmatrix} m_s & 0 \\ 0 & m_{us} \end{bmatrix}\begin{bmatrix} \ddot{x}_s \\ \ddot{x}_{us} \end{bmatrix}+\begin{bmatrix} ab_s & -b_s \\ -b_s & b_s \end{bmatrix}\begin{bmatrix} \dot{x}_s \\ \dot{x}_{us} \end{bmatrix}+\begin{bmatrix} k_s & -k_s \\ -k_s & k_t+k_s \end{bmatrix}\begin{bmatrix} x_s \\ x_{us} \end{bmatrix}=\begin{bmatrix} 0 \\ k_t \end{bmatrix}x_i(t) \tag{2.76}$$

or the general form

$$\mathbf{A}\ddot{\mathbf{x}}+\mathbf{B}\dot{\mathbf{x}}+\mathbf{C}\mathbf{x}=\mathbf{F} \tag{2.77}$$

where $\mathbf{A}$ is the mass matrix, $\mathbf{B}$ is the damping matrix, and $\mathbf{C}$ is the stiffness matrix. If Eqs. (2.75) are reduced to first-order form, then

$$\begin{aligned} \dot{x}_s &= v_s \\ \dot{x}_{us} &= v_{us} \\ \dot{v}_s &= \frac{k_s}{m_s}(x_{us}-x_s)+\frac{b_s}{m_s}(v_{us}-v_s) \\ \dot{v}_{us} &= \frac{k_t}{m_{us}}(x_i-x_{us})-\frac{k_s}{m_{us}}(x_{us}-x_s)-\frac{b_s}{m_{us}}(v_{us}-v_s) \end{aligned} \tag{2.78}$$

and these can be organized into matrix form:

$$\frac{d}{dt}\begin{bmatrix} x_s \\ x_{us} \\ v_s \\ v_{us} \end{bmatrix} = \begin{bmatrix} 0 & 0 & 1 & 0 \\ 0 & 0 & 0 & 1 \\ -\dfrac{k_s}{m_s} & \dfrac{k_s}{m_s} & -\dfrac{b_s}{m_s} & \dfrac{b_s}{m_s} \\ \dfrac{k_s}{m_{us}} & -\left(\dfrac{k_t}{m_{us}} + \dfrac{k_s}{m_{us}}\right) & \dfrac{b_s}{m_{us}} & -\dfrac{b_s}{m_{us}} \end{bmatrix} \begin{bmatrix} x_s \\ x_{us} \\ v_s \\ v_{us} \end{bmatrix} + \begin{bmatrix} 0 \\ 0 \\ 0 \\ \dfrac{k_t}{m_{us}} \end{bmatrix} x_i(t) \tag{2.79}$$

or the general form for linear state equations,

$$\dot{\mathbf{x}} = \mathbf{Ax} + \mathbf{Bu} \tag{2.80}$$

where $\mathbf{x}$ is a vector of state variables, $\mathbf{A}$ is called the *A-matrix* [not to be confused with the mass matrix notation used in Eq. (2.77)], and $\mathbf{B}$ is the *forcing matrix*. When there is only one input, in this example, it is customary to use a lowercase $\mathbf{b}$ to indicate a forcing vector rather than a forcing matrix. It turns out the general form from Eq. (2.77) is a wonderful starting point for vibration analysis, and the general form from Eq. (2.80) is a wonderful starting point for automatic control analysis. It turns out, further, that quite a lot of analytical manipulation can be performed on the general forms of linear equations prior to identifying an actual system or parameters. Some of this development is shown in later chapters. More complete treatment of linear systems analysis may be found in refs. [6] and [7].

Half-Car Model for Vibration Analysis and Control

As a second example of representing a system as linear, consider the half-car model shown in Figure 2.8. This is a side view of a vehicle where the body is represented as a rigid body of mass m, centroidal moment of inertia, I_g, with c.g. location distance a from the front axle and b from the rear axle. The front and rear suspensions and tires are represented identically to the quarter-car example just completed, with the exception of the addition of controllable forces between the sprung masses and the body attachment points. These controllable forces represent force actuators that would be present if we were designing an active suspension for ride quality control or pitch control or perhaps ride height control. If we were able to design an appropriate controller that greatly improved vehicle performance, we would want to know how big the forces need to be and how much power it would take to carry out the control philosophy. All this can be outputted once representative equations are derived. The roadway unevenness is represented by the vertical displacement inputs at the bottom of the front and rear tire springs, $x_{i_f}(t)$ and $x_{i_r}(t)$, respectively.

The coordinates for the problem are shown in the schematic. It is assumed that each coordinate is zero when the vehicle is at rest in a gravity field. As just discussed, we do not need to deal with the gravity force when this reference is used. In this model it

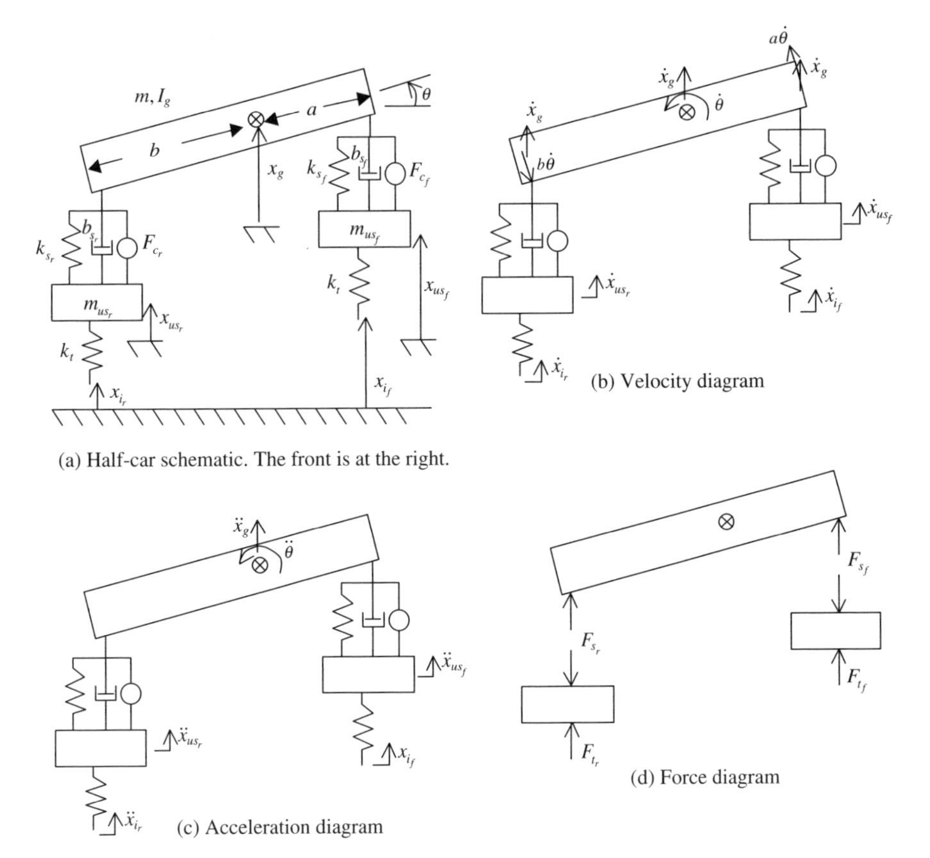

(a) Half-car schematic. The front is at the right.

(b) Velocity diagram

(c) Acceleration diagram

(d) Force diagram

Figure 2.8 Half-car model.

is assumed that the components can only move vertically. The velocity, acceleration, and force diagrams are also shown in Figure 2.8. The velocity components have been transferred from the center of gravity to the right and left attachment points using the velocity transfer relationship. We need the velocity at the attachment points to establish the relative velocity across the dampers.

The equations of motion for the rigid body for translation and rotation are

$$\begin{aligned} F_{s_f} + F_{s_r} &= m\ddot{x}_g \\ aF_{s_f} - bF_{s_r} &= I_g\ddot{\theta} \end{aligned} \tag{2.81}$$

For the unsprung masses at the front and rear, we have

$$\begin{aligned} F_{t_f} - F_{s_f} &= m_{us_f}\ddot{x}_{us_f} \\ F_{t_r} - F_{s_r} &= m_{us_r}\ddot{x}_{us_r} \end{aligned} \tag{2.82}$$

The tire forces at the front and rear come from

$$
\begin{aligned}
F_{t_f} &= k_t(x_{i_f} - x_{us_f}) \\
F_{t_r} &= k_t(x_{i_r} - x_{us_r})
\end{aligned}
\tag{2.83}
$$

The suspension forces at the front and rear require relative displacement across the suspension springs and relative velocity across the dampers. From Figure 2.8 we see that the vertical velocity at the attachment point on the right is

$$
v_{v_f} = \dot{x}_g + a\dot{\theta}\cos\theta \tag{2.84}
$$

and the vertical displacement of the right-side attachment point is

$$
d_{v_f} = x_g + a\sin\theta \tag{2.85}
$$

We are interested in a linear representation of this system. If we restrict our attention to small displacements, and angular displacements, $\cos\theta \approx 1$ and $\sin\theta \approx \theta$. Thus, the velocity and displacement at the attachment point become

$$
\begin{aligned}
v_{v_f} &= \dot{x}_g + a\dot{\theta} \\
d_{v_f} &= x_g + a\theta
\end{aligned}
\tag{2.86}
$$

and the suspension forces at the front and rear become

$$
\begin{aligned}
F_{s_f} &= k_{s_f}[x_{us_f} - (x_g + a\theta)] + b_{s_f}[\dot{x}_{us_f} - (\dot{x}_g + a\dot{\theta})] + F_{c_f} \\
F_{s_r} &= k_{s_r}[x_{us_r} - (x_g - b\theta)] + b_{s_f}[\dot{x}_{us_r} - (\dot{x}_g - b\dot{\theta})] + F_{c_r}
\end{aligned}
\tag{2.87}
$$

Using Eqs. (2.83) and (2.87) in the equations of motion (2.81) and (2.82) yields

$$
\begin{aligned}
&k_{s_f}[x_{us_f} - (x_g + a\theta)] + b_{s_f}[\dot{x}_{us_f} - (\dot{x}_g + a\dot{\theta})] + F_{c_f} \\
&\quad + k_{s_r}[x_{us_r} - (x_g - b\theta)] + b_{s_r}[\dot{x}_{us_r} - (\dot{x}_g - b\dot{\theta})] + F_{c_r} = m\ddot{x}_g \\
&a\left\{k_{s_f}[x_{us_f} - (x_g + a\theta)] + b_{s_f}[\dot{x}_{us_f} - (\dot{x}_g + a\dot{\theta})] + F_{c_f}\right\} \\
&- b\left\{k_{s_r}[x_{us_r} - (x_g - b\theta)] + b_{s_r}[\dot{x}_{us_r} - (\dot{x}_g - b\dot{\theta})] + F_{c_r}\right\} = I_g\ddot{\theta} \\
&k_t(x_{i_f} - x_{us_f}) - \left\{k_{s_f}[x_{us_f} - (x_g + a\theta)] + b_{s_f}[\dot{x}_{us_f} - (\dot{x}_g + a\dot{\theta})] + F_{c_f}\right\} = m_{us_f}\ddot{x}_{us_f} \\
&k_t(x_{i_r} - x_{us_r}) - \left\{k_{s_r}[x_{us_r} - (x_g - b\theta)] + b_{s_r}[\dot{x}_{us_r} - (\dot{x}_g - b\dot{\theta})] + F_{c_r}\right\} = m_{us_r}\ddot{x}_{us_r}
\end{aligned}
\tag{2.88}
$$

These can be put into the general matrix form of Eq. (2.77), with the result

$$
\begin{bmatrix} m & 0 & 0 & 0 \\ 0 & I_g & 0 & 0 \\ 0 & 0 & m_{us_f} & 0 \\ 0 & 0 & 0 & m_{us_f} \end{bmatrix}
\begin{bmatrix} \ddot{x}_g \\ \ddot{\theta} \\ \ddot{x}_{us_f} \\ \ddot{x}_{us_r} \end{bmatrix}
+
\begin{bmatrix} b_{s_f} + b_{s_r} & ab_{s_f} - bb_{s_r} & -b_{s_f} & -b_{s_r} \\ ab_{s_f} - bb_{s_r} & a^2b_{s_f} + b^2b_{s_r} & -ab_{s_f} & bb_{s_r} \\ -b_{s_f} & -ab_{s_f} & b_{s_f} & 0 \\ -b_{s_r} & bb_{s_r} & 0 & b_{s_r} \end{bmatrix}
$$

$$\times \begin{bmatrix} \dot{x}_g \\ \dot{\theta} \\ \dot{x}_{us_f} \\ \dot{x}_{us_r} \end{bmatrix} + \begin{bmatrix} k_{s_f} + k_{s_r} & ak_{s_f} - bk_{s_r} & -k_{s_f} & -k_{s_r} \\ ak_{s_f} - bk_{s_r} & a^2 k_{s_f} + b^2 k_{s_r} & -ak_{s_f} & bk_{s_r} \\ -k_{s_f} & -ak_{s_f} & k_{s_f} + k_t & 0 \\ -k_{s_r} & bk_{s_r} & 0 & k_{s_r} + k_t \end{bmatrix} \begin{bmatrix} x_g \\ \theta \\ x_{us_f} \\ x_{us_r} \end{bmatrix} \tag{2.89}$$

$$= \begin{bmatrix} 0 \\ 0 \\ k_t \\ 0 \end{bmatrix} x_{i_f} + \begin{bmatrix} 0 \\ 0 \\ 0 \\ 1 \end{bmatrix} x_{i_r} + \begin{bmatrix} 1 \\ a \\ -1 \\ 0 \end{bmatrix} F_{c_f} + \begin{bmatrix} 1 \\ -b \\ 0 \\ -1 \end{bmatrix} F_{c_r}$$

These could be reduced further to first-order form, but this will not be done for this example.

Linearization of the Inverted Pendulum

The quarter- and half-car models were, from the beginning, assumed to be linear. Sometimes it is justifiable to "linearize" a nonlinear system to study its behavior for small motions around some equilibrium position. This is demonstrated by returning to the inverted pendulum example shown in Figure 2.6. The nonlinear governing equations were derived in Eqs. (2.63) and are repeated here for convenience:

$$\ddot{x} = \frac{\left(I_g + ma^2\right)\left[F - k(x - x_{fl}) + ma\dot{\theta}^2 \sin\theta\right]}{(m_c + m)I_g + m^2a^2\left[(m_c/m) + \sin^2\theta\right]} - \frac{ma\cos\theta(mga\sin\theta)}{(m_c + m)I_g + m^2a^2\left[(m_c/m) + \sin^2\theta\right]}$$

$$\ddot{\theta} = -\frac{ma\cos\theta\left[F - k(x - x_{fl}) + ma\dot{\theta}^2 \sin\theta\right]}{(m_c + m)I_g + m^2a^2\left[(m_c/m) + \sin^2\theta)\right]} + \frac{(m_c + m)(mga\sin\theta)}{(m_c + m)I_g + m^2a^2\left[(m_c/m) + \sin^2\theta\right]} \tag{2.90}$$

If we restrict our attention to small displacements of the cart and small angular motions of the pendulum, we can use the approximations $\cos\theta \approx 1$ and $\sin\theta \approx \theta$. This reduces (2.90) to

$$\ddot{x} = \frac{\left(I_g + ma^2\right)\left[F - k(x - x_{fl}) + ma^2\dot{\theta}^2\theta\right]}{(m_c + m)I_g + m^2a^2\left[(m_c/m) + \theta^2\right]} - \frac{ma(mga\theta)}{(m_c + m)I_g + m^2a^2\left[(m_c/m) + \theta^2\right]}$$

$$\ddot{\theta} = -\frac{ma\left[F - k(x - x_{fl}) + ma^2\dot{\theta}^2\theta\right]}{(m_c + m)I_g + m^2a^2\left[(m_c/m) + \theta^2\right]} + \frac{(m_c + m)(mga\theta)}{(m_c + m)I_g + m^2a^2\left[(m_c/m) + \theta^2\right]} \tag{2.91}$$

We further neglect terms that are products of variables or products of variables and their derivatives. The argument is that if, say, θ is very small by assumption, then θ^2

or $\theta\dot{\theta}$ is really small and can be neglected. Thus, (2.91) reduces to

$$\begin{aligned}\ddot{x} &= \frac{\left(I_g + ma^2\right)[F - k(x - x_{\rm fl})]}{(m_c + m)I_g + m^2a^2(m_c/m)} - \frac{ma(mga\theta)}{(m_c + m)I_g + m^2a^2(m_c/m)} \\ \ddot{\theta} &= -\frac{ma[F - k(x - x_{\rm fl})]}{(m_c + m)I_g + m^2a^2(m_c/m)} + \frac{(m_c + m)(mga\theta)}{(m_c + m)I_g + m^2a^2(m_c/m)}\end{aligned} \tag{2.92}$$

and the result is a set of linear second-order differential equations. If we let the free length of the spring be zero and perform some simplifications, (2.92) can be written in the matrix form

$$\begin{bmatrix} m_c & 0 \\ 0 & I_g \end{bmatrix}\begin{bmatrix} \ddot{x} \\ \ddot{\theta} \end{bmatrix} + \begin{bmatrix} \dfrac{1 + I_g/ma^2}{D}k & \dfrac{mg}{D} \\ -\dfrac{\dfrac{m}{m_c}a\dfrac{I_g}{ma^2}}{D}k & -\dfrac{\dfrac{m_c+m}{m_c}\dfrac{I_g}{ma^2}}{D}mga \end{bmatrix}\begin{bmatrix} x \\ \theta \end{bmatrix} = \begin{bmatrix} \dfrac{1 + I_g/ma^2}{D} \\ -\dfrac{\dfrac{m}{m_c}a\dfrac{I_g}{ma^2}}{D} \end{bmatrix}F \tag{2.93}$$

and can be further reduced to the first-order form

$$\frac{d}{dt}\begin{bmatrix} x \\ \theta \\ v_x \\ \omega \end{bmatrix} = \begin{bmatrix} 0 & 0 & 1 & 0 \\ 0 & 0 & 0 & 1 \\ \dfrac{-(1/m_c)\left(1 + I_g/ma^2\right)}{D}k & -\dfrac{1}{m_c}\dfrac{mg}{D} & 0 & 0 \\ \dfrac{\dfrac{1}{I_g}\dfrac{m}{m_c}a\dfrac{I_g}{ma^2}}{D}k & \dfrac{\dfrac{1}{I_g}\dfrac{m_c+m}{m_c}\dfrac{I_g}{ma^2}}{D}mga & 0 & 0 \end{bmatrix}\begin{bmatrix} x \\ \theta \\ v_x \\ \omega \end{bmatrix} + \begin{bmatrix} 0 \\ 0 \\ \dfrac{\dfrac{1}{m_c}\left(1 + I_g/ma^2\right)}{D} \\ -\dfrac{\dfrac{1}{I_g}\dfrac{m}{m_c}a\dfrac{I_g}{ma^2}}{D} \end{bmatrix}F \tag{2.94}$$

where

$$D = 1 + \frac{(m_c + m)}{m_c}\frac{I_g}{ma^2} \tag{2.95}$$

Once parameters are identified, Eq. (2.94) can be used to develop a control law for the force input F that will stabilize the inverted pendulum.

2.6 SUMMARY

In this chapter we dealt with deriving equations of motions for dynamic systems which, if solved, would predict the motion–time history of the system. An important part of the procedure is to cast equations into a form suitable for computation. Although the procedure to accomplish this has some ad hoc aspects, certain steps are taken in every case.

1. A system model is proposed and important dynamic components are identified. Springs, dampers, masses, rigid bodies, inputs, and their connections are shown schematically. This step is actually quite difficult outside the classroom. How would the reader know that the quarter-car example shown schematically in Figure 2.7 has anything to do with the performance of a real car? Experience is the only teacher for knowing what is important and what can be neglected in any given model.
2. Coordinates must be selected that describe all the possible positions that a dynamic system can attain during the course of its motion. This step is also difficult outside the classroom, and when several choices are available there can be a big difference in derivation effort between one selection and another.
3. Velocity and acceleration diagrams are drawn where arrows and symbols are used to identify important velocity and acceleration components. In general, the acceleration of all mass particles and centers of gravity of all rigid bodies must be derived.
4. An accompanying force diagram is needed, using arrows and symbols, that exposes all the forces as orthogonal components acting wherever components physically make contact with each other or with constraining barriers.
5. Newton's laws for translation and rotation are applied appropriately in orthogonal directions by having forces in one direction create accelerations in that same direction.
6. Kinematic constraints must be recognized.
7. The time-varying variables and the number of equations are counted up to assure that there are sufficient equations for the number of variables.

When these steps have been completed, all dynamic principles have been used, and for some homework problems this is all that is asked of you. If you are actually making this model in order to understand what the system does or to design a device, the next steps must also be carried out.

When the foregoing steps are complete, there will typically exist a set of coupled, nonlinear second-order differential equations. The coordinate variables will appear, as will some additional symbols for forces and moments that were introduced during the modeling process. The next steps are taken to reduce the sets of equations into a form suitable for computation, and that will typically be the explicit first-order state equations indicated generally in Eq. (2.40). To reduce sets of nonlinear equations to explicit first-order form can take considerable algebra, and the appropriate substitutions are not always obvious. Here are some general things to try.

1. Express spring and damper forces in terms of the coordinates and their derivatives and substitute for these forces in the equations.
2. Try to manipulate the second derivative terms to the left-hand side of the equations and have all other lower-order terms on the right.
3. Try to solve for remaining forces and moments in terms of the coordinates and make substitutions to eliminate the force and moment symbols.
4. Use the kinematic constraints to eliminate some of the coordinate variables.
5. If multiple second-order terms remain in several equations, matrix inversion will probably be needed. This is easiest to carry out using the computer rather than by hand.
6. Ultimately reduce second derivatives to first derivatives using the definitions demonstrated in the examples.

Note that if the system is linear by assumption or linearized after equation derivation, all these steps are much easier.

Let's declare ourselves competent at deriving equations of motion for fairly complex planar systems. Chapter 3 deals with solving the equations using a computer.

REFERENCES

[1] Matlab and Simulink software, Mathworks Co., Natick, Mas.

[2] ACSL software, AEgis Technologies Group, Huntsville, AL.

[3] Easy-5 software, MSC Corporation, Santa Ana, CA.

[4] 20-Sim software, Controllab Products, Enschede, The Netherlands.

[5] B., Carnahan, H., Luther, and J., Wilkes, *Applied Numerical Methods*. Wiley, New York, 1969.

[6] D.C., Karnopp, D.L., Margolis, and R.C. Rosenberg, *System Dynamics: Modeling and Simulation of Mechatronic Systems*, 4th ed., Wiley, Hoboken, NJ, 2005.

[7] L. Meirovitch, *Analytical Methods in Vibrations*. Macmillan, New York, 1967.

PROBLEMS

2.1 The physical system from Problem 1.1(a) is repeated here with some reasonable coordinates identified for the system. (Figure P2.1) In part *a* inertial coordinates are used for both mass elements. In part *b* the coordinate for the top mass is measured from an inertial reference, and the coordinate for the lower mass is measured relative to the top mass. Using these coordinates, make velocity, acceleration, and force diagrams and derive the equations of motion.

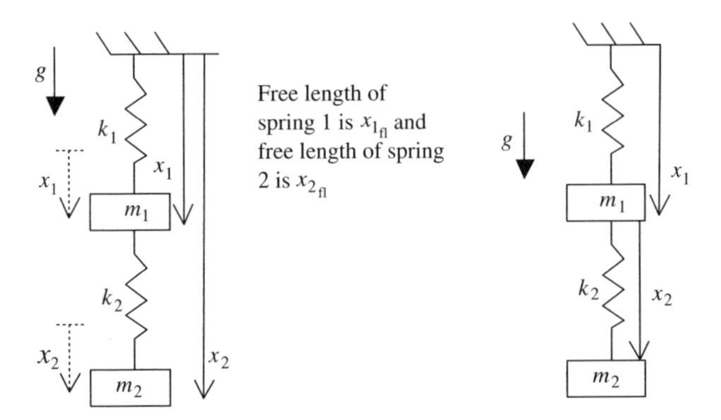

(a) System only moves vertically. Coordinates are measured from an inertial reference.

(b) Coordinate x_1 measured from an inertial reference, coordinate x_2 measured relative to.

Figure P2.1

There is no need to reduce these to a computational form at this time. Ensure that there are sufficient equations for the unknowns.

For part (a) of this problem, the inertial coordinates could be measured from equilibrium in a gravity field. These coordinates are shown on the left side of Figure P2.1a. How would the equations of motion change if such coordinates are used?

2.2 Problem 1.1(b) is repeated here with some coordinates shown for this two-degree-of-freedom system (Figure P2.2). The coordinate Y is an inertial coordinate measured from the base of the spring, and the inertial coordinate y is measured from the equilibrium position in a gravity field. Using the Y,θ coordinates, make velocity, acceleration, and force diagrams and derive the

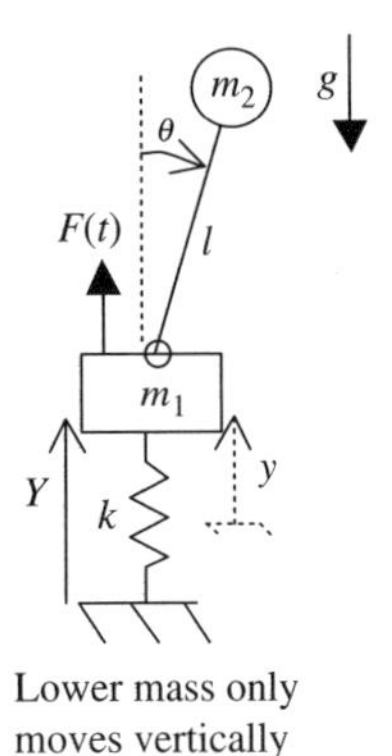

Figure P2.2

equations of motion. How do the equations change if the y-coordinate measured from equilibrium is used in place of the Y-coordinate? There is no need to reduce the resulting equations to a computational form. But ensure that there are sufficient equations for the number of unknowns. If you need to identify the spring free length, call it l_0.

2.3 This system is taken from Problem 1.1(d). It is a double pendulum that moves under the influence of gravity. In Figure P2.3a the coordinates are the angle of each pendulum measured from the inertial vertical direction, and in part b the lower pendulum is located by its angular position measured relative to the upper pendulum. For each set of coordinates, make velocity, acceleration, and force diagrams and derive the equations of motion in mutually perpendicular directions. There is no need to reduce the equations to computational form at this time, but ensure that you have sufficient equations for the number of unknowns.

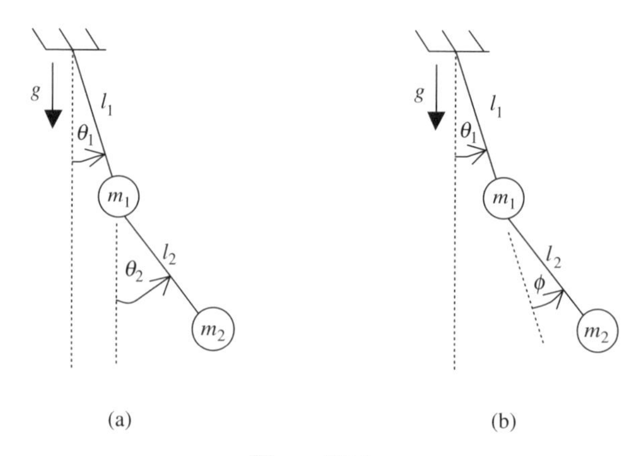

Figure P2.3

2.4 This problem is taken from Problem 1.1(e). It is a double pendulum where the two pendulums are connected by a torsional spring of torsional spring constant k_τ. The rods l_1 and l_2 are rigid. The spring generates a torque proportional to the relative twist angle across it. The positive torque direction is indicated in Figure P2.4. Using the coordinates shown, construct velocity, acceleration, and force diagrams for this system and derive the equations of motion. You do not need to reduce your equations to a computational form, but do ensure that you have sufficient equations for the unknowns.

2.5 This problem is taken from Problem 1.1. This is a two-degree-of-freedom system where the variable y_1 determines the location of m_1 as well as the angle of the lever and the position of the base of the spring. The variable y_2 determines the location of the second mass. The system configuration when these coordinates are zero is described in Figure P2.5, If you feel you need the free length of the spring, call it l_0. Show velocity, acceleration, and force

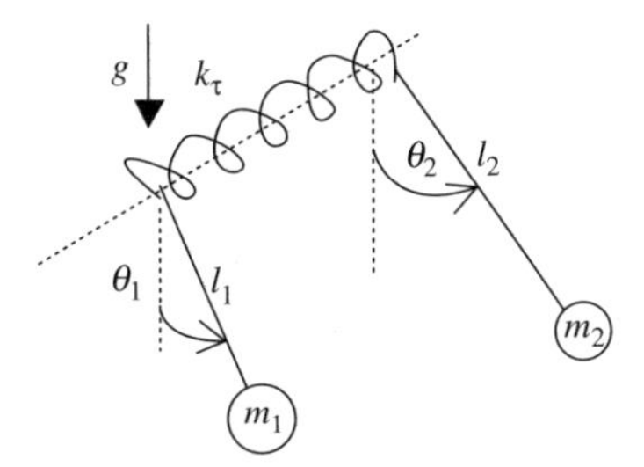

Torsional spring with pendula at its ends. The torque in the spring, positive as indicated, is $\tau = k_\tau(\theta_1 - \theta_2)$.

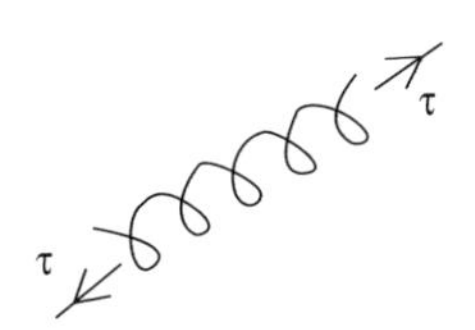

Figure P2.4

diagrams for this system and derive the equations of motion. You can assume that the lever moves through small angles. Ensure that you have sufficient equations for the number of unknowns you introduce.

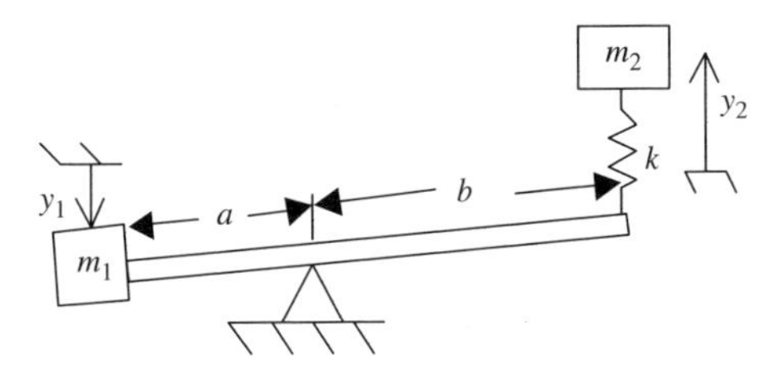

Lever is massless, pin joint is frictionless, m_2 moves vertically. When $y_1 = 0$ and $y_2 = 0$, the lever is horizontal and m_2 is at equilibrium in a gravity field.

Figure P2.5

2.6 Figure P2.6 shows a pendulum made from a general-shaped rigid body of mass m and centroidal moment of inertia I_g. It is pinned without friction at the top. The single-degree-of-freedom is indicated by the single coordinate θ.

(a) Show velocity acceleration and force diagrams for this system. Label the forces at the pin joint F_x in the horizontal direction positive to the right and F_y in the vertical direction positive upward. Derive the equations of motion in the horizontal and vertical directions and write the rotational equation by summing moments about the center of gravity.

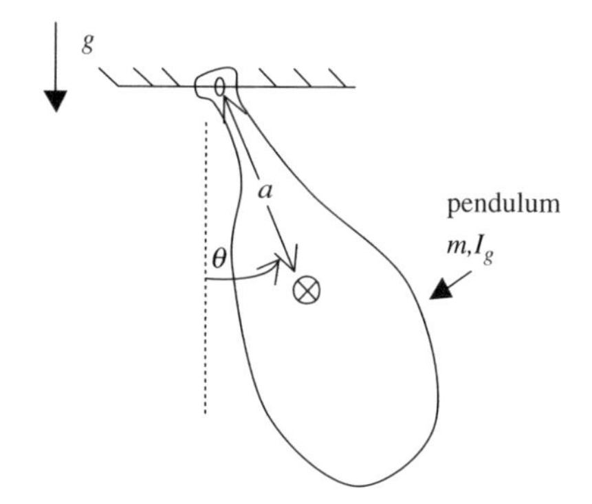

Figure P2.6

(b) Substitute for the force components and express your result in terms of only θ and its derivatives.

(c) From your knowledge regarding writing the moment equation about a fixed point, write the moment equation about the pin joint. Remember that the moment of inertia must now be with respect to the pin joint, I_0.

(d) From your knowledge of the parallel axis theorem (see Appendix C), are the results from (b) and (c) the same?

2.7 This problem is repeated from Problem 1.5. A rigid body has mass m and c.g. moment of inertia I_g (figure P2.7). The inertial coordinates indicated locate the center of gravity and the rotation of the body. The springs that are horizontal remain horizontal and the springs that are vertical remain vertical regardless of the motion of the rigid body.

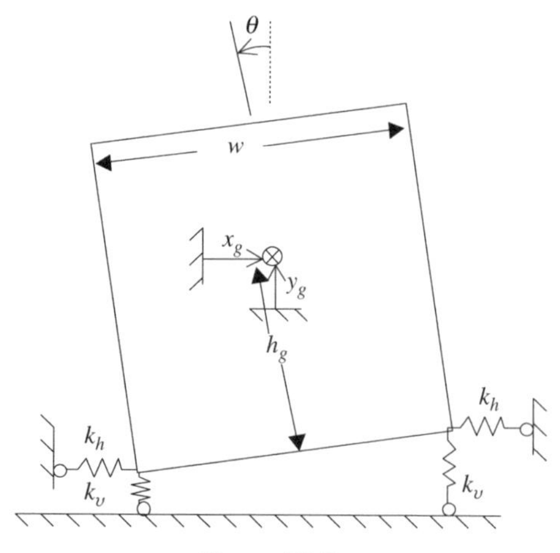

Figure P2.7

(a) Construct velocity, acceleration, and force diagrams for this system using the coordinates given. You will have to give symbolic names to the horizontal and vertical spring forces.

(b) Transfer the velocity components from the center of gravity to the spring ends and align in horizontal and vertical directions. Write expressions for the velocity input to each spring.

(c) Derive the equations of motion for this system. Cast your equations into a final form where the highest-order derivatives are on the left side of the equations. Try to reduce this final form to a set of first-order state equations.

2.8 This problem is taken from Problem 1.6. A cylinder of radius R rolls without slip on a horizontal surface (Figure P2.8). Because it rolls without slip it requires only one coordinate to describe its location. Thus, either the position of the center x or the angular position of the offset center of gravity θ can be used. The coordinate frame is defined such that when $x = 0$, the angle θ also equals zero. A prescribed force $F(t)$ acts on the center of the cylinder.

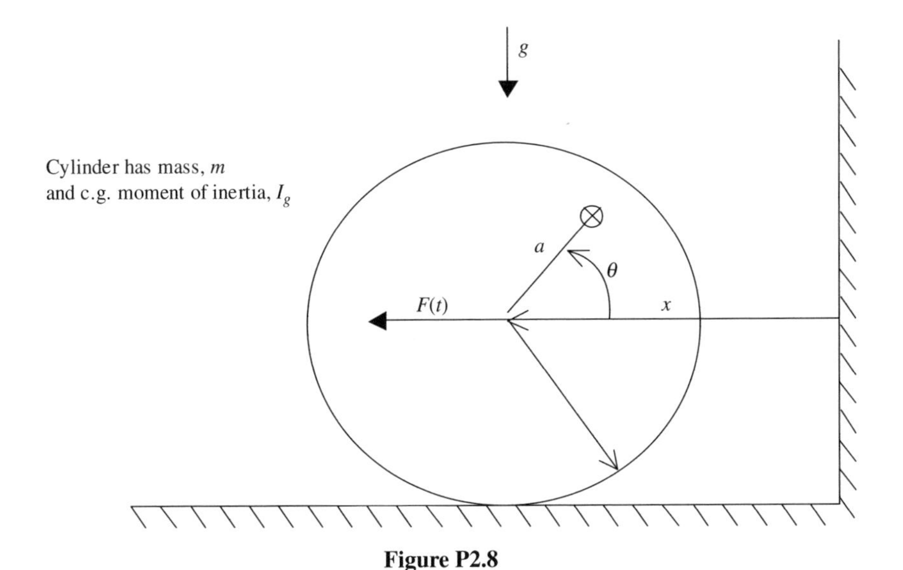

Figure P2.8

(a) Construct velocity, acceleration, and force diagrams for this system. Note that a horizontal force is needed at the base of the cylinder even though no sliding occurs.

(b) Derive the equations of motion writing component equations in the horizontal and vertical directions. Cast these equations into a final form where the highest-order derivatives are on the left side of the equations.

(c) Try to reduce your equations to a first-order state equation form.

2.9 This problem is taken form Problem 1.7. A cylinder with center of gravity in the center is free to roll without slip on a horizontal surface (Figure P2.9). Either the variable X that locates the center of the cylinder or the angle θ that indicates the angular orientation of the cylinder can be used to describe the cylinder motion. A mass particle m is free to slide without friction on the inner surface of the cylinder. The mass particle is located by the angle ϕ measured with respect to an inertial vertical line. Some axes are attached to the cylinder to assist you in expressing the acceleration of the mass particle.

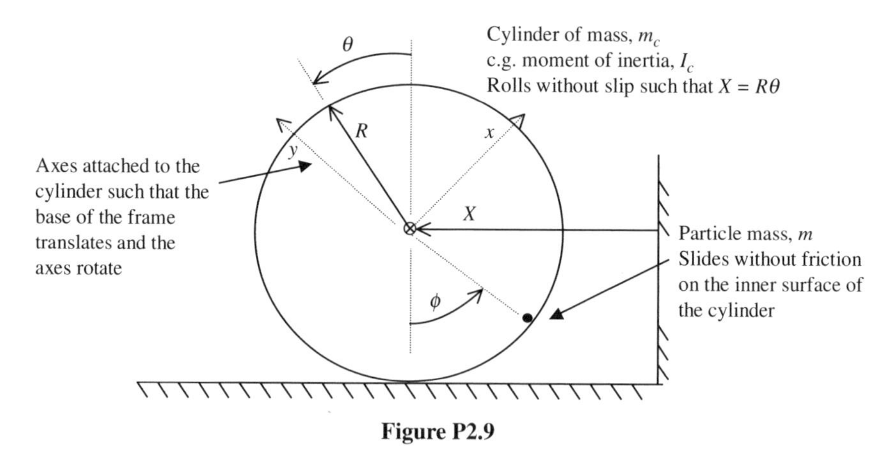

Figure P2.9

(a) Construct velocity, acceleration, and force diagrams for this system. Use arrows and symbols to show velocity and acceleration components. You will have to assign symbolic names to force components. Note that there is a normal force between the mass particle and the cylinder wall. Also, do not forget the gravity force acting on the mass particle.

(b) Derive the equations of motion for the system. Substitute for the force symbols and have final equations in terms of only X or θ and ϕ. In your final result, have the highest derivatives on the left side of equations.

2.10 This problem is a repeat of Problem 1.8 and is similar to Problem 2.9. A big cylinder rolls without slip on a horizontal surface (Figure P2.10). On the inner surface of the big cylinder is a small cylinder that rolls without slip. This is a two degree-of-freedom system and you can use either X or θ to locate the big cylinder and ϕ measured from vertical to locate the small cylinder. Expressing the velocity and acceleration of the centers of gravity of the rigid bodies is similar to Problem 2.9. This problem is a bit difficult in that the moment equation for the small cylinder requires the angular acceleration to be expressed in terms of the coordinates given. This requires derivation of a kinematic relationship.

(a) Derive the kinematic relationship relating the absolute angular velocity of the small cylinder ω_c to the coordinates θ and ϕ. A good starting point is first to imagine a coordinate frame attached to the big cylinder at its center

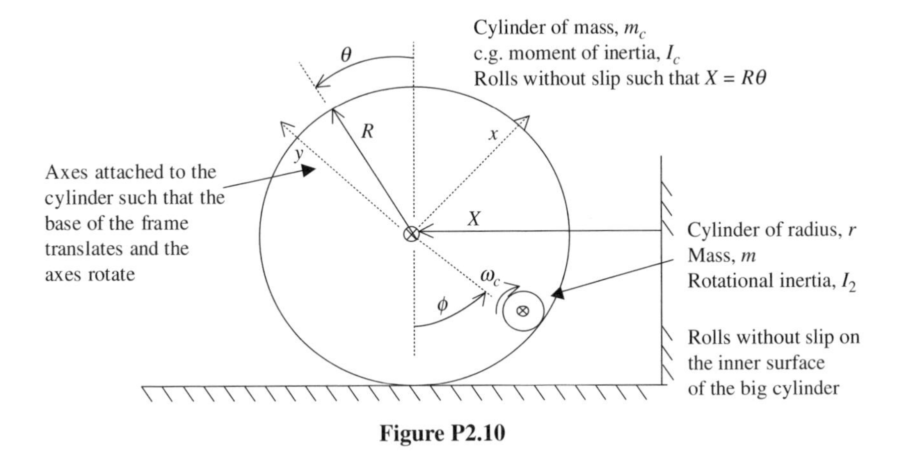

Figure P2.10

of gravity, and show a velocity diagram of the center of the small cylinder and the velocity components at the bottom of the small cylinder. Then imagine a coordinate frame attached to the small cylinder at its contact point with the large cylinder and construct a velocity diagram for the center of the small cylinder. Equate proper components from the two diagrams and solve for ω_C.

(b) Construct acceleration and force diagrams for the system and derive the equations of motion. Try to substitute for the forces and end up with equations in terms of only the coordinates and their derivatives.

2.11 This is the same problem as Problem 1.9. A cylinder of radius R rolls without slip on a horizontal surface (Figure P2.11). There is a string wrapped around an inner radius r (some what like a spool of thread), and the string is being pulled horizontally by the force F. For this single-degree-of-freedom system, construct velocity, acceleration, and force diagrams and derive the equations of motion. Try to obtain an equation in terms of only θ and its derivatives.

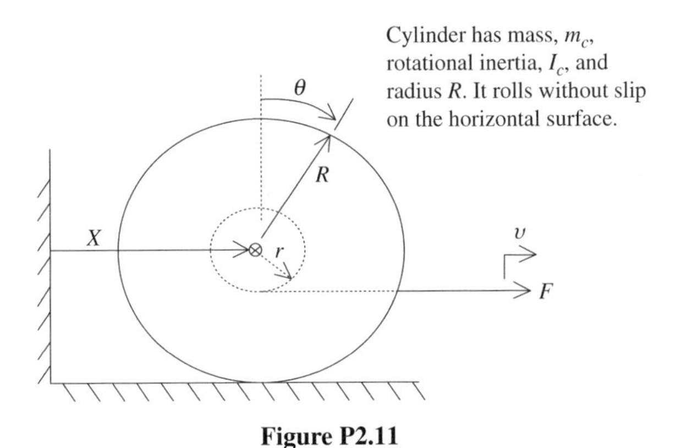

Figure P2.11

2.12 This system is the same as in Problem 1.10. A cylinder rolls without slip on a horizontal surface and has an offset center of gravity (Figure P2.12). Across from the center of gravity a spring is attached as shown. This is a single-degree-of-freedom system and we will use the coordinate θ to locate the cylinder. The setup is such that when $x = 0$, $\theta = 0$ and the spring is relaxed with no force. In this configuration the left wall is a distance D from the center of the cylinder. The mass of the cylinder is m and the c.g. moment of inertia is I_g. Construct velocity, acceleration, and force diagrams and derive the equations of motion in terms of the variable θ and its derivatives.

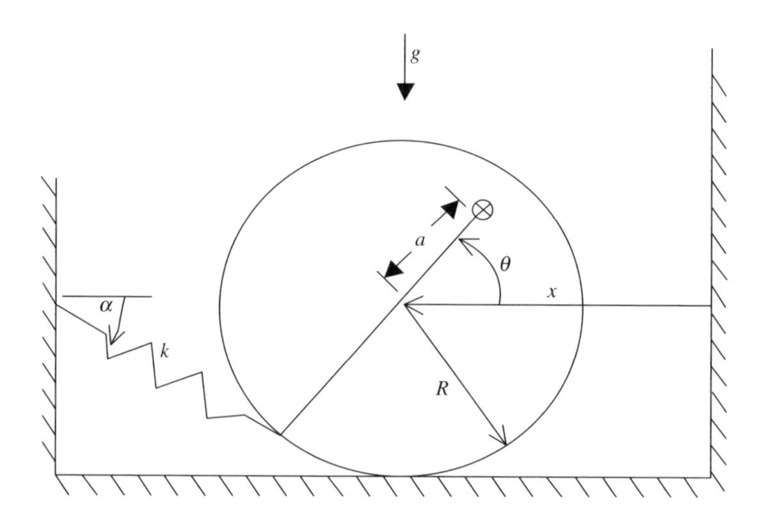

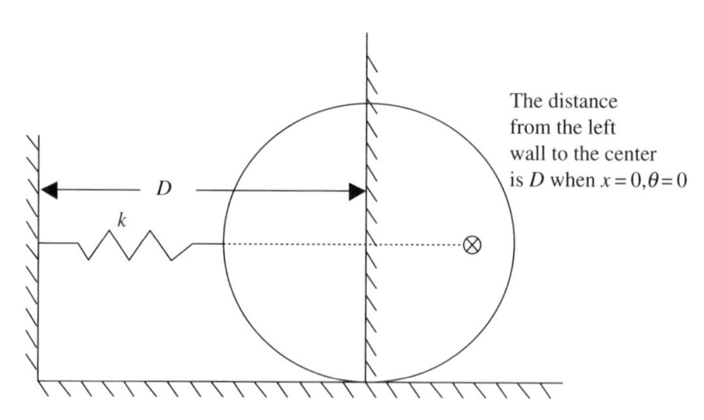

Figure P2.12

2.13 This system is the same as that in Problem 1.11. Figure P2.13 shows the top view of a car that can move in the XY plane. We will use body-fixed coordinates for the problem and there are axes attached to the car at its center of gravity

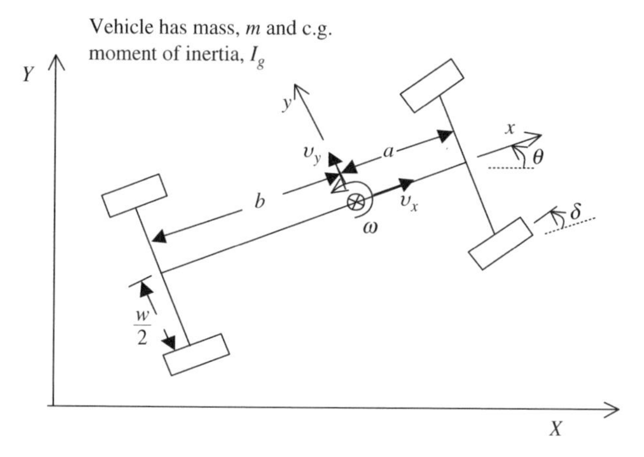

Figure P2.13

and aligned in the body-fixed xy directions. The velocity components in the body-fixed directions are indicated.

(a) Using body-fixed coordinates, construct velocity and acceleration diagrams for the center of gravity of the vehicle. Transfer the velocity components to the center of each wheel and show the components on the diagram.

(b) Show a force diagram using arrows and symbols and realize that there are force components at the contact point of each tire. You will have to assign names to these forces.

(c) Derive the equations of motion. In a later chapter you will learn how the forces are related to the body-fixed velocity components. For now you can leave the equations in terms of the forces and the body-fixed acceleration.

2.14 This is a repeat of Problem 2.7, but this time we are using body-fixed coordinates. Axes are attached to the body at the center of gravity and aligned as shown in Figure P2.14. The absolute instantaneous velocity vector has been broken into two perpendicular components aligned in the body-fixed directions. The springs at the base are constrained to remain horizontal and vertical, respectively, regardless of the motion of the body.

(a) Construct velocity, acceleration, and force diagrams for this system. Transfer the velocity components to the attachment points of the springs and resolve these components to be along the various spring directions.

(b) Derive the equations of motion for this system using body-fixed coordinates for the c.g. motion. Write additional first-order equations for the displacement of each spring. For example, for the horizontal spring on the right with displacement x_R,

$$\frac{dx_R}{dt} = (v_x + h_g\dot{\theta})\cos\theta - \left(v_y + \frac{w}{2}\dot{\theta}\right)\sin\theta$$

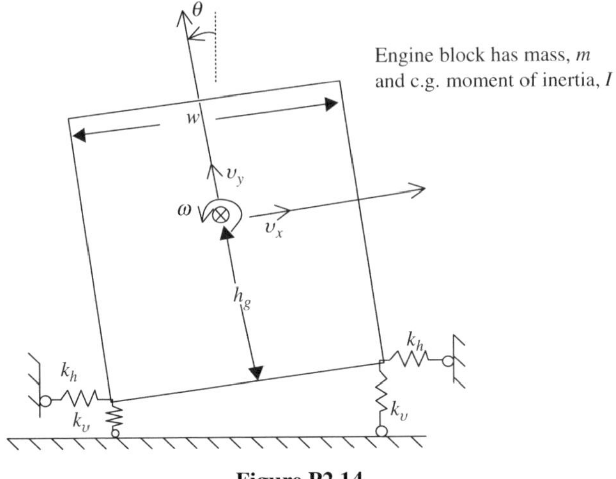

Figure P2.14

The force F_{H_R} in this spring would then be given by $F_{H_R} = k_h x_R$. Note that in a simulation model these additional first-order equations would be solved along with the equations of motion to yield the motion–time history of the system.

2.15 This is the system from Problem 1.15. Two mass particles are at the ends of an inextensible string (Figure P2.15). The upper mass is on a horizontal tabletop and might be moving in the r and θ directions. The lower mass moves only

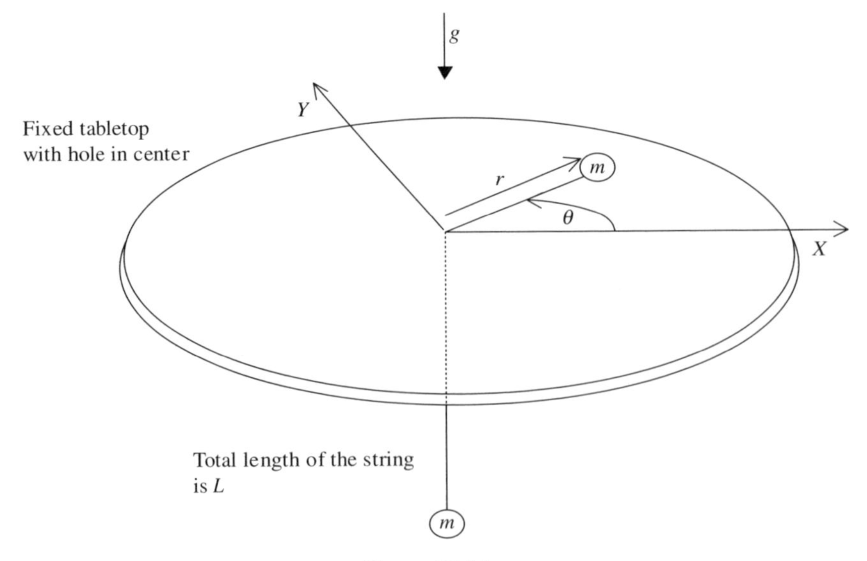

Figure P2.15

vertically and would experience the same r-direction motion as the upper mass. This is a two-degree-of-freedom system in the two coordinates indicated.

(a) Show velocity, acceleration, and force diagrams for this system.

(b) Derive the equations of motion for the two masses. Your final result should be in terms of only the coordinate variables.

2.16 This is the system from Problem 1.16. A rigid body rotates without slip on a fixed rigid cylinder (Figure P2.16). The springs at the ends of the body constrained to remain vertical regardless of the body motion. This is a single-degree-of-freedom system and the variable θ is the single coordinate required. Note that the distance from the center of gravity to the contact point on the cylinder is $R\theta$.

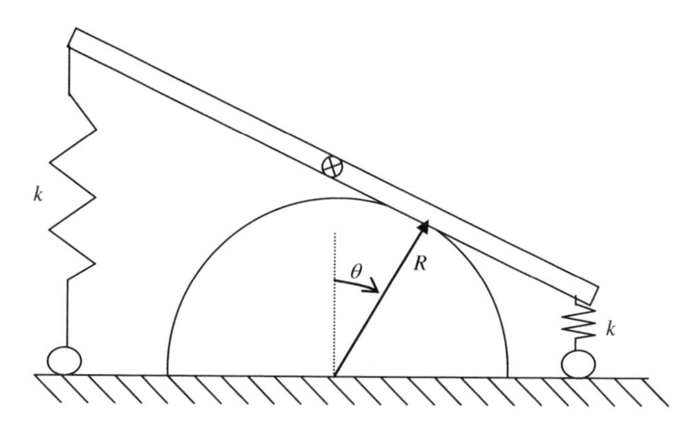

Figure P2.16

(a) Construct velocity, acceleration, and force diagrams for this system and show the velocity components at the ends of the body.

(b) Write expressions for the velocity components in the spring directions.

(c) Derive the equations of motion for the system in terms of θ and its derivatives and the symbols you used for the spring forces. Write additional first-order equations relating the time rates of change of the spring displacements to the appropriate velocity components.

2.17 The sticklike rigid body is free to rotate about the pin joint with no friction. Gravity is acting.

(a) Construct velocity, acceleration, and force diagrams for this system.

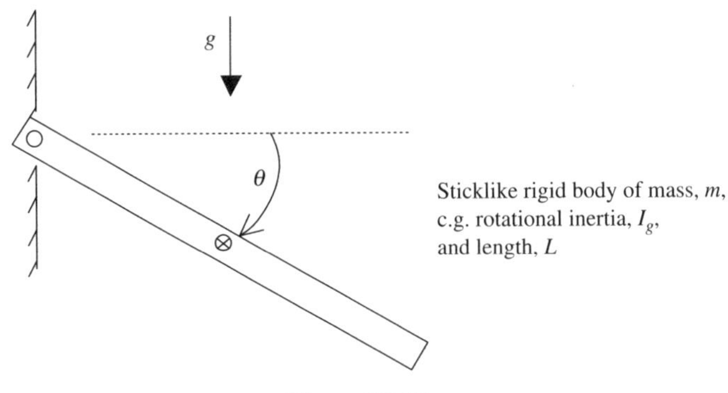

Figure P2.17

(b) Derive the equations of motion, writing the moment equation about the center of gravity. This is a single-degree-of-freedom system and the only variable in your final result should be θ.

(c) From your knowledge of expressing the angular momentum with respect to a fixed point, derive the equations of motion, writing the moment equation about the pin joint. Are your results from part (b) the same as those here?

(d) It is interesting to derive the acceleration of the end of the stick just as it is released from a horizontal position. To do this, let θ be zero in your result from part (c) and solve for the angular acceleration $\ddot{\theta}$. The acceleration at the end of the stick at this instant is $L\ddot{\theta}$. For a stick, the moment of inertia about its end (see Appendix B) is $I_0 = m(l^2/3)$. Use this moment of inertia and see if your result for the acceleration at the end is at all surprising.

2.18 This is a very realistic engineering problem. A company that makes office workspaces needed a design for a damper that would let a shelf door open easily when lifted but close very smoothly when dropped from a horizontal position. This problem presents a model that would allow such a design to be done. The system is identical to that in Problem 2.17, with the addition of

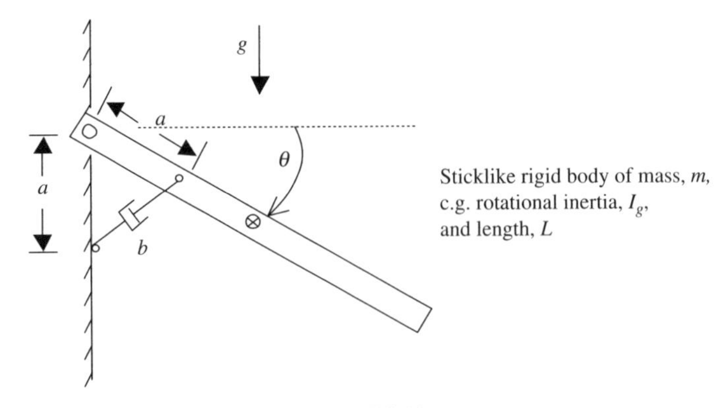

Figure P2.18

a damper that is pinned to the vertical wall and to the "door." The ultimate goal would be to specify the force/velocity characteristics of the damper to accomplish the design goal. Here the damper is linear with damper constant b and we just want to derive some equations of motion.

(a) In Problem 2.17 you needed to construct velocity, acceleration, and force diagrams for a virtually identical system. The force diagram from Problem 2.17 must be modified to include the additional force from the damper. The damper force is proportional to the velocity inline with the damper. Derive the kinematic relationship relating the velocity across the damper to the coordinate θ and its derivatives.

(b) When writing the moment equation either with respect to the center of gravity or the fixed point, you will need the moment arm of the damper force. Derive an expression for the moment arm of the damper force in terms of the coordinate θ.

(c) Derive the equations of motion for this system and ensure that you have sufficient equations for the unknowns.

2.19 Shown in Figure P2.19 is a very simple model of a building with horizontal base excitation. Such a model might be used to initiate a study of how buildings react to earthquakes. The building is represented as a rigid body, and its attachment to ground is represented by the horizontal spring k and the torsional spring k_τ. The torsional spring generates an opposing moment proportional to the angle θ. The earthquake input is represented by the specified input displacement $x_{in}(t)$.

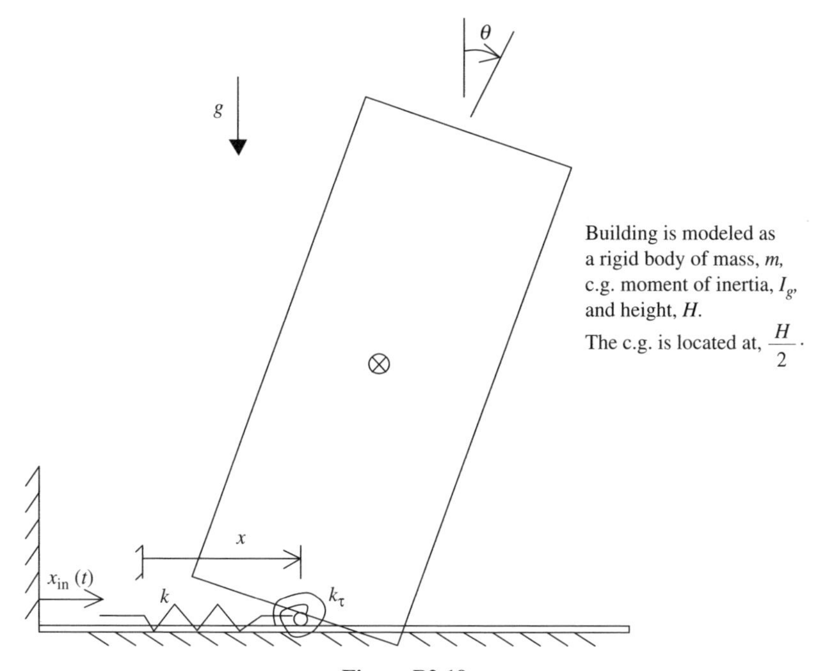

Figure P2.19

For this problem the motions of x and θ are small and we desire linear equations of motion. This is a two-degree-of-freedom system and the coordinates are x and θ, as indicated.

(a) Generate velocity, acceleration, and force diagrams for this system, allowing only small motions of the coordinates.

(b) Derive the linear equations of motion for this system. Put the resulting second-order equations into the matrix form of Eq. (2.77).

(c) Reduce the resulting second-order equations to first-order form and put into the standard matrix form of Eq. (2.80).

2.20 This is a repeat of Problem 2.1(a), but an external force $F(t)$ is acting on the lower mass (Figure P2.20).

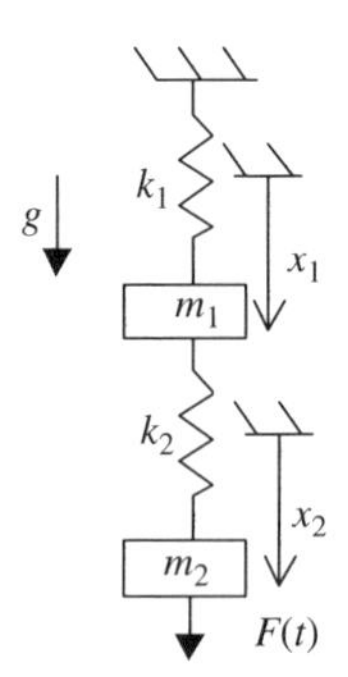

Figure P2.20

The coordinates are measured from equilibrium, so the gravity forces will not be included in the force diagram. For this two-degree-of-freedom system, derive the equations, of motion and put into the matrix form of Eq. (2.77). Reduce to first-order form and put into the form of Eq. (2.80).

2.21 Figure P2.21 is from Problem 2.3(a). For small angular motions, derive the linear equations of motion for this system in terms of the coordinates indicated.

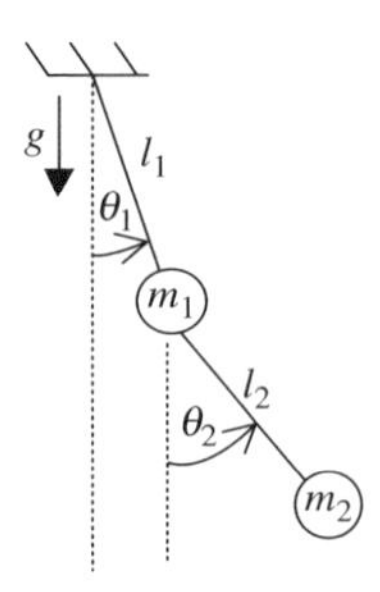

Figure P2.21

For the second-order form of the equations, write them in the matrix form of Eq. (2.77). Reduce the equation to first-order form and express them in the matrix form of Eq. (2.80).

2.22 Figure P2.22 is repeated from Problem 2.4. The two pendulums are connected by the torsional spring shown. For small angular motions, derive the linear equations of motion for the two-degree-of-freedom system. For the second-order form of the equations, write your result in the matrix form of Eq. (2.77). Reduce your equations to first-order form and cast into the matrix form of Eq. (2.80).

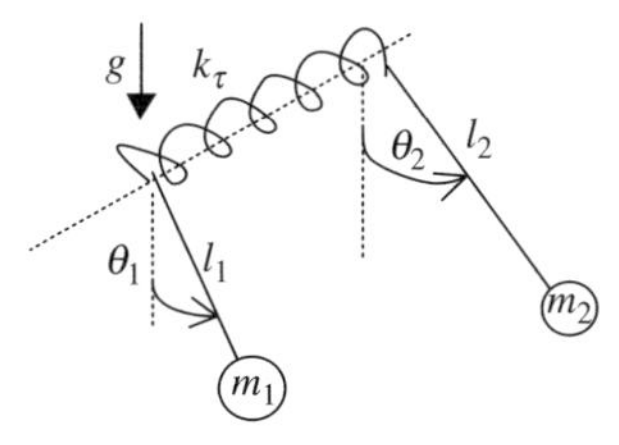

Torsional spring with pendula at its ends. The torque in the spring, positive as indicated is $\tau = k_\tau(\theta_1 - \theta_2)$.

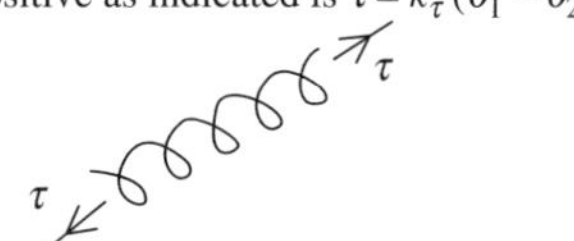

Figure P2.22

2.23 The cylinder shown in Figure P2.23 has an offset center of gravity located distance a from the center. The angle θ locates the center of gravity relative to the vertical axis. Gravity is acting. Derive the linear equations that would predict the motion of this system for small angular motions. You will, of course, need velocity, acceleration, and force diagrams. This is a

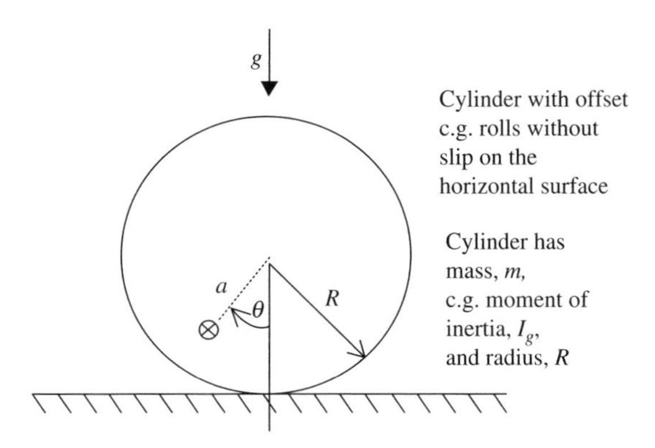

Figure P2.23

single-degree-of-freedom system, so only the variable θ and its derivatives are permitted in your final equations. Reduce your result to first-order form and express in the standard matrix form of Eq. (2.80).

2.24 This is a repeat of Problem 2.9 where a mass particle slides without friction on the inner surface of a cylinder and the cylinder rolls without slip on a horizontal surface. This is a two-degree-of-freedom system and we use the variable θ to locate the cylinder and the variable ϕ to locate the mass particle. For small motions of both θ and ϕ, derive the linear equations of motion for the system. Reduce your result to first-order form and express in the standard matrix format of Eq. (2.80).

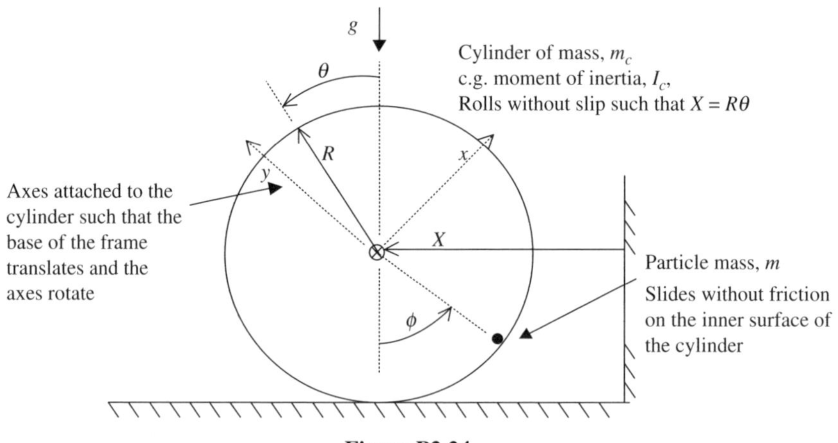

Figure P2.24

2.25 The system is a rigid body that is attached to ground at its left end through a translational and torsional spring (Figure P2.25). The torsional spring

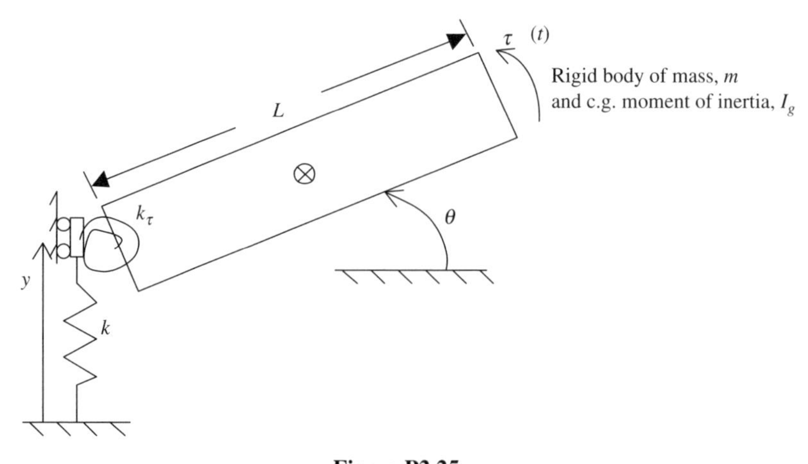

Figure P2.25

generates a torque or moment that opposes the angle θ. An external moment $\tau(t)$ acts at the right end. The two-degree-of-freedom system and the coordinates to be used are the location y of the left end and the angle θ. For small motions of y and θ, derive the linear equations of motion for this system. Express your result in the matrix form of Eq. (2.77), then reduce it to first-order form and express in the matrix form of Eq. (2.80). You will need to construct velocity, acceleration, and force diagrams as part of your solution.

2.26 Two mass elements are connected to the ends of a lever through the springs and damper shown in Figure P2.26. The mass on the right is acted upon by a prescribed force $F(t)$. The moment of inertia of the lever about its pivot is I_0. The coordinates that locate the two mass elements are y_1 and y_2 measured with respect to the equilibrium positions of the two masses. The angle θ indicates the angular rotation of the lever. The coordinates are defined such that when $y_1 = y_2 = \theta = 0$ there is no force in either spring.

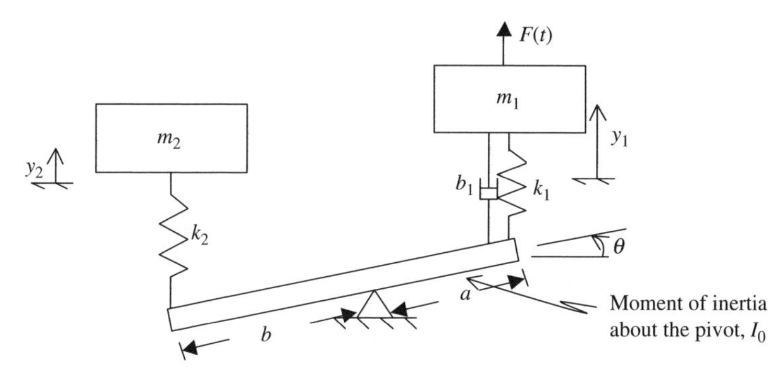

Figure P2.26

(a) Show velocity, acceleration, and force diagrams for this system and derive the small-motion linear equations of motion.

(b) Write your equations in the matrix format of Eq. (2.77). Reduce your second-order equations to first-order form. You can put your result into first-order matrix format if you wish.

3

COMPUTER SOLUTION OF EQUATIONS OF MOTION

In Chapters 1 and 2 we developed the fundamental dynamic principles and showed how to apply them to mass particles and rigid bodies. The result of constructing velocity, acceleration, and force diagrams and using Newton's laws is always sets of second-order differential equations. For most engineering systems these equations are nonlinear, and the nonlinearities typically result from geometric considerations of the system motion and from nonlinear spring and damper force relationships. Sometimes we can justify assuming that the system is linear, which means that the governing equations of motion are sets of linear second-order differential equations. When the equations of motion are linear, an analytical solution can sometimes be pursued. For most systems a computational solution is required whether governed by linear or nonlinear equations of motion.

In Chapter 2 the idea of reducing systems of equations to first-order form was developed. The explicit first-order form is the most appropriate for simulation, and systems of equations should be put into explicit first-order form if at all possible. In this chapter the computational solution of equations of motion is developed. Some concepts for analysis of linear systems are also shown. Time-step simulation is described and some simulations are demonstrated.

3.1 TIME-STEP SIMULATION OF NONLINEAR EQUATIONS OF MOTION

When the first-order form of equations was presented in Section 2.3, the manner in which computers march out solutions to sets of equations was described. With explicit first-order form the equations have the general appearance

$$
\begin{aligned}
\dot{x}_1 &= f_1(x_1, x_2, \ldots, x_n, u_1, u_2, \ldots, u_r) \\
\dot{x}_2 &= f_2(x_1, x_2, \ldots, x_n, u_1, u_2, \ldots, u_r) \\
\dot{x}_3 &= f_3(x_1, x_2, \ldots, x_n, u_1, u_2, \ldots, u_r) \\
&\vdots \\
\dot{x}_n &= f_n(x_1, x_2, \ldots, x_n, u_1, u_2, \ldots, u_r)
\end{aligned}
\tag{3.1}
$$

where the x_i are state variables and the u_j are prescribed inputs. Many algorithms are available to integrate equations of this form starting from some initial conditions for each state variable and marching out the solution until a prescribed final time. What varies among the various software packages is the interface between the user and the program and the manner in which the system is described to the program.

For specialized systems such as magnetic circuits, electrical circuits, and structures there are specialized programs that use graphical input to describe the system, and the equations are derived automatically. Reference [1] is an example of such a specialized program. In other specialized systems the structure of the system is so repetitive that the equations can be preprogrammed and the user need only supply the data that describe the system. Finite-element programs are examples of this [2]. For most dynamic analyses that are done for prototype design or control design or physical system understanding, the system models do not lend themselves to using specialized programs and the user must come up with the equations of motion by applying the physical principles.

There are two fundamental ways to describe equations to an equation solver. One way is to enter the equations almost as they appear from the derivation. The other is to use computational "blocks," where signals are added, multiplied, put through nonlinear functions, and integrated according to the dictates of the equations. References [3] and [4] allow both types of equation input format. In what follows several examples are developed into computational format and some simulation outputs are presented.

Example 3.1 Spring–Pendulum The spring–pendulum was developed in Chapter 2 and is shown in Figure 2.1. The governing first-order equations were derived previously and are repeated here as Eqs. (3.2). The reader will recognize these as explicit first-order state equations ready for simulation. Table 3.1 gives some parameters for this system.

$$
\begin{aligned}
\frac{dr}{dt} &= v_r \\
\frac{d\theta}{dt} &= \omega \\
\frac{dv_r}{dt} &= r\omega^2 + g\cos\theta - \frac{k}{m}(r - l_0) \\
\frac{d\omega}{dt} &= -\frac{g}{r}\sin\theta - \frac{2v_r\omega}{r}
\end{aligned}
\tag{3.2}
$$

TABLE 3.1. Parameters for Example 3.1

$m = 5.0$ kg
Define a frequency $f_n = 1.0$ Hz Calculate the spring stiffness: $k = m(2\pi f_n)^2$, N/m Free length of the spring, $l_0 = 0.5$ m
$g = 9.8$ m/s^2, acceleration of gravity
Initial conditions: $r_{\text{ini}} = l_0,\ 1.5l_0,\ 2l_0$ $\theta_{\text{ini}} = (20^\circ, 45^\circ, 90^\circ)\dfrac{\pi}{180}$ rad $v_{r_{\text{ini}}} = 0$ $\omega_{\text{ini}} = 0$ Other ICs can be tried as well.

Regardless of the format being used to enter the equations, there will be an initial program section where parameter values are defined and initial conditions are assigned. There will additionally be needed a final time t_f when the simulation terminates and, in most programs, some initial value for the time step dt.

In equation format the input file would look as follows:

```
dr=v_r
dtheta=omega
dv_r=r*omega^2+g*cos(theta)-k/m*(r-l_0)
domega=-g/r*sin(theta)-2.*v_r*omega/r
```
(3.3)

When these equations are delivered to the equation solver, it will automatically carry out the integration steps from such statements as

```
r=integ(dr, r_ini)
theta=integ(dtheta, theta_ini)
v_r=integ(dv_r, v_r_ini)
omega=integ(domega, omega_ini)
```
(3.4)

These equations state that each state variable is obtained by integrating their respective derivative starting from some initial value.

Notice that the state variables are automatically available as outputs from the program. We may want to write additional output equations to make other outputs available for plotting. For example, to generate the angular position in degrees rather than radians (computation is *always* done using angles measured in radians), we would write

```
theta_deg=theta*180./pi
```
(3.5)

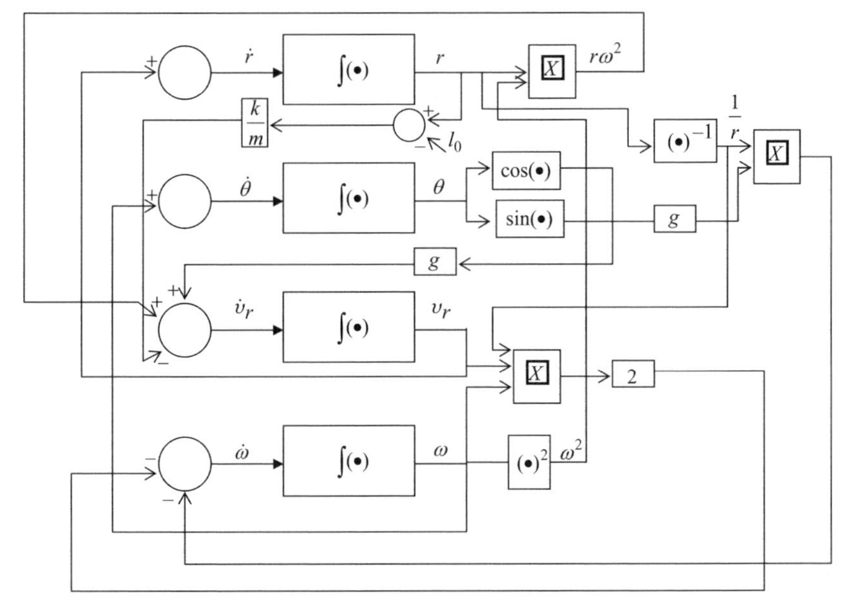

Figure 3.1 Block representation of Example 3.1.

If we desired the force in the spring, we would write

$$\texttt{F\_s} = \texttt{k}^*\texttt{(r-l\_0)} \tag{3.6}$$

and the force becomes an available output.

The graphical type of input for this problem is shown in Figure 3.1. There is a block that represents an integrator, and there is a summing junction at the inlet of each integrator. One such construction exists for each state equation. The derivative of each state variable is an input to an integrator, and the state variable is an output from its respective integrator. The remainder of the diagram is reconstructing the derivatives according to Eqs. (3.2). There are blocks that multiply and function blocks that output sines and cosines and other needed operations. The input to each block is operated on according to the function of that block, and an output results from each operation. The reader should verify that the operations indicated in Figure 3.1 do, in fact, reconstruct the equations of motion.

The equation format of (3.3) and the block form of Figure 3.1 will vary somewhat from program to program, but these are the primary ways to input equation information to commercial equation solvers. One of the commercial solvers was used to produce the simulation results presented next.

When doing an engineering problem that involves modeling, equation derivation, and simulation, it is always a good idea to test the system for special cases where the result might be known or expected. For example, for the spring–pendulum, if the spring were infinitely stiff, we would expect the system to behave like a simple

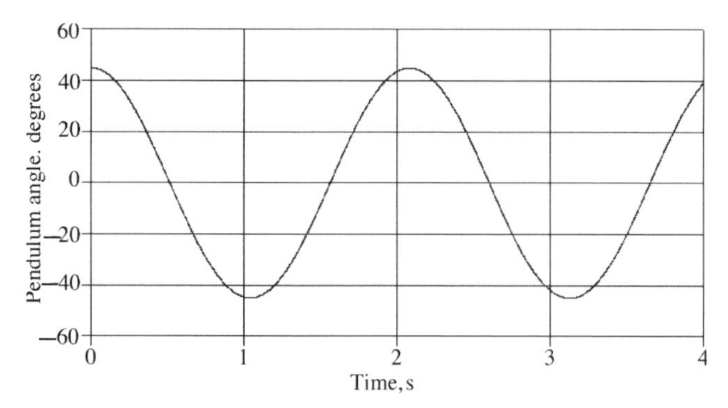

Figure 3.2 Pendulum angle when a spring is very stiff.

pendulum and simply swing back and forth. To test this, the frequency f_n in Table 3.1 was set to 1000 Hz and the spring stiffness was calculated for use in the simulation. The results of this are shown in Figures 3.2 and 3.3. The first figure shows the pendulum angle, and the second shows the pendulum length. The pendulum swings back and forth and the length is virtually fixed (note that the vertical axis in Figure 3.2 indicates no change in length). The frequency f_n is returned to the value in Table 3.1 and the simulation is run again with the results shown in Figures 3.4 and 3.5. The motion is presented as length versus angle rather than using time for the horizontal axis. As indicated in the figure, different initial conditions are used to produce Figures 3.4 and 3.5, and the motion is quite different as a result.

Example 3.2: Pinned Disk with a Slot and Mass Particle A pinned disk with a mass particle moving in a slot in the disk (Example 2.3), is shown schematically in Figure 2.3. The equations of motion were derived and then reduced to first-order form in Eqs. (2.54) and (2.55), equations repeated here for convenience. In this example the final explicit form of the equations is not required for simulation. The third of Eqs.

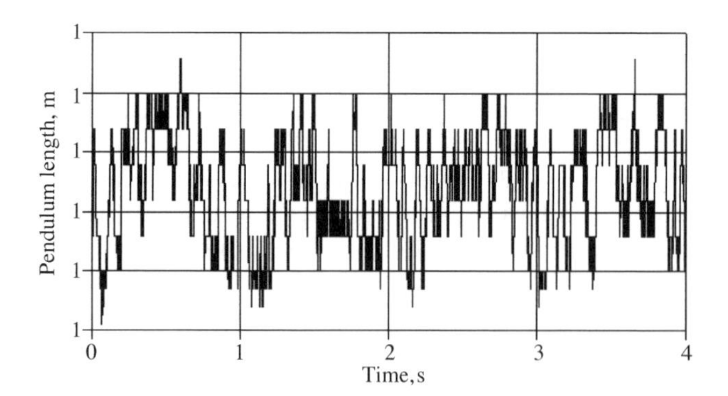

Figure 3.3 Pendulum length when a spring is very stiff.

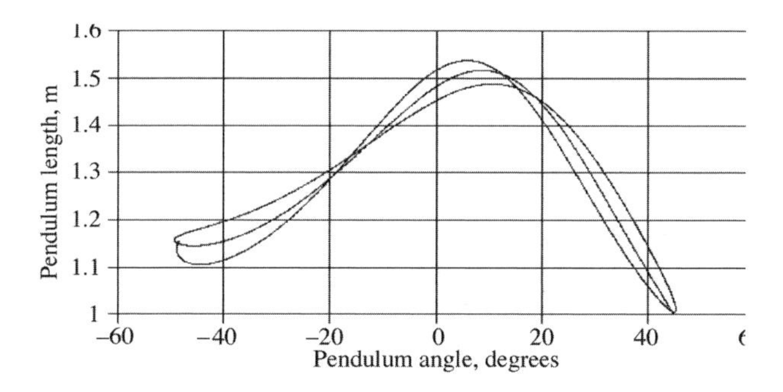

Figure 3.4 Pendulum length versus angle for $\theta_{\text{ini}} = 45^\circ$ and $r_{\text{ini}} = l_0$.

(3.7) could be substituted into the fourth equation to generate the explicit form, but this is not required.

$$
\begin{aligned}
\dot{\theta} &= \omega \\
\dot{y} &= v_y \\
\dot{\omega} &= -\frac{may\omega^2}{I_d + my^2} - \frac{2my\omega v_y}{I_d + my^2} + \frac{kay}{I_d + my^2} + \frac{\tau_i(t)}{I_d + my^2} \\
\dot{v}_y &= -a\dot{\omega} + y\omega^2 - \frac{k}{m}y
\end{aligned}
\tag{3.7}
$$

In equation format the simulation file would look something like

```
dtheta=omega
dy=v_y
domega=(-m*a*y*omega^2-2*m*y*omega*v_y+k*a*y+tau_in)/(I_d+m*y^2)
dv_y=-a*domega + y*omega^2-k/m*y                                  (3.8)
```

In graphical input format, the simulation file would appear something like that shown in Figure 3.6. Notice that the output from the integrator for θ is not connected

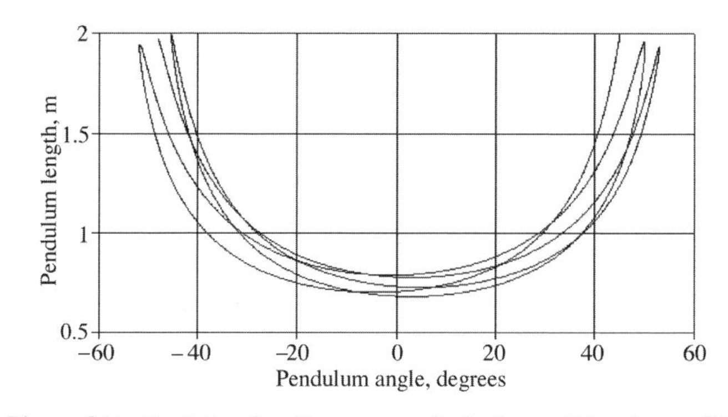

Figure 3.5 Pendulum length versus angle for $\theta_{\text{ini}} = 45^\circ$ and $r_{\text{ini}} = 2l_0$.

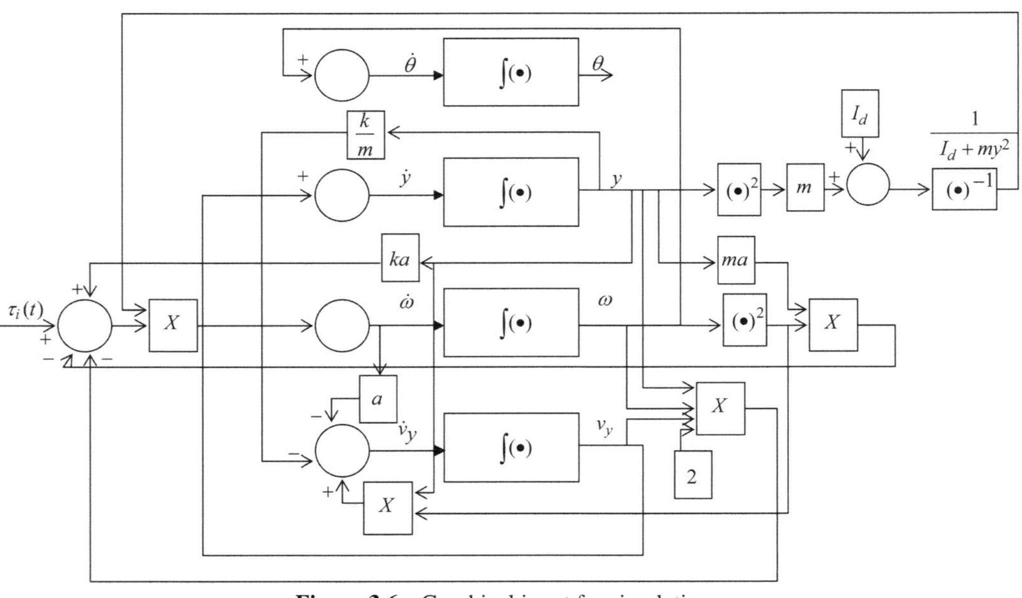

Figure 3.6 Graphical input for simulation.

to any other output. This is because the dynamics of the system do not depend on angular position. Also notice that the last of Eqs. (3.7) is realized by using $\dot{\omega}$ as an input to the final integrator. The reader should confirm that the block diagram of Figure 3.6 reconstructs the equations for the system.

This system was simulated using a commercial equation solver using the parameters from Table 3.2. Some simulation results are presented. In this example there is an external input torque $\tau_i(t)$ that excites the system. This will have to be programmed along with the equations of motion. For the equation input format of Eqs. (3.8), the programming steps would look like

```
if t<=0.5
tau=tau_0
else
tau=0
end
```
(3.9)

This logic would have to be built into the block diagram format of equation entry.

As a first test the spring stiffness was set to be very high by setting the defined frequency f_n to be 1000 Hz. If the mass particle is constrained not to move in the slot, the disk should respond to the input torque by accelerating angularly while the torque is acting and then remain at constant angular velocity when the torque is

TABLE 3.2. Parameters for Example 3.2

Mass of the disk, $m_d = 5.0$ kg Radius of the disk, $R = 0.5$ m Calculate the moment of inertia, $I_d = m_d \dfrac{R^2}{2}$, kg · m^2
Mass of particle, $m = 2.5$ kg Define a frequency $f_n = 1.0$ Hz
Calculate the spring stiffness: $k = m(2\pi f_n)^2$, N/m
Distance to the slot, $a = 0.4$ m
Initial conditions: $y_{\text{ini}} = 0$ $\theta_{\text{ini}} = 0$ $v_{y_{\text{ini}}} = 0$ $\omega_{\text{ini}} = 0$
The input torque is a pulse input, $\begin{cases} \tau_0 & \text{for} \quad 0 \le t \le 0.5\text{ s} \\ 0 & \text{for} \quad t > 0.5\text{ s} \end{cases}$ where $\tau_0 = 8.0$ N·m

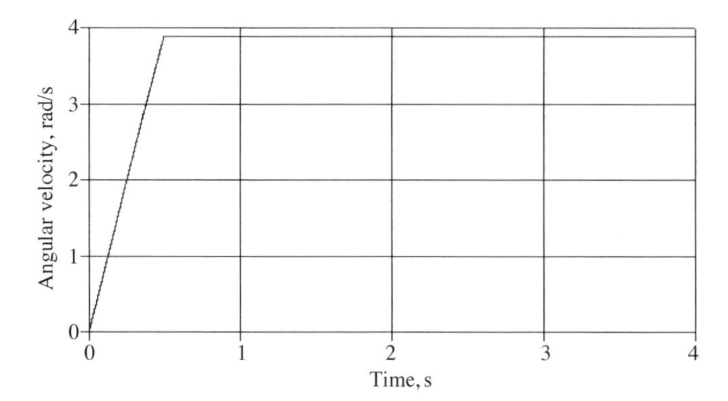

Figure 3.7 Disk angular velocity when the spring stiffness is very high.

removed. Figures 3.7 and 3.8 show the disk angular velocity and the position of the mass particle, respectively. The position response is on the order of 10^{-7} m, which is virtually no motion at all, and the angular velocity response is as expected. This test does not prove that the equations are correct, but it suggests that the equations are correct.

The frequency f_n is set back to its Table 3.2 value and the simulation is run again. Figures 3.9 and 3.10 show the disk angular velocity and the position of the mass particle, respectively. This time the spring force affects the disk motion and the disk motion affects the particle motion. It is truly an interactive dynamic system.

Example 3.3: Stabilizing an Inverted Pendulum In Example 2.4, the equations of motion for a rigid-body inverted pendulum pinned to a cart and spring were derived. The system is shown in Figure 2.6. This was an example where, after applying Newton's laws, the resulting second-order equations had second-order derivatives of the

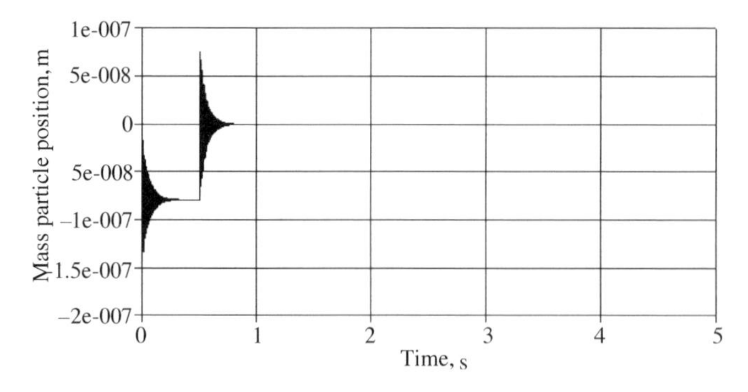

Figure 3.8 Mass particle position when the spring stiffness is very high.

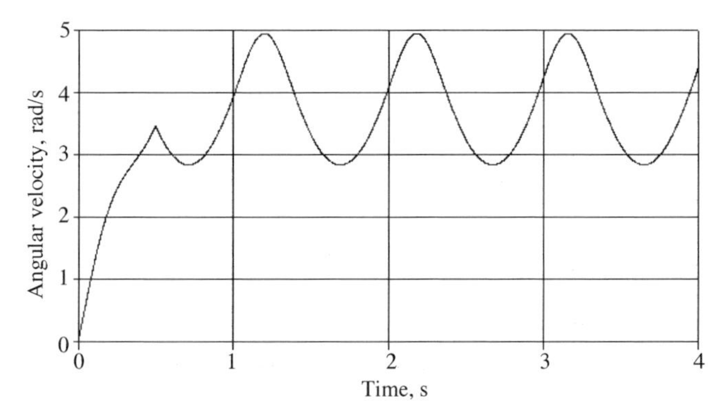

Figure 3.9 Disk angular velocity when the spring stiffness is as given in Table 3.2.

two coordinates x and θ in each of the equations of motion. These were put into a matrix form in Eqs. (2.61) and are repeated here:

$$\begin{bmatrix} m_c + m & ma\cos\theta \\ ma\cos\theta & I_g + ma^2 \end{bmatrix} \begin{bmatrix} \ddot{x} \\ \ddot{\theta} \end{bmatrix} = \begin{bmatrix} F - k(x - x_{\text{fl}}) + ma\dot{\theta}^2 \sin\theta \\ mga\sin\theta \end{bmatrix} \tag{3.10}$$

Because only two coordinates are involved, the inverse of the 2×2 matrix on the left side of (3.10) can be taken analytically with the result of Eqs. (2.63), modified here and put into first-order form as

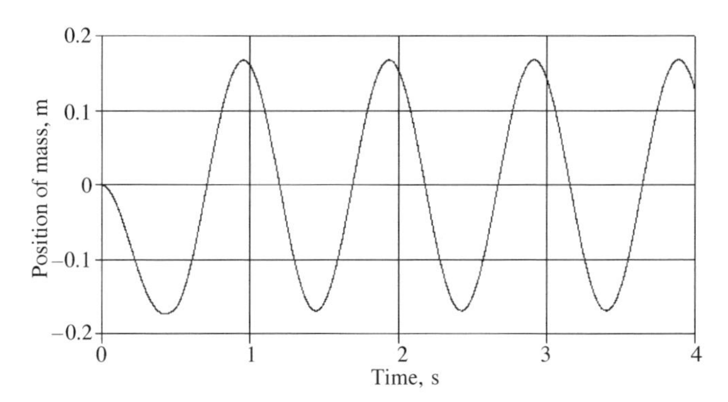

Figure 3.10 Mass particle position when the spring stiffness is as given in Table 3.2.

$$
\begin{aligned}
\dot{\theta} &= \omega \\
\dot{x} &= v_x \\
\dot{v}_x &= \frac{(I_g + ma^2)[F - k(x - x_{\mathrm{fl}}) + ma\dot{\theta}^2 \sin\theta]}{(m_c + m)I_g + m^2a^2\left[(m_c/m) + \sin^2\theta\right]} \\
&\quad - \frac{ma\cos\theta(mga\sin\theta)}{(m_c + m)I_g + m^2a^2\left[(m_c/m) + \sin^2\theta\right]} \\
\dot{\omega} &= -\frac{ma\cos\theta[F - k(x - x_{\mathrm{fl}}) + ma\ddot{\theta}^2 \sin\theta]}{(m_c + m)I_g + m^2a^2\left[(m_c/m) + \sin^2\theta\right]} \\
&\quad + \frac{(m_c + m)(mga\sin\theta)}{(m_c + m)I_g + m^2a^2\left[(m_c/m) + \sin^2\theta\right]}
\end{aligned}
\tag{3.11}
$$

Table 3.3 shows the parameters for this model. In addition to simulating this system, we will also design a simple controller for the control force F and see if we can stabilize the pendulum. From Figure 2.6, if the angle θ is positive as shown and F is positive in the direction shown, the effect will be to decrease θ. Perhaps a reasonable starting point for a controller is to have the control force proportional to the angle. If the pendulum tends to fall over in the positive θ direction, the control force will act to move the angle in the opposite direction. In addition, it is also a good idea to have some control action that reacts to retard the angular velocity of the pendulum. It is proposed to try

$$F = K_p\theta + K_d\dot{\theta} \tag{3.12}$$

TABLE 3.3. Parameters for Example 3.3

Mass of the pendulum,
$m = 2.5$ kg
Length of the pendulum,
$l = 1.0$ m
Calculate the centroidal moment of inertia:
$I_g = m\,\frac{l^2}{12}$, kg · m^2
Length to the center of gravity from the pin joint,
$a = \frac{l}{2}$, m
Free length of the spring,
$x_{\mathrm{fl}} = 0$, or we are measuring x from the free-length position
Mass of the cart
Define the ratio, $m_c/m = 1$ and calculate:
$m_c = \frac{m_c}{m}$, m · kg
Define a frequency
$f_n = 1.0$ Hz
Calculate the spring stiffness:
$k = m_c\,(2\pi f_n)^2$N/m

If this works to stabilize the pendulum, it might be necessary to add additional control action to limit the motion of the cart. The control gains K_p and K_d must be determined through numerical experimentation.

Equations (3.11) were solved using a commercial equation solver with equation entry. The graphical entry can also be used, but the block diagram is quite cumbersome. The equation entry looks like

```
dx=v_x
dtheta=omega
dv_x=((I_g+m*a^2)*(F-k*x+m*a*omega^2*sin(theta))...
-m*a*cos(theta)*(m*g*a*sin(theta)))...
/((m_c+m)*I_g+m^2*a^2*(mcm+sin(theta)^2))                     (3.13)
domega=(-m*a*cos(theta)*(F-k*x+m*a*omega^2*sin(theta)))...
+(m_c+m)*(m*g*a*sin(theta))...
/((m_c+m)*I_g+m^2*a^2*(mcm+sin(theta)^2))
```

```
x=integ(dx, 0)
theta=integ(dtheta, theta_ini)
v_x=integ(dv_x, 0)                                            (3.14)
omega=integ(domega, 0)
theta_deg=theta*180/pi
```

```
F = Kp*theta+Kd*dtheta                                        (3.15)
```

For the first simulation the control gains are set to zero, the cart spring is made very stiff by setting the defined frequency f_n to 1000 Hz, and the initial pendulum angle is set to $\theta_{\text{ini}} = 1°$. In this test the cart should hardly move at all and the pendulum should swing back and forth through a large angle. Figures 3.11 and 3.12 show the angular motion and cart motion, respectively. The pendulum moves as expected, and the cart has virtually no motion.

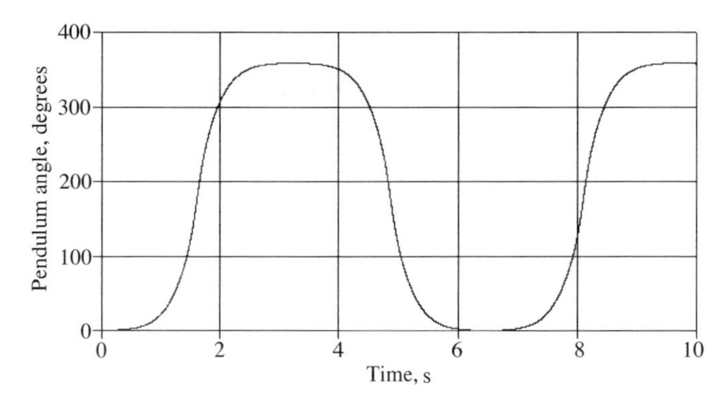

Figure 3.11 Pendulum angle when the cart spring is very stiff and the control force is zero.

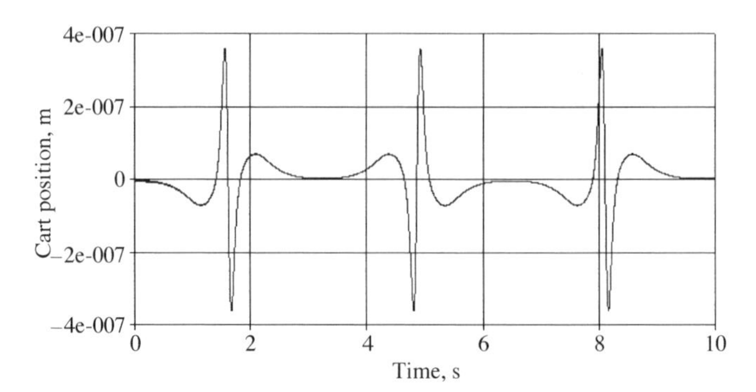

Figure 3.12 Cart position when the cart spring is very stiff and the control force is zero.

In Figures 3.13 and 3.14 the cart spring is back to the prescribed value in Table 3.3 and the control force remains zero. This time the interaction between the cart and the pendulum is quite apparent. The cart moves back and forth clearly influenced by the rotating rigid-body pendulum. The pendulum starts 1° from vertical, and the interaction with the cart affects its angular trajectory. This interaction would be very difficult to observe without the modeling, equation derivation, and simulation steps taken to get to this point.

Just for fun, let's try to stabilize the pendulum using some numerical experimentation and the control equation (3.12). This simulation runs virtually instantly, and multiple runs can be made in a very short time. The proportional gain K_p was varied first and then the derivative gain K_d was varied. The pendulum angle was initialized to 5° and the time response was observed after each run. In a very brief time the response shown in Figure 3.15 was achieved. The control force is capable of stabilizing the pendulum and returning it from its initial offset angle back to vertical. However, notice that the cart position is drifting as shown in Figure 3.16. We can try to fix this problem by using additional control action that tries to maintain the cart position. It

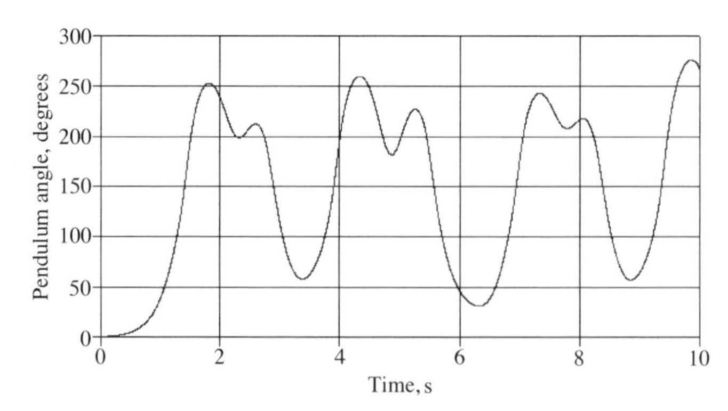

Figure 3.13 Pendulum angle when the cart spring is as specified in Table 3.3 and the control force is zero.

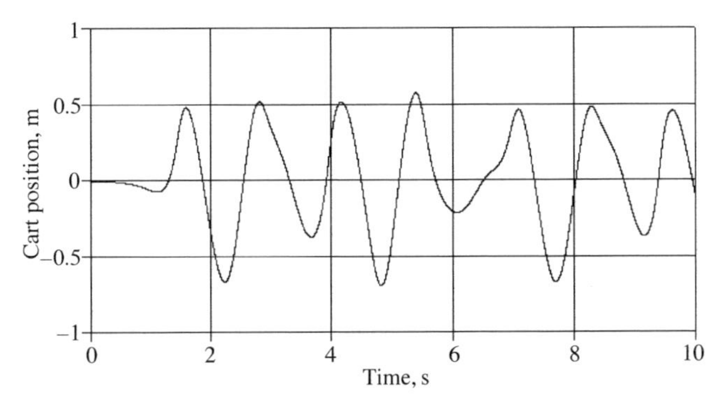

Figure 3.14 Cart position when the cart spring is as specified in Table 3.3 and the control force is zero.

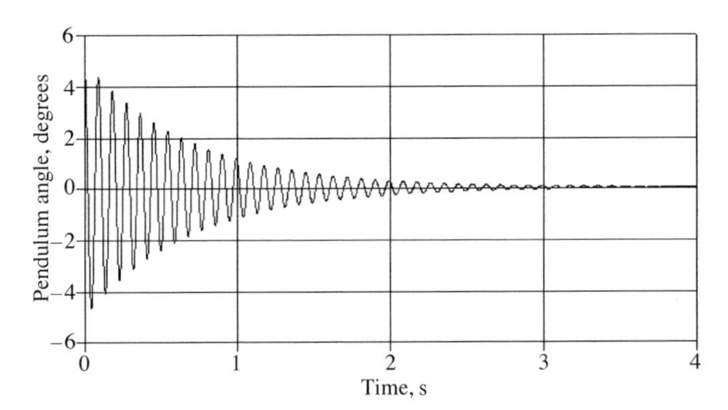

Figure 3.15 Pendulum angle when control is used. For this simulation, $K_p = 10{,}000$ and $K_d = 6$ and the initial angle is $\theta_{\text{ini}} = 45°$.

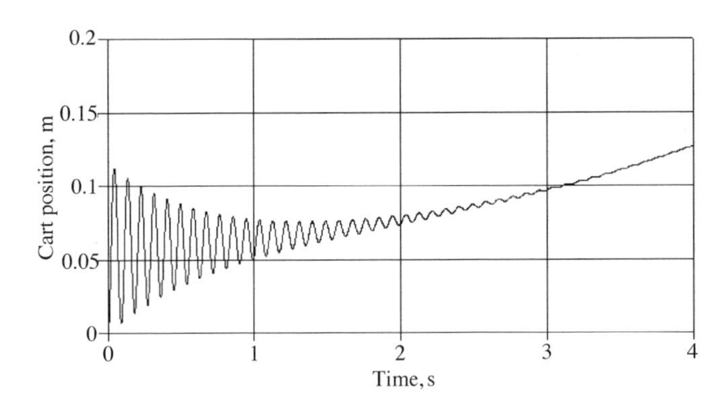

Figure 3.16 Cart position when control is used. The control gains are given in Figure 3.15.

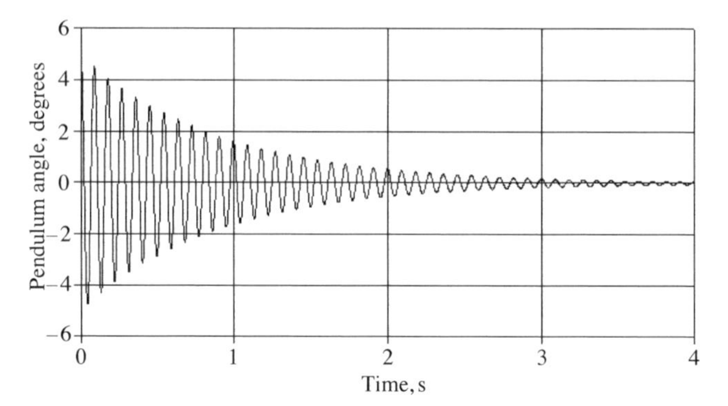

Figure 3.17 Pendulum angle when position control is added to the control action. For this simulation, $K_p = 10{,}000$, $K_d = 20$, $K_{px} = 100$, and $K_{dx} = 2$.

is proposed to modify the control equation from that used in (3.12) to

$$F = K_p\theta + K_d\dot{\theta} + K_{px}x + K_{dx}\dot{x} \tag{3.16}$$

This was done straightforwardly in the simulation, and some additional numerical experimentation was performed. The results of adding some position control to the cart is shown in Figures 3.17 and 3.18. The pendulum angle is still controlled, but this time the cart acquires a steady-state position.

This is just an example of the utility of solving the equations of motion to predict the overall performance of a system. With some background in automatic control one might try linearizing the equations of motion and using some controller design tools to come up with an "optimized" controller. Perhaps we want the system to respond faster or damp out quicker. Perhaps we want to look at the control force and see what size actuator is needed to build the actual system. Once the model is formulated and

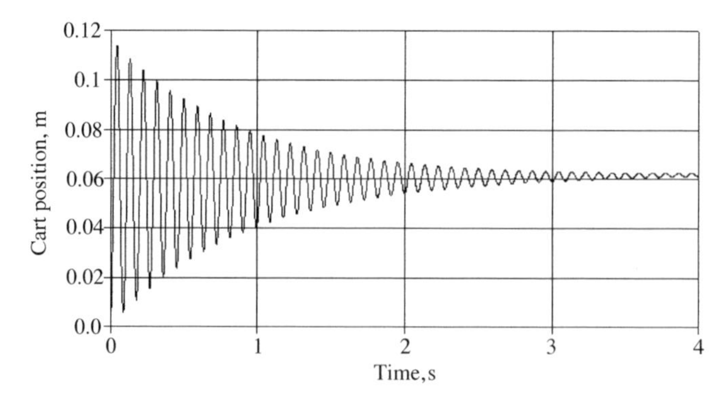

Figure 3.18 Cart position when position control is added to the control action. The control gains are given in Figure 3.17.

a simulation model is built, there really is no end to the studies that can be performed to assist in design of the real system.

3.2 LINEAR SYSTEM RESPONSE

In Chapter 2 it was shown that if the system is assumed to be linear from the start of the modeling process or is linearized after deriving the nonlinear equations of motion, a matrix form of the first-order equations can be realized as

$$\dot{\mathbf{x}} = \mathbf{A}\mathbf{x} + \mathbf{B}\mathbf{u} \tag{3.17}$$

There are typically additional outputs that we desire that are different from the state variables in (3.17). These outputs are dependent on the state variables and are related to the state variables by the output equation

$$\mathbf{Y} = \mathbf{C}\mathbf{x} \tag{3.17a}$$

where $\mathbf{Y}$ is a vector of desired outputs and $\mathbf{C}$ is a matrix that relates outputs to the state variables. In Eq. (3.17), $\mathbf{x}$ is the vector of state variables that come from reducing sets of second order equations to first-order form and $\mathbf{u}$ is the vector of prescribed inputs. Equations of the form (3.17) lend themselves to an enormous collection of analytical techniques, and there are virtually volumes devoted to linear systems analysis. The subject of automatic control relies heavily on linear system analysis tools. References [6] and [7] are typical of books devoted to linear systems and automatic control.

When it comes to engineering applications of dynamics, there are three very important linear concepts that are used over and over to aid in understanding system physics: eigenvalues, transfer functions, and frequency response. These three important concepts are outlined in this section. It is important to remember that real systems are nonlinear and that linear analysis tools do not apply to real systems. However, system understanding often begins with a linearized system model.

Eigenvalues and Their Relationship to System Stability

Eigenvalues or characteristic values come from first setting the system forcing to zero and starting with

$$\dot{\mathbf{x}} = \mathbf{A}\mathbf{x} \tag{3.18}$$

Thus eigenvalues have nothing to do with the forced response of a system. We next guess that the solution to these unforced equations is

$$\mathbf{x} = \mathbf{X}e^{st} \tag{3.19}$$

where $\mathbf{X}$ is a vector of constants and s is the Laplace variable. Substituting (3.19) into (3.18) yields the matrix equation

$$[s\mathbf{I} - \mathbf{A}]\mathbf{X} = 0 \tag{3.20}$$

where $\mathbf{I}$ is an identity matrix with ones on the diagonal and zeros everywhere else. One uninteresting solution to (3.21) is $\mathbf{X} = 0$. If this is true, the real response variables $\mathbf{x}$ also equal zero and the system just stands still. The possibility of having nonzero values of $\mathbf{X}$ requires that

$$\text{Det}[s\mathbf{I} - \mathbf{A}] = 0 \tag{3.21}$$

If this operation is carried out, the result is an nth-order polynomial of the form,

$$s^n + a_1 s^{n-1} + \cdots + a_{n-1}s + a_n = 0 \tag{3.22}$$

This polynomial equation is called the *characteristic equation*. Solutions to this polynomial are called *eigenvalues*, and there are going to be n of them:

$$s = s_1, s_2, \ldots, s_n$$

and the total solution will be modified from our guess of Eq. (3.19) to

$$\mathbf{x} = \mathbf{X}_1 e^{s_1 t} + \mathbf{X}_2 e^{s_2 t} + \cdots + \mathbf{X}_n e^{s_n t} \tag{3.23}$$

where the vectors $\mathbf{X}_i$ are all vectors of constants. The eigenvalue problem would continue with discussions of the calculation and interpretation of the $\mathbf{X}_i$. Here we simply state that these vectors contain constants that could be determined from initial conditions and other considerations. The important thing is the appearance of all the eigenvalues as arguments of exponential functions.

It is now stated that for engineering systems, the eigenvalues are potentially complex numbers, and if a particular eigenvalue is complex, its complex conjugate is also an eigenvalue. We refer to this occurrence as complex-conjugate pairs of eigenvalues. Thus, let's say that the ith eigenvalue is complex and therefore

$$\begin{aligned} s_i &= \alpha_i + j\omega_i \\ s_{i+1} &= \alpha_i - j\omega_i \end{aligned} \tag{3.24}$$

and the contribution of this pair of eigenvalues to the free response of the system comes from (3.23) as

$$\mathbf{x} = \cdots \mathbf{X}_i e^{s_i t} + \mathbf{X}_{i+1} e^{s_{i+1} t} + \cdots \tag{3.25}$$

or

$$
\begin{aligned}
\mathbf{x} &= \cdots \mathbf{X}_i e^{(\alpha_i + j\omega_i)t} + \mathbf{X}_{i+1} e^{(\alpha_i - j\omega_i)t} + \cdots \\
&= \cdots \mathbf{X}_i e^{\alpha_i t} e^{j\omega_i t} + \mathbf{X}_{i+1} e^{\alpha_i t} e^{-j\omega_i t} + \cdots \\
&= \cdots e^{\alpha_i t} \left[\mathbf{X}_i e^{j\omega_i t} + \mathbf{X}_{i+1} e^{-j\omega_i t} \right] \cdots
\end{aligned}
\tag{3.26}
$$

It is now stated without proof that if the vectors of constants $\mathbf{X}_i$ and $\mathbf{X}_{i+1}$ are associated with complex-conjugate eigenvalues, then the associated vectors of constants are themselves complex conjugates; that is,

$$
\mathbf{X}_{i+1} = \mathbf{X}_i^* \tag{3.27}
$$

At this point it is easier to talk about the contribution of this conjugate eigenvalue pair to the response of one of the entries in the solution vector $\mathbf{x}$. Recalling the Euler formulas

$$
e^{j\omega t} = \cos \omega t + j \sin \omega t
$$

and

$$
e^{-j\omega t} = \cos \omega t - j \sin \omega t \tag{3.28}
$$

let's look at the first entry in the solution vector and write

$$
\begin{aligned}
x_1(t) = &\cdots e^{\alpha_i t}[(a_{1_i} + jb_{1_i})(\cos \omega_i t + j \sin \omega_i t) \\
&+ (a_{1_i} - jb_{1_i})(\cos \omega_i t - j \sin \omega_i t)] + \cdots
\end{aligned}
\tag{3.29}
$$

where the complex constants $a_{1_i} + jb_{1_i}$ and $a_{1_i} - jb_{1_i}$ are the first entries in the vectors $\mathbf{X}_i$ and $\mathbf{X}_{i+1}$ respectively. After simplification (3.31) becomes the real contribution,

$$
x_1(t) = \cdots e^{\alpha_i t}[A_{1_i} \cos \omega_i t + B_{1_i} \sin \omega_i t] + \cdots \tag{3.30}
$$

where $A_{1_i} = 2a_{1_i}$ and $B_{1_i} = -2b_{1_i}$.
The contribution of a complex-conjugate pair of eigenvalues to the free response of a linear system can now be interpreted from Eq. (3.30).

1. The real part of an eigenvalue always appears as a real exponential multiplier. If the real part is positive, the term will grow exponentially and will ultimately approach infinity.
2. Stability of a system is dictated completely by the sign of the real parts of eigenvalues. Stable systems require all real parts of all eigenvalues to be negative. If

even one real part of one eigenvalue is positive, it will grow exponentially and ultimately govern the system response as time goes on.

3. The imaginary part of an eigenvalue always appears along with time t as the argument of a sine or cosine function. Therefore, the imaginary part of an eigenvalue is a frequency of oscillation in rad/s.
4. If an eigenvalue is only real and has no imaginary part, that eigenvalue contribution to the response cannot oscillate. It can grow or decay exponentially.
5. If an eigenvalue pair has no real part, the contribution of that eigenvalue pair will be a nondecaying oscillation that will persist forever.

We will make use of these facts about eigenvalues in some of the applications in this book. In the Matlab program [3], the eigenvalues of a linear system of virtually any size can be obtained by the command eig(**A**), where we need the **A**-matrix from first deriving and linearizing the equations of motion. The result of executing this command is a list of the eigenvalues broken into real and imaginary parts. Simply perusing the list tells us quite a lot about the potential system response.

Transfer Functions

A second very important concept in linear system analysis is the *transfer function*, the ratio of an output to an input expressed in terms of the Laplace variable s rather than time. The Laplace transform is used to convert the time-domain equations (3.17) into the *Laplace domain*. The Laplace transform can be found in many references, and it is covered thoroughly in many automatic controls texts (e.g., [8] and [9]. To convert time-domain linear equations into algebraic equations in the s-domain, we really need only one concept from Laplace transforms. For the time function $x(t)$, its Laplace transform is $X(s)$ and we say that

$$L[x(t)] = X(s) \tag{3.31}$$

The actual calculation that converts $x(t)$ into $X(s)$ is

$$L[x(t)] = X(s) = \int_0^\infty x(t)e^{-st}\,dt \tag{3.32}$$

and the reader can find tables of transform pairs where (3.32) has been applied to countless functions [8]. For the application here, it is sufficient to recognize that

$$L\left[\frac{d^n x(t)}{dt^n}\right] = s^n X(s) \tag{3.33}$$

or the Laplace transform of the nth derivative of a time function is s^n times the Laplace transform of the time function. For example, consider the second-order differential

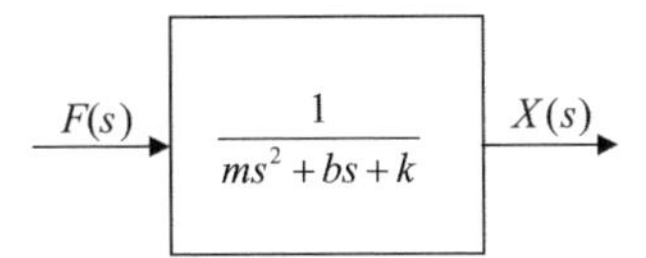

Figure 3.19 Block diagram for a transfer function.

equation

$$m\ddot{x} + b\dot{x} + kx = F(t) \tag{3.34}$$

To convert to the Laplace domain, we take the Laplace transform of the equation, with the result

$$(ms^2 + bs + k)X(s) = F(s) \tag{3.35}$$

We really cannot say anything further about the Laplace transform of the forcing function until the forcing function is specifed.

A *transfer function* is the ratio of output to input in the Laplace domain. Thus, for this example the transfer function becomes

$$\frac{X(s)}{F(s)} = \frac{1}{ms^2 + bs + k} \tag{3.36}$$

In automatic control it is very typical to represent transfer functions using blocks as shown in Figure 3.19. The interpretation is that the input is multiplied by the contents of the block to produce the output.

We now return to the general representation of linear systems from Eq. (3.17). The Laplace transform of this matrix equation is

$$[s\mathbf{I} - \mathbf{A}]\mathbf{X}(s) = \mathbf{B}\mathbf{U}(s) \tag{3.37}$$

or, considering one input at a time,

$$[s\mathbf{I} - \mathbf{A}]\mathbf{X}(s) = \mathbf{b}U(s) \tag{3.38}$$

It is typical to use a lowercase **b** to indicate a column vector when considering only one input at a time. In fact, (3.37) could be written in terms of the separate inputs as

$$[s\mathbf{I} - \mathbf{A}]\mathbf{X}(s) = \mathbf{b}_1 u_1(s) + \mathbf{b}_2 u_2(s) + \mathbf{b}_3 u_3(s) + \cdots \tag{3.39}$$

Also, when considering only one output at a time, it is typical to write Eq. (3.16a) as

$$y = \mathbf{c}^t\mathbf{x} \tag{3.40}$$

where $\mathbf{c}^t$ indicates a vector on its side (i.e., a row vector). The Laplace transform of the output equation becomes

$$Y(s) = \mathbf{c}^t\mathbf{X}(s) \tag{3.41}$$

We can now symbolically exhibit the transfer function between an output and input by starting with (3.38) and writing

$$\mathbf{X}(s) = [s\mathbf{I} - \mathbf{A}]^{-1}\mathbf{b}u(s) \tag{3.42}$$

and then use (3.43), with the result

$$Y(s) = \mathbf{c}^t\mathbf{X}(s) = \mathbf{c}^t[s\mathbf{I} - \mathbf{A}]^{-1}\mathbf{b}U(s) \tag{3.43}$$

Finally, the general expression for a transfer function comes from (3.43) as

$$\frac{Y(s)}{U(s)} = \mathbf{c}^t[s\mathbf{I} - \mathbf{A}]^{-1}\mathbf{b} \tag{3.44}$$

Notice that for each output the $\mathbf{c}^t$ vector will change, and for each input the $\mathbf{b}$ vector will change. There is one further symbolic simplification that can be made to (3.44) to arrive at the final form:

$$\frac{Y(s)}{U(s)} = \frac{\mathbf{c}^t \mathrm{Adj}[s\mathbf{I} - \mathbf{A}]\mathbf{b}}{\mathrm{Det}[s\mathbf{I} - \mathbf{A}]} \tag{3.45}$$

where Adj[·] is the adjoint of the argument matrix.

We would never carry out the operations indicated in (3.45) to derive a transfer function by hand, but it is interesting to note that Det[$s\mathbf{I} - \mathbf{A}$] is the denominator of all input/output transfer functions for a given dynamic system, and this denominator is the characteristic equation that was needed to determine the eigenvalues in Eq. (3.21). For a given dynamic system characterized by the linear equations (3.17), the numerator of any input/output transfer function is different for each input and each output, but the denominator is the same for all transfer functions.

For low-order systems having only a few state equations, transfer functions can be derived by hand using Cramer's rule [10]. To do this we would get to the point of Eq. (3.38) and then solve for the particular entries in the state vector that allow simplest calculation of the output of interest. Assume that we want to solve for $X_i(s)$ in the state vector. Cramer's rule yields

$$\frac{X_i(s)}{U(s)} = \frac{\left|\begin{array}{l}\text{determinant of } s\mathbf{I} - \mathbf{A} \\ \text{having first substituted the } \mathbf{b} \text{ vector} \\ \text{for the } i\text{th column}\end{array}\right|}{|s\mathbf{I} - \mathbf{A}|} \tag{3.46}$$

Consider the second-order equation from (3.34),

$$m\ddot{x} + b\dot{x} + kx = F(t) \tag{3.47}$$

In first-order form this becomes

$$\begin{aligned}\dot{x} &= v_x \\ \dot{v}_x &= \frac{F}{m} - \frac{b}{m} v_x - \frac{k}{m} x\end{aligned} \tag{3.48}$$

or in matrix form,

$$\frac{d}{dt}\begin{bmatrix} x \\ v_x \end{bmatrix} = \begin{bmatrix} 0 & 1 \\ -\dfrac{k}{m} & -\dfrac{b}{m} \end{bmatrix}\begin{bmatrix} x \\ v_x \end{bmatrix} + \begin{bmatrix} 0 \\ \dfrac{1}{m} \end{bmatrix} F \tag{3.49}$$

In the Laplace domain [e.g., Eq. (3.38)], Eq. (3.49) becomes

$$\begin{bmatrix} s & -1 \\ \dfrac{k}{m} & s + \dfrac{b}{m} \end{bmatrix}\begin{bmatrix} X \\ V_x \end{bmatrix} = \begin{bmatrix} 0 \\ \dfrac{1}{m} \end{bmatrix} F \tag{3.50}$$

The transfer function between the output $X(s)$ and input $F(s)$ comes from Cramer's rule as

$$\frac{X(s)}{F(s)} = \frac{\begin{vmatrix} 0 & -1 \\ \dfrac{1}{m} & s + \dfrac{b}{m} \end{vmatrix}}{\begin{vmatrix} s & -1 \\ \dfrac{k}{m} & s + \dfrac{b}{m} \end{vmatrix}} = \frac{1/m}{s^2 + (b/m)s + k/m} \tag{3.51}$$

and the transfer function for the output $V_x(s)$ to the input $F(s)$ is

$$\frac{V_x(s)}{F(s)} = \frac{\begin{vmatrix} s & 0 \\ \dfrac{k}{m} & \dfrac{1}{m} \end{vmatrix}}{\begin{vmatrix} s & -1 \\ \dfrac{k}{m} & s + \dfrac{b}{m} \end{vmatrix}} = \frac{s/m}{s^2 + (b/m)s + k/m} \tag{3.52}$$

It should not be too surprising that the transfer function for the velocity (3.52) is the same as the transfer function for displacement (3.51) except for the additional s in the numerator of (3.52). The velocity is the time derivative of displacement, and in the Laplace domain, time derivatives are indicated by multiplication by s. Thus, since

$$v_x = \frac{dx}{dt}$$

then

$$V_x(s) = sX(s)$$

and the transfer functions are related by

$$\frac{V_x(s)}{F(s)} = s\frac{X(s)}{F(s)} \tag{3.53}$$

This has been a brief introduction to the topic of transfer functions. We will revisit this concept at various places in the book. If we were use a linear analysis program such as Matlab [3], we can determine the transfer functions for very large systems. After formulating a system model, applying physical principles, and deriving a linear set of governing equations in the matrix form of (3.17), we would first create a system using the state-space function with a command such as

```
Systems=ss(A, b, c_t, 0),
```

followed by a command that turns the system into a transfer function,

```
TF_system=tf(System).
```

The result, TF_system, can be viewed symbolically as a ratio of polynomials in the s-domain as well as being used for other analysis applications.

Frequency Response

The third linear analysis concept that is very important in engineering applications is the frequency response. We are interested in the response of a system if an input to the system is a fixed-amplitude sine or cosine function called a *harmonic input*. Such an input is fairly easy to generate in a laboratory environment, and most industries that do any vibration testing will have shaker facilities that can generate harmonic inputs and monitor the response of system components. Automotive companies have shaker facilities for testing suspensions, and even entire vehicles can be shaken in larger facilities. Computer and electronic components are tested using shaker tables that can sweep through many frequencies and determine if components break.

A good guess would be that if the input to a system is a fixed-frequency harmonic, the outputs from that system will also be fixed-amplitude harmonics at the same frequency as the input. It turns out that this is true only for linear systems. For more realistic nonlinear systems the response to a fixed-frequency input can be a very distorted-looking sine wave and can even be at a different frequency than the input. Nevertheless, the fixed-frequency harmonic input is used pervasively in engineering applications and yields useful results even for nonlinear systems.

One way to present a frequency response is to show the time histories of the input and response. For the example system that generated Eq. (3.49), some parameters

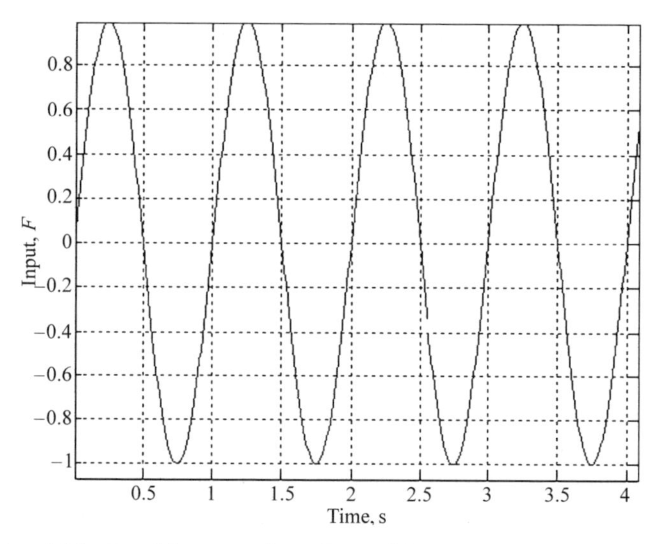

Figure 3.20 Fixed-frequency input force of 1 N amplitude and f_i = 1.0 Hz.

were chosen and the frequency response was generated for input at 1 Hz. The fixed-amplitude input and the corresponding displacement response are shown in Figures 3.20 and 3.21. Notice that after a brief transient the output becomes a fixed-amplitude harmonic with some phase relationship to the input. We could change the frequency and run the simulation again and the response would have a different amplitude and

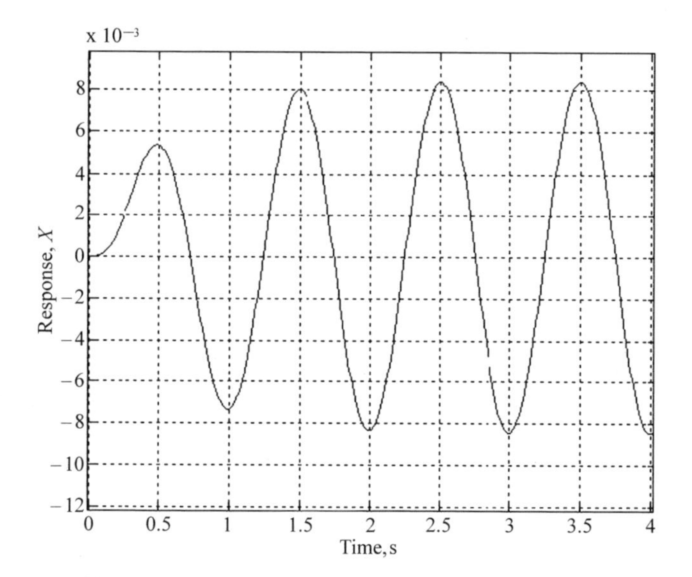

Figure 3.21 Response to a fixed-frequency force input.

phase relationship. We could do this procedure for many frequencies and generate many plots of the time history of the response.

What is of more interest is the response amplitude and phase, as they depend on the forcing frequency. If the input is

$$u(t) = U_0 \cos \omega t \tag{3.54}$$

the steady-state response will be

$$y(t) = Y_0 \cos(\omega t + \phi) \tag{3.55}$$

We are not much interested in plotting the time histories. We would really like to determine how Y_0 and ϕ depend on the forcing frequency ω. When the response amplitude divided by the input amplitude is plotted versus the forcing frequency and the response phase angle is plotted versus the forcing frequency, the result is the *frequency response* of the system. Sometimes such a presentation is called a *Bode plot*. The goal of frequency response analysis is to come up with amplitude and phase dependence on forcing frequency.

One way to determine a frequency response analytically for the single second-order equation (3.34) is to use (3.54) for the input and (3.55) for the response and substitute into the governing equation. This will generate a lot of algebra to solve for Y_0 as a function of ω and ϕ as a function of ω. Another approach is to recognize that

$$u(t) = U_0 \cos \omega t = \text{Re}(U_0 e^{j\omega t}) \tag{3.56}$$

$$y(t) = Y_0 \cos(\omega t + \phi) = \text{Re}(Y_0 e^{j(\omega t + \phi)}) = \text{Re}(Y_0 e^{j\phi} e^{j\omega t}) = \text{Re}(H_y e^{j\omega t}) \tag{3.57}$$

where

$$\begin{aligned} |H_y| &= Y_0 \\ \angle H_y &= \phi \end{aligned} \tag{3.58}$$

We then substitute the final equation of (3.56) and (3.57) into the governing equation and solve for the complex number H_y.

In other words, if we can obtain the complex number H_y, we can recover the response amplitude by taking the magnitude of H_y and the response phase angle by determining the angle of the complex number H_y. The complex number H_y is called the *complex frequency response function*. It turns out that it is incredibly simple to come up with the complex frequency response function and far more difficult to perform the algebra if one starts with the real-time functions from (3.54) and (3.55).

It is now stated without proof that the complex frequency response function is obtained simply by letting $s = j\omega$ in the transfer function between any output and any input. For the transfer function derived in Eq. (3.35), the complex frequency response would be

$$\frac{X(j\omega)}{\mathrm{F}(j\omega)} = \frac{1}{m(j\omega)^2 + bj\omega + k} = \frac{1}{k - m\omega^2 + \ jb\omega} \tag{3.59}$$

The response amplitude divided by the input amplitude comes from the first equation of (3.58) as

$$\left|\frac{X}{F}\right| = \frac{1}{[(k - m\omega^2)^2 + (b\omega)^2]^{1/2}} \tag{3.60}$$

Recall that the magnitude of a complex number comes from the square root of the sum of the squares of the real and imaginary parts. The response phase angle comes from the second equation of (3.58) as

$$\angle\frac{X}{F} = 0 - \tan^{-1}\frac{b\omega}{k - m\omega^2} \tag{3.61}$$

where we have made use of the fact that the associated angle of the ratio of complex numbers is the angle of the numerator complex number minus the angle of the denominator complex number. We have also used the fact that the angle of a complex number is the arctangent of the imaginary part divided by the real part. For this example the numerator has no imaginary part and the arctangent of zero is zero. We could now plot or sketch the frequency response.

The transfer function for the velocity of this system was derived in Eq. (3.52). The complex frequency response function is

$$\frac{V_x(j\omega)}{F(j\omega)} = \frac{j\omega/m}{(j\omega)^2 + (b/m)\, j\omega + k/m} = \frac{(1/m)\, j\omega}{(k/m) - \omega^2 + \ j(b/m)\omega} \tag{3.62}$$

Thus, the magnitude of the frequency response is

$$\left|\frac{V_x}{\mathrm{F}}\right| = \frac{(1/m)\omega}{[((k/m) - \omega^2)^2 + ((b/m)\omega)^2]^{1/2}} \tag{3.63}$$

and the response phase angle is

$$\angle\frac{V_x}{F} = 90^\circ - \tan^{-1}\frac{(b/m)\omega}{(k/m) - \omega^2} \tag{3.64}$$

where this time the numerator of the complex frequency response function (3.62) is purely imaginary with no real part and $\tan^{-1}(\omega/0) = \tan^{-1}\infty = 90\,^\circ$.

For low-order linear systems with only a few governing equations, carrying out the analytical steps shown here is a reasonable exercise. For larger systems we would resort to a computational solution. The commercial program Matlab [3] can produce the frequency response directly from the matrix formulation for linear systems in

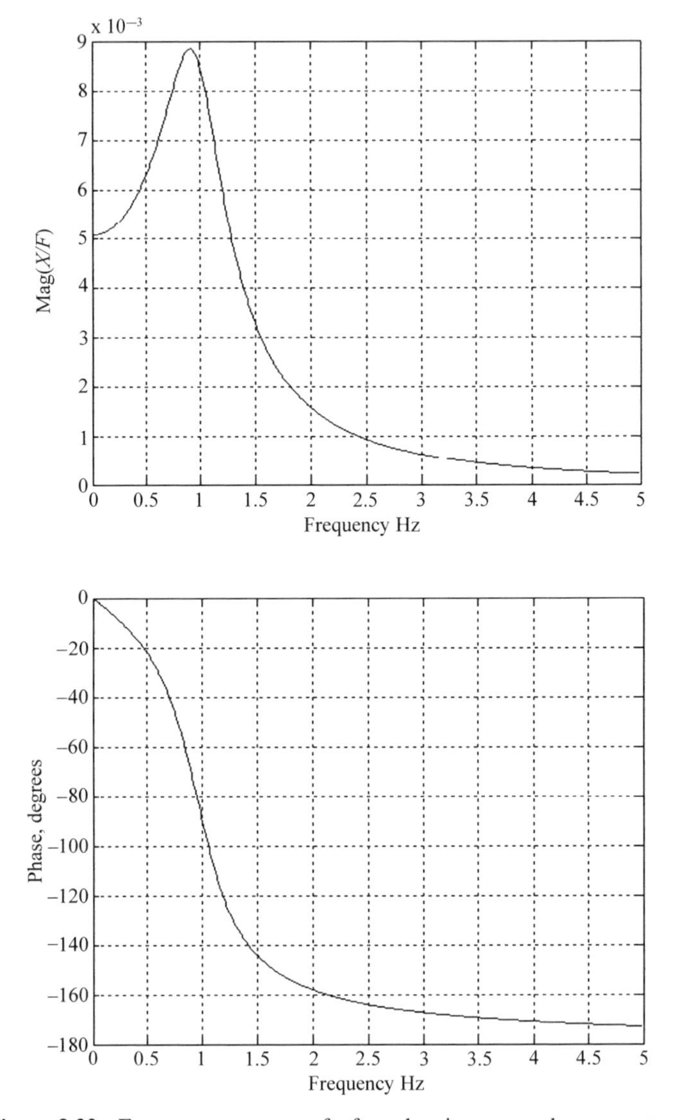

Figure 3.22 Frequency response of a forced spring–mass–damper system.

the form of Eqs. (3.17) and (3.17a). The Matlab program was used to produce the displacement frequency response shown in Figure 3.22. The concept of a frequency response will be used in subsequent chapters.

REFERENCES

[1] *SPICE v 5.0, RD Research*, Norwich, England.

[2] Nastran Finite Element, Analysis software, Noran Engineering, Westminster, CA.

[3] Matlab and Simulink software, Mathworks Co., Natick, MA.

[4] ACSL software, AEgis Technologies Group, Huntsville, AL.

[5] B. Carnahan, H. Luther, J. Wilkes, *Applied Numerical Methods*, Wiley, New York, 1969.

[6] K. Ogata, *System Dynamics* 4th ed., Pearson Prentice Hall, Upper Saddle River, NJ,s 2004.

[7] D.L. Smith, *Introduction to Dynamic Systems Modeling for Design* Prentice Hall, Upper Saddle River, NJ 1994.

[8] F.H. Raven, *Automatic Control Engineering*, 4th ed., McGraw-Hill, New York 1987.

[9] W.J. Grantham, T.L. Vincent, *Modern Control Systems Analysis and Design*, Wiley, New York, 1993.

[10] Wylie, C.R. *Advanced Engineering Mathematics*, McGraw-Hill, New York, 1960.

PROBLEMS

3.1 **(a)** Using any commercial equation solver available to you, simulate the spring–pendulum system from Example 3.1 and reproduce the responses from Figures 3.2 to 3.5.

(b) Run the simulation for the full range of initial lengths and initial angles listed in Table 3.1.

3.2 **(a)** Using any commercial equation solver available to you, simulate the pinned disk with a slot and mass particle from Example 3.2 and reproduce the responses from Figures 3.7 to 3.10.

(b) Change the input torque $\tau_i(t)$ to a harmonic input where $\tau_i(t) = \tau_0 \sin \omega_i t$ and $\omega_i = \omega_n$, $2\omega_n$, and $3\omega_n$, where $\omega_n = 2\pi f_n$. Plot the angular velocity of the disk and the position of the mass particle.

3.3 **(a)** Using any commercial equation solver available to you, simulate the inverted pendulum system from Example 3.3 and reproduce the uncontrolled responses from Figures 3.11 to 3.14.

(b) Add the control as suggested in Eq. (3.16) and use the control gains from Figures 3.15 and 3.17 to generate the outputs shown in Figures 3.15 to 3.18.

(c) Start with an initial angle $\theta_{\text{ini}} = 10^\circ$, 30°, 60° and see if the controller still works. If the controller does not work so well, try some numerical experimentation to determine if better control gains can be obtained.

3.4 Shown in Figure P3.4 is a mass sitting on a nonlinear spring and nonlinear damper. The spring and damper are both positive in compression. The spring force is F_s and the damper force is F_d. There is base excitation indicated by $y_{\text{in}}(t)$ and a prescribed force on the mass $F_{\text{in}}(t)$. The nonlinear spring and damper laws are shown in Figure P3.4. When $y = 0$ and $y_{\text{in}} = 0$, the spring is at its free length and there is no force in the spring. We desire a simulation of this system to specified inputs.

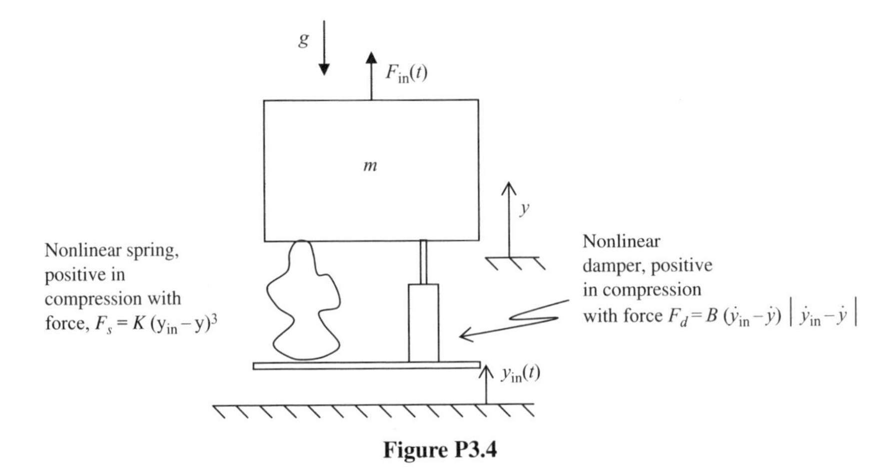

Figure P3.4

(a) Construct velocity, acceleration, and force diagrams and derive $F_s + F_d - mg = m\ddot{y}$, or in first-order form,

$$\begin{aligned} \dot{y} &= v \\ \dot{v} &= \frac{F_s}{m} + \frac{F_d}{m} + \frac{F_{\text{in}}}{m} - g \end{aligned} \tag{P3.4-1}$$

(b) All simulations will start from equilibrium in a gravity field. At equilibrium the spring force will oppose the weight mg. If the equilibrium displacement y_0 is $y_0 = -0.5$ m (note that y is measured positive up) and the mass is $m = 50$ kg, calculate K from $K = mg/(-y_0)^3$. Also, for a relative velocity of $v_{\text{rel}} = 1$ m/s, the damper force is equal to the weight. Calculate the constant B for the damper.

The force input at the top is a step input such that if $t \geq 0$, $F_{\text{in}} = mg$ for one simulation and $F_{\text{in}} = -mg$ for a second simulation. The displacement input at the base is harmonic, such that $y_{\text{in}} = Y_{\text{in}} \sin \omega_{\text{in}} t$ where $Y_{\text{in}} = 0.5\,\text{m}$, $f_{\text{in}} = 1, 5, 10\,\text{Hz}$, and $\omega_{\text{in}} = 2\pi\, f_{\text{in}}$.

(c) Using a commercial equation solver, simulate the system when F_{in} is acting and $y_{\text{in}} = 0$. Try both positive and negative step inputs. Simulate the system when y_{in} is acting and $F_{\text{in}} = 0$. Try the various input frequencies. For all cases, plot the velocity and position of the mass.

(d) Turn both inputs on and simulate the system when the input force magnitude has its negative value and the input frequency is 1.0 Hz. Plot the same results as in part (c).

Some sample responses are shown in Figures P3.4-1 to P3.4-6.

3.5 The bicycle model of a car is shown in Figure P3.5. This is the same schematic as that used in Problem 2.13. Body-fixed coordinates are used with velocity components U oriented longitudinally and V oriented laterally. The yaw rate

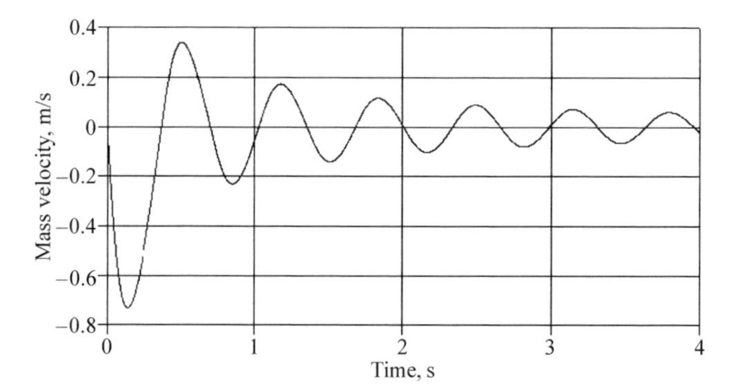

Figure P3.4-1 Mass velocity when the base excitation is zero and the input force is $F_{\mathrm{in}} = -mg$.

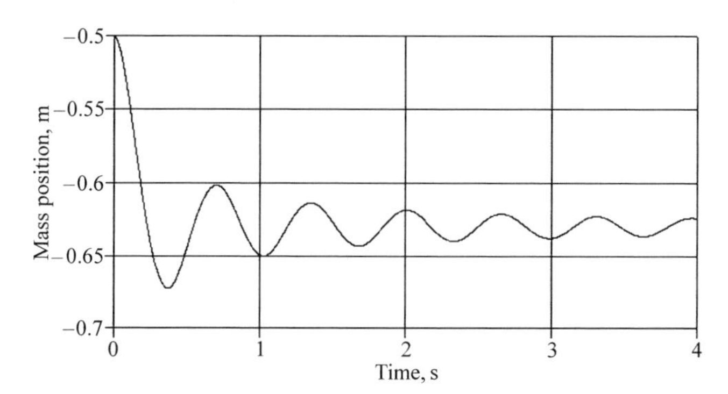

Figure P3.4-2 Mass position when the base excitation is zero and the input force is $F_{\mathrm{in}} = -mg$.

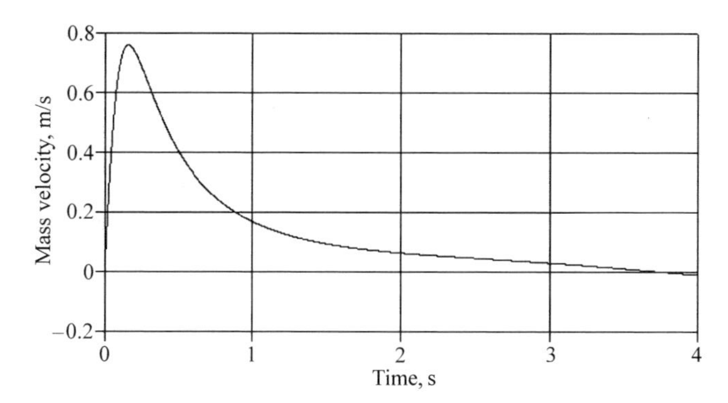

Figure P3.4-3 Mass velocity when the base excitation is zero and the input force is $F_{\mathrm{in}} = mg$.

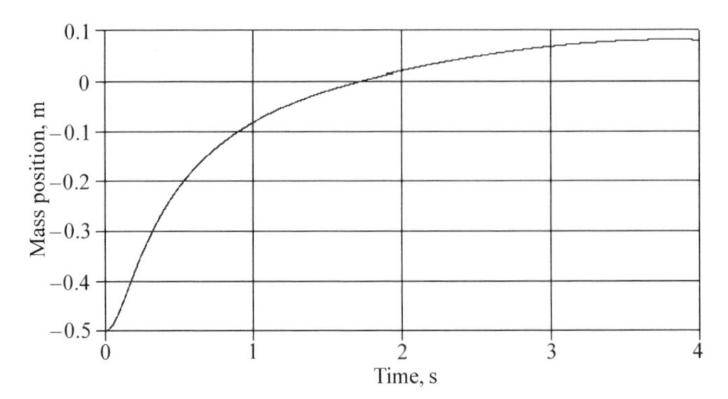

Figure P3.4-4 Mass position when the base excitation is zero and the input force is $F_{\text{in}} = mg$.

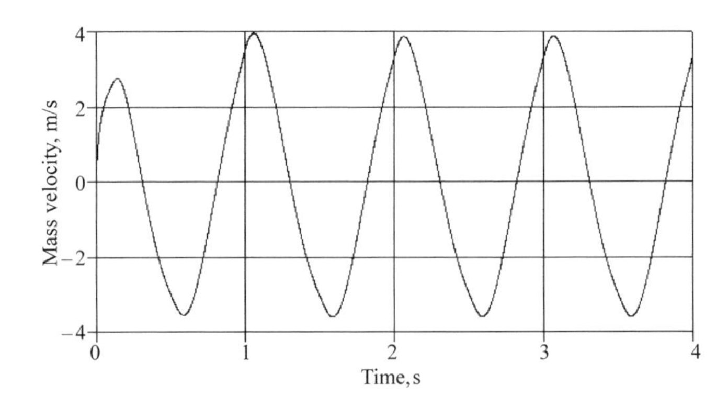

Figure P3.4-5 Mass velocity when the base excitation is $y_{in} = Y_{in} \sin \omega_{in} t$ and the input force is $F_{\text{in}} = 0$.

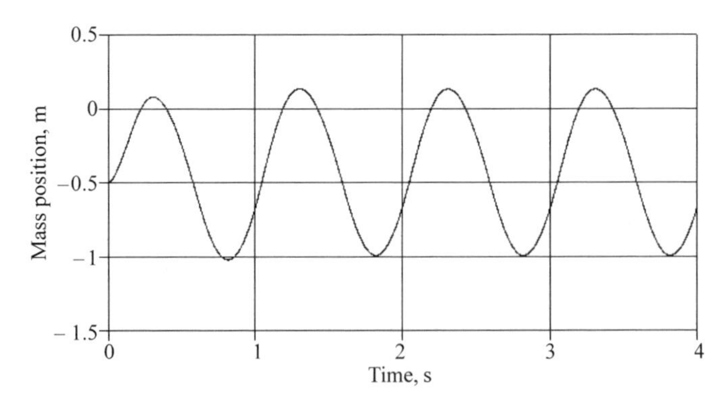

Figure P3.4-6 Mass position when the base excitation is $y_{in} = Y_{in} \sin \omega_{in} t$ and the input force is $F_{\text{in}} = 0$.

is given by ω_y and is positive in the counterclockwise direction. The car can be steered by specifying the steer angle δ. It turns out that the cornering forces F_f and F_r are related to the velocity and angular velocity by

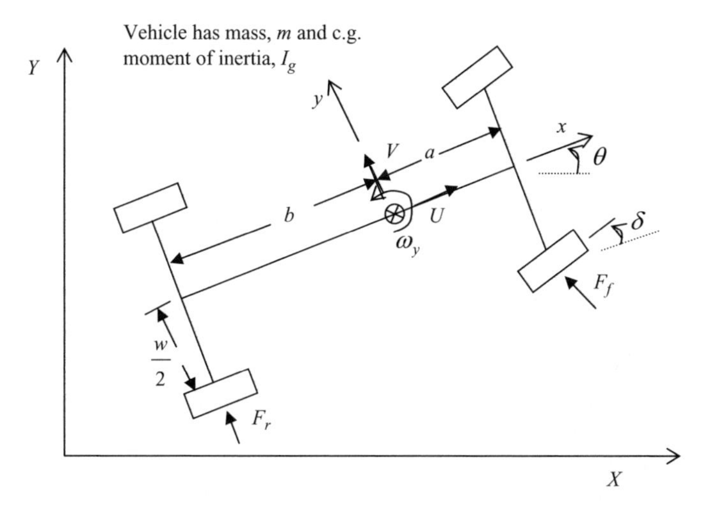

Figure P3.5

$$F_f = \frac{C_f}{U}[U\delta - (V + a\omega_y)]$$
$$F_r = \frac{C_r}{U}(b\omega_y - V) \qquad \text{(P3.5-1)}$$

where C_f and C_r are called *cornering stiffnesses* and are constants that must be given. A way to relate the stiffness parameters is

$$C_r = K\frac{a}{b}C_f \qquad \text{(P3.5-2)}$$

where K is a form of understeering coefficient. If $K > 1$ the car is understeering, if $K < 1$ the car is oversteering, and if $K = 1$ the car is neutral steering. The first-order equations of motion are

$$\dot{V} = -U\omega_y + \frac{F_f}{m} + \frac{F_r}{m}$$
$$\dot{\omega} = \frac{F_f}{I_g}a - \frac{F_r}{I_g}b \qquad \text{(P3.5-3)}$$

We will assume that the forward velocity U is constant for this problem. The parameters for the system are

$$
\begin{aligned}
m &= 1450\,\text{Kg} & I_g &= mab,\ \text{kg--m}^2 \\
L &= 2.2\,\text{m} & C_f &= 50,000\,\text{N/rad} \\
\frac{a}{b} &= 0.9 & C_r &= K\frac{a}{b}C_f \\
b &= \frac{L}{1+a/b},\ \text{m} & U &= 60\,\text{mph} = 26.8\,\text{m/s} \\
a &= L - b,\ \text{m} & K &= 0.8, 0.9, 1.0, 1.1
\end{aligned}
\qquad \text{(P3.5-4)}
$$

(a) Using any commercial equation solver available to you, program Eqs. [P3.5-3] with [P3.5-1] and simulate the motion of the vehicle for the following steering inputs. Run the simulation for the various K values suggested above. Constant steering input; $\delta(t) = \delta_0$, where $\delta_0 = 2^0 = 2(\pi/180)$ rad. Harmonic steering input, $\delta = \delta_0 \sin\omega_i t$, where, $f_i = 1.0$ Hz and $\omega_i = 2\pi f_i$.

(b) Plot the yaw rate ω_y and lateral acceleration a_{lat}, where the lateral acceleration is given by $a_{\text{lat}} = \dot{V} + U\omega$. Plot the lateral acceleration in g's by dividing a_{lat} by g. Plot the trajectory of the vehicle. To do this, inertial direction velocity components are needed. In the x-direction,

$$\dot{x} = U\cos\theta - V\sin\theta \qquad \text{(P3.5-5)}$$

and in the y-direction,

$$\dot{y} = U\sin\theta + V\cos\theta \qquad \text{(P3.5-6)}$$

where

$$\dot{\theta} = \omega_y \qquad \text{(P3.5-7)}$$

Equations [P3.5-5] through [P3.5-7] are simply integrated along with the dynamic equations to provide the desired outputs.

Some sample outputs are shown in Figures P3.5-1 to P3.5-6 for constant steer angle input. It is pretty obvious that the coefficient K plays a large role in determining the response.

3.6 Shown in Figure P3.6 is a linear two-degree-of-freedom system. The connecting rod between the cart and the pendulum mass is of length l and is massless. The variable y_c locates the cart and the variable θ; locates the fixed-length pendulum relative to the vertical direction. Gravity is acting. For small motions

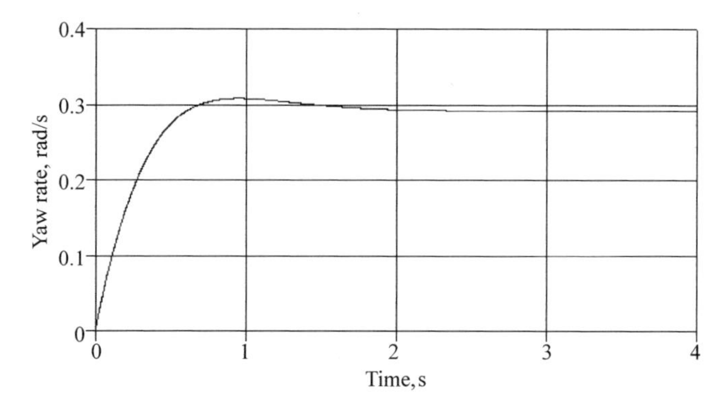

Figure P3.5-1 Yaw rate for constant steer angle input and $K = 1.1$.

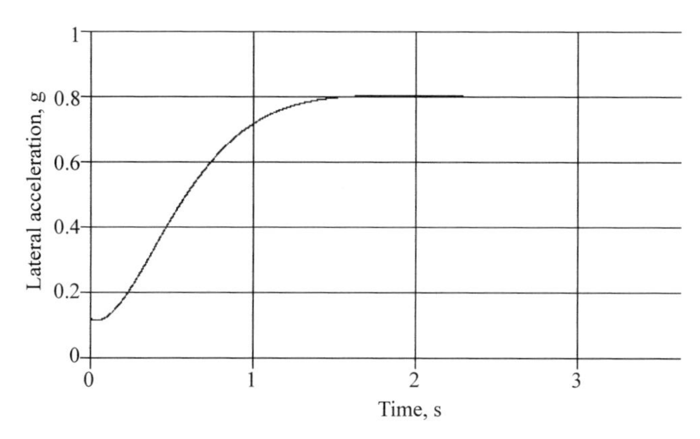

Figure P3.5-2 Lateral acceleration for constant steer angle input and $K = 1.1$.

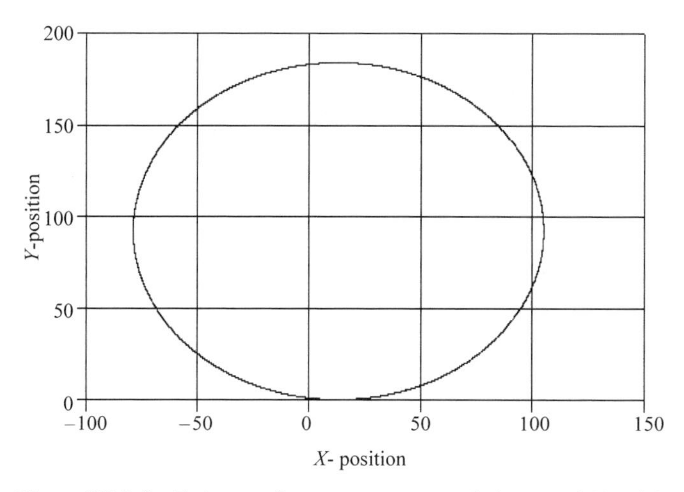

Figure P3.5-3 Trajectory for constant steer angle input and $K = 1.1$.

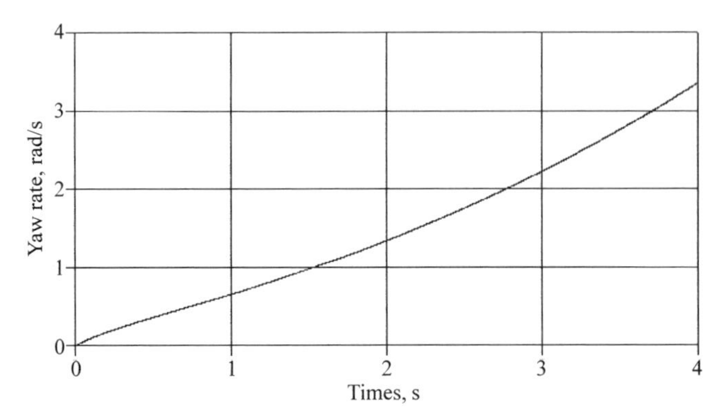

Figure P3.5-4 Yaw rate for constant steer angle input and $K = 0.8$.

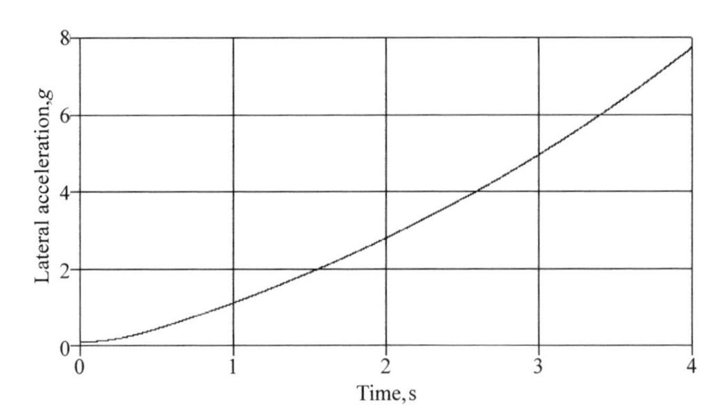

Figure P3.5-5 Lateral acceleration for constant steer angle input and $K = 0.8$.

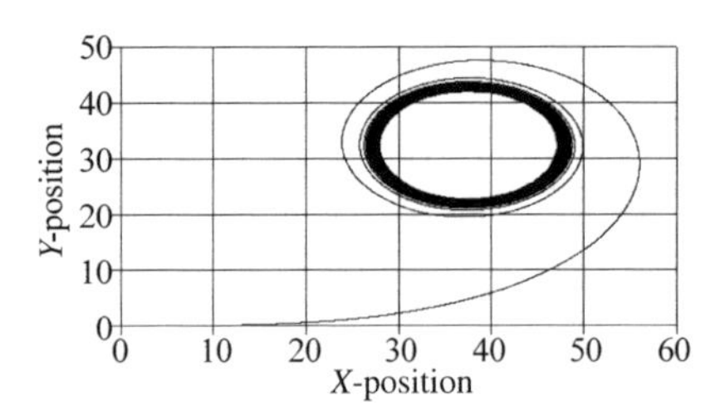

Figure P3.5-6 Trajectory for constant steer angle input and $K = 0.8$.

of y_c and θ, the equations of motion are

$$\begin{aligned} F(t) - ky_c + mg\theta &= m_c\ddot{y}_c \\ -mg\theta &= m(\ddot{y}_c + l\ddot{\theta}) \end{aligned} \tag{P3.6-1}$$

(a) To get one second-order derivative term in each of two equations, solve the first equation of (P3.6-1) for $\ddot{y}_c$ and substitute into the second equation. Use the definitions $\dot{\theta} = \omega$, $\dot{y}_c = v$ and reduce the two second-order equations to four first-order equations. Write the result in the standard matrix format of $\dot{\mathbf{x}} = \mathbf{A}\mathbf{x} + \mathbf{b}u$ from Eq. 2.80.

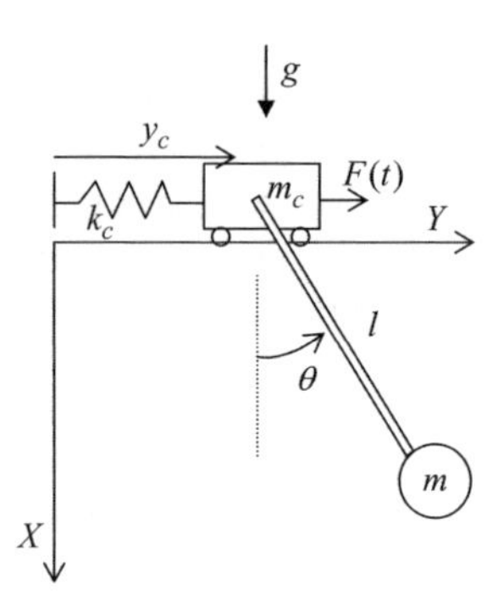

Figure P3.6

(b) Derive the transfer function relating θ to F and y_c to F in the Laplace s-domain.

(c) Write down the characteristic equation from which the eigenvalues would be determined.

3.7 Two identical pendulums are connected by the torsional spring of stiffness k_τ (Figure P3.7). The unforced equations of motion for small motions of the angles θ_1 and θ_2 are

$$\begin{aligned}\ddot{\theta}_1 &= -\left(\frac{g}{l} + \frac{k_\tau}{ml^2}\right)\theta_1 + \frac{k_\tau}{ml^2}\theta_2 \\ \ddot{\theta}_2 &= \frac{k_\tau}{ml^2}\theta_1 - \left(\frac{g}{l} + \frac{k_\tau}{ml^2}\right)\theta_2\end{aligned} \tag{P3.7-1}$$

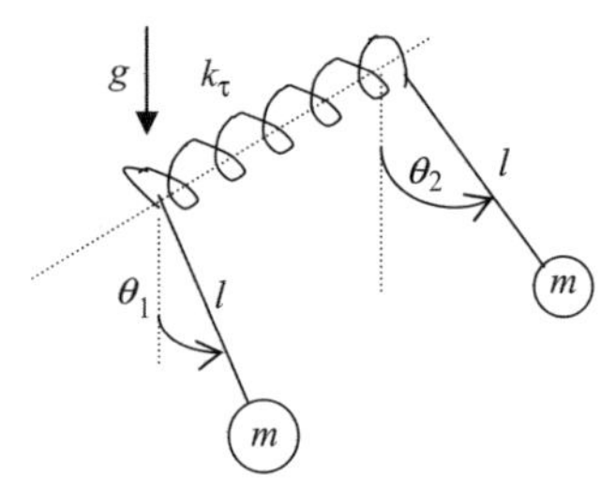

Torsionalspring with pendulums at its ends. The torque in the spring is $\tau = k_\tau(\theta_1 - \theta_2)$

Figure P3.7

(a) Put these equations in first-order matrix form [e.g., Eq. 2.80] and derive the characteristic equation for the eigenvalues.

(b) Solve the biquadratic characteristic equation for the eigenvalues and interpret their meaning.

3.8 Figure P3.8-1 shows a cylinder of radius R, mass m, and c.g. moment of inertia I_g. There is an offset center of gravity and the cylinder rolls without slip on a horizontal surface. An external force $F(t)$ acts through the center of the cylinder. The force diagram (Figure P3.8-2) exposes the forces at the contact point of the cylinder with the ground. You will need to make velocity and acceleration diagrams to derive the equations of motion. For small motions of the angle θ, the linearized equations of motion are

$$\begin{aligned} F - F_f &= m(R\ddot{\theta} - a\ddot{\theta}) \\ F_f(R - a) + Fa - mga\theta &= I_g\ddot{\theta} \end{aligned} \tag{P3.8-1}$$

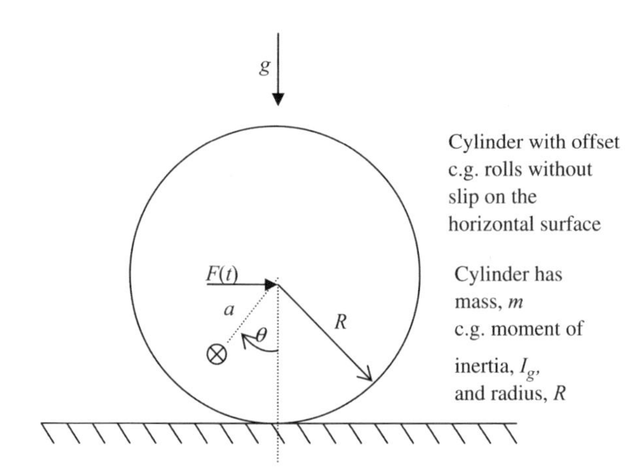

Figure P3.8-1

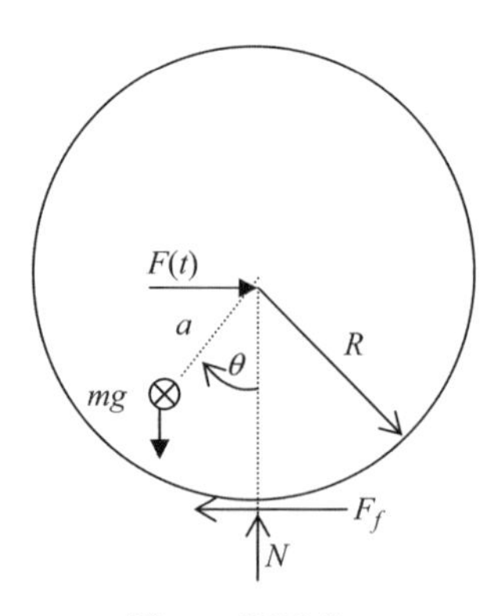

Figure P3.8-2

(a) Solve the first of equations [3.8-1] for F_f and substitute into the second equation and derive a single equation of motion in terms of the single variable θ and its derivatives.

(b) Reduce this equation to first-order form and derive the transfer function between the output $\theta(s)$ and input $F(s)$.

(c) Generate the complex frequency response function and write the expression for the magnitude of the response for a fixed-amplitude harmonic input. Sketch the frequency response. Write an expression for the resonant frequency.

3.9 A massless lever has a mass m attached at one end and a spring–damper attached at the other (Figure P3.9). An external force $F(t)$ acts on the mass. This is a single-degree-of-freedom system and the single variable is θ.

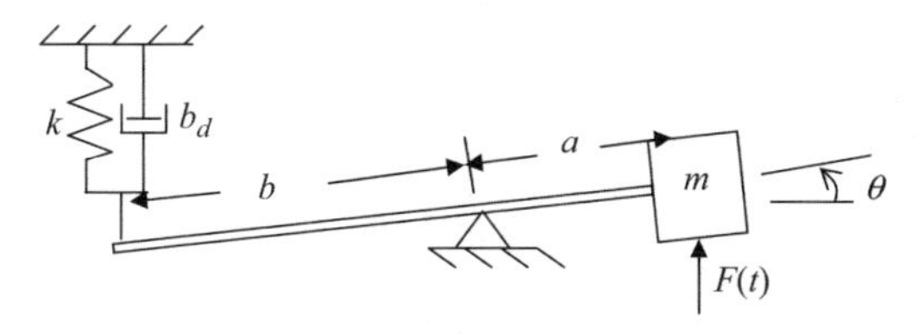

Figure P3.9

(a) Make velocity, acceleration, and force diagrams and derive the small-angle equations of motion for this system.

(b) Cast the resulting second order differential equation into first-order form. Derive the characteristic equation and solve for the eigenvalues. Interpret the eigenvalues.

(c) Derive the transfer function relating the output $\theta(s)$ to the input $F(s)$. Derive the complex frequency response function.

(d) Sketch the magnitude of the frequency response over a range of forcing frequencies.

3.10 Here we revisit Problem 2.19. The system is a rather simple model of a building where the base of the building is attached to ground through a horizontal spring k and a torsional spring k_τ (Figure P3.10). The coordinates for the two degrees of freedom are the position of the base of the building indicated by x and the rotational angle θ. You should derive the equations of motion and make sure that the following small-motion equations of motion are correct.

$$
\begin{aligned}
k(x_{\text{in}} - x) + mg\theta &= m\left(\ddot{x} + \frac{H}{2}\ddot{\theta}\right) \\
-k\left(x_{\text{in}} - x\right)\frac{H}{2} - k_\tau\theta &= I_g\ddot{\theta}
\end{aligned}
\qquad \text{(P3.10-1)}
$$

(a) Solve the second of equations [3.10-1] for $\ddot{\theta}$ and substitute into the first equation. The result will be two equations, each with only one second-

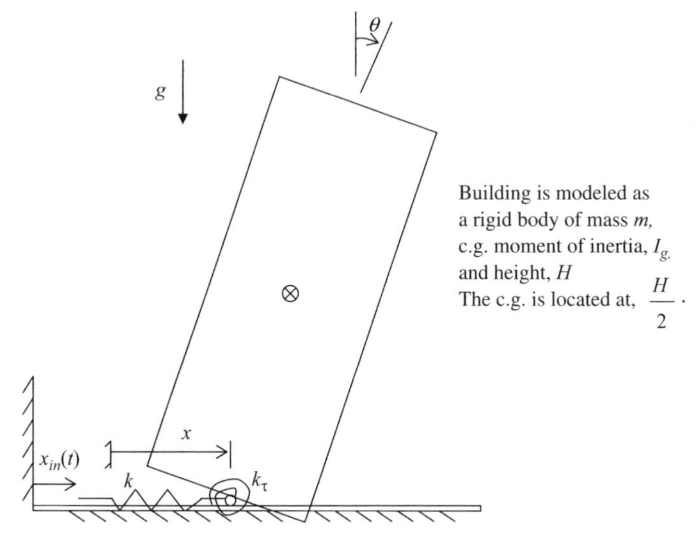

Figure P3.10

order derivative variable. Reduce these second-order equations to first-order form by introducing $\dot{x} = v$ and $\dot{\theta} = \omega$.

(b) Derive the characteristic equation for the eigenvalues. If any coefficient of a power of s is negative, the system will be unstable. Is it possible for any of the coefficients in the characteristic equation to be negative? Discuss this.

(c) Derive the transfer function relating the output $\theta(s)$ to the input $x_{\text{in}}(s)$. Convert this transfer function to a complex frequency response function and sketch the magnitude $|\theta(j\omega)/x_{\text{in}}(j\omega)|$.

3.11 In Problem 3.5 a bicycle model vehicle was presented with steering input. In that problem the objective was to generate a simulation model of the vehicle. The system is repeated here with the objective to derive the eigenvalues and specific input–output transfer functions. The schematic of the vehicle is shown in Figure P3.5 and the linearized equations of motion for constant forward speed U are

$$\begin{aligned} \dot{V} &= -U\omega + \frac{F_f}{m} + \frac{F_r}{m} \\ \dot{\omega} &= \frac{F_f}{I_g}a - \frac{F_r}{I_g}b \end{aligned} \tag{P3.11-1}$$

with

$$\begin{aligned} F_f &= \frac{C_f}{U}[U\delta - (V + a\omega)] \\ F_r &= \frac{C_r}{U}(b\omega - V) \end{aligned} \tag{P3.11-2}$$

(a) Substitute (P3.11-2) into (P3.11-1) and write the equations in the standard matrix format of Eq. (2.80). Use $C_r = K(a/b)C_f$ to introduce the understeer coefficient K into the formulation.
(b) For an input $\delta = 0$, put the equations into the Laplace domain and derive the characteristic equation for the eigenvalues. Solve for the eigenvalues and determine the forward velocity U that will make the constant term of the characteristic equation become negative and thus produce unstable eigenvalues.
(c) Derive the transfer functions relating $\omega(s)$ to $\delta(s)$ and $V(s)$ to $\delta(s)$. The lateral acceleration a_{lat} is given by

$$a_{lat} = \dot{V} + U\omega \qquad \text{(P3.11-3)}$$

Combine the transfer functions already produced and derive the transfer function relating $a_{\text{lat}}(s)$ to $\delta(s)$.

3.12 Figure P3.12 shows a structural mass M acted on by an external force $F_i(t)$. Attached to the structure is an additional spring and mass k_a, m_a. The coordinates that locate the two masses are indicated, and when they are both equal to zero there is no force in the spring.

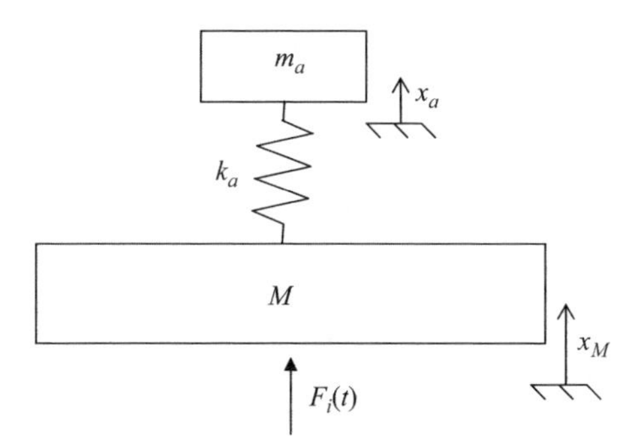

Figure P3.12

(a) Derive the equations of motion for this system.
(b) Using the definitions $\dot{x}_a = v_a$ and $\dot{x}_M = v_M$, put the equations into first-order form and express them in the standard matrix format of Eq. (2.80).
(c) Derive the transfer function relating $v_M(s)$ to $F_i(s)$.
(d) Let $s = j\omega$ and derive the complex frequency response function. Write the expression for the magnitude $|v_M(j\omega)/F_i(j\omega)|$ and sketch the frequency response. Is there an interesting occurrence at the natural frequency of m_a interacting with k_a?

3.13 In this problem, similar to Problem 3.12, a structural mass M is acted on by an external force $F_i(t)$ (Figure P3.13). This time a system is attached to the structure which consists of a massless lever with mass m_a attached at one end

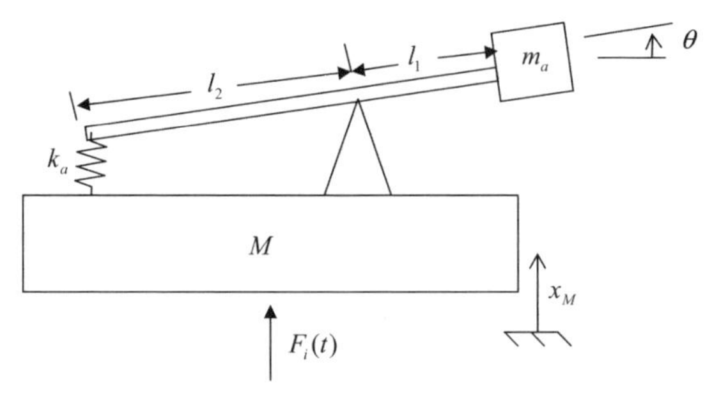

Figure P3.13

and a spring k_a attached at the other. The coordinate for the structural mass is x_M and the coordinate for m_a and the lever is the angle θ.

(a) For small motions of the coordinates, derive the equations of motion for this system. You will need to make velocity, acceleration, and force diagrams for the system. Verify that the equations of motion are

$$
\begin{aligned}
m_a\ddot{x}_M + m_a l_1\ddot{\theta} + \frac{l_2}{l_1}k_a l_2\theta &= 0 \\
M\ddot{x}_M - \frac{l_2}{l_1}k_a l_2\theta &= F_i
\end{aligned}
\tag{P3.13-1}
$$

(b) Use the definitions $\dot{x}_M = v_M$ and $\dot{\theta} = \omega$ and reduce the equations of motion to first-order form and express in the standard matrix format of Eq. (2.80).

(c) Derive the transfer function relating $x_M(s)$ to $F_i(s)$. Let $s = j\omega$ and derive the complex frequency response function. The transfer function for the acceleration of the structure mass can be obtained simply by multiplying the transfer function just derived by s^2. This is because acceleration in the time domain is the second derivative of displacement, and this is indicated in the s-domain by multiplication of $x_M(s)$ by s^2. Derive the transfer function relating the acceleration of the structure mass to the force input.

(d) Sketch the magnitude of the frequency response of the acceleration versus the forcing frequency. You should notice that the frequency where the structure mass acceleration goes to zero is dependent on the length ratio l_2/l_1. What do you think about modifying the lever device such that the fulcrum can be moved to the right and left and thus we can tune the device to provide structure isolation at any input frequency?

3.14 The objective of the vibration isolation systems shown in Figure P3.14 is to isolate the mass m from the ground excitation $x_{\text{in}}(t)$. A conventional isolator is shown in part a of the figure, where whatever material is used between the ground and the mass possesses some stiffness k and damping b properties.

In part b of the figure the damper is between the mass and inertial ground. This type of damping is referred to as *skyhook damping* because to realize it would require an attachment to the sky. If these were springs and shock absorbers associated with vehicle suspensions, then configuration b would be quite impossible to build.

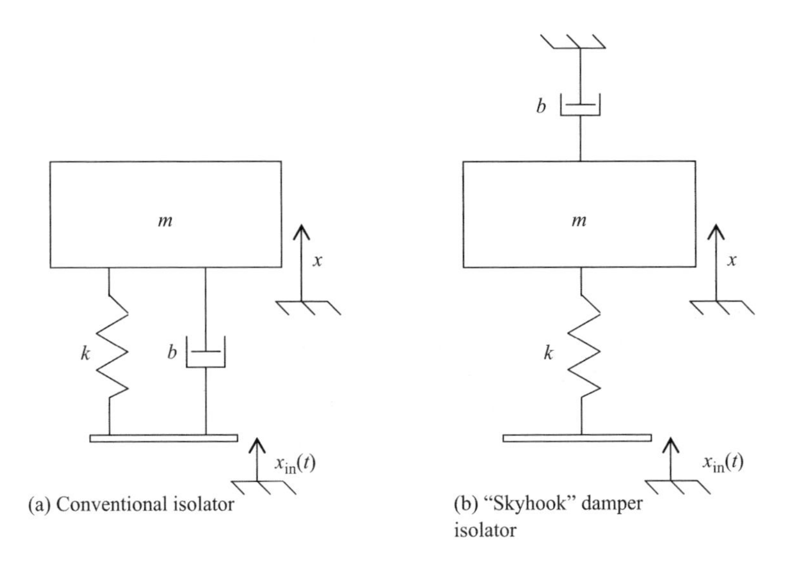

Figure P3.14

(a) Derive the equations of motion for these two systems and derive the transfer functions relating the output $x(s)$ to the input $x_{in}(s)$.

(b) Derive the complex frequency response function and derive the magnitude of the response for these two systems. Sketch the magnitude versus forcing frequency and compare the performance of the two systems.

3.15 This system has an interesting frequency response. When equations and transfer functions have been derived, we will let $m' \to 0$ for the final result. First we need the equations of motion for the system shown in Figure P3.15. When the coordinates are zero there is no force in the spring or damper.

(a) Derive the equations of motion for this system and verify that they are

$$\begin{aligned} k(x' - x_m) &= m\ddot{x}_m \\ b(\dot{x}_i - \dot{x}') - k(x' - x_m) &= m'\ddot{x}' \end{aligned} \tag{P3.15-1}$$

Use the definitions $\dot{x}' = v'$ and $\dot{x}_m = v_m$ and reduce these equations to first-order form and express in the standard matrix format of Eq. (2.80).

(b) Derive the transfer function relating $x_m(s)$ to the input $x_i(s)$. Let $m' \to 0$ and retain only those terms that approach infinity as $m' \to 0$.

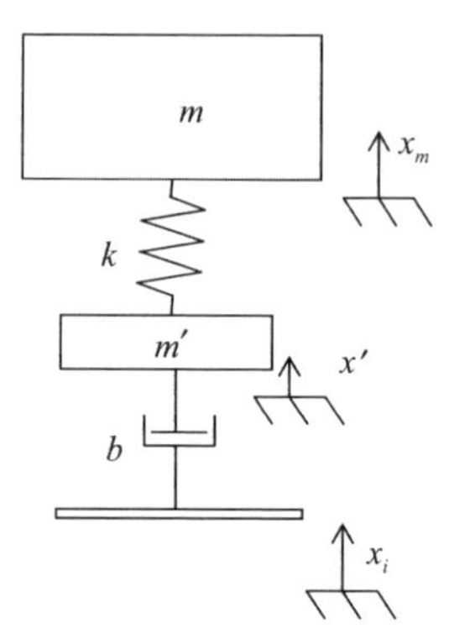

Figure P3.15

(c) Let $s = j\omega$ and derive the complex frequency response function. Sketch the magnitude of the response. Compare the frequency response of this system with those from Problem 3.14. Is the response of this system more like the conventional configuration from Figure P3.14*a* or more like the skyhook configuration from Figure P3.14*b*?

4

ENERGY AND LAGRANGE EQUATION METHODS

Many problems in mechanics can be addressed by considering how energy is stored, conserved, or dissipated. Often, simple restrictions on the possible motions of mechanical systems follow from studies of potential and kinetic energy. In restricted cases, energy methods by themselves can be used to derive equations of motion in ways that are considerably simpler than derivations based on a direct use of basic principles of mechanics.

In Section 4.1 we illustrate how energy considerations can lead to an equation of motion for cases in which the direct application of the laws of mechanics is surprisingly complicated. Unfortunately, for systems with a number of degrees-of-freedom, a study of energy alone is not sufficient to describe the dynamics of the system. We also show how Lagrange's equations extend the use of energy-based methods to formulate equations of motion for systems with multiple degrees-of-freedom. For elementary systems, energy methods are straightforward alternatives to the direct use of Newton's laws, and for complex systems, Lagrange's equations, which use stored energy in a more sophisticated way, are often a practical way to formulate equations of motion for systems that would be very tedious to analyze directly.

Unfortunately, no derivation of Lagrange's equations is easy to follow and, without a good deal of effort, the derivations really do not provide much insight into the reason that they work as an alternative to dynamic principles based on Newton's laws. Virtually every book on advanced mechanics in the reference list has a derivation of Lagrange's equations, but typical introductory texts do not discuss Lagrange's equations at all. A standard derivation of Lagrange's equations starting from Newton's laws may be found in refs. [3], [4], [5], or [6], for example. A different type of derivation based on a generalized Hamilton's principle and variation methods may be found in refs. [2] and [8].

The authors believe that it is perfectly possible to use Lagrange's equations and even to interpret the final results in terms of the laws of mechanics without spending the time to grasp the derivations in detail. Since the emphasis in this book is on applications of dynamics in engineering, Lagrange's equations will simply be stated without derivation. [Many in the academic community may see this as a radical step that promotes a cookbook approach to mechanical analysis, but it is the author's opinion (1) that students can be expert users of Lagrange's equations without studying the derivation of the equations, and that (2) those who become interested in the theory of analytical mechanics have ample opportunity to study the various derivations in numerous books.]

The focus in this book is on the proper use of Lagrange's equations for formulating equations of motion and in the interpretation of terms in the equations using the basic principles of mechanics. In many cases it is found that the simplest route to correct equations of motion is through the use of Lagrange's equations. In most practical applications, the equations of motion turn out to be nonlinear, but now that digital computer simulation is readily available, this is not the barrier to the study of dynamic systems that it once was. Methods for converting the standard-second-order forms for Lagrange's equation to twice as many first-order equations better suited to computer solution are discussed later in the chapter. In fact, one can argue that the ability to formulate correct equations of motion is more important today than it was when Lagrange derived his equations in 1788, since he had little general ability to solve such differential equations at the time.

4.1 KINETIC AND POTENTIAL ENERGY

To begin, let us review the various forms that equations take for kinetic and potential energy. These formulas are derived in practically every book on dynamics listed in the references, including introductory texts, (see, e.g., [1], [7], or [9]). The best known form for kinetic energy is for a particle of mass m moving with vector velocity $\mathbf{v}$:

$$T = \tfrac{1}{2}mv^2 \tag{4.1}$$

where v is the *magnitude* of the vector velocity, $\mathbf{v}$. Another common form involves a rigid body rotating about a *fixed* axis with angular velocity ω. If the body has a moment of inertia I_0 evaluated about the rotation axis, the kinetic energy is

$$T = \tfrac{1}{2}I_0\omega^2 \tag{4.2}$$

For a body in general plane motion, the formula for kinetic energy expression is almost a combination of Eqs. (4.1) and (4.2):

$$T = \tfrac{1}{2}mv_c^2 + \tfrac{1}{2}I_c\omega^2 \tag{4.3}$$

In this formula it is important to remember that v_c is the magnitude of the *velocity of the center of mass* and I_c is the moment of inertia of the body with respect to an *axis perpendicular to the plane of motion and passing through the center of mass*.

For general three-dimensional motion, the formulas for kinetic energy are more complex since the inertia matrix comes into play and the angular velocity is a vector, $\boldsymbol{\omega}$. (These topics were treated in Chapter 1; three-dimensional motion is discussed more extensively in Chapter 7.) The general expression for kinetic energy is

$$T = \tfrac{1}{2}mv_c^2 + \tfrac{1}{2}[\omega]^{\mathrm{T}}[I_c][\omega] \tag{4.4}$$

where $[\omega] = \begin{bmatrix} \omega_x \\ \omega_y \\ \omega_z \end{bmatrix}$ is a column vector of the components of $\boldsymbol{\omega}$ in an orthogonal coordinate system and $[\omega]^{\mathrm{T}} = [\,\omega_x \quad \omega_y \quad \omega_z\,]$ is its transpose. The matrix $[I_c]$ is the inertia matrix referred to the mass center.

For a general x – y – z coordinate system fixed in the body, the inertia matrix has both *moments of inertia* I_{xx}, I_{yy}, and I_{zz} along the main diagonal and *products of inertia* $I_{xy} = I_{yx}$, $I_{xz} = I_{zx}$, and $I_{yz} = I_{zy}$ off the main diagonal:

$$[I_c] = \begin{bmatrix} I_{xx} & I_{xy} & I_{xz} \\ I_{yx} & I_{yy} & I_{yz} \\ I_{zx} & I_{zy} & I_{zz} \end{bmatrix} \tag{4.5}$$

The definitions of moments and products of inertia are given in every reference concerning dynamics, although in introductory texts the discussion is often relegated to an appendix, (see, e.g., ref. [7] or [9]). One should be aware that although the definitions of moments of inertia are consistent among authors, the definitions of products do vary from reference to reference. The differences have to do only with minus signs. When the minus signs are included in the product-of-inertia formulas, the inertia matrix is written as in Eq. (4.5). If the minus signs are not included in the basic formulas, the inertia matrix is shown with a minus sign in front of each of the products of inertia when displayed as in Eq. (4.5). This means that one must be careful to be clear on which definition of products of inertia one is using—with a minus sign or without.

When the axis system is aligned along the *principal directions* 1 – 2 – 3, the inertia matrix is simplified, with only the *principal moments of inertia* appearing. Again, standard books on dynamics describe the eigenvalue problem that arises when one wishes to find the principal directions. For many simple rigid-body shapes, symmetry considerations make the principal directions obvious (see Appendix B for principal moments of inertia for a number of solid bodies):

$$[I_c] = \begin{bmatrix} I_1 & 0 & 0 \\ 0 & I_2 & 0 \\ 0 & 0 & I_3 \end{bmatrix} \tag{4.6}$$

When Eq. (4.4) is written out in components, the general expression using Eq. (4.5) is

$$T = \tfrac{1}{2}mv_c^2 + \tfrac{1}{2}\left(I_{xx}\omega_x^2 + I_{yy}\omega_y^2 + I_{zz}\omega_z^2 + 2I_{xy}\omega_x\omega_y + 2I_{yz}\omega_y\omega_z + 2I_{zx}\omega_z\omega_x\right) \tag{4.7}$$

The result is simpler if the components along the 1–2–3 principal axis system are used:

$$T = \tfrac{1}{2}mv_c^2 + \tfrac{1}{2}\left(I_1\omega_1^2 + I_2\omega_2^2 + I_3\omega_3^2\right) \tag{4.8}$$

In some situations, a body executes general rotation, but a point in the body (or imaginary extension of the body) remains fixed in inertial space. If the body happens to be rotating about a fixed point, 0, and the principal moments of inertia are evaluated about 0, then the kinetic energy has an even simpler form, in which the motion of the center of mass does not need to be taken into account:

$$T = \tfrac{1}{2}\left(I_{01}\omega_1^2 + I_{02}\omega_2^2 + I_{03}\omega_3^2\right) \tag{4.8a}$$

The result for the kinetic energy when there is a fixed point is identical whether Eq. (4.8) or Eq. (4.8a) is used. The different moments of inertia in the two formulas make the two ways to compute the kinetic energy exactly the same. The relations between moments and products of inertia evaluated about the center of mass and about a possible fixed point is the subject of the well-known *parallel axis theorem*, given in Appendix C (again see e.g., ref. [7] or [9]).

The expressions in Eqs. (4.1)–(4.8) are sufficient to cover all the cases of motion of particles or rigid bodies, although a few extra special cases could be added. Using the parallel axis theorem for moments of inertia, one could use Eq. (4.3) to derive Eq. (4.2) as a special case, for example, and the same process could be used to derive a special case for general rotation of a body about a fixed point (see Problem 4.1).

Potential energy expressions are useful for characterizing elements that conserve energy related to deflections rather than velocities. For example, a linear spring that produces a force $F = ke$ directed along its length as a function of a deflection e from the point at which there is no force has a potential energy V given by the familiar

$$V = \tfrac{1}{2}ke^2 \tag{4.9}$$

where k is a spring constant. If the force is not ke but rather some other function of e, $F(e)$, the potential energy is given by the integral

$$V = \int_0^e F(e)\,de \tag{4.10}$$

In either case, it should be noted that the force is given by the derivative of the potential energy:

$$F = \frac{dV}{de} \tag{4.11}$$

Another common use of potential energy arises when a body of mass m has its center of mass a distance z measured upward from an arbitrary fixed location in a uniform gravity field with an acceleration of gravity, g. The potential energy is then simply

$$V = mgz \tag{4.12}$$

Potential energy functions can be found for any number of cases in which energy is a function of position or deflection and is conserved in a cyclic motion. A number of other cases involving elastic elements are discussed in the problems.

4.2 USING CONSERVATION OF ENERGY TO DERIVE EQUATIONS OF MOTION

The conservation of energy is often invoked to solve simple problems that are somewhat more difficult to solve using Newton's laws directly. Galileo, for example, could have calculated how fast a lead ball would be traveling after it was dropped from the leaning tower of Pisa by neglecting air drag and assuming that energy was conserved (assuming, of course, that he had learned how to compute kinetic and potential energy). In this section, however, we illustrate the use of energy methods to formulate an equation of motion. Consider the oscillator formed by a rigid tube of constant cross-sectional area A containing a liquid of mass density ρ in a gravity field with acceleration of gravity g shown in Figure 4.1. The problem is to estimate the natural frequency of oscillations by neglecting viscosity effects and assuming the conservation of energy. (We neglect the effect of friction between the fluid and the walls of the tube.)

The fluid column has a total length of L, so the total mass of the fluid is ρAL. If one assumes that the fluid moves as a single slug, the *magnitude* of the velocity of *all* particles of the fluid is $\dot{x}(t)$, where $x(t)$ is the distance the endpoints of the fluid column have moved from the equilibrium position at any time (see Figure 4.1). (The vector *direction* of the velocities is quite different for the various particles.) This means that the total kinetic energy is simply

$$T = \tfrac{1}{2}\rho AL\dot{x}^2 \tag{4.13}$$

because the kinetic energy depends only on the magnitude of the velocity and not on the direction.

The potential energy requires a special calculation. Suppose that a force F pushes one end of the fluid column down a distance x. (Conceptually, F could be imagined as

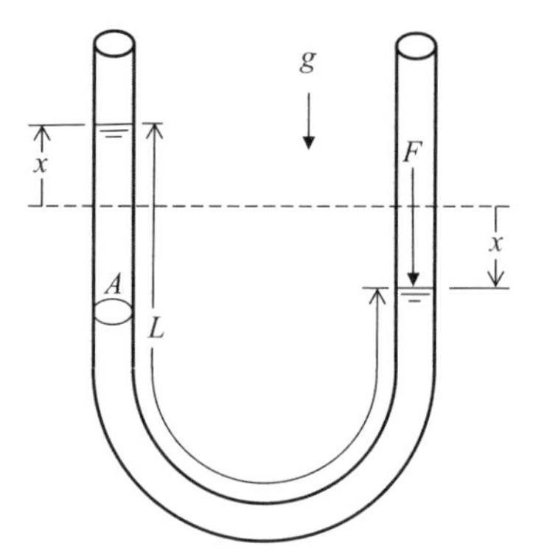

Figure 4.1 U-tube oscillator analyzed using the energy method.

acting upon a massless piston.) We assume that the force gradually increases so slowly that the acceleration of the fluid is very small and that inertia forces are negligible.

As can be seen in Figure 4.1, the force must balance the weight of slug of fluid of length $2x$ because the weight of the rest of the fluid balances out:

$$F = 2\rho A g x \tag{4.14}$$

Now V can be computed using Eq. (4.10):

$$V = \int_0^x F(x)\,dx = \int_0^x 2\rho A g x\,dx = \rho A g x^2 \tag{4.15}$$

Neglecting viscous losses, the conservation of total energy expression is

$$T + V = \tfrac{1}{2}\rho A L \dot{x}^2 + \rho A g x^2 = \text{constant} \tag{4.16}$$

An equation of motion results when Eq. (4.16) is differentiated with respect to time. The result of the time differentiation of Eq. (4.16) is

$$\rho A L \dot{x}\ddot{x} + \rho A g 2 x \dot{x} = 0 \quad \text{or} \quad \ddot{x} + \frac{2g}{L}x = 0 \tag{4.17}$$

where $\dot{x}$ can be divided out whenever it is not exactly zero.

This is a single-degree-of-freedom system vibratory system as discussed in elementary dynamics texts. (More general vibratory systems are discussed in Chapter 8.) Such books analyze a simple mass–spring oscillator with an equation of the form $m\ddot{x} + kx = 0$ and conclude that it will oscillate with a circular undamped natural frequency of $\omega_n = \sqrt{k/m}$ in radians per second.

Since Eq. (4.17) has the form of a mass–spring oscillator equation with $m = 1$ and $k = 2g/L$, the undamped natural frequency is readily seen to be

$$\omega_n = \sqrt{\frac{2g}{L}} \tag{4.18}$$

This example brings out several important points about the use of energy methods. The first is that energies are scalars, while Newton's laws deal with vectors such as forces, velocities, and accelerations. The fluid particles in the lower part of the tube in the example have accelerations that point in various directions, and the forces on the particles come from the tube as well as from neighboring particles. The result is that a direct application of Newton's laws is quite complex but when done correctly will yield the simple result of Eq. (4.17).

In fact, the energy method shows that the shape of the lower part of the tube, shown as a semicircle in Figure 4.1, is actually irrelevant to the final result even though the vector acceleration of particles in the fluid does change if the tube shape is changed. Trying to prove this independence of tube shape using only Newton's laws is challenging, yet it is quite easy using the conservation of energy.

A second point is that is that while the deriving of the energy expressions in Eq. (4.16) is straightforward, the equation of motion involves a time differentiation that must be accomplished using the chain rule. In this case, the required manipulations are elementary,

$$\frac{dx^2}{dt} = \frac{dx^2}{dx}\frac{dx}{dt} = 2x\dot{x}, \qquad \frac{d\dot{x}^2}{dt} = \frac{d\dot{x}^2}{d\dot{x}}\frac{d\dot{x}}{dt} = 2\dot{x}\ddot{x} \tag{4.19}$$

Similar but more complex operations will be encountered in using Lagrange's equations.

Finally, a major limitation of the use of the conservation of energy to derive equations of motion should be evident. The conservation of energy can provide only a single equation. Thus, if kinetic and potential energy can be expressed in terms of a single variable and its derivative—x and $\dot{x}$ in the example—then differentiating the conservation of energy expression will give an equation of motion. Often, one can add a power loss or gain to complete the equation in the case when energy is conserved by inertial and elastic elements, but there is power loss or gain due to friction or applied forces. When more than one variable is required to compute the energy functions, Lagrange's equations can be used to derive equations of motion often with the same advantages over the direct use of Newton's laws that were illustrated above.

4.3 EQUATIONS OF MOTION FROM LAGRANGE'S EQUATIONS

Generalized Coordinates

There are many practical cases in dynamics in which a relatively small number of variables suffice to determine the position of every particle in the system. If this is the case, these variables and their derivatives can, in principle, be used to determine the velocity and acceleration of each particle. A rigid body, for example, may be imagined to have a continuous distribution of mass or an indefinite number of particles. However, if the position of a single point in the body is determined by three linear position variables and the angular position by three angular position variables, the position of the infinite number of points in the body is determined by just six variables. Similarly, velocities and accelerations of the infinite number of points in the body are determined by the position variables and their derivatives.

We say that a rigid body moving in three dimensions has just six *degrees-of-freedom*. For a rigid body we then derive just six basic equations of motion. Three equations relate the net vector force to the acceleration of the center of mass, and three relate the net vector torque to the rate of change of angular momentum. When the center-of-mass velocity and the angular velocity are expressed in terms of the six degrees of freedom, the kinetic energy can be determined using Eq. (4.4). [There is a good deal of theoretical effort required to derive Eq. (4.4) and to define the inertia matrix, but this is included in every basic exposition of dynamics and we assume that the reader is familiar with these operations.]

Lagrange's equations apply to *holonomic systems*, for which a finite number of *generalized coordinates* $q_1, q_2, q_3, \ldots, q_n$ suffice to describe the position of all points in the system. If there are elastic elements or other elements that have potential energy functions, the system potential energy is also determined by the generalized coordinates.

In dynamic studies, the generalized coordinates are functions of time, and the coordinates, together with their derivatives, determine the velocities of all points in the systems and thus the kinetic energy of all inertia elements. These statements will be illustrated by analyzing a simple example. The system of Figure 4.2 illustrates that generalized coordinates are typically a combination of linear and angular position variables. The massive block is acted on by a spring and carries a simple pendulum. At first all friction will be neglected. The generalized coordinate q_1 is the position of the block measured from the point in inertial space at which the spring is unstretched. The coordinate q_2 is the angular position of the pendulum measured from the vertical. Given values of q_1 and q_2, the position of any point in the system can be determined, so the system has just two degrees-of-freedom.

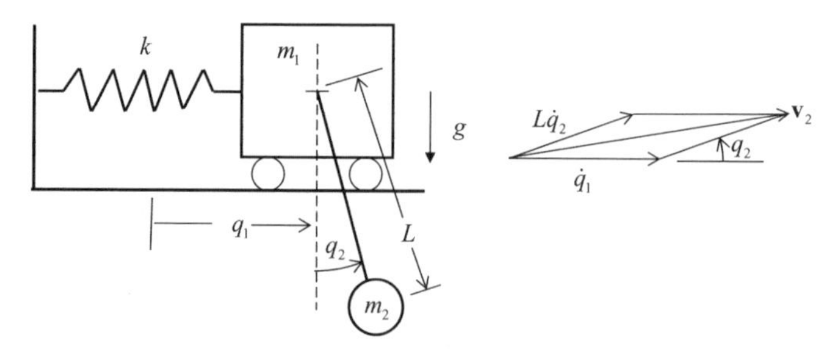

Figure 4.2 System with two generalized coordinates.

The potential energy is easy to evaluate in terms of q_1 and q_2, in this case using Eqs. (4.9) and (4.12):

$$V = \tfrac{1}{2}kq_1^2 + mg(-L\cos q_2) \tag{4.20}$$

[Note that using the center of the block as the reference point for the height of m_2 in the gravity field (positive upward), z in Eq. (4.12) is actually $-L\cos q_2$. The height of m_2 could be measured from any point in inertial space other than the pivot point. This would have the effect of adding only a constant to the expression in Eq. (4.20), but since only derivatives of V are used in Lagrange's equations, an extra possible constant will prove to be irrelevant.]

Now, given values for q_1 and q_2 as well as the *generalized velocities* $\dot{q}_1$ and $\dot{q}_2$, the velocity of all points in the system can be determined. Then, assuming that both masses shown in Figure 4.2 are considered to be particles, the kinetic energy is simply given by Eq. (4.1). For the block, $\dot{q}_1$ is directly the velocity magnitude v_1 that will be used in Eq. (4.1). A complication is that the velocity magnitude for the pendulum mass has to be found by combining the block velocity $\dot{q}_1$ with the velocity component due to the pendulum angular velocity $L\dot{q}_2$ in a vectorial manner, as shown by the diagram on the right side of Figure 4.2. (One way to see this is to consider a coordinate system with origin at the pendulum pivot point and rotating with the pendulum. Then, applying the kinematic relation for the velocity of a point in a rotating coordinate system given in Chapter 1, $\mathbf{v}_p = \mathbf{v}_0 + \boldsymbol{\omega} \times \mathbf{r}_{0p} + \mathbf{v}_{\text{rel}}$, where $v_A = \dot{q}_1$, $\omega \times r_{AB} = L\dot{q}_2$, and $\mathbf{v}_{\text{rel}} = 0$, the vector diagram results.)

Since the square of the velocity magnitude v_2 is required for T, one can either sum the squares of orthogonal components using the Pythagorean theorem or use the law of cosines directly on the vector triangle in Figure 4.2. to evaluate v_2^2:

$$\begin{aligned} v_2^2 &= (\dot{q}_1 + L\dot{q}_2\cos q_2)^2 + (L\dot{q}_2\sin q_2)^2 \\ &= \dot{q}_1^2 + (L\dot{q}_2)^2 + 2\dot{q}_1 L\dot{q}_2\cos q_2 \end{aligned} \tag{4.21}$$

(The search for the square of the magnitude of a velocity from a vector diagram such as that shown in Figure 4.2 is a common feature of the use of Lagrange's equations.)

Finally, the kinetic energy is just

$$T = \tfrac{1}{2}m_1\dot{q}_1^2 + \tfrac{1}{2}m_2\left(\dot{q}_1^2 + L^2\dot{q}_2^2 + 2\dot{q}_1 L\dot{q}_2\cos q_2\right) \tag{4.22}$$

This example clearly shows that the potential and kinetic energy functions are determined by the two variables q_1 and q_2 as well as their time derivatives. Neglecting any losses, expressing the fact that the total energy is constant will give only a single equation when we need two equations for this two-degree-of-freedom system. Lagrange's equations based on these energy functions will yield the two equations required.

Lagrange's Equations

The shortest version of the Lagrange equation for the generalized variable q_i is often given in terms of the Lagrangian L:

$$L = T - V \tag{4.23}$$

The form of the Lagrange equation for the generalized coordinate q_i is

$$\frac{d}{dt}\frac{\partial L}{\partial \dot{q}_i} - \frac{\partial L}{\partial q_i} = Q_i, \qquad i = 1, 2, 3, \ldots, n \tag{4.24}$$

There will be one version of this equation for each of the n generalized coordinates q_i describing an n-degree-of-freedom system, $i = 1, 2, 3, \ldots, n$. A somewhat more explicit version of Lagrange's equations results from the recognition that the potential energy is not a function of any of the *generalized velocities*, $\dot{q}_i$. This is the form used in the examples that follow:

$$\frac{d}{dt}\frac{\partial T}{\partial \dot{q}_i} - \frac{\partial T}{\partial q_i} + \frac{\partial V}{\partial q_i} = Q_i, \qquad i = 1, 2, 3, \ldots, n \tag{4.25}$$

The *generalized forces* Q_i arise from elements in the system that cannot be represented by energy functions. Examples of effects not represented in T and V are friction forces and externally applied forces or torques. The manner in which the generalized forces Q_i are determined will be shown by example.

Let us now apply Lagrange's equations in the form of Eq. (4.25) to the expressions for T in Eq. (4.22) and V in Eq. (4.20). Starting with the first generalized coordinate, q_1, the first step is to evaluate the partial derivative terms:

$$\frac{\partial T}{\partial \dot{q}_1} = m_1\dot{q}_1 + m_2(\dot{q}_1 + L\dot{q}_2\cos q_2) \tag{4.26}$$

Note that in this calculation, we treat q_1, q_2, $\dot{q}_1$, and $\dot{q}_2$ as if they were four independent variables when taking the partial derivative. The two other partial derivatives in Eq. (4.25) are quite simple:

$$\frac{\partial T}{\partial q_1} = 0 \tag{4.27}$$

$$\frac{\partial V}{\partial q_1} = kq_1 \tag{4.28}$$

The tricky part comes in finding the *total time derivative* of Eq. (4.26) to use in Eq. (4.25). We see that $\partial T/\partial \dot{q}_1$ is a function of $\dot{q}_1$, $\dot{q}_2$, and q_2. Considering these three variables as functions of time, the chain rule of differentiation is used to find the *total time derivative* as follows:

$$\begin{aligned}\frac{d}{dt}\frac{\partial T}{\partial \dot{q}_1} &= \frac{\partial}{\partial \dot{q}_1}\left(\frac{\partial T}{\partial \dot{q}_1}\right)\frac{d\dot{q}_1}{dt} + \frac{\partial}{\partial \dot{q}_2}\left(\frac{\partial T}{\partial \dot{q}_1}\right)\frac{d\dot{q}_2}{dt} + \frac{\partial}{\partial q_2}\left(\frac{\partial T}{\partial \dot{q}_1}\right)\frac{dq_2}{dt} \\ &= \frac{\partial}{\partial \dot{q}_1}\left(\frac{\partial T}{\partial \dot{q}_1}\right)\ddot{q}_1 + \frac{\partial}{\partial \dot{q}_2}\left(\frac{\partial T}{\partial \dot{q}_1}\right)\ddot{q}_2 + \frac{\partial}{\partial q_2}\left(\frac{\partial T}{\partial \dot{q}_1}\right)\dot{q}_2\end{aligned} \tag{4.29}$$

Although this expression seems very complicated, when the actual expression in Eq. (4.26) is inserted, the result is not as complex as might be feared:

$$\frac{d}{dt}\frac{\partial T}{\partial \dot{q}_1} = (m_1 + m_2)\ddot{q}_1 + (m_2 L \cos q_2)\ddot{q}_2 + m_2 L\dot{q}_2(-\sin q_2)\dot{q}_2 \tag{4.30}$$

Anyone who can understand the procedure for finding the total time derivative of expressions such as Eq. (4.26) using the chain rule as in Eq. (4.29) to generate the results such as Eq. (4.30) is well on the way to becoming a master of Lagrange's equations.

Now Eqs. (4.30), (4.27), and (4.28) can be substituted into Eq. (4.25) with $Q_1 = 0$ to find the first equation of motion:

$$(m_1 + m_2)\ddot{q}_1 + (m_2 L \cos q_2)\ddot{q}_2 + m_2 L\dot{q}_2(-\sin q_2)\dot{q}_2 + kq_1 = 0 \tag{4.31}$$

The same pattern of operations is used to find the second equation of motion for q_2.

$$\frac{\partial T}{\partial \dot{q}_2} = m_2(L^2\dot{q}_2 + L\dot{q}_1 \cos q_2) \tag{4.32}$$

$$\frac{\partial T}{\partial q_2} = m_2 L\dot{q}_1\dot{q}_2(-\sin q_2) \tag{4.33}$$

$$\frac{\partial V}{\partial q_2} = -mgL(-\sin q_2) \tag{4.34}$$

$$\frac{d}{dt}\frac{\partial T}{\partial \dot{q}_2} = m_2 L^2\ddot{q}_2 + (m_2 L \cos q_2)\ddot{q}_1 + [m_2 L\dot{q}_1(-\sin q_2)]\dot{q}_2 \tag{4.35}$$

Finally, combining Eqs. (4.35), (4.23), and (4.44) according to Eq. (4.25), the result is

$$m_2 L^2\ddot{q}_2 + m_2 L(\cos q_2)\ddot{q}_1 + mgL(\sin q_2) = 0 \tag{4.36}$$

One might note that in evaluating terms in Lagrange's equations, it often happens that some cancellation occurs. In this case, the last term in Eq. (4.35) is canceled by Eq. (4.33), which enters the Lagrange equation with a negative sign. A reason for this somewhat annoying feature of Lagrange's equations is presented in ref. [6].

The two equations (4.31) and (4.36) are the equations of motion for the system of Figure 4.2. These are surprisingly complex nonlinear differential equations for what may seem to be a simple mechanical system, but it is in the nature of mechanical systems to exhibit geometric nonlinearities even when Newton's law seems to be linear, and in this case, the spring is also assumed to be linear. It certainly is possible to derive these equations using only elementary dynamic principles, but the derivation is not simple and there is a large potential for mistakes when setting up free-body diagrams and defining internal forces between the block and the pendulum. In Section 4.4 we show how Lagrange equations can be validated for special cases to check for errors and to make a connection to basic dynamic principles.

Generalized Forces

Lagrange equations are particularly useful for handling systems in which there are geometric constraints among inertial elements and some forces can be found from potential energy functions. *Generalized forces* are used when forces exist in the system that cannot be represented by kinetic or potential energy functions. Energy dissipation due to friction or energy input due to imposed forces can be represented by generalized forces, for example.

Consider the generalized forces Q_i present on the right sides of Eqs. (4.24) and (4.25). Suppose, for example, that there is a friction force acting on the block shown in Figure 4.2. This friction force can be represented by the generalized force Q_1. Assume that the friction force F_f is defined to be a function of the block velocity by one of two possible expressions. The first formula represents a viscous friction assumption and the second represents a Coulomb friction assumption:

$$F_f = B\dot{q}_1 \quad \text{or} \quad F_f = \mu mg \frac{\dot{q}_1}{|\dot{q}_1|} = \mu mg \operatorname{sgn} \dot{q}_1 \tag{4.37}$$

The constant B in the first version of Eq. (4.37) is a viscous friction coefficient, and μ in the second version of Eq. (4.37) is a coefficient of friction that multiplies the normal force mg. In both cases the friction force is defined to be positive when the velocity $\dot{q}_1$ is positive and negative when the velocity is negative. [Although viscous friction is sometimes assumed, it is rarely a good assumption in mechanical systems, and the second formula represents a constant magnitude Coulomb friction force but arranged to switch direction when the velocity changes direction. The force magnitude μmg is multiplied by the velocity divided by the absolute value of velocity, which is either $+1$ or -1. This is one way to define the *signum function*, $\operatorname{sgn}(\dot{q}_1)$, which switches between $+1$ and -1 depending on whether the velocity is positive or negative.]

Physically, it is clear that friction forces generally remove energy from a system, and in this case the friction force always acts to oppose the direction of motion of the block; that is, if the block is moving to the right, the force acts to the left, and if the block moves to the left, the force acts to the right. The definitions in Eq. (4.37) imply that the force is positive (meaning acting to the left) when $\dot{q}_1$ is positive and negative (acting to the right) when $\dot{q}_1$ is negative. This change of sign is automatic for the viscous friction law in Eq. (4.37) when the velocity changes sign, but the Coulomb friction law needs the signum function to provide a sign switch when $\dot{q}_1$ changes sign. So far, the first equation of motion, Eq. (4.31), has no generalized force, but now with friction on the block, a generalized force as in Eq. (4.26) is needed to represent the friction force. The way to find the value of Q_1 is to imagine the incremental work δW done *by* F_f *on* the system when the generalized coordinate q_1 is imagined to increase an infinitesimal amount δq_1 and all other generalized coordinates are held fixed. This change in q_1 is called *virtual variation* since it is only imagined to occur under the special condition that the other generalized coordinates remain constant. A virtual variation is in the nature of a thought experiment and does not actually occur when the system is moving dynamically.

If the force adds energy to the system, the work is counted positive, and if the force takes energy from the system, the work is considered negative. With the friction forces as defined in Eq. (4.37), the virtual work is the product of the force times the infinitesimal displacement of the block:

$$\delta W = Q_1\,\delta q_1 = -F_f\,\delta q_1 \tag{4.38}$$

and the generalized force is just the coefficient (including sign) of δq_1. After the discussion above, it should be clear that when the block moves to the right an amount δq_1, the work done by the friction force is negative. Using the two example friction laws in Eq. (4.37), the first generalized force is given by

$$Q_1 = -F_f, \qquad Q_1 = -B\dot{q}_1 \quad \text{or} \quad Q_1 = -\mu m g\,\operatorname{sgn}\dot{q}_1 \tag{4.39}$$

Friction forces generally lead to negative generalized forces because friction removes energy from a system no matter how it moves. On the other hand, an applied force positive in the direction of q_1 would be represented by a positive generalized force Q_1 just equal to the applied force. Since q_2 is an angle and the work term $\delta W = Q_2\,\delta q_2$ has the dimensions of work or energy, the generalized force Q_2 for this generalized variable is a torque. The friction force on the block does not contribute to Q_2 because if q_2 changes by an amount δq_2 while all other variables remain fixed, the block does not move and the friction force does no work.

Suppose, however, that there is a friction torque τ_0 at the pendulum pivot. The virtual work expression for a variation in q_2 would be

$$\delta W = Q_2\,\delta q_2 = -\tau_0\,\delta q_2, \qquad Q_2 = -\tau_0 \tag{4.40}$$

and one would have to find a formula for how the friction torque was related to the angular velocity $\dot{q}_2$ analogous to the laws for the friction force in Eq. (4.37). If an external torque were applied to the pendulum, positive in the direction of increasing angle, the generalized force would be the applied torque but with a positive sign.

When a force or a torque is assumed to be applied to some part of a system and is also assumed to be a given function of time, the Lagrange equation may have a time function in the generalized forces. Then the Lagrange equations can be used in a simulation program to determine the response of the mechanical system to the forces or torques imposed. This procedure is discussed in Section 4.6.

Imposed Motion

Another way in which mechanical systems are forced or excited is by prescribed motion of some part of the system instead of imposed forces or torques. As an example, suppose that the wall on the left side of Figure 4.2 were not fixed in inertial space but rather, had a motion $X(t)$ positive in the direction of q_1 and independent of the motion of the system. (Some agent must impose this motion—an hydraulic actuator, for example—but it is assumed that the motion is a function of time given independent of any forces that may be required. For now, it is assumed that q_1 is measured from a point in inertial space such that when both X and q_1 are zero, the spring is unstretched.)

This imposed motion would modify Eq. (4.31) because the potential energy in Eq. (4.20) would become

$$V = \tfrac{1}{2}k\,[q_1 - X(t)]^2 + mg(-L\cos q_2) \tag{4.41}$$

and the Lagrange equation for q_1, Eq. (4.31), would have a term $k[q_1 - X(t)]$ instead of just kq_1, so that $X(t)$ would be an input time function for the equation of motion. When the wall is assumed to move, Eq. (4.31) changes to

$$(m_1 + m_2)\ddot{q}_1 + m_2 L\cos q_2\ddot{q}_2 + m_2 L\dot{q}_2(-\sin q_2)\dot{q}_2 + k[q_1 - X(t)] = 0 \tag{4.42}$$

As it happens, the modified expression for V in Eq. (4.41) does not involve q_2 so the Lagrange equation for q_2, Eq. (4.36) does not change in this case.

As another example of how to incorporate imposed motion, suppose that the position q_1 and the velocity $\dot{q}_1$ in Fig. 4.2 were not measured with respect to inertial space but rather with respect to the moving wall. If q_1 represents the stretch of the spring, the potential energy expression in Eq. (4.20) would not have to change. However, the inertial velocity of the block would no longer be simply $\dot{q}_1$ but rather $\dot{q}_1 + \dot{X}(t)$. This would change the calculation of the kinetic energy from Eq. (4.22) to

$$T = \tfrac{1}{2}m_1\left[\dot{q}_1 + \dot{X}(t)\right]^2 + \tfrac{1}{2}m_2\left\{\left[\dot{q}_1 + \dot{X}(t)\right]^2 + L^2\dot{q}_2^2 + 2\left[\dot{q}_1 + \dot{X}(t)\right]L\dot{q}_2\cos q_2\right\} \tag{4.43}$$

[Although this result can be derived by a simple substitution of $\dot{q}_1 + \dot{X}(t)$ for $\dot{q}_1$ in Eq. (4.22), it would also result from the beginning if the diagram on the right of Figure 4.2 had been analyzed with the block velocity shown as $\dot{q}_1 + \dot{X}(t)$ instead of $\dot{q}_1$.]

Using Eq. (4.43), Eq. (4.26) then changes to

$$\frac{\partial T}{\partial \dot{q}_1} = m_1\left[\dot{q}_1 + \dot{X}(t)\right] + m_2\left[\dot{q}_1 + \dot{X}(t)\right] + m_2 L\dot{q}_2 \cos q_2 \tag{4.44}$$

Finally, Eq. (4.30) becomes

$$\frac{d}{dt}\frac{\partial T}{\partial \dot{q}_1} = (m_1 + m_2)\left[\ddot{q}_1 + \ddot{X}(t)\right] + m_2 L\ddot{q}_2 \cos q_2 + m_2 L\dot{q}_2(-\sin q_2)\dot{q}_2 \tag{4.45}$$

and the Lagrange equation for q_1, Eq. (4.31) becomes

$$(m_1 + m_2)\left[\ddot{q}_1 + \ddot{X}(t)\right] + (m_2 L \cos q_2)\ddot{q}_2 + m_2 L\dot{q}_2(-\sin q_2)\dot{q}_2 + kq_1 = 0 \tag{4.46}$$

Under this new definition of q_1, the steps leading to the Lagrange equation for q_2 also change. Instead of Eqs. (4.32)–(4.36), we have the following:

$$\frac{\partial T}{\partial \dot{q}_2} = m_2\left[L^2\dot{q}_2 + L(\dot{q}_1 + \dot{X}(t)\cos q_2\right] \tag{4.47}$$

$$\frac{\partial T}{\partial q_2} = m_2 L\left(\dot{q}_1 + \dot{X}(t)\right)\dot{q}_2\,(-\sin q_2) \tag{4.48}$$

$$\frac{\partial V}{\partial q_2} = -mgL(-\sin q_2) \tag{4.49}$$

$$\frac{d}{dt}\frac{\partial T}{\partial \dot{q}_2} = m_2 L^2\ddot{q}_2 + m_2 L(\cos q_2)\left[\ddot{q}_1 + \ddot{X}(t)\right] + m_2 L\left[\dot{q}_1 + \dot{X}(t)\right](-\sin q_2)\dot{q}_2 \tag{4.50}$$

Finally, Eq. (4.36) changes to

$$m_2 L^2\ddot{q}_2 + m_2 L(\cos q_2)\left[\ddot{q}_1 + \ddot{X}(t)\right] + mgL(\sin q_2) = 0 \tag{4.51}$$

It is important to note that the imposed motion of the wall (with q_1 measured in inertial space) that results in Eq. (4.42) made $X(t)$ the input function. The same imposed wall motion together with the definition of $\dot{q}_1$ as the block velocity *relative* to the wall results in a change from Eq. (4.31) to Eq. (4.45) and Eq. (4.36) to Eq. (4.51) in which the acceleration $\ddot{X}(t)$ plays the role of input function. This illustrates that a change in the definition of the generalized coordinates can change the motion input from position to acceleration. The assumption that $X(t)$ is a known function of time implies that the function $\ddot{X}(t)$ is also known, so that the actual behavior of the system will be predicted correctly no matter which definition of the general coordinates is chosen.

If two the versions of the equations of motion were simulated, both would represent the motion of the system correctly. But the two versions of the generalized coordinate q_1 would not be identical, since one version represents the block motion with respect to inertial space and the other with respect to the moving wall. As we have just seen, if one were to ask how the moving wall changes the equations of motion from the case when the wall is stationary, the correct answer is: "It depends." Depending on the definition of q_1, the equations may have either the wall position $X(t)$ or the wall acceleration $\ddot{X}(t)$ as an input time function. For more complex systems, there may be a number of possible choices for generalized coordinates. For every choice there will be a set of equations of motion and they may appear to be quite different. Unless there is a mistake, each set of equations will describe the motion of the system. It is just that the system is described using different variable choices and thus the equations of motion are different.

4.4 INTERPRETATION OF LAGRANGE'S EQUATIONS

In principle, every term in a Lagrange equation can be interpreted using the principles of mechanics directly. Often, however, the interpretation is not particularly easy when complex nonlinear functions are involved. On the other hand, it is often quite easy to check the correctness of at least some terms in Lagrange equations in special cases by recognizing that if a generalized coordinate is a linear position variable, the Lagrange involves a force balance, and if the generalized coordinate is an angular position, the equation involves a torque balance. It is often helpful to think of special simplifying cases to check the correctness of some term in the general equations. For example, in Eq. (4.31), if the pendulum were fixed to the block such that $\ddot{q}_2 = \dot{q}_2 = 0$, the equation would simplify to

$$(m_1 + m_2)\ddot{q}_1 + kq_1 = 0 \tag{4.52}$$

which is clearly the correct force balance equation for a mass–spring oscillator with the fixed pendulum. Under the same conditions, Eq. (4.42) becomes

$$(m_1 + m_2)\ddot{q}_1 + kq_1 = -kX(t) \tag{4.53}$$

which is the equation for a mass–spring oscillator with q_1 representing the mass motion relative to inertial space and with a moving wall spring connection. Finally, with the pendulum still fixed but using the definition of q_1 to be the block position *relative* to the moving wall, Eq. (4.45) changes to

$$(m_1 + m_2)\ddot{q}_1 + kq_1 = -(m_1 + m_2)\ddot{X}(t) \tag{4.54}$$

which is the equation of a mass–spring oscillator in an accelerated coordinate system since the absolute acceleration of the masses would be $\ddot{q}_1 + \ddot{X}(t)$ in this case.

Another special case can be devised to check the equation for q_2. If the block were fixed in space with $\ddot{q}_1 = 0$, Eq. (4.36) would reduce to

$$m_2 L^2 \ddot{q}_2 + mgL(\sin q_2) = 0 \tag{4.55}$$

which is the correct torque balance equation for a pendulum with a fixed pivot. This equation is easy to derive for a simple pendulum using Newton's law or the torque angular momentum relationship. When q_1 represents the relative displacement of the block with respect to the wall, and $\ddot{q}_1$ is assumed to be zero, Eq. (4.51) becomes

$$m_2 L^2 \ddot{q}_2 + mgL(\sin q_2) = -m_2 L(\cos q_2)\ddot{X}(t) \tag{4.56}$$

which is the equation of a pendulum mounted on an accelerating block. [One way to see this is to consider $-m_2\ddot{X}(t)$ as a d'Alembert force, which when multiplied by the moment arm $L \cos q_2$ yields an effective moment around the pivot point.]

The coupling terms that we have just eliminated by assuming one or the other of the generalized coordinates were fixed are present in the equations to represent how the pendulum motion affects the force balance of the block and how the block motion influences the torque balance of the pendulum. The beauty of Lagrange equation formulation is that the effect of *constraint forces* such as the forces at the pendulum pivot are incorporated automatically in the final equations.

Constraint forces (the force at the pendulum pivot in the example system) are not involved in the stored energy functions, nor are they involved in the generalized forces. This is because the pivot force acting on the pendulum is equal and opposite to the pivot force acting on the block. (This is Newton's law of action and reaction.) Thus, in any virtual variation of the generalized coordinates the virtual work on the total system due to the equal and opposite constraint forces vanishes. In contrast, when using Newton's laws directly, all forces must be included in the free-body diagrams for the parts of the system even though the constraint forces can be eliminated algebraically in the final formulation of the equations of motion. This is a major reason why Lagrange's equations are particularly useful for complex interconnected systems of inertial elements.

The coupling terms in Lagrange's equations could have been found by separating the system into free-body diagrams for the block and the pendulum (including the interaction forces at the pivot point), writing the appropriate dynamic equations and then combining the equations in such a way that the pivot forces were eliminated and only two equations of motion equivalent to the two Lagrange equations were left. Often called the *direct method,* this is the only way to derive equations of motion using Newton's laws alone.

Unfortunately, the calculations of accelerations, the use of multiple free-body diagrams, and the algebraic manipulations required to derive equations of motion for many dynamic systems leave room for many errors. In contrast, Lagrange equations require only correct expressions for kinetic and potential energy and the ability to differentiate properly to derive equations of motion for holonomic systems even in complicated situations.

4.5 NONLINEAR KINEMATICS AND LAGRANGE'S EQUATIONS

It is often true that the use of Lagrange equations for an interconnected system leads to equations of motion more quickly and with less likelihood of error than by the use of free-body diagrams. In this section we explore the options inherent in the use of Lagrange equations for several cases of kinematic constraints between inertial elements. For clarity, the examples will be restricted to a flywheel rotating about a fixed axle and a massive piston moving in a straight line. A kinematic constraint relates the position of the piston to the angle of the flywheel, so that from the point of view of Lagrange's theory, the system is holonomic and there is a single generalized coordinate.

Consider that the flywheel connected to a piston in such a way that the position of the piston, x, is related to the angle of the flywheel, θ, by the relation

$$x = f(\theta) \tag{4.57}$$

If the piston has mass m and the flywheel has moment of inertia I, the kinetic energy of the combination is

$$T = \tfrac{1}{2}m\dot{x}^2 + \tfrac{1}{2}I\dot{\theta}^2 \tag{4.58}$$

If θ is chosen as the generalized coordinate, Eq. (4.57) can be used to modify Eq. (4.58) to involve only $\dot{\theta}$ as follows:

$$\dot{x} = \frac{df(\theta)}{d\theta}\frac{d\theta}{dt} = \frac{df(\theta)}{d\theta}\dot{\theta}, \qquad T = \frac{1}{2}m\left[\frac{df(\theta)}{d\theta}\right]^2\dot{\theta}^2 + \frac{1}{2}I\dot{\theta}^2 \tag{4.59}$$

The Lagrange equation, with Q_θ the generalized force (which in this case is an applied moment, Q_θ), is relatively simple since there is no potential energy:

$$\frac{d}{dt}\left(\frac{\partial T}{\partial\dot{\theta}}\right) - \frac{\partial T}{\partial\theta} = Q_\theta \tag{4.60}$$

Terms in the equation are evaluated below:

$$\frac{\partial T}{\partial\dot{\theta}} = m\left[\frac{df(\theta)}{d\theta}\right]^2\dot{\theta} + I\dot{\theta} \tag{4.61}$$

The total time derivative of this quantity is a little complicated. First, we consider the time derivative of the terms involving $\dot{\theta}$ as if $df(\theta)/d\theta$ were constant. Then we treat $\dot{\theta}$

as if it were constant and compute the time derivative of $[df(\theta)/d\theta]^2$ using the chain rule:

$$\frac{d}{dt}\left(\frac{df}{d\theta}\right)^2 = 2\left(\frac{df}{d\theta}\right)\left(\frac{d^2f}{d\theta^2}\right)\frac{d\theta}{dt} = 2\left(\frac{df}{d\theta}\right)\left(\frac{d^2f}{d\theta^2}\right)\dot{\theta}$$

When these operations are applied to Eq. (4.61), the final result is

$$\frac{d}{dt}\left(\frac{\partial T}{\partial\dot{\theta}}\right) = m\left(\frac{df}{d\theta}\right)^2\ddot{\theta} + I\ddot{\theta} + 2m\left(\frac{df}{d\theta}\right)\left(\frac{d^2f}{d\theta^2}\right)\dot{\theta}^2 \tag{4.62}$$

The second term in Eq. (4.60), using the kinetic energy form, Eq. (4.59), is

$$\frac{\partial T}{\partial\theta} = m\left(\frac{df}{d\theta}\right)\left(\frac{d^2f}{d\theta^2}\right)\dot{\theta}^2 \tag{4.63}$$

Combining Eqs. (4.62) and (4.63), according to Eq. (4.60), the resulting Lagrange equation is more complex than one might at first imagine:

$$\frac{d}{dt}\left(\frac{\partial T}{\partial\dot{\theta}}\right) - \frac{\partial T}{\partial\theta} = m\left(\frac{df}{d\theta}\right)^2\ddot{\theta} + m\left(\frac{df}{d\theta}\right)\left(\frac{d^2f}{d\theta^2}\right)\dot{\theta}^2 + I\ddot{\theta} = Q_\theta \tag{4.64}$$

Note that Lagrange's equation generally requires two differentiations of the kinematic relation Eq. (4.57). In some cases this requirement is not especially difficult, but in other cases it can be very complex. We consider next three specific cases of increasing difficulty.

Example 4.1 First, assume that the piston is moved by a rack-and-pinion system in which the pinion has a contact radius of R (see Figure 4.3.) In this case, Eq. (4.57) has the simple linear form

$$x = f(\theta) = R\theta + \text{constant} \tag{4.65}$$

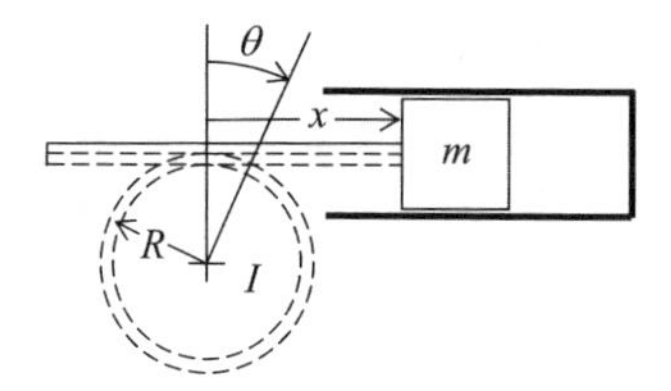

Figure 4.3 Rack-and-pinion system.

[Note that the constant in Eq. (4.65) depends on an arbitrary choice of the fixed point at which $x = 0$. If $x = 0$ when $\theta = 0$, the constant vanishes.] Now

$$\frac{df}{d\theta} = R \quad \text{and} \quad \frac{d^2 f}{d\theta^2} = 0 \tag{4.66}$$

and Eq. (4.64) assumes the simple form

$$(mR^2 + I)\ddot{\theta} = Q_\theta \tag{4.67}$$

This result could have been derived using Newton's equations directly rather than using the Lagrange equation. But this case does provide an indication of the fact that when kinematic relations are linear forms such as Eq. (4.65), the use of Lagrange equations is simple and straightforward. Instead of using the general form of Eq. (4.64), the differentiation of Eq. (4.65) to find $\dot{x} = R\dot{\theta}$ and substitution into Eqs. (4.58) and (4.59) would lead to Eq. (4.67) immediately. When several generalized variables and *linear* constraints are involved, the use of Lagrange formalism often turns out to be more useful than this simple example might indicate.

Example 4.2 As a second example, The Scotch yoke mechanism shown in Figure 4.4 has the nonlinear kinematic relation

$$x = f(\theta) = R \sin\theta + \text{constant} \tag{4.68}$$

In Eq. (4.68), if the constant vanishes, x is measured from the point the edge of the piston assumes when the crank pin is in the vertical position. The derivatives of $f(\theta)$ are fairly easy to evaluate in this case:

$$\frac{df}{d\theta} = R\cos\theta, \qquad \frac{d^2 f}{d\theta^2} = -R\sin\theta \tag{4.69}$$

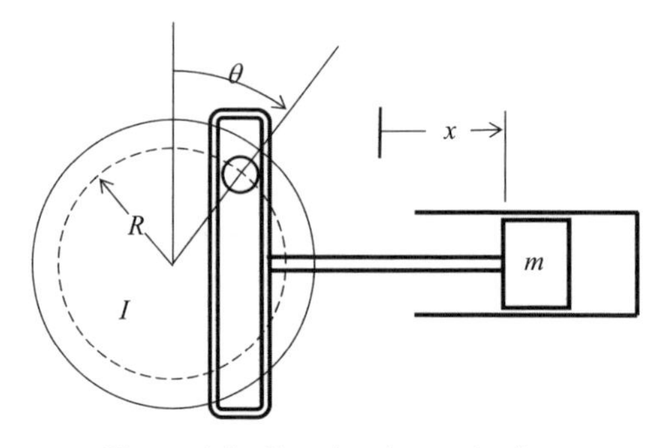

Figure 4.4 Scotch yoke mechanism.

Then Eq. (4.64) becomes

$$m(R^2\cos^2\theta)\ddot{\theta} + m(R\cos\theta)(-R\sin\theta)\dot{\theta}^2 + I\ddot{\theta} = Q_\theta \tag{4.70}$$

This equation is considerably more complicated than Eq. (4.67), and it is not as easily derived using free-body diagrams. Thus, the use of a Lagrange equation instead of basic Newton's laws is probably worthwhile for this example. Furthermore, if an equation of motion had been derived using the direct method and it turned out to be the same as Eq. (4.70), it would be a strong indication that no mistake had been made. Since the two methods of deriving equations are so different, they do serve as excellent checks on each other.

Example 4.3 In some practical cases, the differentiation of kinematic relations is not at all simple. As a third example, consider now that a connecting rod of length L connects the piston to a crankshaft with a pin on the flywheel at radius R, as shown in Figure 4.5. Using geometry (see Problem 4.10), the kinematic constraint is found to be

$$f(\theta) = R\sin\theta + (L^2 - R^2\cos^2\theta)^{1/2} + \text{constant} \tag{4.71}$$

The constant vanishes in Eq. (4.71) if x is measured from the crankshaft center. One differentiation of this $f(\theta)$ is fairly straightforward:

$$\frac{df(\theta)}{d\theta} = R\cos\theta - \frac{R^2\cos\theta\sin\theta}{\left(L^2 - R^2\cos^2\theta\right)^{1/2}} \tag{4.72}$$

but the next differentiation to find $d^2f(\theta)/d\theta^2$ becomes very complex. Using the result in Eq. (4.64) will result in very complex equations of motion that may be hard to check for correctness. In such cases, it is often practical to consider other methods to describe the dynamics of the situation.

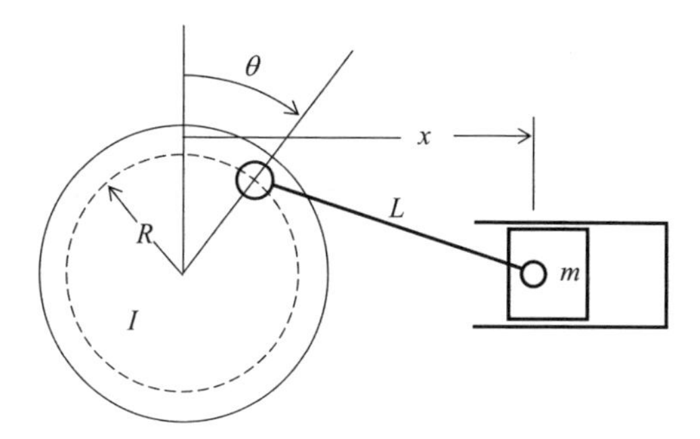

Figure 4.5 Crank and connecting rod mechanism.

Approximate Method for Satisfying Constraints

(Although this is a method generally well known to people working in modeling and simulation, some researchers have given it the fanciful name the *Karnopp–Margolis method* (see [11] and [12]). If the aim of a dynamic analysis is to produce equations that can be solved on a computer, it is sometimes easier to enforce kinematic relationships approximately by imagining stiff springs (and perhaps dampers in parallel) at connection points rather than to assume ideally rigid connections. The springs and dampers sometimes represent real compliance elements at connection points but they may also be *virtual elements* used only to aid in the construction of a computable model of the system. Suppose, for example, that the piston position x does not exactly equal the value given in Eqs. (4.57) and (4.71), but rather, there is a stiff spring with constant k and deflection $x - f(\theta)$ connecting the piston with the end of the connecting rod. This is shown schematically in Figure 4.6. The concept is that by adjusting the value of k, one may make the deflection sufficiently small during a computer simulation such that the kinematic constraint is effectively satisfied.

A linear spring would produce a force $F_s = k[x - f(\theta)]$ that will act to reduce the distance between the piston x and the end of the connecting rod $f(\theta)$. Increasing k will tend to reduce the position error, but very large values of k will have the effect of allowing high-frequency vibrations between the piston and the end of the connecting rod in this approximate model. This will require small time steps for the integration routine used to solve the equations of motion to avoid numerical instabilities and hence long times for simulation. Therefore, a compromise value for the spring constant will have to be found, perhaps by trial and error during the simulation phase. We will see that if this scheme is used, there will be two generalized coordinates rather than a single one as outlined for Example 4.3, but the equation formulation is simplified considerably.

The kinetic energy remains exactly as given in Eq. (4.58), but now there is also a potential energy

$$V = \tfrac{1}{2}k\,[x - f(\theta)]^2 \tag{4.73}$$

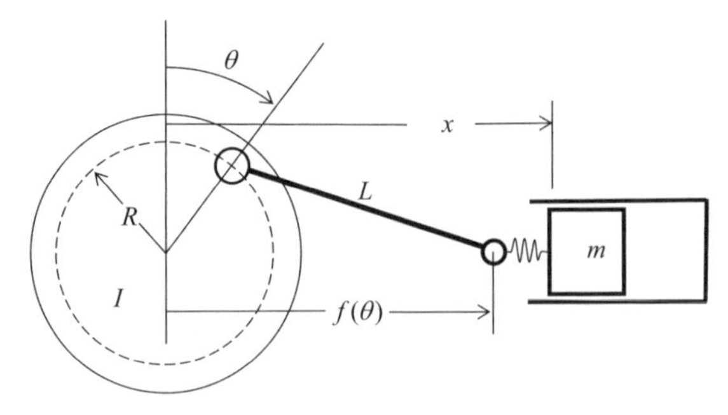

Figure 4.6 System of Figure 4.5 with spring-enforced constraint between piston and connecting rod.

and there are two Lagrange equations to be evaluated. The equations are

$$\frac{d}{dt}\left(\frac{\partial T}{\partial \dot{\theta}}\right) - \frac{\partial T}{\partial \theta} + \frac{\partial V}{\partial \theta} = Q_\theta \tag{4.74}$$

$$\frac{d}{dt}\left(\frac{\partial T}{\partial \dot{x}}\right) - \frac{\partial T}{\partial x} + \frac{\partial V}{\partial x} = Q_x \tag{4.75}$$

where Q_x is a generalized force associated with the piston and Q_θ is a possible torque applied to the crankshaft. (A virtual damper in parallel to the virtual spring to reduce the possible oscillations between the flywheel and the piston mass could also be incorporated in the generalized forces.) Since the kinetic energy now has the simple form of Eq. (4.58) and does not depend on either x or θ, both Eqs. (4.74) and (4.75) are simple to evaluate:

$$I\ddot{\theta} + k[x - f(\theta)]\left[-\frac{df(\theta)}{d\theta}\right] = Q_\theta \tag{4.76}$$

$$m\ddot{x} + k[x - f(\theta)] = Q_x \tag{4.77}$$

The only complicated terms in these equations involve $f(\theta)$ and $df(\theta)/d\theta$, which are given in Eqs. (4.71) and (4.72). The generally more complex term $d^2f(\theta)/d\theta^2$, which is necessary in Eq. (4.64) when the constraint is used directly to achieve a single-degree-of-freedom formulation is not required.

It should be clear from physical considerations on the basis of Figure 4.6 or from mathematical considerations of Eqs. (4.76) and (4.77) that the system so far has no energy dissipation unless some form of damping is to be incorporated in the generalized forces Q_θ and Q_x. Unless the system is to be coupled with other elements that do provide a damping effect, the introduction of a stiff spring to approximately enforce the geometric constraint will lead to the possibility of undamped oscillations between the flywheel and the piston.

These oscillations may be damped out by imagining that a damper is inserted in parallel with the spring and incorporating the damper force in the generalized forces. The relative velocity between the end of the connecting rod and the piston is

$$\frac{d}{dt}[x - f(\theta)] = \dot{x} - \frac{df(\theta)}{d\theta}\dot{\theta}$$

and the damping force F_d is related to this relative velocity. For example, if this virtual damper is assumed to be linear,

$$F_d = b\left[\dot{x} - \frac{df(\theta)}{d\theta}\dot{\theta}\right] \tag{4.78}$$

where b is a damping coefficient and when F_d is positive, it represents a tension force between the piston and the end of the connecting rod.

The work done on the system by the damping force during virtual variations in x and θ is

$$\delta W = -F_d \delta x + F_d \frac{df(\theta)}{d\theta} \delta\theta \tag{4.79}$$

where $[df(\theta)/d\theta]\delta\theta$ represents the virtual motion to the right of the end of the connecting rod when θ is given a virtual increment $\delta\theta$. The generalized forces to be used with Eqs. (4.76) and (4.77) are then

$$Q_x = -b\left[\dot{x} - \frac{df(\theta)}{d\theta}\dot{\theta}\right], \qquad Q_\theta = +b\left[\dot{x} - \frac{df(\theta)}{d\theta}\dot{\theta}\right]\frac{df(\theta)}{d\theta} \tag{4.80}$$

Note that even if we decide to add damping to the spring, we still do not need to evaluate the second derivative if the constraint $f(\theta)$. Just as is the case for finding a suitable spring constant for simulation purposes, a certain amount of trial-and-error simulating may be required to find a suitable damping coefficient.

The two equations of motion, Eqs. (4.76) and (4.77), are remarkably simple compared to the single equation (4.64) mainly because they involve just a single differentiation of the kinematic relation between x and θ rather than two differentiations. Thus, in the last example, one would not have to proceed with differentiation of the constraint beyond Eq. (4.72). Particularly when fairly complex kinematic relations are involved, this technique has much to recommend it and, despite its seeming inelegance, it turns out to be remarkably practical in many cases.

Although trial-and-error adjustment of the spring and damper values during simulation is generally necessary to achieve sufficient accuracy without resorting to very small time steps and long simulation times, it is often possible to decide on rough initial parameters by considering some simplifying linearized special cases. See Problems 4-35 and 4-36, for example.

4.6 FIRST-ORDER FORMS FOR LAGRANGE'S EQUATIONS

As pointed out in Chapter 3, when the goal of a dynamic system study is to simulate the response of a mechanical system, it is useful to transform the equations of motion from second-order form to first-order form. Typically, a state-space first-order form can be achieved from second-order differential equations using velocities and positions as state variables.

Because of the special nature of Lagrange's equations, there is another way to put the equations in first-order form using the *generalized momentum variables*. The resulting equations are sometimes called *Hamilton's equations*, although this term also has a more specialized meaning (see [10)]. The first-order form will be a state space of order $2n$ with state variables consisting of the original n generalized coordinates

$q_1, q_2, q_3, \ldots, q_n$ and n generalized momentum variables $p_1, p_2, p_3, \ldots, p_n$. The state equations are just $2n$ expressions for the first derivatives of the p's and the q's.

The first step seems trivial. We consider a holonomic system with n generalized coordinates. The n generalized momentum variables, p_i, corresponding to the generalized coordinates, q_i, are simply defined by

$$p_i \equiv \frac{\partial T}{\partial \dot{q}_i}, \qquad i = 1, 2, 3, \ldots, n \tag{4.81}$$

Then there are n first-order equations for the rates of change of the momentum variables that can be found by substituting Eqs. (4.81) into either of the two versions of Lagrange equations of Eqs. (4.24) and (4.25):

$$\dot{p}_i = \frac{\partial L}{\partial q_i} + Q_i = \frac{\partial T}{\partial q_i} - \frac{\partial V}{\partial q_i} + Q_i, \qquad i = 1, 2, 3, \ldots, n \tag{4.82}$$

These equations are half of the $2n$ first-order equations equivalent to the n second-order Lagrange equations. The remaining n equations have to do with the generalized velocities, $\dot{q}_i$.

It is shown in ref. [5] that the kinetic energy depends on the generalized velocities, the generalized coordinates, and possibly time in a special way:

$$T = T_2 + T_1 = T_0 \tag{4.83}$$

The term T_2 is a homogeneous quadratic function of the generalized velocities but with coefficients that may depend on the generalized coordinates and time:

$$T_2 = \tfrac{1}{2}\sum_{i=1}^{n}\sum_{j-1}^{n} m_{ij}\dot{q}_i\dot{q}_j = \tfrac{1}{2}[\dot{q}]^T[m(q, t)][q] \tag{4.84}$$

where the "mass" matrix is symmetrical and the elements in the matrix may be constants or functions of the q and t. The term T_1, if it exists, is linear in the generalized velocities

$$T_1 = \sum_{i=1}^{n} a_i(q, t)\dot{q}_i \tag{4.85}$$

and occurs only when the kinetic energy is a function of time explicitly as, for example, when there are imposed motions in the system. The term T_0 involves only the generalized coordinates and time and thus has no influence on the momentum variables as defined in Eq. (4.81).

Using Eqs. (4.84) and (4.85), the definition of the generalized momentum variables, Eq. (4.81), can be written out in detail using vector matrix notation:

$$[p] = [m(q,t)][\dot{q}] + [a(q,t)] \tag{4.86}$$

The second set of n first-order equations to supplement Eqs. (4.79) comes directly from Eq. (4.86). Unfortunately, a matrix inversion of the mass matrix is necessary in general:

$$[\dot{q}] = [m(q,t)]^{-1}[p - a(q,t)] \tag{4.87}$$

In many cases, the mass matrix has a number of zero terms and the individual $\dot{q}$'s in Eq. (4.87) can be found without having to accomplish a formal matrix inversion. In the worst case, one might have to perform a numerical matrix inversion at each time step during a simulation of the system.

Example System

As an example of the procedure, consider the system of Figure 4.2 that has been analyzed previously. The kinetic energy is given by Eq. (4.22), and we see that in this case only the term T_2 exists. The momentum variables p_1 and p_2 have actually been evaluated in Eqs. (4.26) and (4.32), respectively. In matrix form the result is

$$\begin{bmatrix} p_1 \\ p_2 \end{bmatrix} = \begin{bmatrix} m_1 + m_2 & m_2 L \cos q_2 \\ m_2 L \cos q_2 & m_2 L^2 \end{bmatrix} \begin{bmatrix} \dot{q}_1 \\ \dot{q}_2 \end{bmatrix} \tag{4.88}$$

Notice that the mass matrix is symmetrical (as it must be) and contains a nonlinear function of the generalized coordinate q_2 and that there are no a_i functions since T_1 is missing for this example.

The next step is to invert Eq. (4.88) to find the differential equations for the generalized coordinates. This can be done analytically for the example because the mass matrix is only 2×2 and the formula for the inversion of a 2×2 matrix is

$$\begin{bmatrix} a & b \\ c & d \end{bmatrix}^{-1} = \frac{\begin{bmatrix} d & -b \\ -c & a \end{bmatrix}}{ad - bc} \tag{4.89}$$

The set of first-order equations equivalent to Eq. (4.87) for the generalized coordinates are then

$$\begin{bmatrix} \dot{q}_1 \\ \dot{q}_2 \end{bmatrix} = \frac{\begin{bmatrix} m_2 L^2 & -m_2 L \cos q_2 \\ -m_2 L \cos q_2 & m_1 + m_2 \end{bmatrix} \begin{bmatrix} p_1 \\ p_2 \end{bmatrix}}{(m_1 + m_2) m_2 L^2 - (m_2 L \cos q_2)^2} \tag{4.90}$$

To complete the example, we first evaluate Eq. (4.82) using Eqs. (4.27), (4.28), (4.33), and (4.34) to find the differential equations for the generalized momenta:

$$\begin{aligned} \dot{p}_1 &= -kq_1 \\ \dot{p}_2 &= m_2 L \dot{q}_1 \dot{q}_2 (-\sin q_2) - mgL \sin q_2 \end{aligned} \tag{4.91}$$

Equations (4.90) and (4.91) are now in a state-space form that can readily be programmed for machine simulation. But even for this relatively simple example system, it is clear that nonlinear mechanical systems are neither easy to formulate nor easy to put into a desired form for simulation. Experience has shown, however, that an engineer who has the skills required to use Lagrange's equations has a better chance of deriving equations of motion for complex systems successfully than does one who must rely on free-body diagrams and the basic principles of mechanics.

Comments Regarding the Use of p and q Variables in Simulation

One advantage of Hamilton's first-order equations over the second-order Lagrange's equations is that it is not necessary to perform the total time derivative of $\partial T/\partial \dot{q}_i$ as indicated in Eqs. (4.24) and (4.25). As was pointed out in Section 4.5, this operation is sometimes quite difficult to carry out. Equation (4.82) is quite a bit simpler in many cases than the equivalent Lagrange equation. There does remain another possible algebraic problem: the inversion of a mass matrix that may have coefficients that are functions of the generalized coordinates.

For high-order systems, the mass matrix inversion in Eq. (4.87) may be analytically straightforward or it may be impossible. In some cases, the mass matrix may have few coupling terms and it may be possible to invert the matrix analytically as was done in the example. In other cases, the matrix may have only constant terms, so that a single inversion can be accomplished once and for all numerically. (This is often the case in linear vibratory systems.) In the worst case, the matrix can be inverted numerically after each time step in the integration as the system is being simulated. This slows down the simulation significantly. In any case, the inversion of the mass matrix represents a problem no matter what technique is chosen to convert Lagrange second-order equations of motion to first-order form.

A subtle point is that while Eq. (4.87) expresses $[\dot{q}]$ in terms of p and q variables (and perhaps t), Eq. (4.82) expresses $[\dot{p}]$ not only in terms of p and q variables but sometimes also in terms of some $\dot{q}$ variables. In principle, Eq. (4.87) could be used to eliminate any $\dot{q}$'s in the $[\dot{p}]$ equations in terms of p's and q's, thus achieving a true state-space set of differential equations. But for simulation purposes, this is not really necessary. If Eqs. (8.87) are evaluated in a simulation program *before* Eqs. (4.82) are evaluated, the $[\dot{q}]$ values will be known at the time an integration step is made for Eq. (4.82), and the step can be made without the necessity of eliminating any $\dot{q}$ variables in the $[\dot{p}]$ equations. Thus, the $[\dot{p}]$ equations can be used for simulation just as they are given in Eqs. (4.82). In general, this procedure for converting second-order Lagrange equations into first-order form for simulation has proven to be useful. This point is discussed at length in ref. [10].

4.7 NONHOLONOMIC SYSTEMS

This chapter has dealt with holonomic systems that can be treated with Lagrange's equations. Any geometric constraints in such systems can be represented as relations between linear or angular position variables. A set of generalized variables assures that the constraints are always satisfied. There are, however, a number of special system constraints that cannot be expressed as relations among position variables. Generalized variables cannot be defined for such systems, so the Lagrange technique cannot be applied. Examples of constraints that render a system nonholonomic are those constraints that are expressed not as equations but as inequalities or constraints that involve velocities as will as positions. Books on advanced dynamics, such as refs. [2], [4], [5], and [6], treat various forms of nonholonomic dynamic systems. An interesting example of a nonholonomic system is discussed in Problem 7.15. This problem cannot be treated with Lagrange's equations, but, of course, it is still possible to write the dynamic equations directly, and the constraints can be incorporated in the final formulation.

4.8 SUMMARY

The study of energy and power flow in mechanical systems often leads to insights that are not obvious when only Newton's laws are considered. Unfortunately, the idea of writing equations of motion using the conservation of energy or an extended principle that the rate of change of stored energy is equal to the dissipation power can be used only if a single-degree-of-freedom system is involved.

Lagrange's equations can be considered to be the extension of the energy method to holonomic systems with several degrees of freedom. Among the advantages to the use of Lagrange's equations for deriving equations of motion is that only scalar functions involving velocity and position are required to express the kinetic and potential energies as functions of the generalized coordinates and velocities. When Newton's laws are used directly, vector accelerations must generally be found and this is often quite difficult the do correctly.

Another advantage to having familiarity with Lagrange equations techniques is that they provide a completely different method to derive equations than direct use of the principles of dynamics. Thus, one can check the correctness of an analysis by using the two completely different formulation methods. Redoing an analysis using the same technique to check that no error has been made has the danger of simply repeating the mistake a second time.

Finally, Lagrange's equations can be put into a form useful for computer simulation using generalized momentum variables rather than generalized velocities as state variables. Quite apart from the theoretical justification of this procedure, it often provides practical benefits in reducing the algebraic manipulations required to achieve a usable simulation program.

REFERENCES

[1] F. P. Beer, E. R. Johnson, Jr., and W. E. Clausen, *Vector Mechanics for Engineers: Dynamics*, 8th ed., McGraw-Hill, New York, 2007.

[2] S. H. Crandall, D. C. Karnopp, E. F. Jr., Kurtz, Jr., and D. C. Pridmore-Brown, *Dynamics of Mechanical and Electromechanical Systems*, McGraw-Hill, New York, 1968. (Reprint edition, Krieger Publishing, Melbourne, FL.)

[3] J. H. Ginsberg, and J. Genin, *Dynamics*, 2nd ed., Wiley, New York, 1983.

[4] H. Goldstein, *Classical Dynamics*, Addison-Wesley, Reading, MA, 1959.

[5] D. T. Greenwood, *Principles of Dynamics*, Prentice Hall, Englewood Cliffs, NJ, 1988.

[6] E. J. Haug, *Intermediate Dynamics*, Prentice Hall, Englewood Cliffs, NJ, 1992.

[7] R. C. Hibbler, *Engineering Mechanics Dynamics*, 11th ed., Pearson Prentice Hall, Upper saddle River, NJ, 2007.

[8] C. Lanczos, *The Variational Principles of Mechanics*, University of Toronto Press, Toronto, Ontario, Canada, 1949.

[9] J. L. Meriam, L. G. Kraige, and W. J., Palm, III. *Engineering Dynamics*, 5th ed., Wiley, Hoboken, NJ, 2003.

[10] T. M. Vance, and A. Sitchin, Numerical solution of dynamical systems by direct application of Hamilton's principle, *Int. J. Numer. Methods Eng.*, Vol. 4, pp. 207–216, 1972.

[11] A. Zeid, Bond graph modeling of planar mechanisms with realistic joint effects, *Trans. ASME Dyn. Syst. Meas. Control*, Vol. 111, pp. 15–23, 1989.

[12] A. Zeid, and C.-H. Chung, Bond graph modeling of multibody systems: a library of three-dimensional joints, *J. Franklin Inst*, Vol. 329, pp. 605–636, 1992.

[13] J. P. Den Harlog, *Mechanical Vibrations*, 3rd ed., McGraw-Hill, N. Y., 1947.

PROBLEMS

4.1 A flywheel has been mounted incorrectly such that its axis of rotation is a distance r_0 from the center of mass as shown in Figure P4.1 The flywheel has

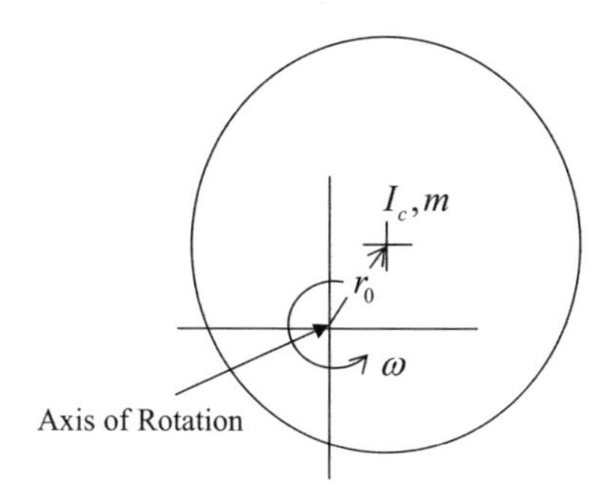

Figure P4.1

a mass m and a centroidal moment of inertia I_c about an axis parallel to the rotation axis. If the flywheel is spinning about the axis of rotation with angular speed ω, show that the same expression for kinetic energy results from using either Eq. (4.2) or (4.3) and the parallel axis theorem.

4.2 If a spring produces a force F as a function of deflection x that can be approximated by the formula $F = Ax^3$, where A is a constant, find an expression for the potential energy.

4.3 A spring generates a force-deflection relation which is shown as a broken line in Figure P4.3. According to Eq. (4.10), the potential energy function should be the area under the force-deflection curve. Since there is no simple analytical expression for this nonlinear spring it is necessary to express the energy and the force by different expressions for three ranges of the deflection x. (In a computer program, such nonlinear functions are typically represented by subroutines.)

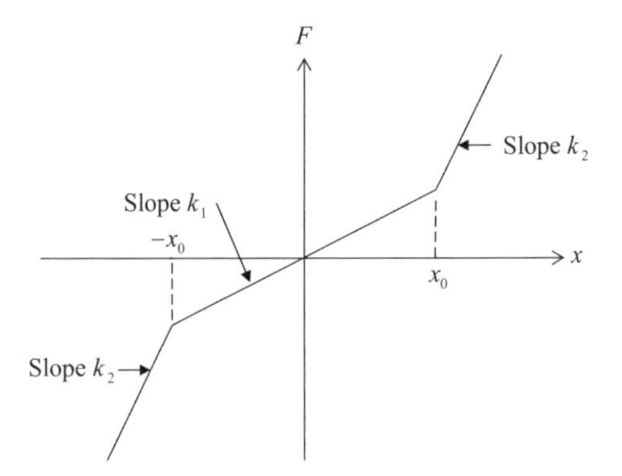

Figure P4.3

Consider three expressions for the potential energy:

$$V = \begin{cases} k_1 x_0 x + \dfrac{k_2}{2}(x - x_0)^2, & x > x_0 \\ \dfrac{k_1}{2} x^2, & -x_0 < x < +x_0 \\ -k_1 x_0 x + \dfrac{k_1}{2}(x + x_0)^2, & x < -x_0 \end{cases}$$

Use Eq. (4.11) to show that these expressions are compatible with the force-deflection law shown in the figure.

4.4 A thin uniform rod of length l, mass m, and centroidal moment of inertia $\frac{1}{12}ml^2$ is suspended by a frictionless pivot at its upper end and acts as a physical pendulum in a gravity field (Figure P4.4). Assume that the pendulum swings in a fixed plane.

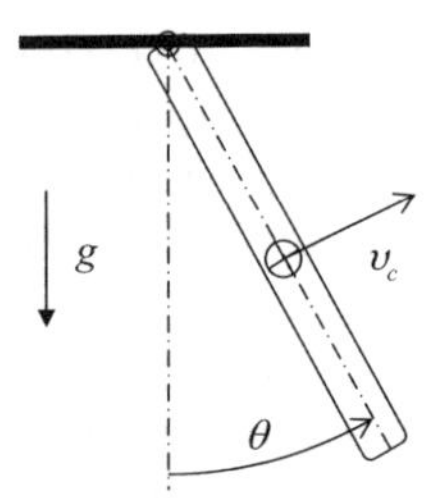

Figure P4.4

(a) Find an expression for the kinetic energy T.

(b) Find an expression for the potential energy V.

(c) Write the equation of motion for θ using the principle of conservation of energy.

4.5 If the pendulum in Problem 4.4 is assumed to have a friction torque $\tau(t) = \tau_0(\dot{\theta}/|\dot{\theta}|) = \tau_0 \operatorname{sgn} \dot{\theta}$, where τ_0 is a constant torque magnitude and $\operatorname{sgn}(\cdot)$ is the *signum function,* which is simply $+1$ when the argument is positive and -1 when the argument is negative. This torque definition assumes that when $\tau(t)$ is positive, it is removing energy from the system. Modify the conservation of energy expression to express the idea that the rate of change of the stored energy is equal to the power dissipated by friction. This should allow you to modify the equation of motion found in Problem 4.4 to include the friction effect.

4.6 The mass m moves on a frictionless *horizontal* plane under the action of a spring with spring constant k and free length l_0 (Figure P4.6).

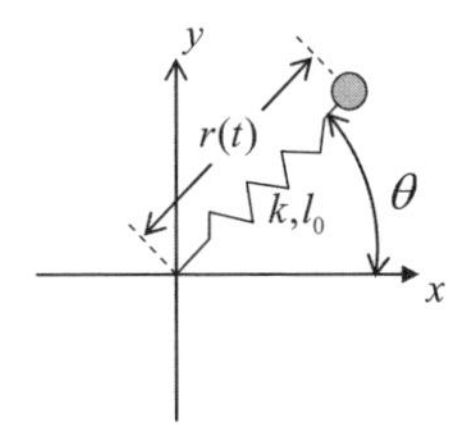

Figure P4.6

(a) Write expressions for the kinetic and potential energies using r and θ as generalized coordinates.

(b) Write Lagrange's equations for the two coordinates and give a physical interpretation of each.

4.7 Imagine that the system shown in Figure 4.2 has an additional force $F(t)$ acting on the pendulum mass. The force is horizontal and acting toward the right. The problem is to determine the generalized forces Q_1 and Q_2 to be used in the

Lagrange equations for the generalized coordinates q_1 and q_2 that will be necessary because of this applied force.

(a) Write the work expression δW that expresses the work done *on the system* by $F(t)$ when the position coordinate q_1 is incremented by the infinitesimal amount δq_1 while q_2 is held constant.

(b) Write the work expression for the case that the angular variable q_2 is incremented by the amount δq_2 while q_1 is held constant. Note that the work is the force times the distance the pendulum mass moves *in the direction of the force*.

(c) Write the expressions for Q_1 and Q_2 due to the presence of $F(t)$.

4.8 A solid uniform disk with radius R, mass M, and centroidal moment of inertia $MR^2/2$ (see Appendix C) rolls along the ground without slipping (Figure P4.8). Attached to the disk is a point of mass m a distance r from the center. Because of the no-slip assumption, the angle θ can function as the single generalized coordinate needed to describe the motion. The gravity force on m will cause the disk to roll with an unsteady angular velocity or to oscillate back and forth.

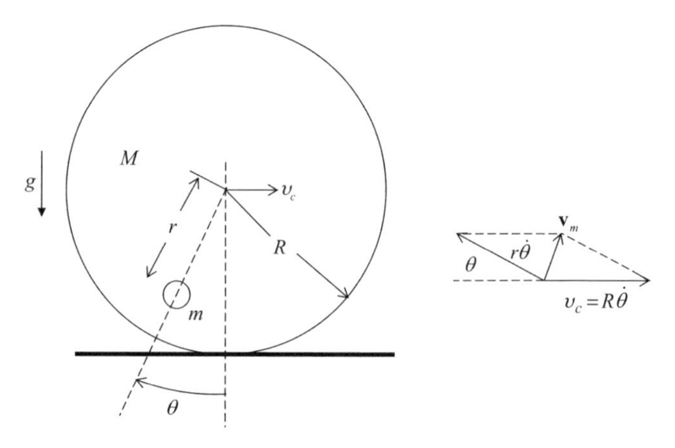

Figure P4.8

(a) Find an expression for the kinetic energy of the system. The diagram on the left shows how the vector velocity $\mathbf{v}_m$ for m is composed from the velocity of the disk center and the velocity with respect to the center due to $\dot{\theta}$. See Figure 4.2 and Eqs. (4.21) and (4.22) for a similar situation.

(b) Find the potential energy for the system.

(c) Find the equation of motion using Lagrange's equation.

(d) Is there a friction force between the bottom on the disk and the ground? If so, does one need to use a generalized force to represent it?

4.9 A simple pendulum with mass m on a massless rod of length l is pivoted on a block that is forced to move up and down in a gravity field with a position $X(t)$ (Figure P4.9). (The actuator responsible for the block motion is not shown.) On the right, the components of the mass velocity are shown. Neglect friction.

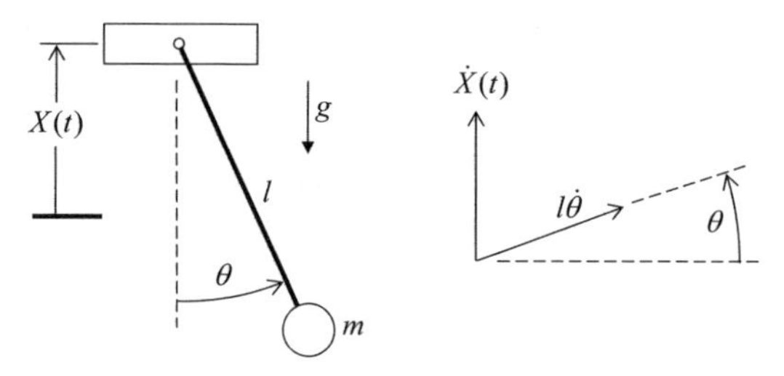

Figure P4.9

(a) Derive an expression for the kinetic energy.
(b) Derive an expression for the potential energy.
(c) Write the Lagrange equation for the angle θ.
(d) Show that for the case when θ is assumed to be small so that $\sin\theta \approx \theta$, the equation has the form of the *Matthieu equation* $m\ddot{x} + (k + \Delta k \sin \omega t) = 0$ if $\ddot{X}(t) = A \sin \omega t$. This equation is discussed at length in ref. [13]. The pendulum of this problem is actually stable upside down if A and ω have certain specific values.

4.10 Make a simplified sketch of the geometry of Figure 4-5 and use it to derive Eq. (4.71).

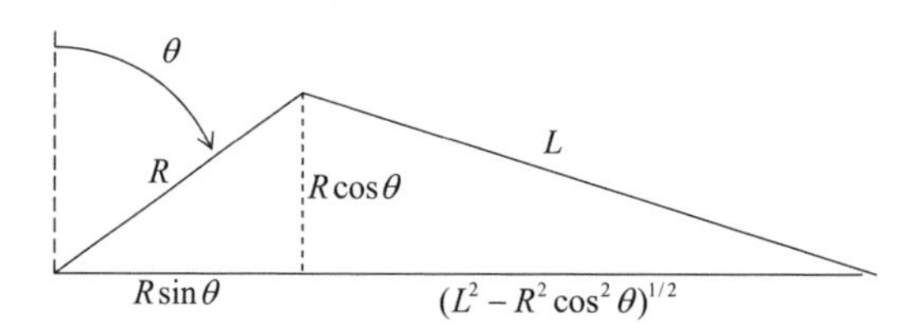

Figure P4.10

4.11 A bicycle salesman claims that the weight of the two wheels on a bicycle is twice as important as the weight of the frame. In Figure P4.11, M is the mass of the frame, m is the mass of each wheel, and for bicycle wheels most of the mass of each wheel is concentrated near the wheel radius, so the centroidal moment of inertia for each wheel is approximately $I_c \approx mR^2$. Assuming that the wheels do not slip, write an expression for the kinetic energy of the entire system, and use this expression to determine how the wheel mass would affect the acceleration of the system. Does the salesman's statement make any sense?

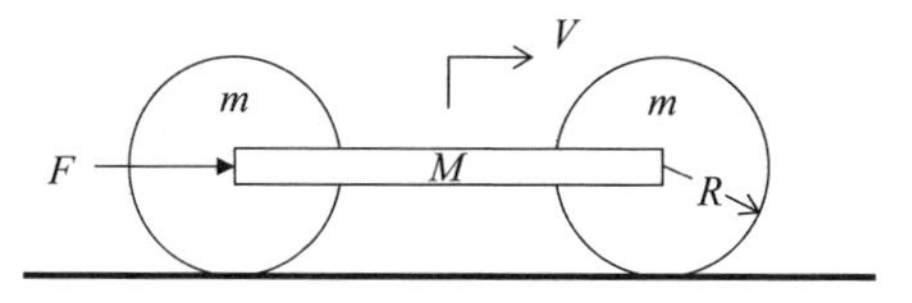

Figure P4.11

4.12 A massive block attached to a beam assumed to be massless vibrates with small amplitudes and with negligible gravity effects (Figure P4.12). The block mass is M and its centroidal moment of inertia is I_c. The system is symmetrical and moves only in the plane. The two generalized coordinates to be used are the tip deflection of the beam y and the angle of the beam end and block θ.

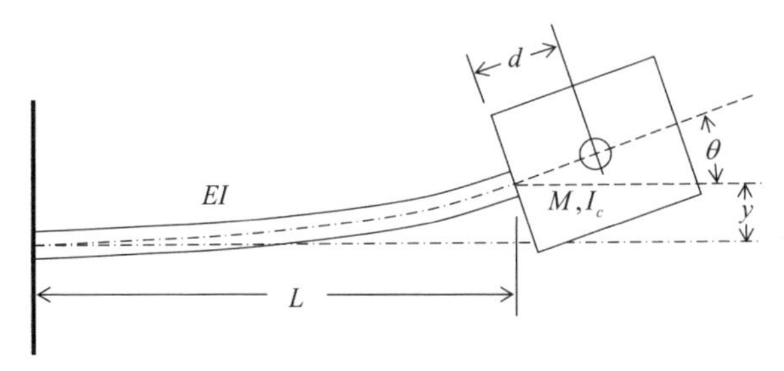

Figure P4.12

(a) Write an expression for the kinetic energy of the block. Note that if θ is assumed to be a small angle, the velocity of the center of mass of the block is approximately $v_c \approx \dot{y} + d\dot{\theta}$.

(b) If the beam is acted on by a force f in the y-direction and a torque τ in the θ-direction, it is a standard exercise in beam theory to find a spring constant matrix relating the force and torque to the deflections y and θ (see [2]).

$$\begin{bmatrix} f \\ \tau \end{bmatrix} = \begin{bmatrix} k_{11} & k_{12} \\ k_{21} & k_{22} \end{bmatrix} \begin{bmatrix} y \\ \theta \end{bmatrix}$$

where $k_{11} = 12EI/L^3$, $k_{12} = k_{21} = -6EI/L^2$, and $k_{22} = 4EI/L$. When a spring constant matrix is involved, the expression for potential energy involves the same type of operation as is found for the kinetic energy in Eqs. (4.4) and (4.7). In the present case

$$V = \frac{1}{2} [\, y \quad \theta \,] \begin{bmatrix} k_{11} & k_{12} \\ k_{21} & k_{22} \end{bmatrix} \begin{bmatrix} y \\ \theta \end{bmatrix}$$

Write out the potential energy in terms of the k's and y and θ using this matrix formula.

(c) Write out the equations of motion using Lagrange's equations for y and θ and put the result in a matrix form in which a symmetric mass matrix and the (symmetric) stiffness matrix should appear.

4.13 Figure P4.13 shows the rear axle of a truck seen from above. The angular positions of the two wheels are called θ_1 and θ_2, and the angle of the drive shaft is θ_3. These angles are shown as if they were vectors using the right-hand screw rule. In this problem we do not consider the translational movement of the truck but rather, imagine the truck to be fixed in space with spinning tires. Forces F_1 and F_2 are friction forces on the bottom of the tires from the ground. When multiplied by the tire radius R, these forces exert torques on the wheels. The gearbox exerts a torque $\tau_3(t)$ on the drive shaft in a direction that supplies power to the system. Three moments of inertia are present: I_1 and I_2 at the wheels and I_3 at the driveshaft.

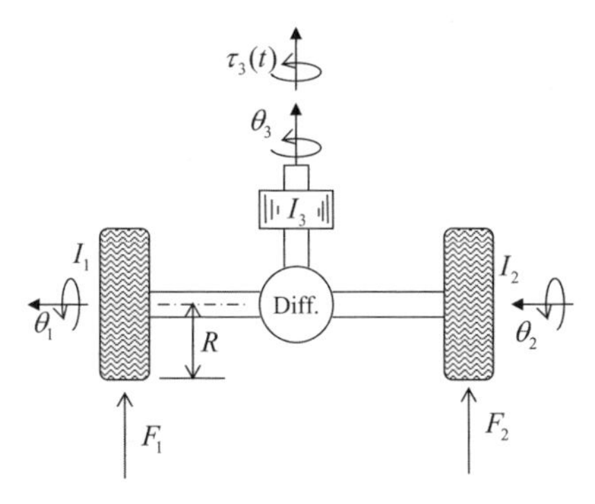

Figure P4.13

The differential constrains the three angles by the expression $\theta_3 = (G/2)(\theta_1 + \theta_2)$, where the constant G is the rear axle ratio. (If both wheels were turning the same speed, then the drive shaft would be turning G times faster.) Because of this constraint, the system has only two degrees of freedom. Use $\theta_1 = q_1$ and $\theta_2 = q_2$ as the two generalized coordinates for a Lagrange equation analysis.

(a) Write the expression for the kinetic energy.

(b) Find the generalized forces Q_1 and Q_2 in terms of F_1, F_2, and $\tau_3(t)$ and the parameters R and G.

(c) Find the equations of motion and put them in matrix form.

4.14 A bar is suspended on two springs with spring constants k_1 and k_2 and vibrates vertically with small motions around a nominally horizontal position (Figure P4.14). Since the springs are linear, it is permissible to neglect the effects of gravity in finding equations of motion in order to compute the natural frequencies of oscillation. The bar has mass m and centroidal moment of inertia I_c. The center-of-mass location is given by the distances a and b with respect to the spring locations. The equations are somewhat simplified if the distance $l = a + b$ is used. Use the spring deflections x_1 and x_2 as generalized coordinates.

(a) Write the velocity of the center of mass in terms of $\dot{x}_1$ and $\dot{x}_2$, remembering that the angular motions are assumed to be small.

(b) Find the angular velocity of the bar.

(c) Find the kinetic and potential energies.

(d) Write the equations of motion and display them in matrix form.

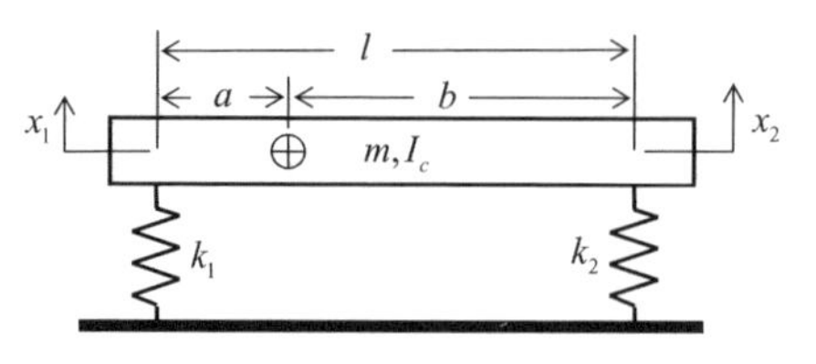

Figure P4.14

(e) In order to add damping to the system, suppose one assumed that dashpots were connected in parallel to the springs generating forces $F_1 = B_1\dot{x}_1$ and $F_2 = B_2\dot{x}_2$, where B_1 and B_2 are viscous damping coefficients. Find generalized forces Q_1 and Q_2 to add the damping to the system an then modify the equation in part (d) to include these damping forces.

4.15 Figure P4.15 shows two ways to represent a simple pendulum. On the left, a point mass is imagined to be mounted on a rigid massless rod of length l. On the right the rod is represented as a spring with free length l and spring constant k. The force-deflection law for the spring in the radial direction is

$$F = k(r - l) = k\left[(x^2 + y^2)^{1/2} - l\right]$$

The idea is that as the spring constant is increased, the spring will begin to have the effect of the rigid rod. On the other hand, the system on the left has a single degree of freedom and θ can function as the generalized coordinate, while the system on the right needs two generalized coordinates, such as x and y. Even though in some circumstances, both system models can produce almost identical motions of the mass, the equations of motion are surprisingly different.

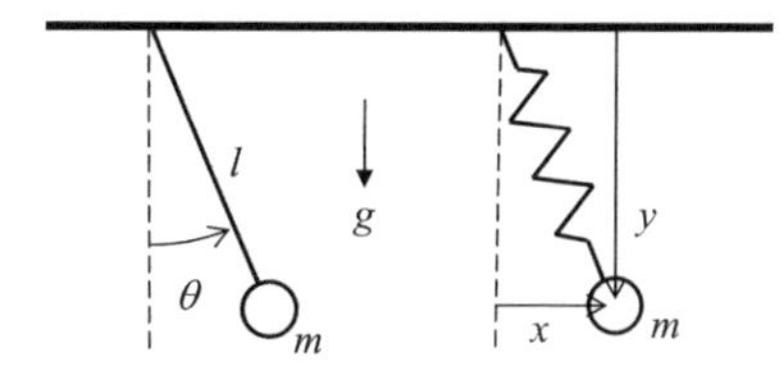

Figure P4.15

(a) For the system on the left, write expressions for the kinetic and potential energies and derive the equation of motion.

(b) For the system on the right, also write expressions for the kinetic and potential energy and derive the two equations of motion.

(c) Try to rearrange the x equation to interpret it as implying the $m\ddot{x} =$ force in the x-direction.

4.16 Figure P4.16 shows part of a machine in which a slotted bar is forced to oscillate by a small pin mounted on a rotating disk. The pin is at a radius r from the center of the disk and the pivot point of the bar is a distance L from the center of the disk. If the disk and the bar are considered to be rigid bodies, the angle of the bar φ is related to the angle of the disk θ and the system has a single degree-of-freedom.

(a) Considering the sketch of the geometry of the system shown at the right of the figure, show that the relation between the two angles can be written $\tan\varphi = r\sin\theta/(L - r\cos\theta)$.

(b) Although this is a relationship between φ and θ, it is not easy to put it into the form of Eq. (4.57) or to evaluate the terms in Eq. (4.64). Instead, you are to try the idea leading to Eq. (4.73) of imagining a stiff spring between the pin and the bar instead of a rigid connection. Show that if the two angles do not exactly satisfy the constraint, there will be a vertical gap of magnitude $(L - r\cos\theta)\tan\varphi - r\sin\varphi$ between the pin and the slot in the bar. If a spring of constant k has this gap as its deflection, compute the resulting potential energy in the spring. (Don't worry about why the spring reacts only to the vertical gap when the bar generally has an angle. This is a virtual spring anyway.)

(c) Now consider that the disk has a moment of inertia I_1 and the bar has a moment of inertia about its pivot point of I_{02}. Assume that there is a torque $\tau(t)$ on the disk and write Lagrange's for θ and φ, which are now considered independent. *Hint:* In case you don't remember, $d\tan\varphi/d\varphi = \sec^2\varphi = 1/\cos^2\varphi$.

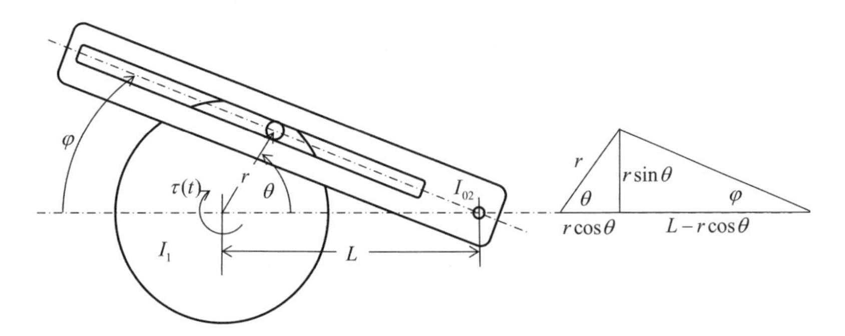

Figure P4.16

4.17 A flywheel with moment of inertia I spins without friction about a vertical axis with angular speed $\dot{\varphi}$ (Figure P4.17). Attached to the flywheel and spinning with it is a shaft upon which a long thin rod with mass m and length l is pivoted. The rod can swing with an angle θ from the vertical so that the angular velocity of the rod has to do with both $\dot{\theta}$ and $\dot{\varphi}$. The principal moments of inertia, evaluated about the pivot point 0, are $I_{01} \approx 0$, $I_{02} = I_{03} = \frac{1}{3}ml^2$. Note that

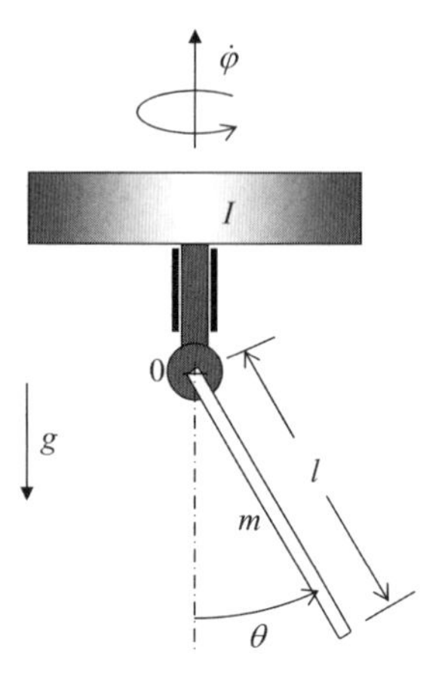

Figure P4.17

point 0 is fixed in space. The center of gravity of the rod is a distance $l/2$ from the pivot point.

(a) Compute the angular velocity components about appropriate 1 – 2 – 3 axes and then use Eq. (4.8a) to write the kinetic energy.

(b) Find the potential energy.

(c) Write the Lagrange equations for φ and θ.

(d) It seems obvious that the system could spin in a steady way with $\dot{\varphi} = \omega_0 =$ constant and with $\theta = \theta_0 =$ constant. Verify from the equations of motion that this is possible and find the relationship between θ_0 and ω_0.

4.18 Figure P4.18 shows a shaft inside a machine that spins at an angular rate $\dot{\varphi}$. Attached to the shaft is an arm with a pivot a distance L from the shaft centerline on which a thin rod is attached. The rod has mass m, length l and principal centroidal moments of inertia $I_1 \approx 0,\ I_2 = I_3 = \frac{1}{12}ml^2$, and its angular position is θ, as shown.

(a) Write an expression for the kinetic energy of the rod.

(b) First assume that $\dot{\varphi} = \omega_0 =$ constant and that there is no friction at the pivot. Write an equation of motion for θ and decide whether or not the rod could generally spin at a constant rate.

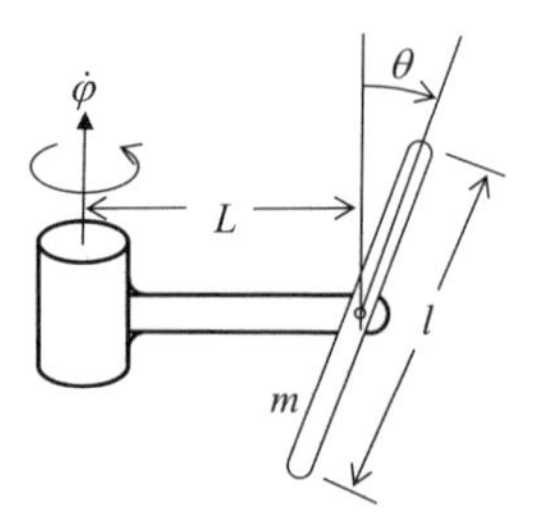

Figure P4.18

(c) It may be that a torque $\tau(t)$ would have to be exerted on the shaft in order for it to spin at a constant speed ω_0 because of the motion of the rod. One way to find the torque is to consider that φ is another generalized coordinate and to set up another equation of motion. Then $\tau(t)$ will be a generalized force and the equation will involve both φ and θ. Finally, in the equation, if $\dot{\varphi} = \omega_0 =$ constant is then assumed, an expression for the torque will result. Carry this operation out assuming that the moment of inertia of the shaft and arm is I.

4.19 Certain insects apparently use vibrating rods as rate gyroscopes. A point mass mounted on a beam spring mounted on a slowly rotating turntable (Figure P4.19) may help to explain how this is possible.

(a) Write T and V for the mass and spring in terms of x, $\dot{x}$, y, $\dot{y}$, and ω_0 using the coordinates rotating with the turntable. (This is the crucial step!)

(b) Write the equations of motion and put them in matrix form if possible.

(c) If the insect could excite the x-motion and could sense the resulting y-motion, could it determine the magnitude and sign of ω_0? Indicate how this would be possible.

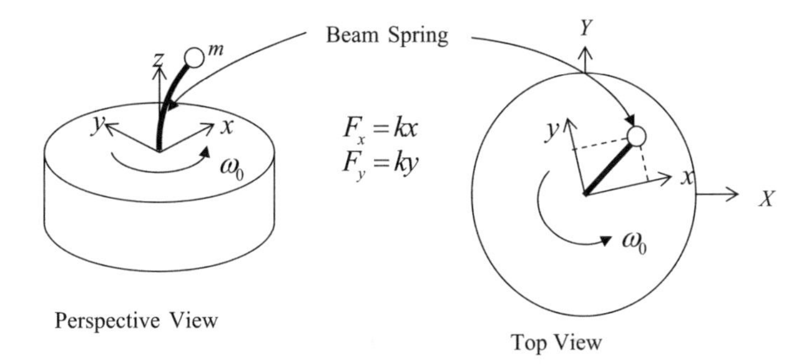

Figure P4.19

4.20 In the system of Problem 4.18, there is a limit to the angular speed ω_0 that would result in satisfactory behavior. What is the limit, and what would go wrong if the limit were exceeded?

4.21 Consider an n-degree-of-freedom vibratory system with generalized coordinates $q_1, q_2, \ldots, q_n$. If T and V are quadratic functions, Lagrange's equations will produce linear equations of the form $[M][\ddot{q}] + [B][\dot{q}] + [K][q] = [F]$, where $[q] = \begin{bmatrix} q_1 \\ q_2 \\ \vdots \\ q_n \end{bmatrix}$. The mass and spring constant matrices can be found from the kinetic and potential energies. The general formulas for the matrix elements derived in books on vibrations are $m_{ij} = \partial^2 T/\partial\dot{q}_i\partial\dot{q}_j$ and $k_{ij} = \partial^2 V/\partial q_i \partial q_j$. These results show that when Lagrange's equations are used to derive linear vibration equations, the mass and spring constant matrices will always be symmetrical. (This might not be the case if Newton's laws are used directly.)

Usually, the damping matrix $[B]$ and the forcing vector $[F]$ are found by considering generalized forces for all forces that act on the system but are not represented in T and V. In Problem 4.18, however, an antisymmetric $[B]$ matrix was found derived from a kinetic energy function that contained quadratic terms involving products of generalized coordinates multiplied by generalized velocities. The resulting terms in the equations of motion, called *gyroscopic coupling terms,* do not represent damping forces. The gyroscopic coupling terms can be found by the formula $b_{ij} = \partial^2 T/\partial \dot{q}_i \partial q_j - \partial^2 T/\partial \dot{q}_j \partial q_i$. Notice that $b_{ii} = 0$ and $b_{ij} = b_{ji}$. Consider terms in T of the form $\alpha \dot{q}_i q_j + \beta \dot{q}_j q_i$. Show that $b_{ij} = \alpha - \beta$. Then, using Lagrange's equation for q_i, show that the term resulting in the q_i equation is indeed $(\alpha - \beta)\dot{q}_j$, showing that the formula is correct.

4.22 Figure P4.22 shows a mechanism in which the only significant inertia effects are associated with the two point masses m_1 and m_2. The vertical shaft and attached arm are acted on by a torsional spring with constant k_t that exerts a torque on the shaft of $k_t\alpha$ in addition to the applied moment $M(t)$. The mass m_2 swings as a pendulum from a pivot on m_1 in such a way that the vertical shaft, the arm, m_1 and m_2 remain in a single plane. The angle between the bar connecting m_1 and m_2 with the vertical is θ.

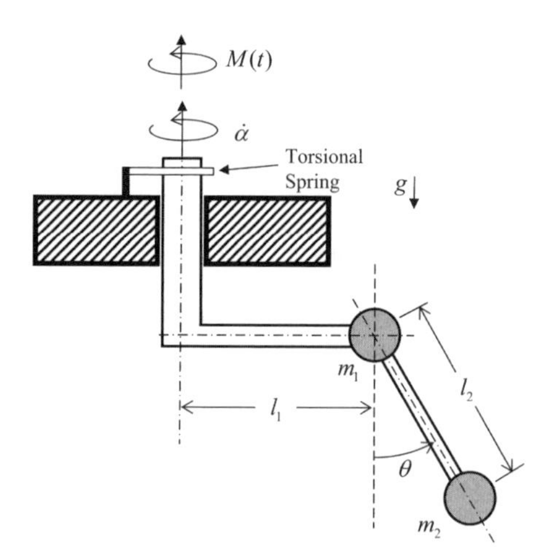

Figure P4.22

(a) Write expressions for the kinetic and potential energies and work expressions for determining the generalized forces.

(b) Write Lagrange equations for α and θ neglecting all friction effects.

4.23 Figure P4.23 shows two masses attached to a cable and capable of vibrating under the influence of the cable tension. The cable has been stretched tightly and has a tension T_0 that is assumed to remain essentially constant when the masses move small distances in the vertical direction, y_1 and y_2. Because of the high tension level, the effects of gravity are neglected. The lower part of the figure shows the geometry of the situation. The potential energy in the cable changes when the masses move from their equilibrium positions $y_1 = y_2 = 0$ because the length of the cable becomes $S_1 + S_2 + S_3$ instead of $3L$. The potential energy is then the (constant) tension times the stretch:

$$V = T_0[(S_1 - L) + (S_2 - L) + (S_3 - L)] + \text{constant}$$

Noting that the first segment obeys $S_1^2 = y_1^2 + L^2$ or $S_1 = (L^2 + y_1^2)^{1/2} = L[1 + (y_1/L)^2]^{1/2}$ and since we assume that $y_1/L \ll 1$, the formula $(1 + \varepsilon)^{1/2} \approx 1 + \varepsilon/2$ if $\varepsilon \ll 1$ can be used to derive the approximate relation $S_1 \approx L\left[1 + \frac{1}{2}(y_1/L)^2\right] = L + y_1^2/L$. A similar analysis can be applied to the remaining two segments of the cable.

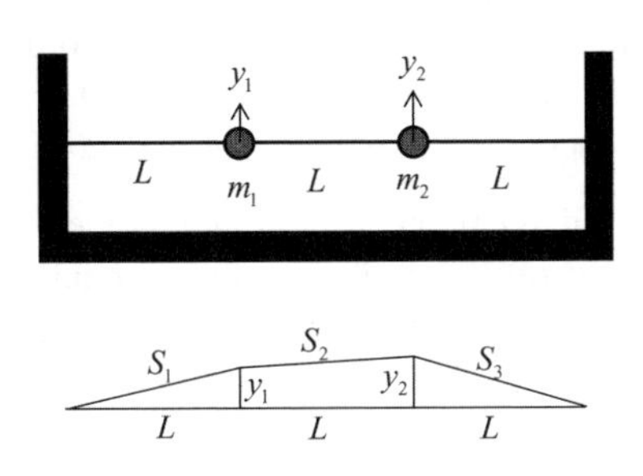

Figure P4.23

- **(a)** Write expressions for the kinetic and potential energies for the system.
- **(b)** Using Lagrange's, write the equations of motion and put them in matrix form.

4.24 Find the mass matrices for the systems shown in Figures 4.3 to 4.6 (The first three systems have only a single degree of freedom, so the matrices are actually scalars. The last system has two degrees of freedom, so the matrix is 2×2.)

4.25 For the system of Problem 4.6:

- **(a)** Find the generalized momenta p_θ and p_r corresponding to the generalized coordinated θ and r.
- **(b)** Using the methods of Section 4.6, write four first-order equations for the system. (Note that $\dot{\theta}$ occurs in the $\dot{p}_r$ equation and in this case could easily be eliminated analytically in favor of p_θ.)

4.26 For the vibratory system of Problem 4.12, derive the first-order equations of motion after defining generalized momentum variables p_y and p_θ. Since this is only a two-degree-of-freedom system, the results can be put in an explicit vector matrix form.

4.27 Convert the differential equations for the solution to Problem 4.13 to first-order form by defining the generalized momentum variables p_1 and p_2 corresponding to θ_1 and θ_2.

4.28 Modify the equations in Problem 4.14 (e) to first-order form using the generalized coordinates p_1 and p_2 corresponding to the generalized coordinates x_1 and x_2. Write the equations in an order such that a computer program could evaluate them step by step in a simulation even though the generalized velocities appear in the equations for $\dot{p}_1$ and $\dot{p}_2$.

4.29 Is there any problem inverting the mass matrix in Problem 4.16? If not, why not?

4.30 Consider the system of Problem 4.22.

(a) Find the mass matrix relating the generalized momentum variables p_α and p_θ to the generalized coordinates α and θ and find the inverse of this matrix.

(b) Write the four first-order state equations for the system using the generalized momentum variables and generalized coordinates. Put the equations in an order such that they could be evaluated by a computer without eliminating any generalized velocities that may appear.

4.31 A rigid wheel has a mass M and a centroidal moment of inertia J_c and a radius R (Figure P4.31). It rolls along the ground without slipping. A point mass m is attached to the disk at radius r and because of gravity causes the disk to roll in a nonuniform manner. In reality, the ground is not perfectly rigid and does absorb energy as the wheel rolls. Suppose that you try to represent this effect by assuming that there is a moment, M, exerted at the contact point which removes energy from the system as it rolls. Define M to be positive in the direction of positive θ. (Actually, a dissipative moment will always act to oppose the instantaneous motion with a form such as $-B\dot{\theta}$ if it were assumed to be linear.) You are to find a differential equation for the motion as described by the variable $\theta(t)$.

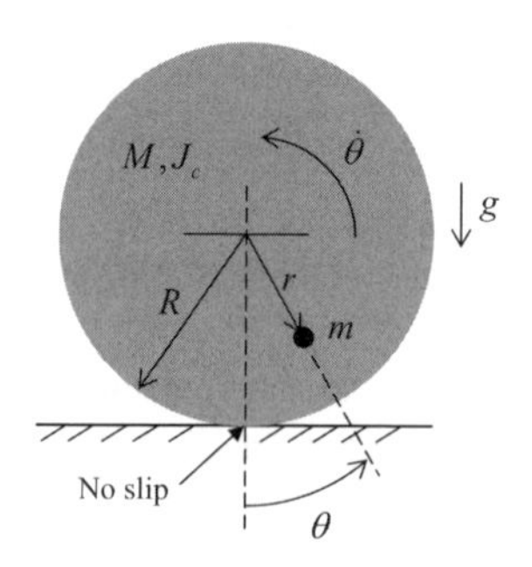

Figure P4.31

(a) Find the velocity of the center of mass of the disk and express the kinetic energy of the disk.

(b) Show a sketch of the velocity of the mass m and write its kinetic energy.
(c) Find the potential energy of the mass m.
(d) Using the Lagrange equation for θ, find the equation of motion including M as a generalized force.

4.32 Suppose that the system of Problem 4.31 now also has a force, $F(t)$, acting toward the left on the center of the disk.
(a) How does that modify the equation of motion?
(b) As a check on the correctness of the equation of motion, first imagine that the force is adjusted to make the angular acceleration $\ddot{\theta} = 0$. This means that the disk rolls to the left with constant speed and constant angular velocity $\dot{\theta}$. It also means that the disk is in static equilibrium. Find an expression for the force required to accomplish this by specializing the equation of motion.
(c) Make a sketch showing that the expression derived in part (b) can be interpreted as a moment equilibrium involving F, the gravity force mg, and the "centrifugal force" $mr\dot{\theta}^2$ using d'Alembert's principle.

4.33 This problem concerns the experimental determination of the moment of inertia of a wheel. Suppose that the mass m of the wheel can be found by weighing it and that the wheel is hollow so that it can be hung on a pivot as shown in Figure P4.33. When it is disturbed from its equilibrium position, it will oscillate as a physical pendulum with the center of mass a distance r from the pivot point. By writing an equation of motion for the system and finding the natural frequency of small oscillations, determine the moment of inertia from the mass, m, the internal radius r, and the period of the oscillation T, which is easily determined by timing a large number of oscillations.

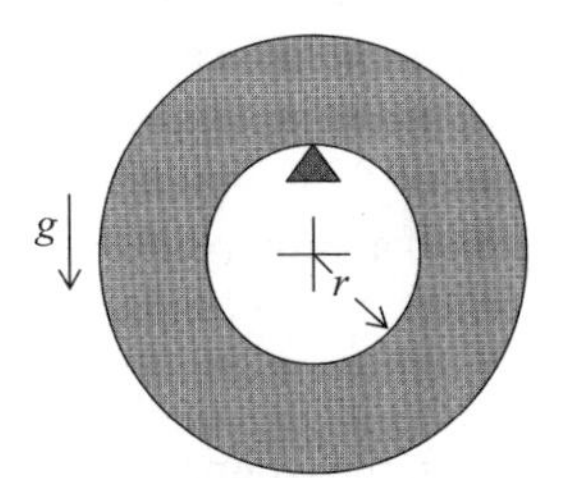

Figure P4.33

4.34 Figure P4.34 shows a heavy block attached to a lightweight beam. The task is to set up equations of motion useful for studying the small-amplitude vibrations of this system. The mass of the beam will be neglected, so it will be analyzed as a static beam. Since this will be a linear analysis, the deflection due to gravity can be neglected when studying vibrations. The block has a mass of M and a moment of inertia relative to the center of mass of I_c. The distance from the beam attachment point to the center of mass is l. The beam has length L, elastic modulus E, and an *area* moment of inertia I. (Don't confuse the area moment of inertia used in beam theory with the mass moment of inertia used in dynamics.) The beam can be represented by a spring constant matrix relating the force and torque acting on the end of the beam to the linear and angular deflections at the point where the beam is attached to the block: $\begin{bmatrix} f \\ \tau \end{bmatrix} = \begin{bmatrix} k_{11} & k_{12} \\ k_{21} & k_{22} \end{bmatrix} \begin{bmatrix} y \\ \theta \end{bmatrix}$. The corresponding potential energy function is $V = k_{11}y^2/2 + k_{12}y\theta + k_{22}\theta^2/2$. (This works because the conservative nature of elastic elements means that all spring constant matrices are symmetric. In this case, $k_{12} = k_{21}$.)

(a) Write the kinetic energy of the block in terms of $\dot{y}$ and $\dot{\theta}$.

(b) Books dealing with beam deflections often give superposition tables that express deflections as functions of imposed forces and torques. In this case the result is $\begin{bmatrix} y \\ \theta \end{bmatrix} = \begin{bmatrix} \frac{L^3}{3EI} & \frac{L^2}{2EI} \\ \frac{L^2}{2EI} & \frac{L}{EI} \end{bmatrix} \begin{bmatrix} f \\ \tau \end{bmatrix}$. This *compliance matrix* is obviously the inverse of the spring constant matrix discussed above. Verify that the spring constant matrix for this problem is $\begin{bmatrix} \frac{12EI}{L^3} & \frac{-6EI}{L^2} \\ \frac{-6EI}{L^2} & \frac{4EI}{L} \end{bmatrix}$ by multiplying the compliance matrix with this spring constant matrix to get the unit matrix $\begin{bmatrix} 1 & 0 \\ 0 & 1 \end{bmatrix}$.

(c) Find the equations of motion in matrix form using Lagrange's equations.

4.35 Recall that an equation of the form $m\ddot{x} + b\dot{x} + kx = F$ represents an oscillator with an undamped natural frequency (in radians/second) of $\omega_n = \sqrt{k/m}$ and a damping ratio of $\zeta = b/2\sqrt{mk}$. Consider the example system of Figure 4.6 with the functions $f(\theta)$ and $df(\theta)/d\theta$ given by Eqs. (4.71) and (4.72). Write out the equation for x, Eq. (4.77), using the generalized force from Eq. (4.80) and under the assumption that θ is *constant* at $\pi/2$. Note that in this condition, $f(\theta) = R + L$. Knowing m and choosing desired values for ω_n and ζ, show how to compute values for k and b that might form starting values for spring and damper parameters that could approximately enforce the geometric constraint for a simulation.

4.36 Suppose that one wanted to perform an analysis similar to that done in Problem 4.35 for the other equation in the example of Figure 4.6, Eq. (4.76), with the generalized force in Eq. (4.80). This will require a linearization about some specific angle and the assumption that the mass is held fixed while θ executes small motions. Let the steady angle be $\theta_s = 0$ and let $\theta(t)$ be a small angle away from the zero steady angle.

(a) Show that under these conditions, $f(\theta) \approx R\theta + (L^2 + R^2)^{1/2}$ and $df/d\theta \approx R$.

(b) Using these values and letting x be constant at the value $(L^2 - R^2)^{1/2}$, write out Eq. (4.76) using the generalized force from Eq. (4.80).

(c) Knowing I and picking values for ω_n and ζ, find another set of k and b values as a starting point for spring and damper parameters. (The values found in Problems 4.35 and 4.36 are only approximate since in both calculations the coupling between the two equations has been neglected. However, when using a trial-and-error method, it is better to have some ideas about starting points, however crude, than to pick values blindly for parameters in the range 0 to ∞.)

5

NEWTON'S LAWS IN A BODY-FIXED FRAME: APPLICATION TO VEHICLE DYNAMICS

This chapter has two goals. First, we aim to introduce an interesting and important application of dynamics: study of the dynamics of vehicles. Second, we use this application to show how the principles of dynamics appear when they are described in a noninertial rotating coordinate system. The use of such coordinate systems is rarely discussed in elementary texts, but these types of coordinate systems are nonetheless of great utility in certain applications. The study of vehicle dynamics is an interesting and important application of Newton's laws [3]. Although vehicles do move in three dimensions, many useful analyses actually deal with plane motion subsets of the possible motions and are therefore quite easy to understand. In this chapter the focus is on plane motion and equations of motion that can be used to study the stability of vehicles. Whereas in this chapter we concentrate on the plane motion of ground vehicles, similar studies can be found for the longitudinal dynamics of aircraft (see, e.g., [1]).

Many modern vehicle dynamic studies use a coordinate system attached to the vehicle. Because this type of coordinate system is rarely discussed in basic dynamics texts, it is worthwhile to devote some time to deriving equations of motion for a rigid body in body-fixed coordinates. Since such a coordinate system rotates with the vehicle and is thus not an inertial frame, the acceleration terms assume somewhat unusual forms. On the other hand, the use of body-centered coordinates often results in simpler equations than is the case when an inertial coordinate system is used. In this chapter the plane motion of ground vehicles is discussed and the contrast between the use of body fixed and inertial coordinates is pointed out. The general three-dimensional form of the dynamic equations for rigid bodies in a body-fixed coordinate system is presented in Appendix A.

5.1 THE DYNAMICS OF A SHOPPING CART

Many ground vehicles use pneumatic tires, but it is not easy to start immediately to analyze such vehicles without beginning with a discussion of the means of characterizing the generation of lateral forces by tires [2, 3]. On the other hand, it is possible to introduce some of the basic ideas of vehicle dynamics if a ground vehicle can be idealized in such a way that the tire characteristics do not have to be described in any detail. In the introductory example of an idealized shopping cart, we replace actual tire characteristics with simple kinematic constraints. This idealization actually works quite well for the hard rubber tires typically used on shopping carts—up to the point at which the wheels actually are forced to slide sideways. This type of idealization allows one to focus on the dynamic model of the vehicle itself without worrying about details of how the tires generate side forces.

Most courses in dynamics mainly discuss inertial coordinate systems when applying the laws of mechanics to rigid bodies. This approach is logical when studying some vehicle types, but there is another way to write equations for freely moving vehicles such as automobiles and airplanes that turns out to be simpler and is commonly used. This second description of vehicle motion involves the use of a coordinate system attached to a body and moving with it. This coordinate system describes the motion of a vehicle as it would appear to an observer riding in the vehicle rather than to an observer on the ground. In this shopping cart example, the two ways of writing dynamic equations are presented as they apply to the special case of plane motion.

Most supermarket shopping carts have casters in the front that are supposed to swivel so that the front can be pushed easily in any direction, and have a back axle with fixed wheels that are supposed to roll easily and to resist sideways motion. Anyone who has actually used a shopping cart realizes that real carts often deviate significantly from these ideals. The casters at the front often do not pivot well, so the cart is hard to turn, and one is often forced to skid the back wheels to make a sharp turn. Furthermore, the wheels often don't roll easily, so quite a push is required to keep the cart moving. On the other hand, a mathematically ideal cart serves as a good introduction to a study of ground vehicles in general.

The interaction of shopping cart tires with the ground will be highly simplified. We assume that the front wheels generate no side force at all because of the casters. In fact, the front of a cart can be moved many direction with no force at all because the casters are supposed to swivel without friction and the wheels are also assumed to roll without friction. At the rear, the wheels are assumed to roll straight ahead in the direction in which the cart is headed, without friction, and the rear wheels are assumed to allow no sideways motion at all. Obviously, if the rear wheels are pushed sideways hard enough, they will slide sideways, but if the side forces are small enough, real wheels roll almost exactly in the direction in which they are pointed.

Inertial Coordinate System

In the first analysis we consider the motion of a shopping cart to be described in an x–y inertial reference frame. Figure 5.1 shows a view of a cart seen from above. The

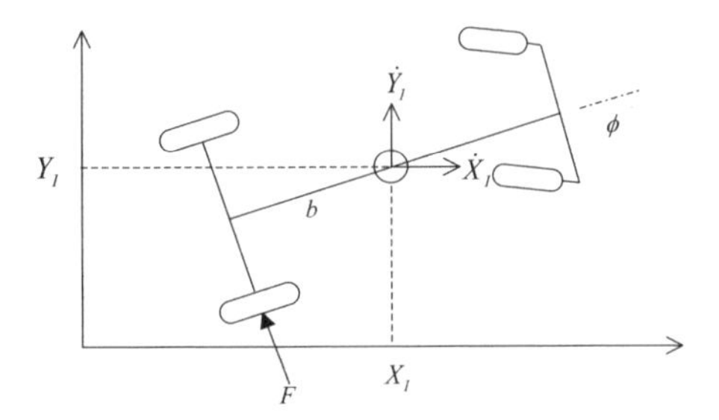

Figure 5.1 Shopping cart described in an inertial coordinate frame.

coordinates X_I and Y_I locate the center of mass of the cart with respect to the ground, and ϕ is the angle of the cart centerline with respect to a line on the ground parallel to the X_I-axis. The X_I–Y_I axes are supposed to be neither accelerating nor rotating, and thus they are an inertial frame in which Newton's laws are easily written.

In what follows, there is a distinction between physical constants such as dimensions and inertial constants and variables that describe how the motion evolves in time. We will call the physical constants *parameters*. In Figure 5.1, the parameters of the cart are a and b, the distance from the center of mass to the front and rear axles, respectively, m the mass, and I_z the moment of inertia about the mass center with respect to the vertical axis. A *basic motion* that the cart could execute is to move along the x-axis with constant speed U.

The dynamic analysis of the shopping cart will concentrate on the *stability* of the cart motion. The question to be answered is whether a cart moving initially in a straight line will continue in essentially the same path after a small disturbance, which tends to move it off the original path. The disturbance could be a small force or unevenness in the ground that temporarily causes a minor change in the motion. For the analysis of stability, we consider a small *perturbation* of the basic motion [3]. The *perturbed motion* is described by the variables $Y_I(t)$ and $\phi(t)$ shown in the figure. [In principle, $X_I(t)$ is also a variable, but for the basic and the perturbed motion, it is assumed that $\dot{x} \cong U$, a constant, so $X_I(t) \cong U_t +$ constant. Thus, in the analysis the speed U will play the role of a parameter rather than that if a variable.] In the basic motion, $\dot{Y}_I$, ϕ, and $\dot{\phi}$ all vanish and the cart proceeds along the x-direction at constant speed U.

For the perturbed motion, $\dot{Y}_I$ is assumed to be small compared with U, and ϕ and $\dot{\phi}$ are also small. (To be precise, $b\dot{\phi}$ is small compared to U.) This means that the small-angle approximations $\cos\phi \cong 1$ and $\sin\phi \cong \phi$ can be used when considering the perturbed motion. If the perturbed motion turns out to be unstable, ϕ will not remain small and after a time these approximations will not be valid. The side force at the rear is F, and since the rear wheels are assumed to roll without friction, there is no force in the direction the rear wheels are rolling. At the front axle there is no force at all in the x–y plane because of the casters. (There are, of course, vertical forces at both axles that are necessary to support the weight of the cart, but they will play essentially no role in the analysis of the lateral dynamics of the cart.)

Because the cart is described in an inertial coordinate system and is executing plane motion, three equations of motion are easily written: Force equals mass times acceleration in two directions, and moment equals rate of change of angular momentum around the z-axis. Using the fact that the angle ϕ is assumed to be very small for the perturbed motion, the equations of motion are

$$m\ddot{X}_I = -F\sin\phi \cong -F\phi \cong 0 \tag{5.1}$$

$$m\ddot{Y}_I = F\cos\phi \cong F \tag{5.2}$$

$$I_z\ddot{\phi} = -Fb \tag{5.3}$$

Because of the assumption that ϕ is small, we assume that $\sin\phi \approx \phi$ and $\cos\phi \approx 1$. In Eq. (5.1), the acceleration in the X_I-direction is seen not to be exactly zero. However, we consider that the forward velocity is initially large and does not change much as long as the angle ϕ remains small. From Eqs. (5.2) and (5.3) we see that the acceleration in the Y_I-direction and the angular acceleration are large compared to the X_I-direction acceleration as long as ϕ remains small.

Considering the velocity of the center of the rear axle, one can derive the condition of zero sideways velocity for the rear wheels. The basic kinematic velocity relation for two points on the same rigid body (see Chapter 1) is

$$\mathbf{v}_p = \mathbf{v}_0 + \boldsymbol{\omega} \times \mathbf{r}_{0p} \tag{5.4}$$

where $\boldsymbol{\omega}$ is the angular velocity of the body and $\mathbf{r}_{0p}$ is the vector distance between points 0 and p. In the case at hand, let $\mathbf{v}_0$ represent the velocity of the mass center and let $\mathbf{v}_p$ represent the velocity of the center of the rear axle. Then $\boldsymbol{\omega}$ is a vector in the z-direction perpendicular to the x–y plane. The magnitude of $\boldsymbol{\omega}$ is $\dot{\phi}$. The magnitude of $\mathbf{r}_{0p}$ is just the distance b between the center of gravity and the center of the rear axle.

Figure 5.2 shows the velocity components involved when Eq. (5.4) is evaluated. The component $b\dot{\phi}$ represents the term $\boldsymbol{\omega} \times \mathbf{r}_{0p}$ in Eq. (5.4). From Figure 5.2 one can see that the side velocity (with respect to the centerline of the cart) at the center of the rear axle is

$$\dot{Y}_I\cos\phi - b\dot{\phi} - U\sin\phi \tag{5.5}$$

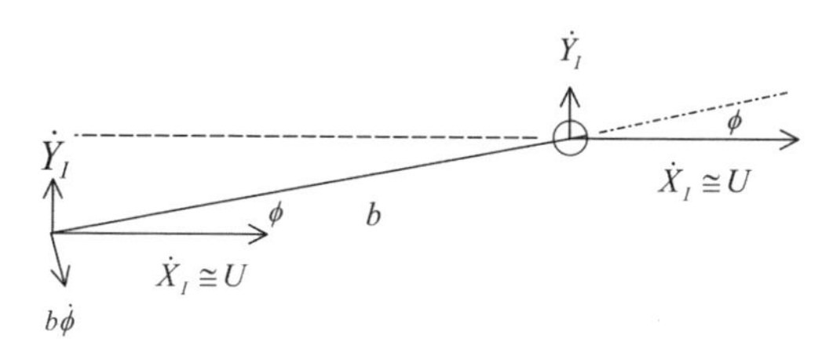

Figure 5.2 Velocity components at the center of the rear axle.

Using the small-angle approximations for the cosine and the sine, the final relation needed is simply a statement that this sideways velocity of the center of the rear axle should vanish:

$$\dot{Y}_I - b\dot{\phi} - U\phi = 0 \tag{5.6}$$

[It is true, but perhaps not completely obvious at first, that if the two rear wheels cannot move sideways but can only roll in the direction in which they are pointed instantaneously, any point on the rear axle also cannot have any sideways velocity. This means that the kinematic constraint of Eq. (5.6) derived for the center of the axle properly constrains the variables such that the two wheels also have no side velocity.]

Now, combining Eqs. (5.2) and (5.3), one can eliminate F, yielding a single dynamic equation. Then, after differentiating Eq. (5.6) with respect to time, $\ddot{Y}_I$ can be eliminated from this dynamic equation to yield a single equation for ϕ. The result is

$$\left(I_z + mb^2\right)\ddot{\phi} + mbU\dot{\phi} = 0 \tag{5.7}$$

Equation (5.7) may appear to be of second order since it involves a term containing $\ddot{\phi}$, but because ϕ itself is missing, a first-order version of Eq. (5.7) can be studied instead.

The angle ϕ actually has no particular significance. It is just the angle the cart makes with the x-axis, which itself is a line on the ground in an arbitrary direction. Since the angle ϕ itself does not appear in Eq. (5.7), it is logical to consider the angular rate $\dot{\phi}$ as the basic variable rather than ϕ. The angle ϕ is called the *yaw angle*, and it is common to call the angular rate $\dot{\phi}$ the *yaw rate* and to give it the symbol r in vehicle dynamics. Standard symbols commonly used in vehicle dynamics are presented in Appendix A.

In terms of yaw rate, the Eq. (5.7) can be rewritten in first-order form. Noting that $\dot{\phi} = r, \ddot{\phi} = \dot{r}$, the result, after dividing out the term $I_z + mb^2$, is

$$\dot{r} + \frac{mbU}{I_z + mb^2} r = 0 \tag{5.8}$$

This equation is a linear ordinary differential equation with constant coefficients and is of the general form

$$\dot{r} = Ar, \quad \text{with } A = -mbU/\left(I_z + mb^2\right) \tag{5.9}$$

The assumption that it is possible for stability analysis to consider small perturbations from a basic motion generally leads to linear differential equations. In the present case, the nonlinear equations, Eqs. (5.1), (5.2), and (5.5), became linear because of the small perturbation assumption.

The analysis of the stability of the first-order equation, Eq. (5.9), is elementary. The solution to this linear equation can be assumed to have an exponential form such as $r = \mathrm{R}e^{st}$. Then $\dot{r} = s\mathrm{R}e^{st}$, which when substituted into Eq. (5.9), yields $s\,\mathrm{R}e^{st} = A\mathrm{R}e^{st}$ or

$$(s - A)\mathrm{R}e^{st} = 0 \tag{5.10}$$

This result is the basis of an *eigenvalue* analysis. For a complete discussion of eigenvalue analysis see, for example, [3]. For now it is enough to note that when three factors must multiply to zero, as in Eq. (5.10), the equation will be satisfied if any one of the three factors vanishes. For example, if R were to be zero, the product in Eq. (5.10) would certainly be zero. This represents the *trivial solution*. The assumed solution would then simply imply that $r(t) = 0$, which should have been an obvious possibility from the beginning. Another possible factor to vanish in Eq. (5.10) is e^{st}. Not only is the vanishing of this factor not really possible, but also even if it were, the result would again be the trivial solution. The only important condition is the vanishing of the final factor in Eq. (5.10), which happens when

$$s = A \tag{5.11}$$

Thus, we have determined that the quantity s takes on the value A, which is called the *eigenvalue*. The only nontrivial solution has the form

$$r = \mathrm{R}e^{At} \tag{5.12}$$

in which R is an arbitrary constant determined by the initial value of r at time $t = 0$.

Figure 5.3 shows the nature of the solutions for the cases when A is greater than and less than zero. When A is negative, the system is stable and the yaw rate will return to zero after a disturbance. If A is positive, the yaw rate increases in time, the cart begins to spin faster and faster, and the system is unstable. In the case of the shopping cart with

$$A = \frac{-mbU}{I_z + mb^2} \tag{5.13}$$

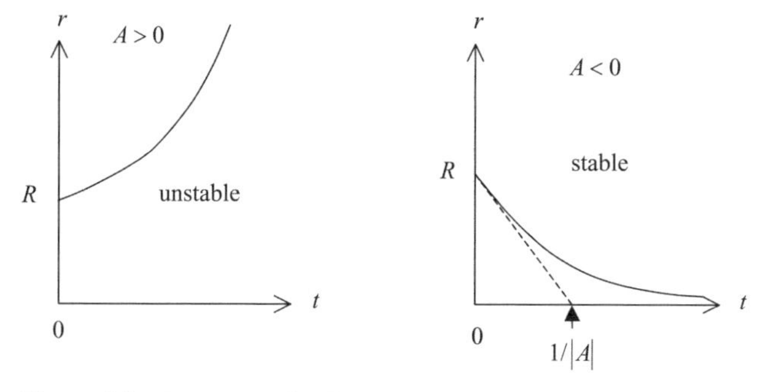

Figure 5.3 Responses for first-order unstable and stable systems.

it is clear that A will normally be negative since all the parameters are positive and there is a negative sign in front of the expression. The motion will then be stable as long as U is positive, as shown in Figure 5.2, since all other parameters are inherently positive. This means that if a shopping cart is given a push in the normal direction but with an initial yaw rate, the yaw rate will decline toward zero exponentially as time goes on. The cart will roll with an ever-decreasing yaw rate until ultimately, it is moving in essentially a straight line. In this case, the shopping cart moving in the forward direction is considered to be *stable*.

On the other hand, Eq. (5.6) remains valid whether U is considered to be positive in the sense shown in Figure 5.2 or negative, and the same is true of Eq. (5.13). (The proof of this statement requires a redrawing of Figure 5.2.) Thus, if the cart is pushed backward, one can simply consider U to be negative in the relations derived so far. In this case, the combined parameter A in Eq. (5.13) will be positive, and if the yaw rate is given any initial value, however small, the yaw rate will increase exponentially in time. For backward motion, the cart is then *unstable*. In fact, the backward-moving cart will eventually turn around and travel in the forward direction even though the linearized dynamic equation we have been using cannot predict this, since the small-angle approximations are no longer valid after the cart has turned through a large angle.

Another interesting aspect of first-order equation response is the speed with which the stable version returns to zero. If the response is written in the form

$$r = \mathrm{Re}^{-t/\tau} \tag{5.14}$$

where τ is defined to be the *time constant*

$$\tau = \frac{1}{|A|} \tag{5.15}$$

it is clear that the time constant is just the time at which the initial response decays to $1/e$ times the initial value for the stable case. For the shopping cart,

$$\tau = \frac{I_z + mb^2}{mbU} \tag{5.16}$$

Figure 5.3 shows that the time constant is readily shown on a plot of the response in the stable case by extending the slope of the response from an initial point (assuming that $t = 0$ at the initial point) to the zero line. The time constant is the time when the extended slope line intersects the zero line.

From Eq. (5.16) it can be seen that the time constant depends on the physical parameters of the cart and the speed. Increasing the speed decreases the time constant and increases the quickness with which the raw rate declines toward zero for the stable case. Pushing the cart faster in the backward direction also quickens the unstable growth of the yaw rate.

Body-Fixed Coordinate System

The use of an inertial coordinate system to describe the dynamics of a vehicle may seem reasonable at first, but in fact many analyses of vehicles use a moving coordinate system attached to the vehicle itself. The vehicle motion is described by considering linear and angular velocities rather than positions. (In the previous stability analysis of the shopping cart, the angular position of the cart turned out to be less important than the angular rate.) To introduce this idea we now repeat the previous stability analysis using a coordinate system attached to the center of mass of the cart and rotating with it. This will require a restatement of the laws of dynamics, taking into consideration the noninertial moving coordinate system. The general case of rigid-body motion described in a coordinate system attached to the body itself and executing three-dimensional motion is presented in Appendix A. Here we present the simpler case of plane motion appropriate for the shopping cart.

Figure 5.4 shows the moving coordinate system and the velocity components associated with it. The x–y–z coordinate system is now attached to and moves with the cart. The view of the cart is from above and the z-axis points down, so that the x–y–z system is right-handed. The side force, now called Y, at the rear axle always points exactly in the y-direction, and again, there is no side force at the front axle. In this description of the motion, the basic motion consists of a constant forward motion with velocity U, which again will function as a parameter. The motion variables for the cart are now the side or lateral velocity V, the angular velocity, and the yaw rate r. This notation is in conformity with the general notation presented in Appendix A. For the basic motion, the side velocity and angular velocity are zero, $V = r = 0$. The perturbed motion has a small lateral velocity and yaw rate $V \ll U$, $r \ll U/b$. Again, the parameters of the cart are a, b, m, I_z, and U.

We are now in a position to write equations relating the force to the mass times the acceleration of the center of mass and the moment to the change of angular momentum. Because the x–y–z axis is rotating with the body, we must find the absolute acceleration in terms of components in a rotating frame. The fundamental way to find the absolute rate of change of any vector $\mathbf{A}$, measured in a frame rotating with angular velocity $\boldsymbol{\omega}$

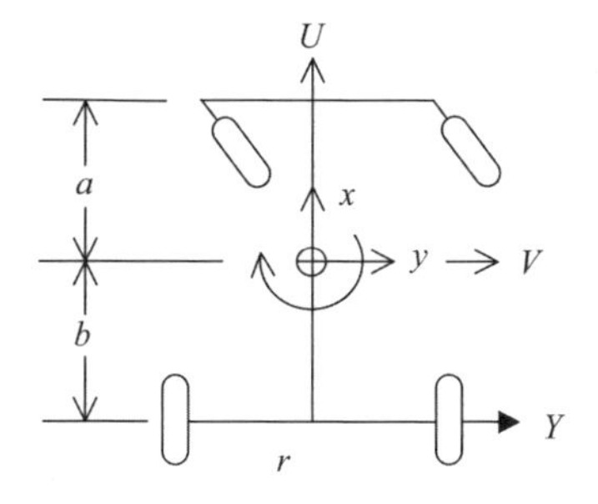

x, y, z coordinates are attached at the center of mass and move with the cart

U,V are velocity components with respect to x, y

r is the yaw rate (angular velocity around the z-axis)

Y is the force on the rear axle

Figure 5.4 Coordinate system attached to the shopping cart.

is given by the formula (see Chapter 1)

$$\frac{d\mathbf{A}}{dt} = \left.\frac{d\mathbf{A}}{dt}\right|_{\text{rel}} + \boldsymbol{\omega} \times \mathbf{A} \tag{5.17}$$

The first term on the right of Eq. (5.17) represents the rate of change of the vector as seen in the coordinate system itself, and the second term corrects for the effects of the rotation of the coordinate system. In the case at hand, we are interested in the acceleration, so **A** represents the velocity with components U and V, and the only component of $\boldsymbol{\omega}$ is the z-component, r. In Appendix A, general formulas for accelerations in vehicle-centered coordinate systems are derived, but for now the relevant expressions resulting from use of the general formula will simply be stated. (See Problem 5.1 for a derivation of these equations.) The acceleration components in the x- and y-direction upon application of Eq. (5.17) are

$$a_x = \dot{U} - rV \tag{5.18}$$

$$a_y = \dot{V} + rU \tag{5.19}$$

Since the side force Y is the only force on the body, the equations of motion then are

$$m(\dot{U} - rV) = 0 \tag{5.20}$$

$$m(\dot{V} + rU) = Y \tag{5.21}$$

$$I_z\dot{r} = -bY \tag{5.22}$$

where the moment about the center of mass in the positive z-direction is due only to the force on the back axle and the moment is negative when the force Y is positive, as shown in the sketch. These equations have the same meaning as Eqs. (5.1)–(5.3), but they are expressed in the body-fixed coordinate system attached to the shopping cart. From Eq. (5.20), one can see that even when the x-force is zero, the forward velocity U is not really constant. In fact, $\dot{U} = rV$. For the basic motion, both r and V are zero and U is exactly constant. For the perturbed motion, both V and r are assumed to be small, $\dot{U} \cong 0$, and U is nearly, but not exactly, constant.

To complete the analysis, we still have to put in the fact that the side velocity at the rear axle is zero. Using the same kinematic law as used previously, Eq. (5.4), the side velocity at the center point of the rear axle is the side velocity of the center of mass plus the side velocity due to rotation. Considering Fig. 5.4, the side velocity is $V - br$ and the constraint that this velocity is assumed to vanish is

$$V - br = 0 \tag{5.23}$$

This relation corresponds to Eq. (5.6) in the previous formulation using the inertial coordinate system. (Once again it can be verified that if the center point of the axle is constrained to have no side velocity, there is no side velocity at either wheel attached to the axle.) Using a body-fixed coordinate system there is no need to assume that the yaw angle is small as was required to derive Eq. (5.6). In fact, the yaw angle need not enter the analysis at all with a body-fixed coordinate system.

Now, considering U to be essentially constant, Eqs. (5.21) and (5.22) and the zero side velocity relation, Eq. (5.23), form a set of three equations for the three variables V, r, and Y. It is left as an exercise to show that if V and Y are eliminated from these three equations, the result is exactly the same equation of motion as before, Eq. (5.8). Thus, use of the coordinate system attached to the cart produces the same equation of motion as was produced using the inertial coordinate system. (It would be a great problem for Newtonian mechanics if two alternative ways to set up an equation of motion for a system did not produce an equivalent, if not identical equation.) The stability analysis is then identical for vehicle motion whether an inertial reference frame or a moving frame is used.

It turns out that the use of a coordinate system moving with the vehicle is in many cases the most convenient way to describe the dynamics of vehicles. Freely moving vehicles such as cars and airplanes are often described with the help of a body-fixed coordinate system such as the one just used for the shopping cart. Since this type of coordinate system is rarely discussed in basic dynamics texts, it is worthwhile to devote some time to a study of the general equations of motion for a rigid body using this type of coordinate system, presented in Appendix A.

Connection between Inertial and Body-Fixed Frames

As has been demonstrated above, body-fixed coordinate systems deal only with linear and angular velocities and accelerations and not with positions. In contrast, inertial coordinate systems start with position variables which have as their time derivatives, velocities and accelerations. It is possible, however, to use body-fixed coordinates and yet also find the vehicle position in a computer simulation. Figure 5.5 shows both inertial and body-fixed coordinate system velocity components. In a computer

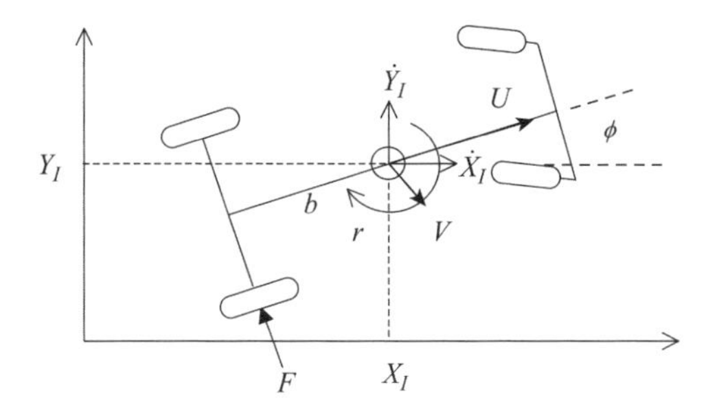

Figure 5.5 Inertial and body-fixed velocity components.

simulation, equations of motion such as Eqs. (5.20)–(5.22) could be used to find the time histories of U, V, and r given the forces in the x- and y-direction and the torque around the z -axis. This may be enough for some purposes, but if it was also desired to find the position in inertial space, then three more equations can be integrated to determine the vehicle position as it was described in Figure 5.1.

$$\dot{X}_I = U\cos\phi + V\sin\phi \tag{5.24}$$

$$\dot{Y}_I = U\sin\phi - V\cos\phi \tag{5.25}$$

$$\dot{\phi} = -r \tag{5.26}$$

Equations (5.24) and (5.25) are simply relations among velocity components in the two coordinate frames. Equation (5.26) is a statement that the yaw rate is just the rate of change of the yaw angle. The minus sign in this equation is necessary because the yaw angle is positive in the counterclockwise sense in Figure 5.1, but the yaw rate is positive in the clockwise direction in Figure 5.4.

With these additional three equations, the body-fixed frame simulation would consist of a total of six first-order equations. This corresponds to the three second-order equations in the inertial frame: for example, Eqs. (5.1)–(5.3), which could also yield the vehicle position in simulation. The shopping cart example showed that for the analysis of stability, the position variables were not needed and that the body-centered coordinate system was then simpler to use than the inertial coordinate system. When some forces are related to *positions*, the simplicity of body-centered coordinate formulations may be lost (see e.g., Problem 5.14).

5.2 ANALYSIS OF A SIMPLE CAR MODEL

Figure 5.6 shows the *bicycle model* of an automobile, in which only a single equivalent front wheel and a single equivalent rear wheel are shown together with the body-centered coordinate system discussed in Section 5.1. (This model cannot be used to model an actual bicycle since it does not allow for a tilt angle.) The model is actually used for three- or four-wheeled vehicles with the assumption that the effect of the roll angle can be neglected. The idea that a single equivalent wheel can be used to represent the wheels at the front and rear axles is based on the observation that in some cases,

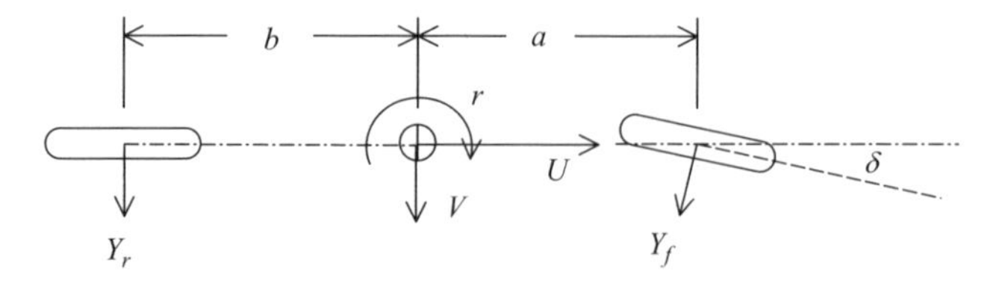

Figure 5.6 Body-centered coordinate system for a bicycle automobile model.

the forward and lateral velocities for the two wheels at an axis are almost identical. This is the case, for example, when the vehicle is traveling at fairly high speed in turns in which the turn radius is considerably longer than the vehicle wheelbase. When this assumption is justified, the bicycle model is a useful simplification and it is often used to study the stability and steering behavior of ground vehicles. A feature not needed for the shopping cart is the steer angle δ, shown at the front in Figure 5.6. This feature is not needed for stability analysis of straight-line motion but is included for later use when steering behavior is discussed. For straight-line stability analysis, the steer angle δ will simply be assumed to be zero.

Again, U and V are velocity components of the center of mass with respect to the body coordinates x and y that are attached to the vehicle body and moving with it. Once again, it will be assumed that the velocity in the forward x -direction, U, is strictly constant for the basic motion and is essentially constant for the perturbed motion, so it is a parameter rather than a variable. The two variables needed to describe the motion are V, the velocity of the center of mass in the y-direction, and r, the angular velocity or the yaw rate about the z-axis, just as they were for the shopping cart.

The lateral tire forces are denoted Y_r and Y_f, although the front lateral tire force does not point strictly in the y-direction if the front wheel is steered as shown. These definitions are in conformity with the general notation introduced in Appendix A. The parameters of the vehicle are a and b, the distances from the center of mass to the axles, m is the mass, and I_z is the centroidal moment of inertia about the z-axis. The equations of motion are specialized from the general equations of Appendix A and are similar to Eqs. (5.21) and (5.22). It is assumed that the steer angle δ is small, so that $\cos\delta \cong 1$. The equation

$$m(\dot{V} + rU) = Y_f + Y_r \tag{5.27}$$

determines the lateral acceleration of the center of mass, and the equation

$$I_z\dot{r} = aY_f - bY_r \tag{5.28}$$

determines the angular acceleration.

The next step is to determine how the front and rear lateral forces are related to the vehicle motion. The detailed characterization of pneumatic tires can be quite complex [2, 3], but here we use the concept of a *cornering coefficient*, which is often used in studies of ground vehicle stability. The idea is that when tires are under a lateral load, they develop a small lateral velocity in addition to their forward or rolling velocity. When the lateral loads are small relative to the normal loads on the tire, there is a small angle between the velocity of the center of the wheel and the direction the wheel is pointing. This angle is called the *slip angle*, although it is due fundamentally to distortion of the tire rather than actual slipping. The assumption we will use is that the lateral forces are related to the slip angle through a constant called the *cornering coefficient*, which we consider a parameter for the purposes of stability analysis.

The calculation of the slip angles is accomplished with the help of the sketches shown in Figure 5.7. Note that β is a sort of slip angle for the center of gravity of the

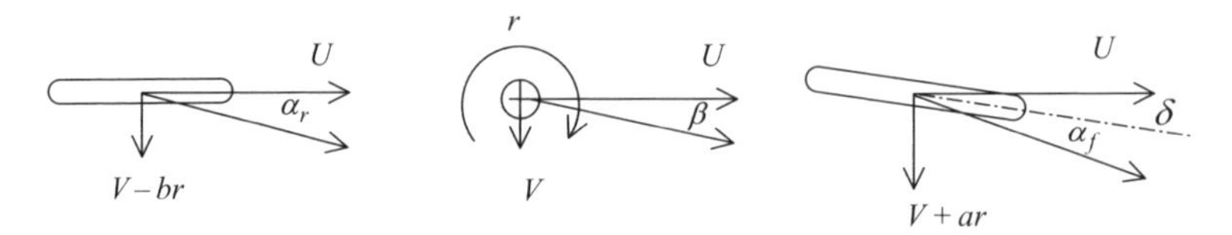

Figure 5.7 Velocity components for slip angle calculations.

vehicle itself. When cornering, a car does not generally go in exactly the direction it is pointed, and thus $\beta \cong V/U$ is not generally zero. The velocities at the wheels are found by adding the velocity of the center of mass to components induced by the angular velocity r in the fashion of Section 5.1 and as discussed in Chapter 1. In this case, the extra components, ar and br, are easily seen to add as shown in Figure 5.7.

Using the sketches and assuming small lateral velocities and thus small angles, the slip angles at the front and rear, α_f and α_r, are readily expressed as the ratios of the lateral velocities (with respect to the wheel pointing direction) to the rolling velocities.

$$\alpha_r = \frac{V - br}{U}, \qquad \alpha_f = \frac{V + ar}{U - \delta} \tag{5.29}$$

Note how the steer angle δ affects the front slip angle. The car is steered by changing the steer angle. This changes the slip angle, and when the slip angle is changed, the lateral force at the front wheels changes and begins the process of changing the direction in which the car moves. Comparing Figures 5.6 and 5.7, it may become clear that if the slip angles are positive as defined by Figure 5.7 and Eqs. (5.29), the forces *on the vehicle from the ground* would actually be in the negative Y-direction. This is a common occurrence in vehicle dynamics and arises simply when, as in Figures 5.6 and 5.7, both lateral velocities and the corresponding lateral forces on the body are defined as positive in the positive Y-direction.

We assume that cornering coefficients C_f and C_r relate lateral forces on the front and rear axles to the slip angles at the front and rear axles. This is a linearization of the tire force laws valid only for small lateral forces and slip angles [3]. For a four-wheeled car, the cornering coefficients represent the combined effects of the two front and two rear tires. The cornering coefficients for the front and rear axles are not simple to estimate because they depend on many factors, such as the normal forces at the wheels, the nature of the surface of the ground, and the particular tires used at the front and rear. Since the purpose here is to illustrate the use of rigid-body dynamics using body-fixed frames, the detailed description of lateral force generation is left to the many reference books on ground vehicle dynamics (see, e.g., [2] or [3]).

Since it is preferable to use positive cornering coefficients rather than negative ones, the linearized force laws will be written with a built-in negative sign so that a positive slip angle produces a negative force:

$$-Y_f = C_f\alpha_f, \qquad -Y_r = C_r\alpha_r \tag{5.30}$$

where C_f and C_r are the cornering coefficients at the front and rear axles, respectively, with units of N/rad. For a particular situation, these coefficients can be considered to be constant parameters. Combining Eqs. (5.27)–(5.30), the equations of motion then become

$$m(\dot{V} + rU) = -\frac{(C_f + C_r)V}{U} - \frac{(aC_f - bC_r)r}{U} + C_f\delta \tag{5.31}$$

$$I_z\dot{r} = -\frac{(aC_f - bC_r)V}{U} - \frac{\left(a^2C_f + b^2C_r\right)r}{U} + aC_f\delta \tag{5.32}$$

These two equations can be put into several possible matrix forms. For example,

$$\begin{bmatrix} m & 0 \\ 0 & I_z \end{bmatrix}\begin{bmatrix} \dot{V} \\ \dot{r} \end{bmatrix} + \begin{bmatrix} \dfrac{C_f + C_r}{U} & \dfrac{aC_f - bC_r}{U} + mU \\ \dfrac{aC_f - bC_r}{U} & \dfrac{a^2C_f + b^2C_r}{U} \end{bmatrix}\begin{bmatrix} V \\ r \end{bmatrix} = \begin{bmatrix} C_f \\ aC_f \end{bmatrix}\delta \tag{5.33}$$

or, by dividing out the inertia coefficients,

$$\begin{bmatrix} \dot{V} \\ \dot{r} \end{bmatrix} = -\begin{bmatrix} \dfrac{C_f + C_r}{Um} & \dfrac{aC_f - bC_r}{Um} + U \\ \dfrac{aC_f - bC_r}{UI_z} & \dfrac{a^2C_f + b^2C_r}{UI_z} \end{bmatrix}\begin{bmatrix} V \\ r \end{bmatrix} + \begin{bmatrix} \dfrac{C_f}{m} \\ \dfrac{aC_f}{I_z} \end{bmatrix}\delta \tag{5.34}$$

Equation (5.34) is in a form that could be used in a computer simulation to study how the vehicle would respond to steer angle inputs as outlined in Chapter 3.

5.3 VEHICLE STABILITY

Linearized dynamic equations are often used to determine vehicle stability properties. As an example, Eq. (5.33) will be used to determine the stability of a vehicle when running straight ahead. For the *basic motion* many variables vanish, $\delta = V = r = Y_f = Y_r = \alpha_f = \alpha_r = 0$, and U is strictly constant. For the perturbed motion, we assume that $\delta = 0$, that U is essentially constant, and that the variables $V(t)$ and $r(t)$ take on small enough values that the slip angles remain small. Then Eq. (5.33) becomes

$$\begin{bmatrix} m & 0 \\ 0 & I_z \end{bmatrix}\begin{bmatrix} \dot{V} \\ \dot{r} \end{bmatrix} + \begin{bmatrix} \dfrac{C_f + C_r}{U} & \dfrac{aC_f - bC_r}{U} + mU \\ \dfrac{aC_f - bC_r}{U} & \dfrac{a^2C_f + b^2C_r}{U} \end{bmatrix}\begin{bmatrix} V \\ r \end{bmatrix} = \begin{bmatrix} 0 \\ 0 \end{bmatrix} \tag{5.35}$$

Now an eigenvalue analysis is started by assuming that the two variables have an exponential form as a function of time:

$$\begin{bmatrix} V \\ r \end{bmatrix} = \begin{bmatrix} \overline{V} \\ \overline{r} \end{bmatrix} e^{st} \tag{5.36}$$

where $\overline{V}$ and $\overline{r}$ are assumed to be unknown constants and s is a generally complex number that will take on values called *eigenvalues* that will determine whether the vehicle is stable or unstable.

From Eq. (5.36) it is seen that the derivatives are also exponential functions.

$$\begin{bmatrix} \dot{V} \\ \dot{r} \end{bmatrix} = \begin{bmatrix} \overline{V} \\ \overline{r} \end{bmatrix} s e^{st} \tag{5.37}$$

If Eqs. (5.37) and (5.34) are substituted into Eq. (5.35), the result is

$$\begin{bmatrix} ms + \dfrac{C_f + C_r}{U} & \dfrac{aC_f - bC_r}{U} + mU \\ \dfrac{aC_f - bC_r}{U} & I_z s + \dfrac{a^2 C_f + b^2 C_r}{U} \end{bmatrix} \begin{bmatrix} \overline{V} \\ \overline{r} \end{bmatrix} e^{st} = \begin{bmatrix} 0 \\ 0 \end{bmatrix} \tag{5.38}$$

The present eigenvalue analyses is a slightly more complex version of the analysis of the first-order equation for the shopping cart, Eq. (5.10). Equation (5.38) is a linear algebraic equation set for two unknowns with the trivial solution that both V and r could be zero.

A fundamental theorem of algebra states that the solution of such a set of linear algebraic equations is unique if the determinant of the coefficient matrix is *not* zero. Since we are interested in solutions such as Eq. (5.36) which are *not* the trivial zero solution, we look for values of s for which the determinant vanishes. Such values of s are called *eigenvalues* and they determine the stability or instability of the differential equations (see [3] or [4]). The condition to be imposed is

$$\det \begin{bmatrix} ms + \dfrac{C_f + C_r}{U} & \dfrac{aC_f - bC_r}{U} + mU \\ \dfrac{aC_f - bC_r}{U} & I_z s + \dfrac{a^2 C_f + b^2 C_r}{U} \end{bmatrix} = 0 \tag{5.39}$$

When the determinant is written out and set equal to zero, the *characteristic equation*, which is the equation that determines the eigenvalues, emerges explicitly. (Evaluating the determinant is something of an algebraic challenge even though it involves only a 2×2 matrix.) The result of evaluating the determinant in Eq. (5.39) is

$$mI_z s^2 + \left[\frac{m(a^2C_f + b^2C_r)}{U} + \frac{I_z(C_f + C_r)}{U} + maC_f G\right]s$$
$$+ \frac{(C_f + C_r)(a^2C_f + b^2C_r)}{U^2} - \frac{(aC_f - bC_r)^2}{U^2} - m(aC_f - bC_r) = 0 \tag{5.40}$$

If the terms in Eq. (5.40) not involving s are combined algebraically, the characteristic equation becomes a little simpler:

$$mI_z s^2 + \left[\frac{m(a^2C_f + b^2C_r)}{U} + \frac{I_z(C_f + C_r)}{U}\right]s$$
$$+ \frac{(a+b)^2 C_f C_r}{U^2} + m(bC_r - aC_f) = 0 \tag{5.41}$$

The two values of s that satisfy Eq. (5.41) (the eigenvalues) will generally be complex numbers with real and imaginary parts. The basic idea of stability for linear systems is easily stated: If *all* the eigenvalues of a system have negative real parts, the system is stable. If one or more of the system eigenvalues has a positive real part, the system is unstable. If a stable system is perturbed, then the variables will eventually decay to zero. In contrast, after a perturbation, the variables of an unstable system will grow with time, and eventually the linearized equations of motion will no longer be a valid representation of the system dynamics.

Equation (5.41) is a second-order characteristic equation that will have two eigenvalues. (An nth-order system will have a characteristic involving s^n and n eigenvalues.) Although Eq. (5.41) has complicated coefficients, it can be compared to the characteristic equation for a mass–spring–damper system. If m, b, and k are the mass, damping, and spring parameters for such a system, the dynamic equation of motion is of the form

$$m\ddot{x} + b\dot{x} + kx = 0 \tag{5.42}$$

and after assuming an exponential form for x, the characteristic equation is

$$ms^2 + bs + k = 0 \tag{5.43}$$

This equation is of the same form as Eq. (5.41) but much easier to understand, since the coefficients are not complicated functions of many parameters.

If m, b, and k are all positive, it is intuitively obvious that the system represents a damped oscillator and is stable. Since Eq. (5.43) is only second order, the quadratic formula can be used to show that the eigenvalues will indeed have negative real parts under this condition. On the other hand, a negative damping coefficient or negative spring constant will make the system unstable and lead to positive real parts for the eigenvalues. (Although physically the mass cannot be negative, the linearized values for b and k can, under some circumstances, be negative.)

A close inspection of Eq. (5.41) shows that the coefficients of s^2 and s—corresponding to the mass and damping coefficient in Eq. (5.43)—are inherently positive but the last coefficient—corresponding to the spring constant in Eq. (5.43)—could possibly be negative. Thus, the criterion for stability of the vehicle can be written

$$\frac{(a+b)^2 C_f C_r}{U^2} + m(bC_r - aC_f) > 0 \tag{5.44}$$

If the inequality in Eq. (5.44) is true, the car is stable; if not, it is unstable.

5.4 STABILITY, CRITICAL SPEED, UNDERSTEER, AND OVERSTEER

Several interesting facts can be seen easily from the simplified stability criterion, Eq. (5.44). First, if the speed is sufficiently low, the first term in Eq. (5.44) involving $1/U^2$ will be a large enough positive number that the criterion will be satisfied for any parameter set. This confirms the obvious fact that all automobiles are stable at very low speeds. (Early automobiles that were only capable of very low speeds did not need to be analyzed for stability and handling properties. Later, as engines became more powerful, allowing high speeds, vehicle dynamics and stability became important areas of vehicle design.)

Second, if the second term in Eq. (5.44) is positive, the car will surely be stable at any speed. The second term is positive when

$$bC_r > aC_f \tag{5.45}$$

This condition is described by the term *understeer*. Clearly, this term relates to the steering properties of the car. For now, all we know is that an understeering vehicle is stable even at high speeds, where the $1/U^2$ term in the stability criterion, Eq. (5.44), becomes small.

Finally, if

$$bCr < aC_f \tag{5.46}$$

we see that the second term in the stability criterion, Eq. (5.44), is negative. This means that if the speed U is gradually increased from zero, the positive first term will decrease to a point at which the first term will just balance the negative second term. Above this speed, called the *critical speed*, U_{crit}, the stability criterion will no longer be satisfied and the car will become unstable. The critical speed is determined by equating the positive and negative terms in Eq. (5.44) and solving for the speed:

$$U_{\text{crit}}^2 = \frac{-(a+b)^2 C_f C_r}{m(bC_r - aC_f)} \tag{5.47}$$

Note that this expression yields a positive number for U_{crit}^2 because we assume that $bC_r < aC_f$. This situation, described by the term *oversteer*, means not only that the car is unstable for speeds greater that the critical speed but also that the speed drastically affects the steady-state cornering behavior of the vehicle. Steering behavior is studied in the next section.

5.5 STEERING TRANSFER FUNCTIONS

In the preceding section, the stability of an automobile traveling in a straight path with a zero steering angle was studied. We now extend our study to include steering dynamics using the same vehicle model. When linear tire characteristics are assumed, transfer functions relating steering inputs to various response quantities are a convenient way to represent the steering dynamics. Transfer functions relate input and output variables for linear systems and are described in automatic control texts such as [4].

To find transfer functions, we take the Laplace transform of Eqs. (5.33) or (5.34). The Laplace transforms of time-dependent variables are, in principle, new variables which are functions of the transform variable s. In conformance with common practice, we now let the variable names V, r, and δ also stand for the Laplace transforms of the time-dependent functions. The context in which these variables appear will make clear whether the variable symbol stands for a function of time or of the Laplace transform of the variable.

In deriving transfer functions, zero initial conditions are assumed. The only difference between the time-dependent Eq. (5.33) or (5.34) and the transformed version is that $\dot{V}$ and $\dot{r}$ become sV and sr, where s is the Laplace variable. In matrix form, Eq. (5.33) then transforms into a set of algebraic rather than differential equations:

$$\begin{bmatrix} ms + \dfrac{C_f + C_r}{U} & mU + \dfrac{aC_f - bC_r}{U} \\ \dfrac{aC_f - bC_r}{U} & I_z s + \dfrac{a^2 C_f + b^2 C_r}{U} \end{bmatrix} \begin{bmatrix} V \\ r \end{bmatrix} = \begin{bmatrix} C_f \\ aC_f \end{bmatrix} \delta \tag{5.48}$$

The transfer functions are found by solving these algebraic equations for variables such as V and r given δ as an input variable. This can be done by using Cramer's rule from the theory of linear algebra.

Cramer's rule states that every variable in a set of linear algebraic equations can be expressed as the ratio of two determinants. The denominator in the ratio is, in every case, the determinant of the array at the left side of Eq. (5.48). When written out, this determinant is exactly the *characteristic polynomial* that we have encountered previously in the eigenvalue analysis. When the characteristic polynomial is set to zero, the result is just the *characteristic equation*, which was used previously to determine the eigenvalues and the stability of the vehicle. Since this polynomial in s will be used frequently, we give it the symbol Δ. The last term in the polynomial is given in the simplified form as it was given in the characteristic equation in Eq. (5.41):

$$\Delta = mI_z s^2 + \left[\frac{m(a^2C_f + b^2C_r)}{U} + I_z\frac{(C_f + C_r)}{U}\right]s + \frac{(a+b)^2C_fC_r}{U^2} + m(bC_r - aC_f) \tag{5.49}$$

The numerator determinant in Cramer's rule has to do with which variable of the system is of interest. The numerator determinant is the determinant of the array to the left in Eq. (5.48) but with the column vector at the right in Eq. (5.48) substituted for one of the columns in the array. The transfer functions are then found by dividing the numerator determinant by Δ. The procedure is best understood by example. The forcing column to be used in the numerator determinant will involve the input variable δ. If this column is substituted in the first column of the array in Eq. (5.48), the result will be the first variable, V. If the forcing column is substituted in the second column, the result will be the second variable, r.

As an example of Cramer's rule applied to Eq. (5.48), when the first column in the coefficient array is substituted by the forcing column on the right-hand side of Eq. (5.48) and the determinant is divided by Δ the result is a relationship between the first variable V and the forcing quantity δ:

$$V = \frac{\det\begin{vmatrix} C_f\delta & mU + (aC_f - bC_r)/U \\ aC_f\delta & I_z s + (a^2C_f + b^2C_r)/U \end{vmatrix}}{\Delta} \tag{5.50}$$

When this result is expressed as a ratio of V to δ, it becomes the transfer function relating the lateral velocity to the front steering angle:

$$\frac{V}{\delta} = \frac{I_zC_f s - maC_fU + (a+b)bC_fC_r/U}{\Delta} \tag{5.51}$$

The other transfer function is derived in a similar manner but with the forcing column at the right of Eq. (5.48) substituted for the second column in the array at the left of Eq. (5.48). The result expressed as a ratio of r to δ is another transfer function.

$$\frac{r}{\delta} = \frac{maC_f s + (a+b)C_fC_r/U}{\Delta} \tag{5.52}$$

One should remember that all transfer functions have the same denominator, Eq. (5.49), and that this polynomial also appears in the characteristic equation used to determine the eigenvalues and the stability of straight-line running. This denominator is second order in the Laplace variable s. The transfer functions above have numerators that are first order in s.

Another transfer function of interest deals with the lateral acceleration of the center of mass, $\dot{V} + rU$. It is derived by combining the transfer functions given above in Eqs. (5.51) and (5.52). In the Laplace domain, the derivative of the lateral velocity is expressed using the Laplace transform variable s, $\dot{V} = sV$. The result of expressing the lateral acceleration as s, $V + rU$ is the transfer function

$$\frac{\dot{V} + rU}{\delta} = \frac{I_z C_f s^2 + (a+b)bC_f C_r s/U + (a+b)C_f C_r}{\Delta} \tag{5.53}$$

This acceleration transfer function is notably different from the transfer functions in Eqs. (5.51) and (5.52). It has a second-order numerator as well as a second-order denominator. This is significant because transfer functions become complex exponential frequency response functions when the variable s is replaced by $j\omega$, where j is the imaginary number and ω stands for the forcing frequency in radians per second.

A transfer function with the same order in s for the numerator and the denominator, as in Eq. (5.53), has a response ratio that approaches a constant as the forcing frequency becomes very large. In contrast, a transfer function with a numerator of lower order than the denominator such as those in Eqs. (5.51 and 5.52) will have a frequency response ratio that becomes small at very high frequencies (see, e.g., [4]). This means that the acceleration response to sinusoidal steering inputs does not fall off at high frequencies, as is the case for the lateral velocity and the yaw response. If you could turn the steering wheel of a car back and a small amount in an approximately sinusoidal fashion, our linear model predicts that the yaw rate, lateral velocity, and lateral acceleration would all respond in a sinusoidal fashion. However, if you were able to move the steering wheel with the same amplitude of motion but with a higher and higher frequency, you would find that the amplitude of the yaw rate and the lateral velocity would decline as the frequency increased. The amplitude of the lateral acceleration would, however, remain nearly constant. (It is quite easy to sense the yaw rate and the lateral acceleration when driving a car, but you should be very careful about trying this experiment on a normal road.)

Another aspect of the difference between the transfer functions for lateral acceleration and those for yaw rate and lateral velocity is that a small sudden change in steer angle will result in an immediate increase in acceleration, but the yaw rate and lateral velocity will take some time to build up. (This is another experiment you probably should not try at home.) This difference in the transfer functions for yaw rate and lateral acceleration also has implications if one were to consider automatic steering control systems. The problem of designing a feedback steering control system is different depending on whether a yaw rate sensor or a lateral accelerometer sensor is used because of the difference in the transfer functions relating steer angle inputs to the yaw rate and to lateral acceleration (see [3]).

Yaw Rate and Lateral Acceleration Gains

When the Laplace variable s is set equal to zero in transfer functions, the steady-state relation of output to input is given (see [4]). For example, the steady yaw rate for a constant front steer angle is found from the r/δ transfer function, Eq. (5.52), by setting $s = 0$. In this case the result is

$$\left.\frac{r}{\delta}\right|_{s=0} = \frac{U}{(a+b) + m(bC_r - aC_f)U^2/(a+b)C_f C_r} \tag{5.54}$$

This expression is simplified by defining the *understeering coefficient*, K, which appears in the transfer functions and plays a prominent role in a discussion of the steering behavior of automobiles. The understeering coefficient is defined in Eq. (5.55):

$$K = \frac{m(bC_r - aC_f)}{(a+b)C_f C_r} \tag{5.55}$$

The sign of K is the same as the sign of the quantity $bC_r - aC_f$. Thus, positive values of K correspond to what we have previously labeled as *understeering*, Eq. (5.45), negative values correspond to *oversteering*, Eq. (5.46), and if K is zero, the car is called *neutral steering*.

It is common to call the transfer functions evaluated at $s = 0$ the *zero-frequency gains*. Here we designate these gains by G with subscript r for yaw and a for acceleration. Using the understeering coefficient, the zero-frequency gains are as follows:

$$\left.\frac{r}{\delta}\right|_{s=0} = G_r = \frac{U}{(a+b) + KU^2}, \qquad \left.\frac{\dot{V} + rU}{\delta}\right|_{s=0} = G_a = \frac{U^2}{(a+b) + KU^2} \tag{5.56}$$

The terms *understeering, neutral steering*, and *oversteering* have significance with respect to stability, but the reason for using this nomenclature may be more obvious in the discussion of cornering behavior below.

Special Case of the Neutral Steering Vehicle

The neutral steering vehicle with $K = 0$ is sometimes thought of as ideal from the point of view of handling. Indeed, when good handling is a main priority, designers often strive for equal weight distribution on the front and rear axles. This, particularly with the same tires and tire pressures front and rear, favors a neutral or nearly neutral steering characteristic. A number of other factors besides weight distribution enter into the determination of the understeering coefficient and hence the handling behavior of a vehicle. These factors, such as the roll stiffness at the front and rear axles of the vehicle, allow a designer to adjust the dynamics of an automobile within limits. Here, we examine some special features of neutral steering vehicles.

In the linearized model, a neutral steering vehicle is one for which $aC_f = bC_r$, so $C_r = aC_f/b$ can be eliminated in favor of C_f in all previous results. When this is done, a number of terms disappear. The denominator of the transfer functions simplifies from Eq. (5.49) to

$$\Delta = \left(ms + \frac{(a+b)C_f}{bU}\right)\left(I_z s + \frac{a(a+b)C_f}{U}\right) \tag{5.57}$$

The factored form of Eq. (5.57) indicates that if this characteristic polynomial were set equal to zero to find the eigenvalues, there would be two real, negative eigenvalues. The car would not only be stable, but would also have no tendency to oscillate at all in this case. In addition, for a neutral steering car, the yaw rate transfer function has an exact cancellation of terms in the numerator and denominator (often called *pole–zero cancellation*) that reduces the transfer function to first order.

$$\frac{r}{\delta} = \frac{aC_f}{I_z s + a(a+b)C_f/U} \tag{5.58}$$

The acceleration transfer function for a neutral steering car remains second order because there is no pole–zero cancellation.

$$\frac{\dot{V} + rU}{\delta} = \frac{I_z C_f s^2 a(a+b)C_f s/U + a(a+b)C_f^2/b}{\Delta} \tag{5.59}$$

A final interesting simplification occurs for a neutral steering vehicle that has the special value of its polar moment of inertia:

$$Iz = mab \tag{5.60}$$

If this is the case, the moment of inertia can be eliminated as an independent parameter. This may seem to be a very special case, but in fact, the relation of Eq. (5.60) is approximately true for many cars. Under this assumption, the two real negative eigenvalues become equal and the denominator of the transfer functions becomes

$$\Delta = ab\left[ms + \frac{(a+b)C_f}{bU}\right]^2 \tag{5.61}$$

The transfer functions simplify quite a bit for a neutral steering car when in addition, Eq. (5.60) is true.

$$\frac{r}{\delta} = \frac{C_f}{b(ms + (a+b)C_f/bU)} \tag{5.62}$$

$$\frac{\dot{V} + rU}{\delta} = \frac{abC_f(ms^2 + (a+b)C_f s/bU + (a+b)C_f/b^2)}{\Delta} \tag{5.63}$$

Although these highly simplified transfer functions for neutral steering cars with a special value for the moment of inertia certainly do not apply in general, they have been used in general studies of handling dynamics and steering control since they contain a reduced number of parameters. Furthermore, neutral steering cars are often thought of as representing a sort of optimum with respect to handling qualities, and these specialized transfer functions can be used to compare the steering response of real vehicles with the response of one version of an ideal vehicle.

5.6 STEADY CORNERING

In previous sections the stability of a simple model of an automobile was studied. The model assumed that the car moved only in plane motion and that the relation between the tire forces and the (small) slip angles was linear. Under these assumptions, it was shown that single equivalent wheels could represent the front and rear axles, and cornering coefficients could represent the total forces acting on the two axles. Although stability was analyzed only for motion in a straight line, the equations of motion in body-centered coordinates were derived, including a small steer angle δ for the front wheels. By allowing δ to have nonzero values, the model can also be used to study cornering behavior. It will be shown that there is an interesting link between the stability properties of a car and the manner in which the steer angle required to negotiate a turn varies with speed.

Description of Steady Turns

To simplify the discussion, we consider *steady turns*, i.e., turns taken at a constant speed and having a constant turn radius, R. Strictly, such turns do not occur in normal driving, but often there is a period of time in actual turns when the car is in nearly a steady state. Furthermore, it is easy for a manufacturer to construct a circular skid pad for testing cars, and thus steady turns at various speeds are often used to characterize a car's steering behavior. The terms *understeering* and *oversteering* which we encountered earlier are most easily related to steady turn behavior.

In Figure 5.8 we consider a steady right-hand turn of radius R. To keep angles small, we assume that the turn radius is large compared to the wheelbase, $R \gg a + b$ and that the lateral acceleration is small enough that the slip angles are not too large. This allows the use of the cornering coefficient approach to tire force generation. These assumptions will not apply to extreme maneuvers that involve actual skidding of the tires. The angles are exaggerated somewhat for clarity in Figure 5.8. The yaw rate (for the body and the entire figure) $r \cong U/R$, and the lateral acceleration of the center of mass is approximately U^2/R. Notice in Figure 5.8 that the tire forces are in the positive direction but the slip angles are negative, as is the lateral velocity V.

The relationships between the steer angle and the yaw rate or the lateral acceleration in a steady turn was derived from the transfer functions evaluated for $s = 0$ in Eqs. (5.56) in terms of the understeer coefficient defined in Eq. (5.55). If the relation $r = U/R$ is substituted in the expression for the yaw rate gain, and the result is solved for

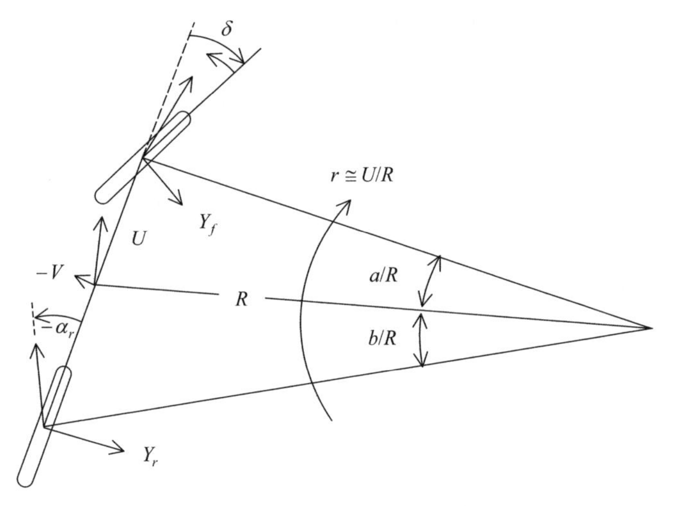

Figure 5.8 Automobile in a steady turn.

the steer angle, the following relationship is found:

$$\delta = \frac{KU^2}{R} + \frac{a+b}{R} \tag{5.64}$$

The understeering coefficient determines how the steer angle changes as the speed or the turn radius changes in a steady turn. Figure 5.9 shows plots of Eq. (5.64) for cases when the vehicle is understeering, oversteering, or neutral (i.e., $K > 0$, $K < 0$, or $K = 0$). Note that U^2/R is the lateral acceleration, so Figure 5.9 indicates how the steer angle must change as the lateral acceleration is increased.

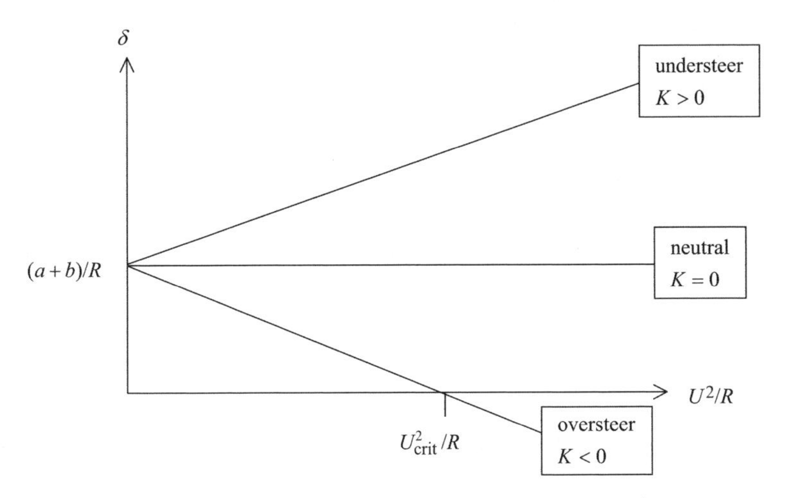

Figure 5.9 Steer angle as a function of lateral acceleration.

Significance of the Understeering Coefficient

Since most cars are designed to understeer, the understeering line in Figure 5.9 shows the common experience that if a constant-radius turn is taken at increasing speeds (and thus increasing lateral acceleration), the driver has to increase the steer angle. Another way to say this is that for an understeering car, the magnitude of the front slip angle increases faster than the magnitude of the rear slip angle for increases in lateral acceleration. For a neutral steering vehicle, the steer angle remains constant with increasing lateral acceleration. (In fact, the slip angles in Figure 5.8 must increase to generate higher forces, but this is accomplished by rotation of the entire car with respect to the turn radius without a change in the steer angle.)

For an oversteering car, it turns out that the critical speed (found from stability analysis) is also the speed at which the steer angle is zero in a steady turn. This can be seen by setting $\delta = 0$ in the expression for steer angle in Eq. (5.64) and solving for the value of U^2 that results in a zero steer angle:

$$U^2 = \frac{a+b}{-K} \tag{5.65}$$

Using the definition of the understeering coefficient in Eq. (5.55), it can be seen that the expression for U^2 in Eq. (5.65) is exactly the formula for critical speed for an oversteering vehicle found in the stability analysis, Eq. (5.47). This means that there are two interesting aspects for the critical speed for an oversteering car.

First, such a car becomes unstable for speeds greater than the critical speed when traveling in a straight line. Second, when negotiating a steady turn at critical speed, the car will turn without any steer angle at all. In fact, if the critical speed is exceeded, not only will the car be unstable, requiring the driver to manipulate the steer angle to stabilize the vehicle, but the steady steer angle will be in the opposite direction from the turn direction. This *countersteering* is peculiar, but it is not as serious for the driver as the necessity for the driver to move the steering wheel continuously to keep the car on track because it is unstable. The driver's impression is that the car will go into a spin if he or she cannot keep it on track by continual motions of the steering wheel.

From Figure 5.8 it can be appreciated that as a car increases its lateral acceleration in a steady turn the rear slip angle must increase in magnitude in order for the rear tires to generate a larger force. This can happen only if the centerline of the car rotates with respect to the turn radius as the lateral acceleration increases. If the steer angle is maintained constant, this change in attitude of the car with respect to the turn radius will increase the magnitude of the front slip angle as much as the rear slip angle.

For a neutral steering car, the forces at the front and rear can be adjusted to give a larger lateral acceleration simply by changing the car attitude without any change in steer angle, as indicated in Figure 5.9. In contrast, an understeering car requires a larger change in slip angle at the front than at the rear to maintain equilibrium for a higher lateral acceleration. This can be accomplished by increasing the steer angle. Just the opposite is true for an oversteering car, for which the steer angle must be reduced to reduce the front slip angle change so that it is less than the rear slip angle change necessary to accommodate to the increased lateral acceleration.

One should keep in mind that the definition of understeering coefficient given here applies strictly only to the linear range of the tire force–slip angle relation. The behavior of pneumatically tired vehicles when the lateral forces become large is significantly nonlinear and quite complex.

Acceleration and Yaw Rate Gain Behavior

The results derived above can be studied to reveal some interesting aspects of the problem facing a driver acting as a controller of an automobile. In short, the response of the vehicle to steering inputs varies drastically with speed and is fundamentally different depending on whether the car is understeering, neutral steering or oversteering. Two particular aspects of steering response will be considered: the yaw rate and the lateral acceleration of the center of mass.

When a driver moves the steering wheel and thus changes the angle of the front road wheels, δ, two easily sensed effects are produced. One is a change in the lateral acceleration and another is a change in the yaw rate. Although the two effects are coupled in a conventionally steered automobile, under differing situations, one is often more important to the driver than the other. For example, if the task is to change lanes on a straight freeway when traveling at high speed, the desire is to accelerate laterally without much change in heading or yaw rate. On the other hand, when rounding a right-angle corner, one must establish a yaw rate in order to change the heading angle by 90°, and the lateral acceleration is simply a necessary by-product during the maneuver. Thus, the driver must use only one control input, the steering angle, to accomplish two quite different types of tasks. Furthermore, the input/output relationships are not constant but vary depending on how fast the vehicle is traveling. It is no wonder that it takes some training and practice to become a good driver.

Consider the lateral acceleration and yaw rate gains defined in Eq. (5.56). These relationships between the yaw rate and lateral acceleration apply directly to steady turn behavior. It is useful to discuss separately the three cases $K > 0$, $K = 0$, and $K < 0$, corresponding to understeering, neutral steering, and oversteering, respectfully. For the neutral case, $K = 0$ and

$$G_a = \frac{U^2}{a + b} \tag{5.66}$$

This means that the acceleration the driver feels per unit steer angle increases as U^2. In this sense, driving a neutral steering car involves controlling a variable gain system. The acceleration the driver experiences due to a change in steer angle varies as the square of the car speed.

For the understeering case, $K > 0$, and it proves to be useful to define a *characteristic speed*, U_{ch}, such that at the characteristic speed the acceleration gain is one half that of a neutral car. Considering Eq. (5.56), this requires that

$$KU_{\text{ch}}^2 = a + b \tag{5.67}$$

With this definition, the (positive) understeering coefficient K can be eliminated algebraically in favor of the characteristic speed in Eq. (5.56).

$$G_a = \frac{U_{\text{ch}}^2}{a+b} \frac{U^2/U_{\text{ch}}^2}{1 + U^2/U_{\text{ch}}^2} \tag{5.68}$$

For the oversteering case, $K < 0$, the *critical speed*, U_{crit}, defined in Eq. (5.47) or Eq. (5.65), can replace the understeering coefficient in Eq. (5.56):

$$G_a = \frac{U_{\text{crit}}^2}{a+b} \frac{U^2/U_{\text{crit}}^2}{1 - U^2/U_{\text{crit}}^2} \tag{5.69}$$

Figure 5.10 shows in general how acceleration gains vary with speed. One should remember that the three cases shown have no particular relation to each other. They are for three separate automobiles. There is no obvious connection between the characteristic speed for an understeering car and the critical speed for an oversteering car in Figure 5.10. On the other hand, it is possible to change a particular car from understeering to neutral steering to oversteering by changing the loading, the tires, or even the tire pressure. Most cars are designed to be as insensitive to these changes as possible, so it usually takes a major change in one of these factors to change the understeering coefficient significantly.

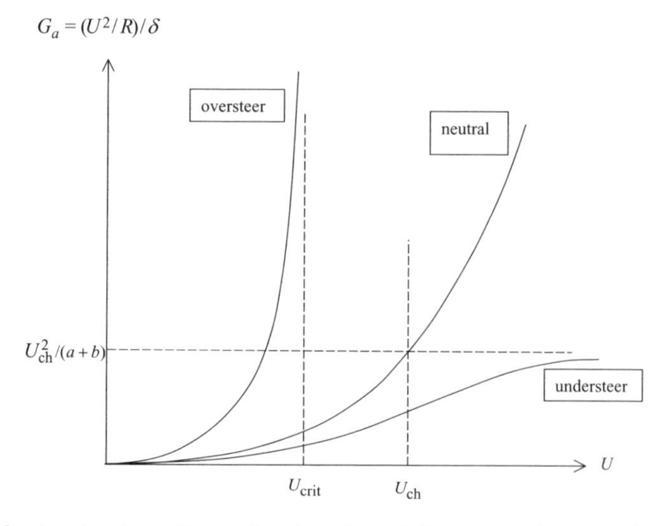

Figure 5.10 Acceleration gain as a function of speed for oversteering, neutral steering, and understeering automobiles.

At very low speeds, the acceleration gains for all three cases, Eqs. (5.66), (5.68), and (5.69), approach identical expressions. The gains simply become the ratio of the square of the speed divided by the wheelbase. Note, however, that the lateral acceleration becomes very sensitive to steer angle for a neutral steering car at high speed. For an oversteering car, this sensitivity even becomes infinite as the finite critical speed is approached. On the other hand, the acceleration gain approaches a finite limit for an understeering car for speeds high compared to the characteristic speed. Thus, the three types of cars behave quite differently from one another as the speed varies.

Another quantity of interest is the yaw rate gain, that is, the steady yaw rate per unit steer angle, G_r, also defined in Eq. (5.56). Again, there are three cases in which the understeering coefficient can be eliminated in favor the characteristic speed and the critical speed: for neutral steering,

$$K = 0, \quad G_r = \frac{U}{a+b} \tag{5.70}$$

for understeering,

$$K > 0, \qquad G_r = \frac{U}{a+b}\frac{1}{1+U^2/U_{\text{ch}}^2} \tag{5.71}$$

and for oversteering,

$$K < 0, \qquad G_r = \frac{U}{a+b}\frac{1}{1-U^2/U_{\text{crit}}^2} \tag{5.72}$$

Figure 5.11 shows the general form of the yaw rate gains for the three cases. Once again one must be careful not to assume that the three cases shown are necessarily related. It is true that the equality of the initial slopes of the three cases plotted in Figure 5.11 implies that the three cars have the same wheelbase, $a + b$. Equations (5.71) and (5.72) approach Eq. (5.70) at low speeds, and in the Figure 5.11 the three curves approach each other at low speeds, but there is no connection between the values of the critical speed for the oversteering car and the characteristic speed for the understeering car.

The oversteering car is on the verge of instability at the critical speed, and at that speed the yaw rate gain becomes infinite. This means that the yaw rate is infinitely sensitive to steering inputs at the critical speed. Above the critical speed, control reversal occurs and the driver must countersteer as well as to attempt to stabilize an unstable vehicle. This effect is not shown in Figure 5.11. The driver feels that the car seems continually to want to slew around and ultimately to travel in reverse. The yaw rate gain for a neutral car increases linearly with speed, so that if a specific yaw rate is desired, smaller steering inputs are required at high speed than at low speed.

For the understeering car, the yaw rate gain initially rises with speed, but the gain reaches a maximum at the characteristic speed and declines at even higher speeds.

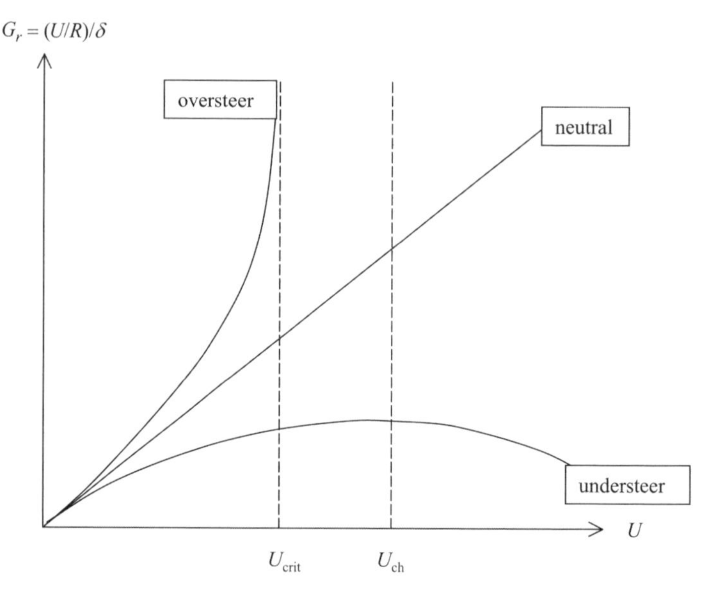

Figure 5.11 Yaw rate gain as a function of speed for oversteering, neutral steering, and understeering automobiles.

The understeering car is stable at high speed but requires large steer inputs to achieve a given yaw rate at very high speeds. This correlates intuitively with the notion that an understeering car is not only stable at high speeds but also becomes less responsive to steering inputs at very high speeds. This is an example of the common property of many vehicles that a high degree of stability is often purchased at the expense of maneuverability.

As Figures 5.10 and 5.11 indicate, the sign of the understeering coefficient has a large effect on the compromise between stability and controllability. A very stable car may not respond to steering inputs very well in the sense of generating lateral acceleration or yaw rate, particularly at high speeds. On the other hand, a car that is very responsive to steering inputs may become unstable at high speed. Neither extreme situation is desirable. An extremely stable car may not be able to swerve to avoid an obstacle, and a car that is unstable at high speed may well be difficult for the driver to control. These plots make plausible the idea that a neutral or nearly neutral steering automobile represents a sound compromise in steering response.

Automobile designers have a number of means at their disposal to adjust the handling of a car. Among these are the location of the center of gravity, the spring rates at the front and rear, the antisway bar rates at the front and rear, the tire sizes and pressures at the front and rear, and even the shock absorber characteristics. These effects are not strictly accounted for in the simple bicycle model. Qualitatively, it is possible to consider that these effects change the effective axle cornering coefficients and thus modify the understeering coefficient. Another complicating feature not taken into account yet has to do with driving and braking. This not only introduces transient effects but also requires the tires to generate longitudinal forces as well as lateral forces. This affects a tire's ability to provide lateral forces significantly and can drastically affect the handling characteristics of a vehicle.

5.7 SUMMARY

In this chapter we introduced the idea of using a noninertial coordinate system to describe the dynamics of rigid bodies. When the coordinate frame is attached to the body and rotating with it, the acceleration assumes a form that is rarely discussed in introductory treatments of dynamics. On the other hand, as the vehicle dynamics examples show, a great deal of interesting information about vehicle dynamics is easily derived using such a coordinate frame. It has been pointed out that in case body positions are of interest, one can still use a body-fixed coordinate system by adding equations that relate inertial velocity components to velocity components in the moving frame. Whether in such a case it is an advantage to use one frame or another is difficult to say in general and usually depends on the details of the problem under study.

In this chapter we focused on plane motion, but Appendix A gives the equations of motion for three-dimensional motion of a rigid body using a body-fixed frame. These general equations will find use in Chapter 7, in which rigid-body motion in three dimensions is discussed.

REFERENCES

[1] A.W. Babister, *Aircraft Dynamic Stability and Response*. Pergamon Press, Oxford, 1980.

[2] T.D. Gillespie, *Fundamentals of Vehicle Dynamics*, Society of Automotive Engineers, Warrendale, PA, 1992.

[3] D. Karnopp, *Vehicle Stability*, Marcel Dekker, New York, 2004.

[4] K. Ogata, *Modern Control Engineering*, Prentice-Hall, Englewood Cliffs, NJ, 1970.

PROBLEMS

5.1 Using Figure P5.1 and the sketches of the body-fixed velocity components of the shopping cart shown in Figure 5.4, derive expressions for acceleration in the x and y directions given in Eqs. (5.18) and (5.19).

5.2 To help understand the expressions for acceleration in a rotating coordinate system consider the situation shown in Figure P5.2. A point mass is positioned at the constant location $X = R_0$ in the inertial X, Y coordinate system and is subject to no force at all. The mass location can also be described using the rotating coordinate system x,y, which can be imagined attached to a turntable rotating with the constant angular rate r. (We call r the yaw rate when describing vehicle motion.) The rotating coordinate system has an angle of rt with respect to the inertial coordinate system. We know that the acceleration of the particle is zero in the inertial system, $\ddot{X} = \ddot{Y} = 0$. In the rotating system, the acceleration is given

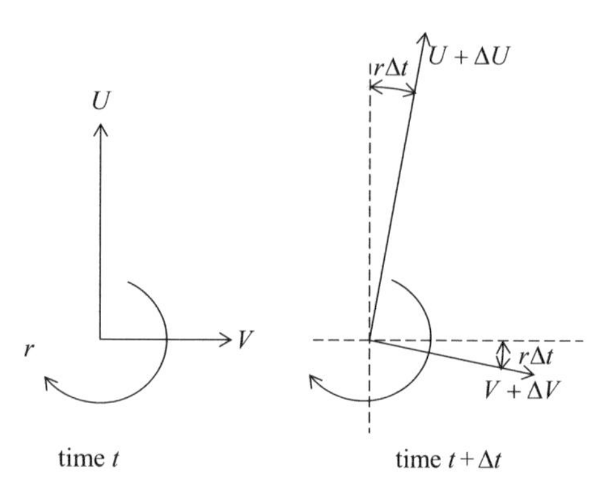

Figure P5.1

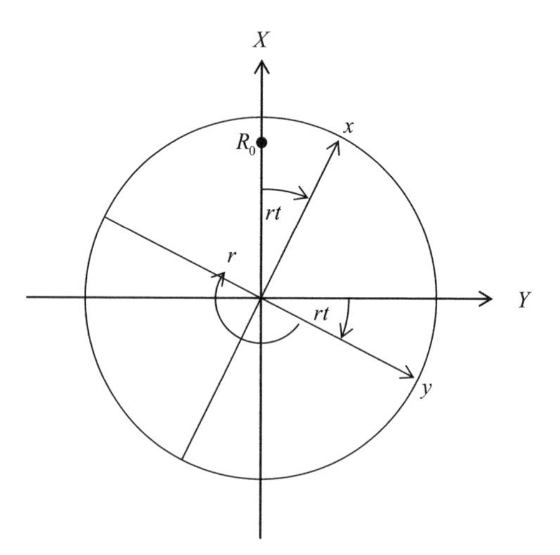

Figure P5.2

by Eqs. (5.18) and (5.19). Show that these equations actually also state that the acceleration is zero by expressing x and y in terms of R_0 and the angle rt and identifying U with $\dot{x}$ and V with $\dot{y}$ and substituting into Eqs. (5.18) and (5.19).

5.3 Derive Eq. (5.8) from Eqs. 5.21.

5.4 Consider the characteristic equation for a mass–spring–damper, $ms^2 + bs + k = 0$.

(a) Show that if all three parameters are positive, the system is stable.

(b) Show that if the spring constant is negative, the system is unstable.

5.5 Derive Eq. (5.64) for a steady turn from Eq. (5.56) using the fact that the steady yaw rate is $r = U/R$, where R is the turn radius.

5.6 Suppose that a point mass is traveling around a circular path at a constant speed U (Figure P5.6). The x–y coordinate system is attached to the point mass and rotates such that the y-direction remains along a radius of the circle. The coordinate system is similar to that used in Figure 5.4. Note that for the x–y–z axis system to be right-handed, the z-axis must point down. The angular velocity around the z-axis is called r and is positive in the direction shown. The radius of the circular path is R and the yaw rate for the coordinate system is actually negative, as indicated in the figure. The lateral velocity in the y-direction remains constant at zero. Using Eq. (5.19), compute the acceleration in the y-direction. Does this result agree with the formula for centripetal acceleration given in books on dynamics?

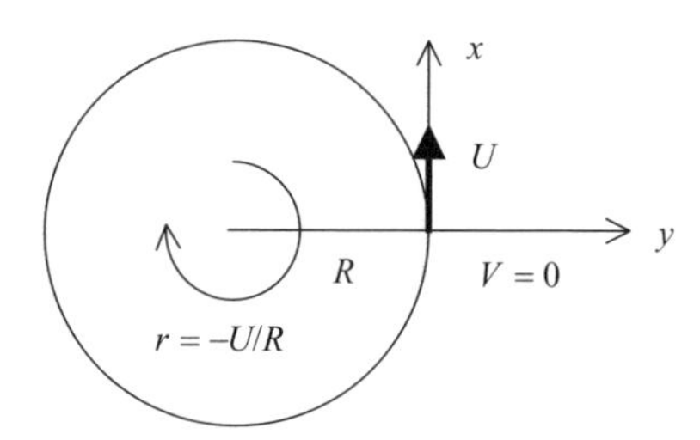

Figure P5.6

5.7 Figure P5.7 is similar to Figure 5.4 except that the separation of the wheels on the back axle of the cart has been designated by the symbol t for "track." The sideways velocity for the center point of the axle was given in Eq. (5.23) as $V - br$. The total velocity of the center of the back axle was found by applying Eq. (5.4) starting from the velocity of the center of mass with components U and V and then adding the term $\boldsymbol{\omega} \times \mathbf{r}_{AB}$, where the magnitude of $\boldsymbol{\omega}$ is r and the

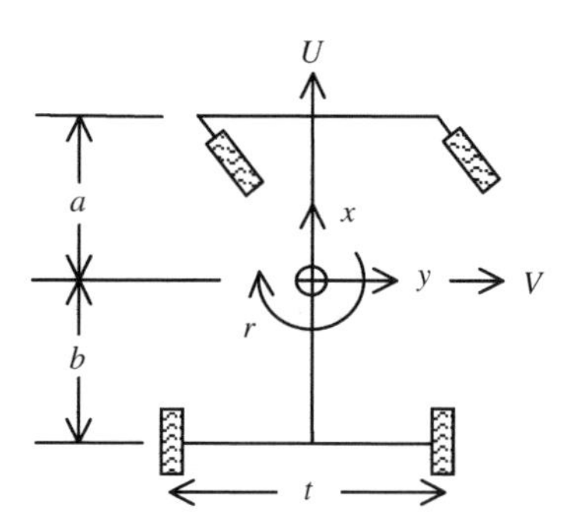

Figure P5.7

length of the vector $\mathbf{r}_{AB}$ is just b. Equation (5.23) considered only the sideways component of the total velocity and constrained it to be zero. To find the total velocity of a wheel, one can again apply Eq. (5.4), but now $\mathbf{r}_{AB}$ is a vector with one component in the minus x-direction of length b and another in the plus or minus y-direction of length $t/2$. Find the total velocity at either wheel and confirm that the sideways velocity component at the wheel location is still $V-br$. As this result is independent of the track, it is confirmed that Eq. (5.23) correctly constrains the wheels to move only in the direction in which they are pointed. It also shows that all points on the axle have the same velocity in the y- direction.

5.8 The claim is made that Eqs. (5.21)–(5.23) can be combined to yield Eq. (5.8). If instead of eliminating Y and V to find an equation for r, one were to find an equation for Y by eliminating V and r, what would the result be? Is it a surprise that it resembles Eq. (5.8)?

5.9 Figure P5.9 shows the motion of an automobile described in inertial coordinates. Although the figure shows four wheels, the mathematical model will be a bicycle model assuming a single slip angle for the two front wheels and a single slip angle for the two rear, as was assumed in Figure 5.6. The sketch is similar to Figures 5.1 and 5.4 for the shopping cart, but with x and y being the position variables for the center of mass and Ψ playing the role of yaw angle.

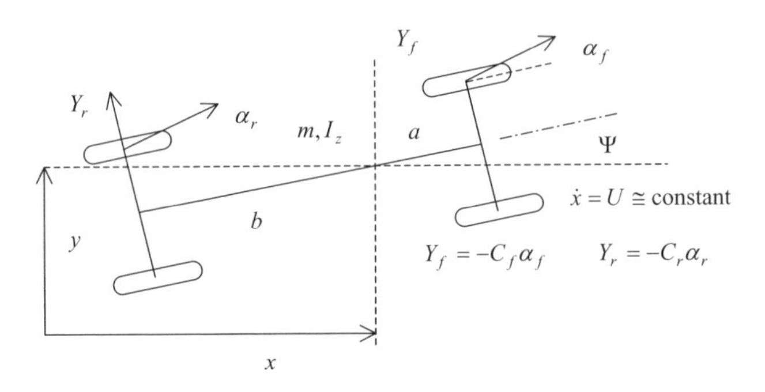

Figure P5.9

With the definitions of slip angles and lateral forces shown in the figure, the slip coefficients are positive. Assume that the slip angles as well as the heading angle Ψ remain small. Also assume that the forward speed U is constant and is essentially the same as the velocity in the x-direction. Derive the following equations of motion for this car model in inertial coordinates y and Ψ, but instead of using Newton's laws as was done in the text, try your hand at the use of Lagrange's equations. The kinetic energy is easily expressed and there is no potential energy, so the only challenge is to find the generalized force for the y-generalized variable and the generalized moment for the Ψ variable.

$$\begin{bmatrix} m & 0 \\ 0 & I_z \end{bmatrix} \begin{bmatrix} \ddot{y} \\ \ddot{\Psi} \end{bmatrix} + \begin{bmatrix} \dfrac{C_f + C_r}{U} & \dfrac{aC_f - bC_r}{U} \\ \dfrac{aC_f - bC_r}{U} & \dfrac{a^2C_f + b^2C_r}{U} \end{bmatrix} \begin{bmatrix} \dot{y} \\ \dot{\Psi} \end{bmatrix}$$

$$+ \begin{bmatrix} 0 & -(C_f + C_r) \\ 0 & -(aC_f - bC_r) \end{bmatrix} \begin{bmatrix} y \\ \Psi \end{bmatrix} = \begin{bmatrix} 0 \\ 0 \end{bmatrix}$$

Derive the characteristic equation for this system by assuming that y and Ψ have an e^{st} time form. You will see that this requires much more algebraic work than the case when body-fixed coordinates are used, Eq. (5.35). If you are able to find the characteristic equation, you will find that the system has two zero-valued eigenvalues and that the two finite eigenvalues are identical to the eigenvalues for the body-centered analysis from Eq. (5.41).

5.10 Verify that Eq. (5.33) results from a combination of Eqs. (5.27)–(5.30) .

5.11 Refer to the section on yaw rate and acceleration gains in steady cornering in Section 5.6. Prove that an understeering vehicle has its maximum yaw rate gain at a forward speed equal to its characteristic speed, as indicated in Figure 5.11. Consider Eqs. (5.56), (5.67), and (5.71) and use calculus to derive an expression for the maximum yaw rate gain.

5.12 Figure P5.12 shows two rear views of a four-wheeled vehicle. In the first instance the vehicle is running in a straight line with no lateral acceleration.

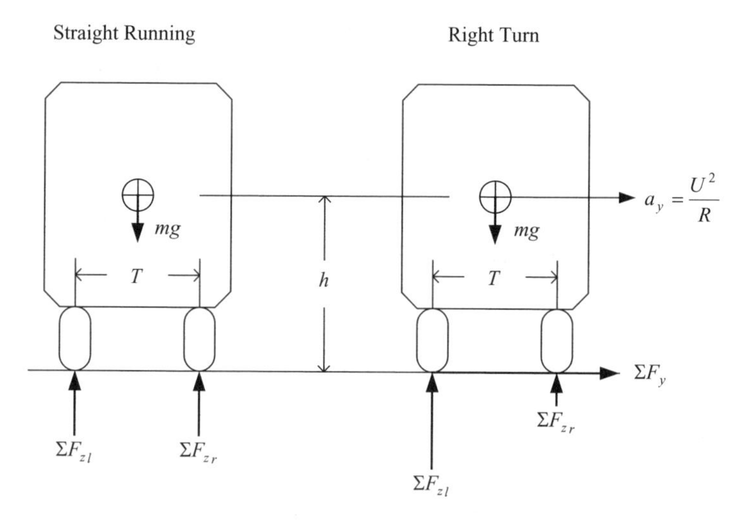

Figure P5.12

The forces shown as $\sum F_{zl}$ and $\sum F_{zr}$ represent the sum of the front and rear normal forces on the tires at the left and right sides, respectively. The distance between the tires, the track T, is assumed to be the same for the front and the rear axles. In this case, the symmetry of the situation makes it obvious that the total normal forces on the two sides of the vehicle must add up to the weight. Thus, $\sum F_{zl} = \sum F_{zr} = mg/2$. In the second part of the figure, the vehicle is executing a steady turn of radius R with speed U so there is a lateral acceleration of $a_y = U^2/R$. There will also be a total lateral force on all the tires of $\sum F_y$. In this case, the height of the center of mass, h, and the track determine how the normal forces on the outside of the turn increase and the forces on the inside of the turn decrease.

(a) Assuming that the vehicle is in a steady state with no angular acceleration around the roll axis, show that the normal forces are given by the following expressions:

$$\sum F_{zl} = m\left(\frac{g}{2} + \frac{hU^2}{TR}\right), \qquad \sum F_{zr} = m\left(\frac{g}{2} - \frac{hU^2}{TR}\right)$$

(b) Show that the vehicle can achieve a lateral acceleration of 1 g only if the center of mass height is equal to or less than half the track.

(c) Suppose that the vehicle has suspension strings that allow it to lean a small amount toward the outside of turns. Qualitatively, how would this change the expressions for $\sum F_{zl}$ and $\sum F_{zr}$ in part (a)?

5.13 In Figure 5.7, note that the lateral velocity at the front can be written as $U(\delta + \alpha_f)$ and at the rear as $U\alpha_r$.

(a) Using these expressions and the lateral velocity of the center of mass expressed as $U\beta$, find an expression for the angle β in terms of the front and rear slip angles and the steer angle.

(b) It has been proposed the a benefit of steering both the front and the rear wheels would be to make β be zero (i.e., to make the center of mass velocity line up with the vehicle centerline during a steady turn). Is this possible without steering the rear wheels at low speeds where the slip angles are very small? (see Figure 5.8.)

(c) Could β ever be zero at high speeds?

5.14 Figure 5.11 shows that the yaw rate gain for an understeering vehicle is a maximum for a speed U that is equal to the characteristic speed U_{ch} as defined in Eq. (5.67). Use calculus to show that the yaw rate gain in Eq. (5.71) is indeed a maximum when $U = U_{\mathrm{ch}}$.

5.15 This problem illustrates that the use of body-centered coordinates can be a little tricky when forces are applied to a body that are related to the position of the body. (The position of the body does not actually enter these equations of motion.) The problem is to set up equations to determine the stability of motion for a trailer being towed at a constant speed by a large vehicle that does not react to forces on it from the trailer (Figure P5.15). The trailer has mass m and moment of inertia I_z. The dimensions a and b locate tha center of mass. This problem is analyzed using inertial coordinates in ref. [3].

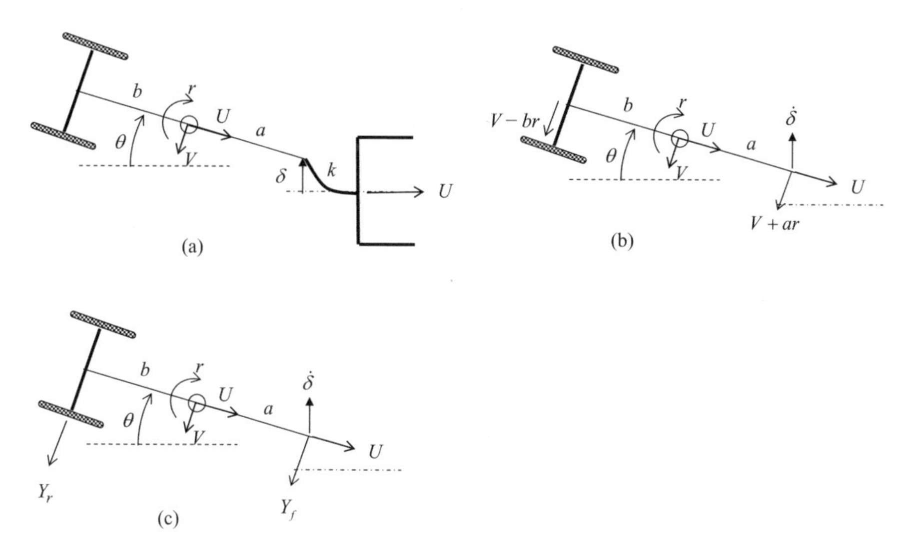

Figure P5.15

We assume that the angle θ is very small for the stability analysis and that the speed of the towing vehicle is much larger than any lateral velocities of the trailer. This means that the constant speed of the towing vehicle and the speed U in the trailer coordinates differ by only a factor of $\cos\theta$, which is nearly unity. (The figure exaggerates the size of the angle θ and the lateral velocity V, for clarity.) The hitch has a flexible connection (shown as a leaf spring) to the nose of the trailer. The deflection of the spring is shown as δ in Figure P5.15a and the spring force is $k\delta$. At the axle, we assume that the lateral velocity shown in part b is nearly zero, as was done for the shopping cart in Section 5.1. (The slip angles of the wheels are assumed to vanish, an assumption that does not give a useful result for the car model in Section 5.2 but simplifies the analysis in this case.)

(a) Noting the similarity between Figure P5.15 and Figure 5.4, write the basic equations of motion for the system. They will be similar to Eqs. (5.21) and (5.22) except that there will be lateral forces at the front and rear of the trailer, shown as Y_f and Y_r, respectively.

(b) Write the expression that states that the lateral velocity at the trailer axle is zero.

(c) Now comes the tricky part. The force from the spring is $k\delta$, so $Y_f \cos\theta = k\delta \simeq Y_f$, but the body-centered coordinate variables really just provide the velocities on the body. So we must use a differentiated form of the spring law $\dot{Y}_f = k\dot{\delta}$. From Figure P5.15*c*, derive the relation $\dot{\delta} = -(V + ar)\cos\theta - U\sin\theta \simeq -(V + ar) - U\theta$. Note that although θ is very small, because U is large, the last term in the $\dot{\delta}$ must be included. Finally, noting that $r = \dot{\theta}$, find the relation for a second derivative of the force in terms of the body-centered coordinate variables r and V in the form $\ddot{Y}_f = -k(\dot{V} + a\dot{r} - Ur)$.

(d) Try your hand at getting a single equation for the yaw rate r by eliminating V using the results of part (b) and taking two derivatives of the equations in part (a) in order to use the results in part (c). The answer should be

$$(I_z + mb^2)\dddot{r} + mbU\ddot{r} + (a+b)^2kr + (a+b)kU = 0$$

(e) The Routh criterion for a third-order equation of the form $a_0\dddot{r} + a_1\ddot{r} + a_2\dot{r} + a_3r = 0$ states that the system will be stable (1) if all the coefficients $a_1 \ldots a_3$ are positive and (2) if $a_1a_2 > a_0a_3$ (see [3] or [4]). Use this criterion to prove that the trailer will be stable if $b > 0$ and $mab > I_z$.

The significance of this is that trailers can be unstable if the center of gravity is behind the axle, or the loaded trailer has a moment of inertia that is too large. If you ever rent a trailer, you should be instructed to load it so that it has "positive tongue weight." This is the common person's way of saying that the center of gravity should be ahead of the axle. There is no simple way of saying that an excessive centroidal moment of inertia can also cause instability, but accidents have resulted when very long, heavy objects have been towed on a trailer.

5.16 Figure P5.16 shows an airplane moving in plane motion. Its motion could be described much as the car was in Section 5.2, but the motion plane is now vertical instead of horizontal. Using the notation in Appendix A, the idea is that the forward velocity is U and the vertical velocity is W, but there is no lateral velocity V. Also, there is a pitch rate q, but the yaw rate and the roll rate are zero, $r = p = 0$. The lift force at the wing L_w and the lift force at the tail L_t act at distances from the center of gravity of a and b, respectively.

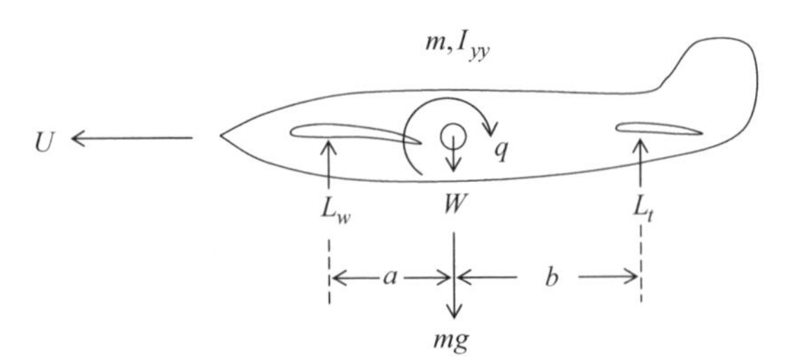

Figure P5.16

The inertia parameters are the mass m and the moment of inertia about the y-axis I_{yy}. Assuming that the forward velocity is constant, write the equations of motion corresponding to Eqs. (5.21) and (5.22) or (5.27) and (5.28) by specializing the equations in Appendix A for this case.

5.17 Figure P5.17 shows the simplified airplane model of Problem 5.16 modified for a stability analysis. The basic motion of the plane is assumed to be horizontal flight at a speed U. For this to be possible, the lift forces at the wing and the tail must balance the weight and also have no net moment about the center of mass. The vertical velocity W and the pitch velocity q are zero for the basic motion. In the figure the variables W and q represent small deviations from zero values.

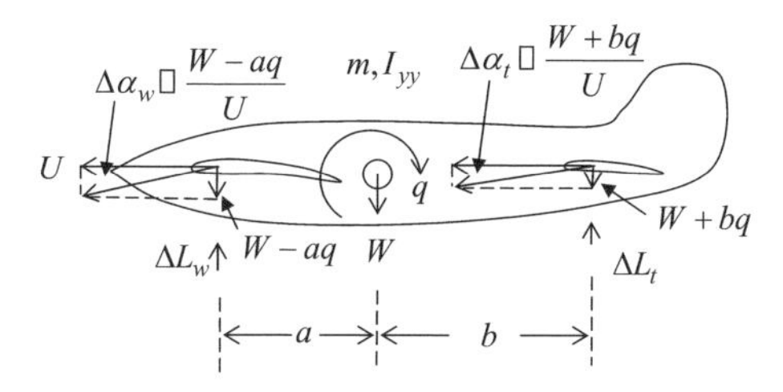

Figure P5.17

The lift forces at the wing and at the tail have to do with the dynamic pressure, $\frac{1}{2}\rho U^2$, the wing and tail areas, S_w and S_t, the slope of the lift coefficients as a function of the angles of attack, $\partial C_{Lw}/\partial\alpha_w$ and $\partial C_{Lt}/\partial\alpha_t$, where the angles of attack α_w and α_t are the angles the relative wind makes with the wing and tail surfaces. For a discussion of these relationships, see ref. [1] or [3]. For the present purposes, the figure leaves out the equilibrium lift forces and the weight and considers only the change in the lift forces due to a change in the angles of attack when the plane is perturbed from the basic motion. It is assumed that the change in lift forces are given by

$$\Delta L_W = \frac{1}{2}\rho U^2 S_W \frac{\partial C_{LW}}{\partial\alpha_W}\Delta\alpha_W, \qquad \Delta L_t = \frac{1}{2}\rho U^2 S_t \frac{\partial C_{Lt}}{\partial\alpha_t}\Delta\alpha_t$$

with the change in angles of attack as given in the figure.

(a) Show that two constant coefficients C_w and C_t can be defined such that the change in lift can be expressed simply as $\Delta L_w = C_w U(W - aq)$ and $\Delta L_t = C_t U(W - aq)$.

(b) Write the equations of motion for the aircraft in a form similar to Eq. (5.33), which was derived for the bicycle model of a car. Instead of the variables V and r, you will use W and q, and although many terms are similar, the signs are not exactly the same.

(c) Now follow the pattern of Eqs. (5.37)–(5.40) to find the characteristic equation of the airplane.

(d) See if the arguments leading to the criterion for stability equation (5.44) are valid in this case. Is there any possibility of a critical speed for this aircraft model?

6

MECHANICAL SYSTEMS UNDER ACTIVE CONTROL

In previous chapters we have emphasized that the equations of motion of mechanical systems are often significantly nonlinear. Many elements of mechanical systems are best described by nonlinear laws—Coulomb friction and bottoming or topping springs are typical examples—but also geometric nonlinearities are common in mechanical systems but not in many other types of dynamic systems. Because of the multiple types of nonlinearity common to mechanical systems, computer simulation is often the only practical option to predict the motion of mechanical systems with fidelity.

On the other hand, linearized versions of nonlinear equations of motion for a mechanical system are very useful when developing an active control system. There are many techniques for finding useful feedback control concepts if the system equations are linear but very few if either the mechanical system or the control system itself is described by nonlinear differential equations. A typical procedure is to develop a concept for control and to study trade-offs using linear techniques, and then to consider nonlinear effects such as saturation of signals, limits of motion, and so on, at a later phase of the design. A final version of the design may be validated by computer simulation of a nonlinear model or by experiment on a physical prototype.

In this chapter, a number of linear techniques typically used in automatic control are discussed as they relate to mechanical dynamic systems. Transfer functions and frequency response functions have already been introduced, and eigenvalues have been found to study the stability of specific example systems. In this chapter, linearized models of mechanical systems are discussed in a more general way. We then present some examples in which active control is used to modify the dynamics of mechanical systems. Block diagrams for the mechanical system and for the feedback control system are used as well as state equations. Another typical use for linearized mechanical system equations is in the field of vibrations, which is developed further in Chapter 8.

6.1 BASIC CONCEPTS

Let us first review some basic concepts of a state-space model of a linearized dynamic system. These concepts are included in almost all books on automatic control, such as [7]. In a number of examples, we have been able to define a state vector $[x]$ with n components, an input vector $[u]$ with m components, and vector–matrix state differential equations of the form

$$[\dot{x}] = [A][x] + [B][u] \tag{6.1}$$

In most cases, the $[A]$ and $[B]$ matrices have constant elements based on parameters of the system, although occasionally the elements can vary with time. In the following we do not deal with the time-varying case.

A set of output variables of interest $[y]$ are defined by the algebraic equations

$$[y] = [C][x] + [D][u] \tag{6.2}$$

where again, we assume that $[C]$ and $[D]$ are constant matrices. Figure 6.1 is a general block diagram of Eqs. (6.1) and (6.2). The solid arrows denote the flow of variables in the mechanical system. The dotted lines denote a feedback control system of a particular type called *state-variable feedback.* If this control scheme is used, the input variables are found from the state variables using a constant matrix $[K]$, $[u] = -[K][x]$. The effect of this type of feedback control on the dynamics of the system is discussed below.

Note in the diagram that, although the notation for the variables has not changed, the variables are, strictly speaking, no longer functions of time but rather, functions of s, which can be considered to be the Laplace transform variable. Use of the same

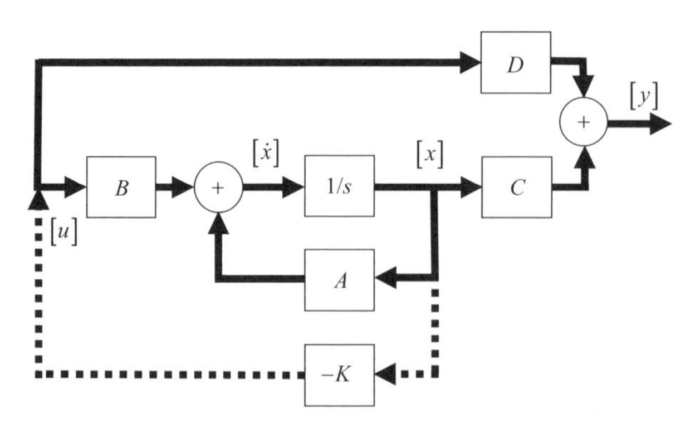

Figure 6.1 General block diagram of a constant-coefficient linear system.

symbol to stand for both a function of time and the equivalent function of s is common in automatic control. [Instead of considering the Laplace transformation of Eqs. (6.1) and (6.2), one can as well assume that the time behavior of all variables is a constant times e^{st}.]

In the s-domain, the equivalent of time differentiation is multiplication by s and time integration is division by s. Thus, the transition from $[\dot{x}]$ to $[x]$ is shown as a multiplication of each variable by $1/s$. In the Laplace domain, the solution of Eq. (6.1) is deceptively easy. With $[I]$ representing the unit matrix (with ones on the main diagonal and zeros elsewhere), Eq. (6.1) becomes

$$[sI - A][x] = [B][u] \tag{6.1a}$$

with the formal solution

$$[x] = [sI - A]^{-1}[B][u] \tag{6.3}$$

and with the output equations

$$[y] = \left[[C][sI - A]^{-1}[B] + [D]\right][u] \tag{6.4}$$

Characteristic Equation

When there are no input variables, $[u] = [0]$, one can see from Eq. (6.1a) that the trivial solution $[x] = [0]$ is always a possible solution. The theory of linear algebra states that this trivial solution will be the only solution unless the determinant of the coefficient matrix is zero. Nontrivial solutions exist only if

$$\det[sI - A] = 0 \tag{6.5}$$

which yields the *characteristic equation* if the determinant is written out. The left side of the characteristic equation is an nth-order polonomial which previously we called the *characteristic polonomial*. The n values of s that satisfy Eq. (6.5) are the *eigenvalues* that we have encountered in specific cases in Chapter 5. The eigenvalues determine the speed and nature of the system response and, in particular, whether the system is inherently stable or unstable. The system is stable if *all* the eigenvalues have negative real parts. The system is unstable if *one* (*or more*) of the eigenvalues has a positive real part.

Transfer Functions

When Eq. (6.3) is solved for the state variable x_i when only the input variable u_j is present, the transfer function $T_{ij}(s)$ is defined as

$$T_{ij}(s) = \frac{x_i}{u_j} \tag{6.6}$$

In a similar way, one can define transfer functions relating the output variables to input variables. Transfer functions have many uses, including the finding of frequency response functions when s is replaced by $j\omega_f$, where ω_f is the forcing frequency when one of the input variables is sinusoidal. Transfer functions are discussed at length in automatic control texts such as [7].

State-Variable Feedback

A powerful idea for a control system is to measure the system state variables and to define a feedback matrix $[K]$ as shown in Figure 6.1. With

$$[u] = -[K][x] \tag{6.7}$$

Eq. (6.1) changes to

$$[\dot{x}] = [A - BK][x] \tag{6.8}$$

and the characteristic equation changes from Eq. (6.5) to

$$\det[sI - A + BK] = 0 \tag{6.9}$$

Much of linear control theory has to do with how effective such a scheme would be in changing the system eigenvalues to more favorable locations or to improving the transfer functions or frequency response functions. Another question that arises has to do with the measurement of the state variables or their estimation based on measurement of related variables. In this chapter we do not attempt to give a comprehensive discussion of linear control theory, but rather, illustrate by example how mechanical systems can be described in ways useful for active control.

6.2 STATE VARIABLES AND ACTIVE CONTROL

Figure 6.2 shows the classical single-degree-of-freedom system often used to introduce basic ideas of vibration control. The figure is typical of sketches used to define single-degree-of-freedom vibratory systems. Here, this elementary system will be used to make some points about the definition of variables and the choice of state variables. The parameters m, b, and k represent the mass, damping coefficient, and spring constant of a linearized version of the actual system, and the force $F(t)$ and base displacement $x_0(t)$ are disturbance inputs to the system. Finally, x is the position of the mass in inertial space, and δ is the relative position of the mass with respect to the base.

Even in this simple system there are some subtle points to be made. The first point has been discussed previously and has to do with constant forces such as gravity forces acting on the system. In Figure 6.2, no gravity force is shown, although the figure seems to imply that the system is in a vertical orientation. If gravity is

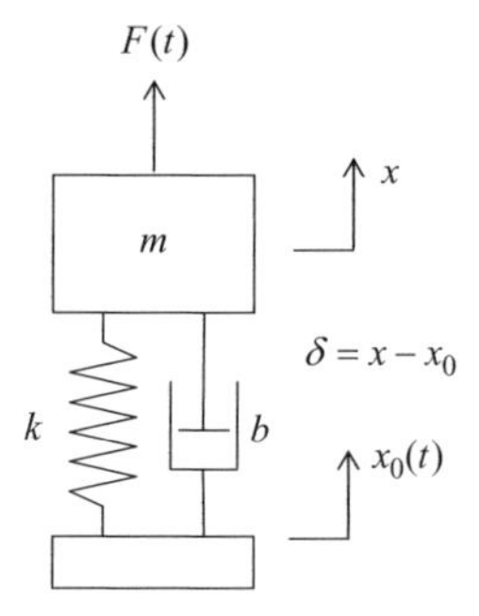

Figure 6.2 Classical single-degree-of-freedom vibration model.

actually present, in the absence of any disturbance, there must be a constant force to hold up the mass. This force can be supplied by the spring if it has a steady deflection.

Suppose that the position of the mass x as well as the position of the base x_0 are measured from positions such that when $x = x_0 = 0$, the spring has a deflection such that the spring force just cancels the gravity force. Then if the spring force is written as $k(x - x_0)$, this expression is really the *extra* spring force that can accelerate the mass when there is a deviation from the equilibrium position. When this is done, the steady spring force in the equilibrium condition and the gravity force will cancel out of the equations of motion. This neglect of the gravity force is justified if the spring is represented by a *linear* force-deflection law but not if the relation is nonlinear. In the nonlinear case, the presence or absence of a gravity force may have a large effect on the natural frequency of small oscillations.

A second point is that the relative deflection δ is actually not well shown in the figure. In the absence of disturbances, if x and x_0 both would vanish, the mass would actually be a fixed distance from the base, but according to the formula in the figure, $\delta = x - x_0$ would also vanish. In fact, δ really represents the relative deflection away from the equilibrium distance between the mass and the base. This equilibrium distance between the base and the isolated mass is often not shown in a sketch of the system. The equilibrium distance does not enter the dynamic equations, so it, like the equilibrium deflection of the spring under the influence of gravity, is usually simply ignored when writing the equations of motion for this linear model.

Writing Newton's law for the mass, the second-order equation of motion is easily written under the assumptions discussed above.

$$m\ddot{x} = F - b(\dot{x} - \dot{x}_0) - k(x - x_0)$$

or

$$m\ddot{x} + b\dot{x} + kx = F + b\dot{x}_0 + kx_0 \tag{6.10}$$

Using the ideas of Chapter 3, a first-order version of Eq. (6.10) almost in the form of Eq. (6.1) can be found by using the velocity $v = \dot{x}$:

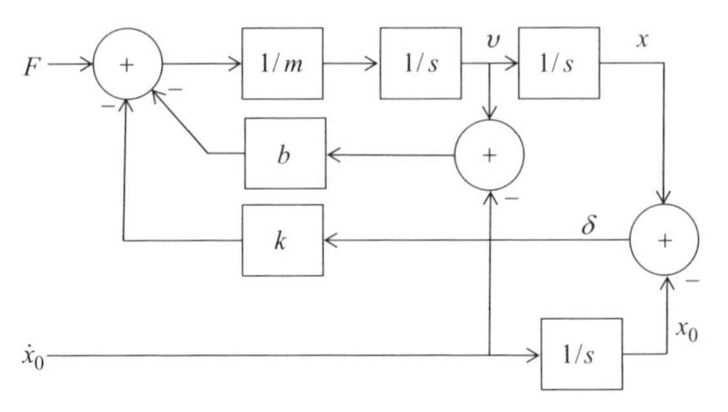

Figure 6.3 Block diagram for Eq. (6.10) or Eqs. (6.11) and (6.12).

$$\begin{bmatrix} \dot{x} \\ \dot{v} \end{bmatrix} = \begin{bmatrix} 0 & 1 \\ -\dfrac{k}{m} & -\dfrac{b}{m} \end{bmatrix} \begin{bmatrix} x \\ v \end{bmatrix} + \begin{bmatrix} 0 & 0 & 0 \\ \dfrac{1}{m} & \dfrac{b}{m} & \dfrac{k}{m} \end{bmatrix} \begin{bmatrix} F \\ \dot{x}_0 \\ x_0 \end{bmatrix} \tag{6.11}$$

Since δ is not one of the states, it can be found as an output variable almost in the form of Eq. (6.2):

$$[\delta] = [\,1 \quad 0\,] \begin{bmatrix} x \\ v \end{bmatrix} + [\,0 \quad 0 \quad -1\,] \begin{bmatrix} F \\ \dot{x}_0 \\ x_0 \end{bmatrix} \tag{6.12}$$

Equations (6.11) and (6.12) have the form of the general state equations, Eqs. (6.1) and (6.2), except that the role of the base motion is unusual. In the general case, the input variables appear alone, but in this case, both the input variable, x_0, and its derivative, $\dot{x}_0$, appear.

Now it is true that in any specific case, if $x_0(t)$ is a known of time, then $\dot{x}_0(t)$ is also known and the converse is also true. A block diagram of Eqs. (6.11) and (6.12) is shown in Figure 6.3, and it should be clear that it does not quite fit the pattern of Figure 6.1. The two input quantities are the disturbance force and the base velocity, but the integral of the base velocity is also needed. There are many known ways to manipulate Eqs. (6.11) and (6.12) into the true state-space form of Eqs. (6.1) and (6.2), but here we show that there is a physically motivated means of choosing another set of state variables rather than the standard means of using mass position and velocity to reduce second-order equations to first-order forms.

Compromises in Passive Vibration Isolation

Before discussing an alternative set of state variables that bring some new insight into the possibilities of active control, let us consider some of the compromises associated with vibration isolation. Suppose that in Figure 6.2, the mass is to be isolated to the extent possible from the effects of both the disturbance force and the base motion by choosing the “best values” of the spring constant and damping coefficient of the isolating structure. Using Eq. (6.10) or the equivalent Eqs. (6.11) and (6.12), it is straightforward to derive several transfer functions of interest:

$$\frac{x}{x_0} = \frac{\dot{x}}{\dot{x}_0} = \frac{\ddot{x}}{\ddot{x}_0} = \frac{bs + k}{ms^2 + bs + k} \tag{6.13}$$

$$\frac{\delta}{x_0} = \frac{-ms^2}{ms^2 + bs + k} \tag{6.14}$$

Equation (6.13) relates the mass position, velocity, and acceleration to the base position, velocity, and acceleration, while Eq. (6.14) relates the relative deflection to the base position.

If one considers the response of the system to the disturbance force, the relative deflection is the same as the absolute mass motion because the base does not move, $x_0 = 0$:

$$\frac{x}{F} = \frac{1}{ms^2 + bs + k} = \frac{\delta}{F} \tag{6.15}$$

Frequency response functions are found from the transfer functions by letting $s = j\omega$, where ω is the forcing frequency for sinusoidal inputs. For example, Eq. (6.13) becomes

$$\frac{x}{x_0} = \frac{j\omega b + k}{(k - m\omega^2) + j\omega b} \tag{6.16}$$

There are some simple results that can be seen even without plotting the amplitude and phase angle of frequency response functions such as Eq. (6.16), as is done in introductory vibration texts (see, e.g., [11]). For example, it is easy to eliminate the motion of the isolated mass completely in response to base motion simply by setting $b = k = 0$ in Eq. (6.16). This is not entirely unrealistic since it means suspending the mass on a spring with negligible friction and using a spring that has a zero spring constant. Such springs can be realized in a number of ways, although they will always have a limited range of deflection. Under these conditions, the isolated mass does not move in response to base motion because there is no isolator force at all (except possibly a constant force counteracting the gravity force if one exists).

On the other hand, Eq. (6.14) shows that when $b = k = 0$, $\delta = -x_0$, meaning that the relative deflection is equal in magnitude to the base position. If the system were to represent the suspension of a car, under these conditions the car would have a smooth ride as long as it never encountered a bump or hill of greater size than the maximum possible deflection of the wheel with respect to the body. Obviously, the idea that isolation from base motion is improved by reducing the damping and the spring stiffness continuously is limited by consideration of the relative suspension deflection, which must also be limited in practice.

One can also give the argument that b and k should have very large values since the deflection in Eqs. (6.14) and (6.16) will be reduced as these parameters are increased.

Also, this will reduce both the isolation and the deflection in response to the disturbance force, as shown by Eq. (6.15). But when the suspension is very stiff, there is no isolation from base motion since Eq. (6.13) shows that x would be almost equal to x_0. The conclusion is that under some conditions, very small values of b and k would give good results, but in other cases very large values of b and k would be best. In general, then, a trade-off is necessary between controlling deflections and reducing the isolated mass accelerations, depending on the input variables to be encountered and the purpose of the suspension system.

A particular problem occurs if any finite value of k is used with a very small value of b. The denominators of all frequency response functions, such as Eq. (6.16), then have very small magnitudes at the undamped natural frequency

$$\omega_n = \sqrt{\frac{k}{m}} \tag{6.17}$$

since at this frequency, the term $k - m\omega^2$ vanishes completely. Then, for a forcing frequency equal to the natural frequency, if b is small,

$$\frac{x}{x_0} \simeq \frac{k}{j\omega_n b} = -\frac{jk}{\omega_n b}$$

which would be a large imaginary number. This is the *resonance problem,* which results in large responses for a lightly damped system any time a disturbance input has a sinusoidal component with a frequency near the natural frequency. Another unfortunate effect of very light damping is that the system will oscillate at a frequency near the natural frequency and will calm down only slowly if it is ever excited by a force or base motion.

The point of this discussion is that a naive optimization often results in a trivial answer that is not practical because in improving one aspect of the system response, another aspect is worsened. A successful optimization results when there is a trade-off between conflicting goals. In that case, there is usually a range of optimum systems, depending on the weight assigned to the conflicting goals.

Active Control in Vibration Isolation

Now, instead of assuming a passive isolator consisting of an elastic element and a damping element in parallel, we consider an active force generator to support the mass to be isolated. Although it may or may not be possible to design a practical actuator to generate the suspension force, this exercise may be instructive in clearly delineating the limitations of more practical passive isolators. Figure 6.4 shows a simple version of an active isolation system. A mystery device is, for now, assumed to be able to produce any control force F_c desired without frequency or magnitude limitations.

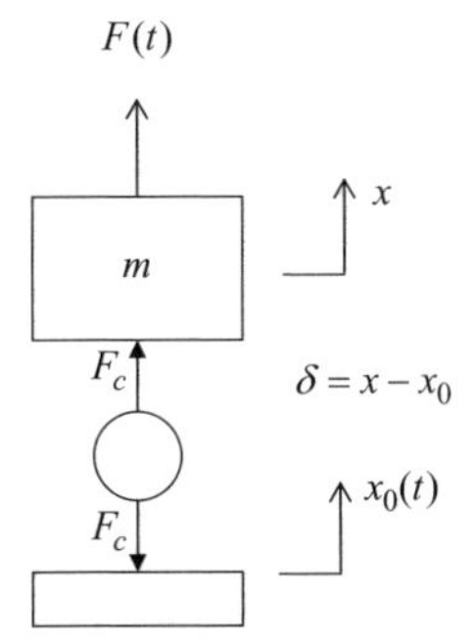

Figure 6.4 An active vibration isolation suspension.

The conventional route to the equation of motion and its conversion to state-space form follows Eqs. (6.10)–(6.12) but contains another surprise.

$$m\ddot{x} = F + F_c \tag{6.18}$$

$$\begin{bmatrix} \dot{x} \\ \dot{v} \end{bmatrix} = \begin{bmatrix} 0 & 1 \\ 0 & 0 \end{bmatrix} \begin{bmatrix} x \\ v \end{bmatrix} + \begin{bmatrix} 0 & 0 \\ 1/m & 1/m \end{bmatrix} \begin{bmatrix} F \\ F_c \end{bmatrix} \tag{6.19}$$

In this formulation, the base motion $x_0(t)$ does not appear at all as an input and does not appear to affect the dynamics of the system.

As we have seen previously, the relative deflection δ is a variable of interest that should be taken into account since any isolation system has only a limited amount of deflection space between the isolated mass and the base. In fact, it would be best to minimize δ if it can be done without too much loss of isolation of the mass from disturbances. One idea is to change the state space so as to include base motion as an input and to include the important variable δ as a state variable. Equation (6.18) can be written as in Eq. (6.19) as

$$\dot{v} = \frac{F}{m} + \frac{F_c}{m} \tag{6.20}$$

and δ can be used instead of x as the other state variable with the equation $\dot{\delta} = v - \dot{x}_0$. The new state equations are then

$$\begin{bmatrix} \dot{\delta} \\ \dot{v} \end{bmatrix} = \begin{bmatrix} 0 & 1 \\ 0 & 0 \end{bmatrix} \begin{bmatrix} \delta \\ v \end{bmatrix} + \begin{bmatrix} 0 & 0 & -1 \\ 1/m & 1/m & 0 \end{bmatrix} \begin{bmatrix} F \\ F_c \\ \dot{x}_0 \end{bmatrix} \tag{6.21}$$

Note that this simple change of state variables has two effects: (1) now only $\dot{x}_0$ is an input variable, not both $\dot{x}_0$ and x_0 as in Eqs. (6.10)–(6.12), and (2), the deflection δ appears directly as a state variable. A comparison of parts a and b of Figure. 6.5 shows how closely related the two state-space formulations are.

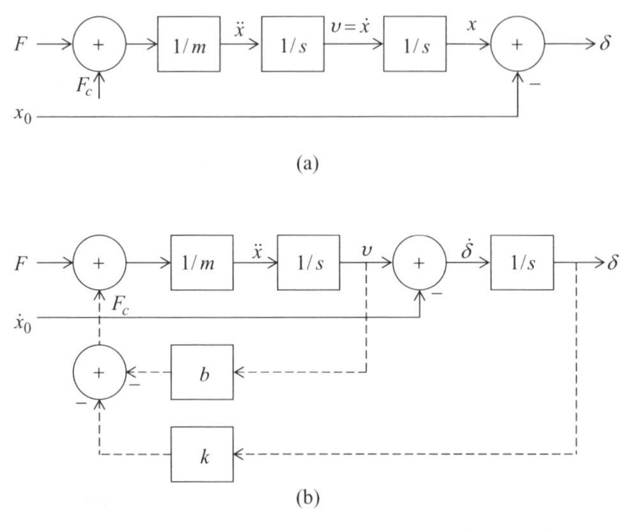

Figure 6.5 Block diagrams corresponding to (*a*) Eq. (6.19) and (*b*) Eq. (6.21).

Figure 6.5*a* makes clear that with the state space of Eq. (6.19), $x_0(t)$ and δ do not affect the dynamics of the state space. Therefore, this state space does not lend itself to the general scheme of Figure 6.1. On the other hand, Figure 6.5*b* has the same form as indicated in Figure 6.1, and a state-variable feedback scheme has been shown in dotted lines as it was shown in general in Figure 6.1. With this state space, the state variables v and δ could be used in a feedback scheme where the feedback coefficients have been called b and k, as if they were damping and spring constants of physical devices.

Optimized Active Vibration Isolator

We have argued previously that passive vibration isolators using springs and dampers must always be designed as a compromise. Stiff springs and dampers lead to small relative deflections but large suspension forces and thus large accelerations of the suspended mass when base motion is considered. Soft springs and dampers lead to low acceleration of the suspended mass but large relative deflections. In any particular case, one can search for an optimum compromise for the suspension parameters. A question then arises: Could an active system, such as that shown in Figures 6.1 and 6.5, provide an even better isolation system in some sense? Control theory provides several answers to this question, but we must first decide how to describe the problem in a precise way and then accomplish some mathematical manipulations that are beyond the scope of this presentation [3, 4]. Here, we present only some interesting basic results.

One idea is to consider that the position of the mass is disturbed from its equilibrium position and ask that the active suspension should return the system to equilibrium while minimizing an integral criterion, J:

$$J = \tfrac{1}{2} \int_0^{\infty} \left(\delta^2 + W F_c^2 \right) dt \tag{6.22}$$

where W is a weighting factor. When W is very large, J will be minimized only if F_c (and hence $\ddot{x}$) is very small. On the other hand, if W is small, J will be minimized when δ (and hence x) returns to zero rapidly using a large control force. In this way a series of optimal suspensions will be found, depending on the value of the weighting factor chosen. Assuming that a means can be found to minimize J for any value of the weighting factor W, we will have found an optimal trade-off between stiff suspensions that rapidly control the deflection and soft suspensions that do not force the isolated mass to have large accelerations.

Another version of the optimal system comes by assuming that the input disturbance is a broadband random process. In the case of Eq. (6.21) and Figure 6.5*b*, the input considered is $\dot{x}_0$. Then a new criterion J' can be defined involving mean-square responses:

$$J' = E(\delta^2 + W F_c^2) \tag{6.23}$$

where E stands for the expected value. The interesting result of the optimization of either Eq. (6.22) or Eq. (6.23) is shown by the dotted feedback signals in Figure 6.5*b*. That is, the optimum control force can be realized by

$$F_c = -bv - k\delta \tag{6.24}$$

for values of b and k that depend on the weight W in the criteria.

A comparison of Figures 6.3 and 6.5*b* shows that the spring in the passive isolator produces a force $k\delta$ just as the optimum active isolator would. This means that one could use a normal spring to generate some of the force the optimum active system would produce. However, in the passive system, the damper reacts to the *relative velocity* $\dot{\delta} = v - \dot{x}_0$, and in the active system a component of the optimum force has only to do with the *absolute velocity* v. Thus, the only part of the optimum active isolator force that really needs to be generated by an actuator is a force proportional to the absolute velocity of the isolated mass. Intuitively, the optimum active system active does not produce a large force when the base velocity is large, but the passive system does. In a sense, the passive system has the damper in the wrong place: between the base and the mass.

The passive system has the transfer function given in Eqs. (6.13) and (6.16), but the active system has a significantly different one. It can be found by writing out the equations corresponding to Figure 6.5*b*:

$$\frac{x}{x_0} = \frac{\dot{x}}{\dot{x}_0} = \frac{\ddot{x}}{\ddot{x}_0} = \frac{k}{ms^2 + bs + k} \tag{6.25}$$

The obvious difference between the transfer function for the passive and active systems is that the passive isolator has a term bs or $j\omega b$ in the numerator of the transfer function, but the active isolator does not. Among other effects of this difference, the response to high-frequency disturbances is very different.

For large values of ω, Eq. (6.16) approaches the value $x/x_0 = -jb/m\omega$, which means that the response of the passive system to base excitation falls off proportional to $1/\omega$. At high frequencies, the frequency response function corresponding to Eq. (6.25) (with $s = j\omega$) approaches the value $x/x_0 = -k/m\omega^2$, which means that the response of the passive system to base motion falls off more rapidly proportional to $1/\omega^2$. The active system can do a better job than the passive system at isolating the mass from high-frequency base motion.

Another interesting detail of the optimum active system is that whereas the parameter k—and hence the undamped natural frequency in Eq. (6.17)—varies with changes in weighting factor W, the parameter b varies in a way that keeps the damping ratio ζ at the value 0.707 independent of the value of the weighting factor:

$$\zeta = \frac{b}{2\sqrt{mk}} = \frac{\sqrt{2}}{2} \tag{6.26}$$

Figure 6.6 shows that the optimum isolator can be realized by two other systems in addition to the purely active system shown in Figure 6.4. In part *a* of the figure, it is shown that a different passive isolator from the one shown in Figure 6.2 would actually give the same results as the optimum active isolator of Figure 6.4. The spring is in the same location as in the passive isolator of Figure 6.2, but the damper has moved in such a way that it reacts to the absolute rather than the relative velocity. The damper is sometimes called the *skyhook damper* because one end is attached to inertial space. Unfortunately, in most practical cases there is no such place to attach the upper end of the damper. Figure 6.6*b* shows a more practical case. The simple mechanical spring actually supplies the control force component in Eq. (6.24), and the active component supplies the force component that has to do with the isolated mass absolute velocity.

A final idea that has been realized in several automobile suspensions and in other isolation suspension systems is the use of a special damper in the location shown in Figure 6.2 . The damper is one that can have its damping changed rapidly in response to an external signal. This is possible using a fast-acting electrohydraulic valve or a magnetorheological fluid that changes its frictional characteristics as a function of a magnetic field. The idea is to mimic the force a skyhook damper would generate whenever possible but using a damper reacting to relative velocity. Although this is not possible at every instant, there are many times when it is. Whenever the passive but variable damper would produce a force of the opposite sign to the force of the skyhook damper, the passive damper would be controlled to produce almost a zero force. Such *semiactive dampers* can have better performance that passive dampers, yet avoid the energy use and complication of a truly active force actuator (see [4]). This example illustrates that simple linear optimal control studies sometimes give insight into new concepts that are not intuitively obvious. The final result, as in the

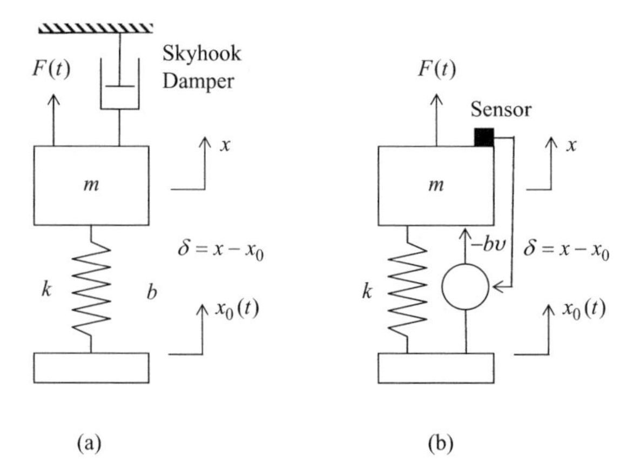

Figure 6.6 Two means of realizing the optimum vibration isolation suspension.

case of semiactive automobile suspensions, often requires extensive development to deal with nonlinear behavior of practical devices.

6.3 STEERING CONTROL OF BANKING VEHICLES

In this section, the problem of stabilizing two-wheeled vehicles such as bicycles and motorcycles through active control of steering is studied. This will be an example of the formulation and linearization of the equations of motion for a mechanical system in order to study the application of a linear control system. There are various ways in which vehicles can be made to lean in curves, but in this section we concentrate on studying the use of steering control to stabilize the lean angle and to follow a curved path. This will provide a simple explanation of how a human operator can control a bicycle or motorcycle or how an automatic control could be designed to balance a tilting vehicle using a steering actuator.

Some tilting trains and commuter vehicles are tilted by direct action of an actuator and an automatic control system that forces the vehicle body to tilt to the inside of a turn against its natural tendency to tilt toward the outside. This requires an active system with an energy supply, but the design of this type of tilting servomechanism is fairly straightforward. Single-track vehicles such as bicycles and motorcycles are tilted in a completely different way by action of the steering system. In this section the dynamics and control of steering-controlled banking vehicles are discussed. These vehicles are generally unstable in the absence of active control of steering by either a human operator or an automatic control system. Although at high-enough speeds, the geometry of the front wheel and fork can be arranged such that "hands-off" stable operation is possible, this phenomenon is not discussed here.

Obviously, steering loses effectiveness as a means of balancing such vehicles at very low speeds or at rest, so another means of achieving balance must be provided.

(Bicycle and motorcycle riders put their feet down for this purpose.) Once a certain speed has been reached, the steering system not only is used to cause the vehicle to follow a desired path, but is also used to stabilize the vehicle and to cause it to bank at the proper angle in a steady turn. The present analysis applies not only to single-track two-wheeled vehicles but also to three-or four-wheeled vehicles with a roll axis near the ground, a high center of gravity, and a suspension with very low roll stiffness. For such vehicles, direct tilt control using an actuator would have to be used to supplement steering control at very low speeds. Here we introduce a number of simplifications to achieve a low-order linear model that is easily understandable, yet illustrates the essential dynamics and control features of steering controlled banking vehicles.

Development of the Mathematical Model

Figure 6.7 shows a number of the dimensions and variables associated with the mathematical model of a tilting vehicle. For a single-track vehicle, Figure 6.7 is to be imagined as existing in the ground plane, although the actual vehicle is tilted with respect to the ground plane. Much of the notation is carried over from the bicycle model of automobiles of Chapter 5. Now, however, the tilting motion of a real bicycle or motorcycle is considered. For a multiwheeled vehicle, the two wheels represent equivalent single wheels for the front and rear axles much as was done for the bicycle

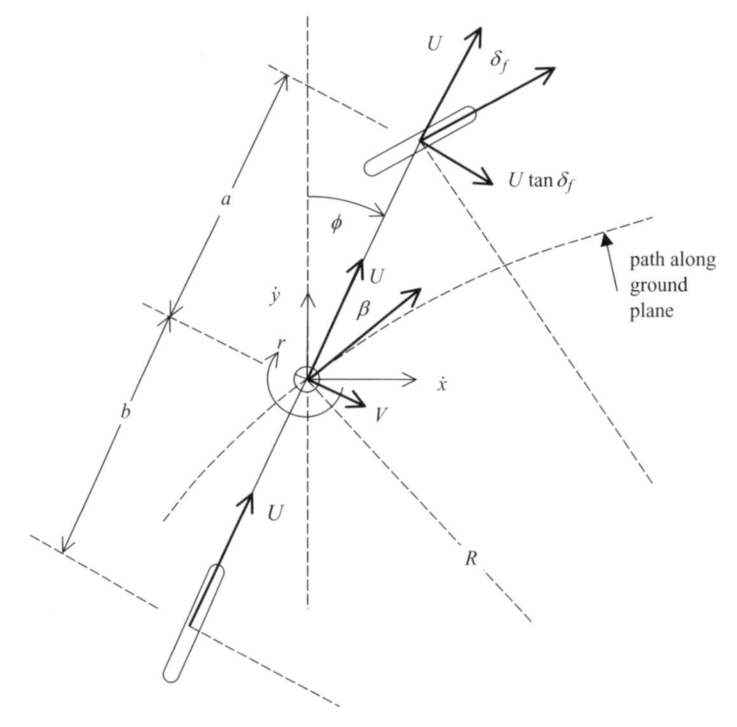

Figure 6.7 Ground plane geometry for a two-wheeled vehicle in a turn.

model for automobiles in Chapter 5. The ground plane would then pass through the roll center of the suspension. Dimensions a and b relate to the distances from the projection of the center of mass to the front and rear axles in the ground plane. The velocity components U and V describe the velocity of the center of mass projection on the ground plane. (Because of the time-varying tilt angle, the center of mass has other velocity components besides these components of the projection in the ground plane.)

The coordinates x and y locate the ground plane projection of the center of mass in inertial space and the angle ϕ, which may be large, represents the orientation of the vehicle with respect to the y-axis. The steer angle δ will be assumed small since the model is intended for use at relatively high speeds when the turn radius, R, is large with respect to the wheelbase, $a + b$. A major simplifying assumption is that the slip angles are negligible. This may seem to be an odd assumption, since in the analysis of automobiles in Chapter 5, slip angles played a major role in the stability analysis. In this section, however, the dynamics of tilting is prominent and the slip-angle effects are not very important as long as the tires do not come close to skidding.

This assumption certainly breaks down at high lateral acceleration, and it precludes the use of nonlinear tire characteristics, but it has the great advantage that no tire characteristics at all are involved in the model. (The introductory example of the shopping cart in Chapter 5 also used this type of assumption.) It is certainly common experience, when riding a bicycle, that it is hard to discern any tire slip angle when riding at moderate lean angles. The wheels appear to roll almost exactly in the direction that they are pointed with no noticeable slip angles.

We also assume that the forward velocity U is constant. With these assumptions, motion in the ground plane is determined purely kinematically. Using the small-angle assumption simple geometric considerations result in an expression for the turn radius (Problem 6.3 involves the derivation of this relationship):

$$\tan\delta \cong \sin\delta \simeq \delta \tag{6.27}$$

$$R = \frac{a+b}{\delta} \tag{6.28}$$

It may be useful to note here that Eq. (6.28) is related to the steer angle relationships encountered previously in the "bicycle model" of an automobile that used finite slip angles and cornering coefficients. If the understeering coefficient should vanish, Eq. (5.64) would match Eq. (6.28). This means that if the tilting vehicle had small slip angles but were a strictly neutral steer vehicle, the relationships would be the same as if the slip angles were actually zero (see Figure 5.8).

The yaw rate r is found easily using Eq. (6.28):

$$r \cong \frac{U}{R} = \frac{U\delta}{a+b} \tag{6.29}$$

The lateral velocity of the center of mass position projected in the ground plane is found again purely kinematically by considering the lateral velocities at the front, $V_f = U\delta$, and the rear, $V_r = 0$:

$$V \cong U\frac{\delta b}{a+b} \tag{6.30}$$

Finally, the slip angle for the center of mass projection in the ground plane, β, is

$$\beta = \frac{V}{U} = \frac{\delta b}{a+b} \tag{6.31}$$

If it is desired to track the location of the center of mass projection in the ground plane during computer simulation, for example, the following equations can be used:

$$\dot{\phi} = r, \qquad \dot{y} = U\cos\phi - V\sin\phi, \qquad \dot{x} = U\sin\phi + V\cos\phi \tag{6.32}$$

With the assumption of zero slip angles for the front and rear wheels, the motion in the ground plane is determined completely by the time histories of the front and rear steer angles. Now the dynamics of the tilting of the vehicle body will be modeled.

Derivation of the Dynamic Equations

Figure 6.8 shows the vehicle body with its center of mass a distance h above its ground plane projection and tilted at the lean angle θ. The angle θ functions as the

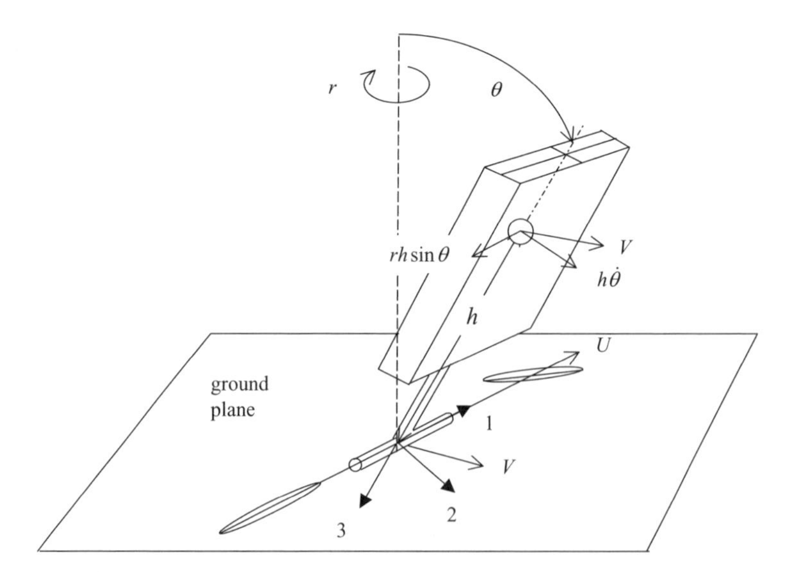

Figure 6.8 The Vehicle body tilting around the ground plane axis.

single geometric degree-of-freedom. The principal axes of the body are assumed to be parallel to the 1–2–3 axis system shown in the ground plane in the figure. The 1-axis lies along the longitudinal axis of the vehicle in the x–y plane, the 3-axis is aligned with the vehicle axis that is vertical when the lean angle is zero, and the 2-axis is perpendicular the 1- and 3-axes. The principal moments of inertia relative to the mass center are denoted I_1, I_2, and I_3. When writing the expression for kinetic energy, one can imagine the 1, 2, 3 axes translated to the center of mass and actually then being the principal axes for the body.

The equation of motion will be derived using Lagrange's equation as described in Chapter 4. In this case of three-dimensional motion, the kinetic energy expression is more complicated than for the plane motion vehicle model, as discussed in detail in Chapter 5. There is now also a potential energy term having to do with the height of the center of mass in the gravity field. Because of the assumption of zero slip angle, the tire lateral forces are perpendicular to the tire velocities in the ground plane and thus do no work. This means that there is no need for a generalized force to represent these forces. The steer angles merely provide a prescribed kinematic motion of the ground plane axis about which the vehicle tilts. The motion variables U, which is assumed to be constant, and V, which is determined by the steer angles through Eq. (6.30), will, however, enter the expression for the kinetic energy.

From Figure 6.8 one can find the square of the velocity of the center of mass, noting that the U and V velocity components lie in the ground plane. The velocity of the center of mass is composed of the velocity of its projection in the ground plane with components U and V added to the components induced by the angular rotation rates r and $\dot{\theta}$. All the components of the center of mass velocity are shown translated to the center of mass location in Figure 6.8. The square of the center of mass velocity can be written as the sum of the squares of two orthogonal components. The first component is the component in the 1-direction, $U - rh\sin\theta$. The second component has to do with the vector sum of V and $h\dot{\theta}$. These two vectors are separated by the angle θ. The square of the vector sum can be found using the law of cosines to be $V^2 + h^2\dot{\theta}_2 + 2Vh\dot{\theta}\cos\theta$. The final expression for the square of the velocity of the center of mass is

$$v_c^2 = (U - rh\sin\theta)^2 + (V^2 + h^2\dot{\theta}^2 + 2Vh\dot{\theta}\cos\theta) \tag{6.33}$$

The angular velocities along the 1, 2, 3 directions are seen to be

$$\omega_1 = \dot{\theta}, \qquad \omega_2 = r\sin\theta, \qquad \omega_3 = r\cos\theta \tag{6.34}$$

Then the kinetic energy expression appropriate for three-dimensional motion as given previously in Chapter 4 is

$$T = \frac{mv_c^2}{2} + \frac{(I_1\omega_1^2 + I_2\omega_2^2 + I_3\omega_3^2)}{2} \tag{6.35}$$

in which Eqs. (6.33) and (6.34) will be substituted. The potential energy expression is just mg times the height in the gravity field,

$$V = mgh\cos\theta \tag{6.36}$$

where in this case V represents the potential energy rather that a lateral velocity.

Finally, substituting Eqs. (6.33) – (6.36) into Lagrange's equation for the single generalized coordinate $\theta(t)$, the resulting equation of motion is found to be

$$(I_1 + mh^2)\ddot{\theta} + (I_3 - I_2 - mh^2)r^2\cos\theta\sin\theta - mgh\sin\theta = -mh\cos\theta(\dot{V} + rU) \tag{6.37}$$

This equation is certainly a more complicated equation of motion than necessary for present purposes since no small-angle approximations have yet been made for the lean angle θ. Only in extreme cases do bicycles or motorcycles achieve large lean angles, so θ will be assumed to be small enough to allow the use of the usual trigonometric small-angle approximations. (Large lean angles for a bicycle riders commonly occur only just before a crash. Motorcycle racers do often achieve lean angles of as much as 45°, but most riders never see such extreme tilt angles.)

The middle term on the left-hand side of Eq. (6.37) will be neglected since it involves the product of the small lean angle and the square of the yaw rate, which is also small for cases of practical interest. After applying the small-angle approximations for the remaining functions of θ, a linearized equation results.

$$(I_1 + mh^2)\ddot{\theta} - mgh\theta = -mh(\dot{V} + rU) \tag{6.38}$$

(This version of the equation can also be derived using Newton's laws and the-small angle approximations from the beginning, but it requires the introduction of tire lateral forces and then their subsequent elimination.)

The terms on the right side of Eq. (6.38) are determined entirely by the time-varying steer angles using the kinematic equations derived above, Eqs. (6.29) and (6.30). After using these relations, the final form of the linearized equation of motion is

$$(I_1 + mh^2)\ddot{\theta} - mgh\theta = -\frac{mh}{a+b}(bU\dot{\delta} + U^2\delta) \tag{6.39}$$

This equation has several interesting features. For example, if the steer angle and its rates were both zero, the equation would describe an upside-down pendulum (for small angles). The term $I_1 + mh^2$ can be recognized as the moment of inertia of the pendulum about a pivot point in the ground plane. This should be no surprise since a motorcycle with locked steering would surely tend to fall over just as an upside-down pendulum would.

The right side of Eq. (6.39) indicates that as $U \to 0$, steering action becomes ineffective in influencing the lean angle as would be expected. What may come as a surprise is that not only does the steer angle influence the lean angle, but the steer angle rate also has an effect. In fact, for low speeds, the steer angle rate is more important than the angle since as the speed decreases, the effectiveness of the rate declines only with U, while the effectiveness of the angle declines more drastically with U^2.

Stability of the Lean Angle

To study the control aspects of banking vehicles, it is useful to define some combined parameters that allow the structure of the dynamic equation to be seen clearly. By dividing Eq. (6.39) by mgh and defining two time constants and a gain, the equation appears in the form

$$\tau_1^2\ddot{\theta} - \theta = -K(\tau_2\dot{\delta} + \delta) \tag{6.40}$$

where

$$\tau_1^2 \equiv \frac{I_1 + mh^2}{mgh},$$
$$\tau_2 \equiv \frac{b}{U},$$
$$K \equiv \frac{U^2}{g(a+b)} \tag{6.41}$$

Note that the time constant τ_1 is determined solely by the physical parameters of the vehicle but that the time constant, τ_2, as well as the gain, K, change with the speed. As the speed increases, the gain K increases, but the variable time constant τ_2 becomes shorter.

For the *basic motion* of a stability analysis, the lean angle, the steer angle, and its derivatives all vanish. This assumption obviously satisfies Eq. (6.40). For the *perturbed motion*, the right-hand side of Eq. (6.40) is still zero, but θ is no longer assumed to vanish. The characteristic equation for the system is then

$$\tau_1^2 s^2 - 1 = 0 \tag{6.42}$$

which, when solved, results in two real eigenvalues,

$$s = \frac{\pm 1}{\tau_1} \tag{6.43}$$

The existence of a positive real eigenvalue confirms that the vehicle is unstable in lean when it is not controlled by a manipulation of the steering angles. (Since the model does not contain any dynamics of the front wheel and fork, it cannot predict the stability of hands-off operation. No active control in the present context means that the steer angle δ simply remains zero.)

Steering Control of the Lean Angle

Suppose that we consider designing a lean control system using steering. The transfer function relating θ to δ can be found in the standard way from Eq. (6.40),

$$\frac{\theta}{\delta} = -\frac{K(\tau_2 s + 1)}{\tau_1^2 s^2 - 1} \tag{6.44}$$

To stabilize the lean angle, let us first try the simplest possible proportional control system,

$$\delta = -G(\theta_d - \theta) \tag{6.45}$$

in which G is the proportional gain, θ_d is a desired lean angle, and the minus sign compensates for the minus sign inherent in the transfer function, Eq. (6.44). [The minus sign on the gain in Eq. (6.45) is not particularly important and is simply the result of the choice of signs for the direction of the steer angles in the derivation of the dynamic equation. The control system is a negative feedback system when all the negative signs are considered.] A block diagram for the controlled system is shown in Figure 6.9.

The transfer function for the closed-loop system can be found by algebraically combining the control law, Eq. (6.45), with the open-loop transfer function, Eq. (6.44), or by block diagram algebra to be

$$\frac{\theta}{\theta_d} = \frac{GK(\tau_2 s + 1)}{\tau_1^2 s^2 + GK\tau_2 s + (GK - 1)} \tag{6.46}$$

The denominator of this transfer function (which is the second-order characteristic polynomial for the closed-loop system) indicates that the system will be stable if all the coefficients of the powers of s are positive. (The characteristic equation has the same form as for the car model analyzed in Sections 5.2 and 5.3.) Only the constant term in the characteristic equation has the possibility of being negative, so stability requires that

$$GK = \frac{GU^2}{g(a+b)} > 1 \tag{6.47}$$

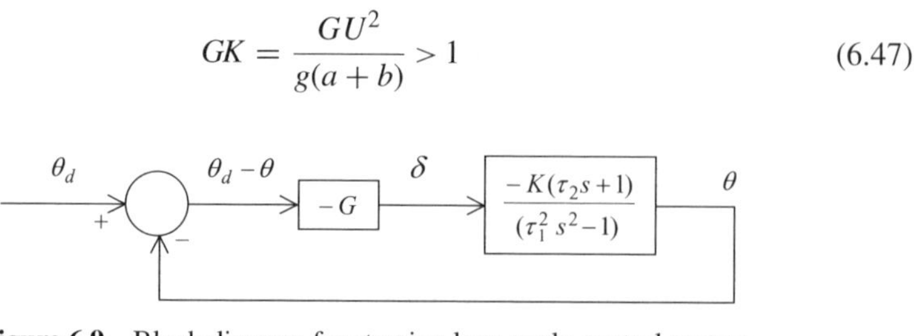

Figure 6.9 Block diagram for steering lean angle control system.

At high speeds, it is no problem to find a value of the gain G to satisfy Eq. (6.47), but as U decreases, G must increase drastically to maintain stability. Practically, it is not possible to increase the gain sufficiently for stability as $U \to 0$.

A bicycle rider slowing to a stop typically steers in ever-wider excursions trying to maintain balance but is ultimately forced to put down a foot to remain upright. In terms of the present model and the proportional control assumed, to maintain stability the rider appears to be trying to increase G as U decreases. A large gain means that the steer angle becomes large even for small deviations of the lean angle from the lean angle desired. If a rider is stopping, the lean angle desired is zero. The rider can react to small lean angles with large steer angles in an attempt to keep the bicycle upright, but eventually the attempt fails as the speed approaches zero. (It is true that because of the geometry of actual bicycle front forks, skilled riders can actually balance at standstill, but this effect is not in the present model. In any event, balancing at zero speed is a difficult trick that most riders can rarely perform.)

By setting $s = 0$ in the closed-loop transfer function, Eq. (6.46), one can see how the steady-state lean angle θ_{ss} is related to the desired lean angle θ_d. Using Eq. (6.41) an expression for the relation as a function of speed can be found:

$$\frac{\theta_{ss}}{\theta_d} = \frac{GK}{GK - 1} = \frac{GU^2/g(a+b)}{\left[GU^2/g(a+b)\right] - 1} \tag{6.48}$$

If the system is stable, $GK > 1$ and the steady lean angle is at least proportional to the lean angle desired. For a given gain G, as the speed increases the steady lean angle approaches the desired lean angle because GK becomes large and the ratios in Eq. (6.48) approach 1. Thus, this simple proportional control system not only stabilizes the system (except at low speeds) but also forces the lean angle to approach any given desired lean angle at high speeds or at moderate speeds if the gain G is large.

Now, by considering the transfer function relating the steer angle and the lean angle, Eq. (6.44), when $s = 0$, it can be seen that in the steady state, the lean angle is related directly to the steering angle δ,

$$\theta_{ss} = K\delta_{ss} = \frac{U^2}{g(a+b)}\delta_{ss} \tag{6.49}$$

The steering angle is kinematically related to the turn radius and to the yaw rate equations (6.29) and (6.30). These relations provide a philosophy for determining the desired lean angle. First the rider (in the case of a bicycle or motorcycle) or an automatic control system (in the case of a controlled tilting vehicle) determines a turn radius or a yaw rate required to follow a desired path. This then determines a steer angle from Eq. (6.29) or Eq. (6.30):

$$\delta = \frac{a+b}{R} \quad \text{or} \quad \delta = \frac{(a+b)r}{U} \tag{6.50}$$

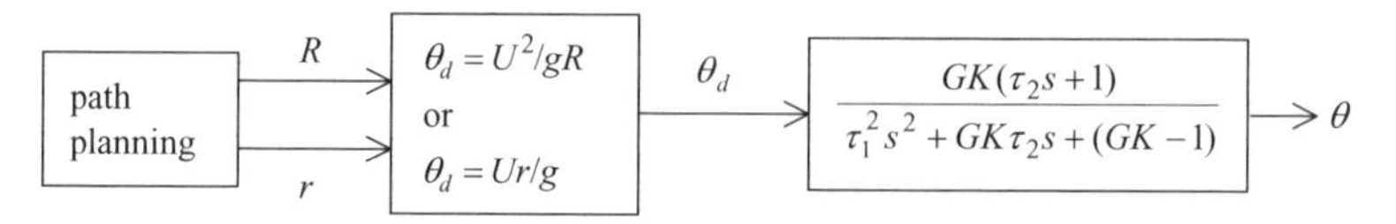

Figure 6.10 How a desired turn radius or yaw rate determines the lean angle desired for a lean control system.

Then, assuming that $\theta_d \cong \theta_{ss}$, the steady-state relation, Eq. (6.49), can be used to derive a desired lean angle given the turn radius or yaw rate desired:

$$\theta_d \cong \theta_{ss} = \frac{U^2}{gR} \quad \text{or} \quad \theta_d \cong \theta_{ss} = \frac{Ur}{g} \tag{6.51}$$

The block diagram in Figure 6.10 indicates the process.

Beginning motorcycle riders are often taught to think about leaning the motorcycle in the direction of a desired turn rather than steering directly in the direction desired as one would in a tricycle or a car. The block diagram makes this advice seem reasonable. The idea is to learn how much lean is required at a given speed either to follow a given turn radius or to achieve a given yaw rate. The peculiarity of tilting vehicles is that just turning the wheel immediately in the turn direction desired will not work, but making the vehicle lean will.

Counter Steering or Reverse Action

There is nothing particularly unusual about the response of the lean angle of the controlled system to a change in desired lean angle as long as the gain G in combination with the speed are large enough to make $GK > 1$, so that the system is stable. The closed-loop transfer function in Eq. (6.46) has a first-order numerator and a second-order denominator just like the transfer function relating yaw rate to steering angle for the automobile model studied in Chapter 5 [Eq. (5.52)]. What may come as a surprise is that the steer angle itself does have unusual dynamics for the closed-loop system.

Using the dynamic and controller equations directly, Eqs. (6.40) and (6.45), or using block diagram algebra on Figure 6.9, one can derive the transfer function relating the steer angle and the desired lean angle:

$$\frac{\delta}{\theta_d} = \frac{-G(\tau_1^2 s^2 - 1)}{\tau_1^2 s^2 + GK\tau_2 s + (GK - 1)} \tag{6.52}$$

For a constant desired lean angle, the steady state steer angle produces the essentially the turn radius specified if Eq. (6.51) is used. The steady steer angle for a constant desired lean angle is found by setting $s = 0$:

$$\delta_{f,\text{ss}} = \frac{G}{GK - 1}\theta_d = \frac{G}{GU^2/g(a+b) - 1}\theta_d \tag{6.53}$$

At high speeds, when $GK >> 1$,

$$\delta_{f,\text{ss}} \rightarrow \frac{g(a+b)}{U^2\theta_d} \tag{6.54}$$

If the lean angle desired is given in terms of the speed and the turn radius from Eq. (6.51), the result is that

$$\delta_{f,\text{ss}} \rightarrow \frac{a+b}{R} \tag{6.55}$$

as expected, considering Eq. (6.29). At lower speeds, or for lower values of G, the steady-state steer angle would not exactly result in the turn radius desired, but a rider or a controller could correct this by adjusting θ_d somewhat if a larger or smaller turn radius would be necessary to follow a desired path.

The initial value of the steer angle to a sudden step change in the lean angle desired is found by letting $s \rightarrow \infty$ in the transfer function (see [7]). When s approaches infinity in Eq. (6.52), the result is

$$\delta(0) = -G\theta_d \tag{6.56}$$

Equations (6.56) and (6.53) mean that for a positive step in the lean angle desired, the steer angle is initially negative but ultimately becomes positive. See the steer angle step response to the change in lean angle desired shown in Figure 6.11. The reason for this peculiar behavior lies in the numerator of the transfer function that has one "zero" in the right half of the s-plane. This means that the transfer function between the desired lean angle and the steering angle is *nonminimum phase* and has a reverse-action steering angle response. (The term *nonminimum phase* is discussed in books on automatic control such as ref. [7]. Here we merely note that the steering transfer function step response has the peculiar reverse action shown in the sketch.)

Motorcycle and bicycle riders recognize this reverse action phenomenon that is required to initiate a sudden turn. The initial phase of steering, which is in the opposite direction to the final steer direction, is called *counter steering*. In order to make a sharp turn in one direction, one initially has to turn the handlebars in the opposite direction. Some motorcycle riding instructors actually encourage their students to practice this counterintuitive counter steering in order to better handle rapid avoidance maneuvers. The reason for counter steering is often explained in less technical terms as follows: To make a right turn, the vehicle must establish a lean to the right. Starting from an upright position, the vehicle is made to lean right only by initially steering left. By steering left, the wheels move to the left relative to the center of mass, thus creating a

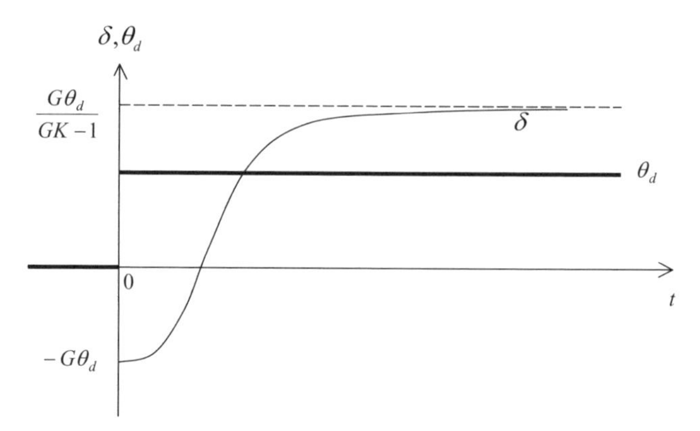

Figure 6.11 Response of the steer angle to a step change in desired lean angle.

lean to the right. As the lean develops, the steer angle must change from left to right to stop the lean angle from increasing too much. In the end the vehicle turns to the right at a constant lean angle with gravity, tending to make the vehicle fall to the inside of the turn and the centrifugal force tending to make the vehicle lean to the outside of the turn. In a steady turn, these two tendencies just balance each other.

Riders of bicycles and motorcycles are often not aware of the need to counter steer if they negotiate only gentle turns. In normal riding, they rarely are perfectly upright and instead, are continually making small steering corrections to keep the lean angle near zero. When it comes time to initiate a turn, the rider simply allows a lean angle in the correct direction to develop without correction and then steers to stabilize the lean angle at a value that corresponds to the speed and the turn radius desired.

This example illustrates that the linearization of a nonlinear mechanical equation is not only a great help in making a preliminary design for a control system but also often provides insight to the dynamic response of the system. Certainly, a more complex mathematical model can be computer simulated and used to refine a proposed control system, but the simpler linearized model often contributes an intuitive understanding of the dynamic behavior of a mechanical system that can be lost if only a complex model is studied.

6.4 ACTIVE CONTROL OF VEHICLE DYNAMICS

In this section we deal with the possibilities of changing the dynamic properties of vehicles through the use of active control. In particular, the idea that the stability of a vehicle can be enhanced by active control means is discussed. In Chapter 5 we showed how a special form of Newton's laws could be applied to ground vehicles and concentrated mainly on analyzing the inherent stability properties of these vehicles. By the term *inherent stability* we mean the ability of the vehicle to return to a basic motion after a disturbance without the aid of a human operator or an automatic control system.

As we have seen, for many ground vehicles stability problems do not arise until a critical speed is exceeded. This fact meant that for most ground vehicles, until power plants were developed that allowed high speeds to be obtained, little attention was paid to vehicle dynamics in general and to stability in particular. Today, however, the dynamics of many ground vehicles are influenced not only by mechanical parameters but also by automatic control systems. Such systems can be designed successfully only by first understanding the dynamics of the mechanical system and then applying the principles of automatic control. Even in mass-produced automobiles, we now see examples of active control of brakes, suspension elements, and steering systems. These active systems consist of sensors, electronic control computers, and actuators and they are capable of significantly modifying the dynamic behavior of vehicles. This can be very useful if the dynamics of the vehicle as determined by the mechanical properties alone are unsatisfactory for some reason.

Stability and Control

In most vehicles, there is a conflict between stability on one hand and controllability or maneuverability on the other hand. In Chapter 5 it has been shown that an understeering car is stable, but its yaw rate gain decreases for speeds higher that the characteristic speed and the lateral acceleration gain approaches a limit at high speeds. This means that a car with a low characteristic speed may be very stable at high speeds, but on the other hand, it may not respond well to steering inputs at high speed. Although stability in a vehicle is generally regarded as a positive attribute, excessive stability usually comes at the price of reduced maneuverability and controllability.

The compromise between stability and controllability has been recognized for many years, and the degree of stability of most vehicles has been tailored to their intended uses. One would not expect a sports car to be as directionally stable as an 18-wheel tractor-trailer, for example. In the first part of the twentieth century, recognition of the importance of stability and controllability mainly affected the physical design of vehicles. Later, as the principles of automatic control became understood and as sensors, actuators, and various forms of computers or controllers were developed, it became clear that stability and automatic control could become linked.

Many innovations involving active stability and control found their first applications in aircraft. The autopilot and the first aircraft antilock brakes were fairly simple automatic control devices concerned with stability and control. The autopilot was able to keep an airplane on course without the attention of the pilot. This is particularly useful for an airplane without a high degree of stability because such airplanes normally require the pilot's attention continuously during gusty weather. Since the attempt to design a high degree of inherent stability into an airplane often compromises other aspects of its performance, the introduction of the autopilot allowed the aircraft designer more freedom to optimize other aspects of performance. Antilock brakes that initially just prevented a pilot from blowing out the tires by overbraking on landing have developed into devices that alter the dynamic behavior of aircraft and ground vehicles in sophisticated ways. To illustrate the basic idea of active control of vehicle dynamics in general and stability in particular, some examples from the automotive area will be studied here. Basic models developed in Chapter 5 will be used even though it must be appreciated that actual production systems are based on more extensive mathematical models as well as on experimental results.

From ABS to VDC

Although there are countless examples of the use of active means to influence vehicle dynamics, we concentrate first on a familiar example from the automobile sector. It is particularly impressive because it involves important safety aspects of a mass-market product, which means that it must be cost-effective and fail-safe. In addition, it deals directly with stability issues in a sophisticated way. What will be described is an active automobile stability control system that is a development of ABS, the *antilock braking system*. Many trade names and special acronyms are used to describe these electronic stability enhancement systems, but a fairly generic term is VDC (vehicle dynamic control) [1,12].

The original antilock brakes for airplanes were intended at first merely to prevent brake locking when landing and thus to lessen the chances of catastrophic tire failure. The idea was to monitor wheel angular velocity and to detect when it began to decrease abruptly, indicating when the wheel was about to lock. When imminent locking was detected, the brake hydraulic pressure was released until the wheel angular speed accelerated again under the influence of the friction force on the tire from the runway. At the time the wheel speed returned to near its original value, brake pressure was reapplied. This resulted in a cycling of the brake pressure and of the wheel angular acceleration until the pilot no longer required maximum braking. The average braking force during the cycling behavior may not have been as large as it would have been if the pilot had been able to exert just enough pressure on the brake to achieve the optimum longitudinal slip, but the important point was to prevent a wheel from locking and the tire from skidding along the runway surface.

In the course of time, the antilock braking control strategy increased in sophistication. By the time it began to be applied to cars, the antilock system was capable not only of preventing wheel lockup, but also of searching for the optimal longitudinal slip for maximal deceleration. This extension of the function of ABS brought along its own stability problems. when a car travels on a *split-μ surface*, a surface with different coefficients of friction on the right and left sides, optimal braking on one side would generate a larger retarding force than on the other side. This would cause a yaw moment on the car and the car would tend to swerve. In this case the system that prevented instability in the wheel angular speeds during braking caused vehicle instability in yaw. This problem was ameliorated but not really solved by adopting strategies such as the *select low principle* at the back axle. "Select low" meant that the ABS action was applied equally to both rear wheels when either one began to lock. Thus, some of the braking force was sacrificed at a rear wheel that was on a high-friction surface in order to reduce the yaw moment associated with unequal maximum braking forces available at the two rear wheels.

The original ABS systems were only capable of releasing and reapplying brake pressure being supplied in response to the driver's foot pushing on the brake pedal. (Of course, many cars had power brakes that provided higher hydraulic pressure than could be generated by the driver's force on the brake pedal. It is still true for such systems that if the driver is not pushing on the brake pedal, there is no pressure to modulate at any of the wheels.) Even for simple ABS systems, however, a number of components must be present. There have to be wheel speed sensors, high-speed electrohydraulic valves, and a digital control system. In addition, there had to be a small pump to replace the brake fluid discharged from the brake lines during an ABS episode.

It was soon realized that most of the ABS equipment could be used to solve another problem. Cars with an open differential on the drive axle deliver essentially the same torque to both wheels. If one wheel happens to be on a surface with poor traction, it cannot generate a large traction force without spinning. In such a situation, better traction at the other drive wheel cannot be utilized because the poor traction at the spinning wheel limits the torque to both wheels. One solution to this problem is a limited slip differential that can increase the torque to the nonspinning wheel. This requires the addition of a more complex and expensive component in place of the simple open differential. Another solution is to apply the service brake at the spinning wheel. This has two effects. A spinning wheel generally produces a smaller traction force than a nonspinning wheel and also is capable of generating a smaller lateral force for a given slip angle. A reduction in the angular speed of a spinning wheel generally helps increase the possible lateral and longitudinal tire forces.

The other effect is to apply the total torque applied to the spinning wheel through the differential to the wheel with better traction. The total torque at the spinning wheel consists of whatever torque is caused by the traction force plus the applied brake torque. Thus, even when one wheel is on ice with an extremely low coefficient of friction, the other wheel can receive a large torque from the differential equal to at least the brake torque applied at the spinning wheel. Thus, the nonspinning wheel can generate as large a traction force as is allowed by the pavement on which it is resting.

Since ABS systems monitor wheel speed, it is possible to detect wheel spin as well as imminent locking. If the system has a source of hydraulic pressure other than that provided by the driver's foot, the controller can be programmed to control wheel spin by applying the brake at the spinning wheel. *Traction control systems* (TCSs) typically use the brakes and ABS components in combination with throttle control. In low-traction situations, the brakes can react more quickly than the throttle control to reduce wheel spin, but on a longer time scale, it is logical to reduce engine power. In any case, a TCS system can make the car react in a more stable and predictable way in low-traction situations.

As ABS and TCS systems were developed into commercially viable products, engineers began working on a true stability enhancement system, *vehicle dynamic control* (VDC). With the addition of a supply of hydraulic pressure independent of the braking action of the driver for use with TCS, it became evident that it was possible to apply any one of the four brakes on a car at any time. This opened up the possibility of monitoring the motion of the car and detecting any unusual situations. Then the brakes could be used to correct any deviations from the motion the driver might have expected under normal conditions.

Model Reference Control

VDC systems have some similarity to model reference control systems used in aviation. In these systems, the pilot's inputs to the airplane are measured and applied to a computer model of a reference airplane. The response of the computer model is computed in real time and compared to the sensed response of the actual airplane. Then deviations between the model response and the actual response are used as error signals and used by a controller to actuate control surfaces in such a way as to reduce these deviations. In rough terms, this is the way a space shuttle is made to fly like a commercial transport aircraft even though the shuttle itself inherently has very different characteristics when it is operating as an airplane in the atmosphere. The general idea of model reference control applied to the steering dynamics of an automobile is shown schematically in Figure 6.12. We return to this example of model reference control in the section on active steering, but now the basic idea is applied to VDC systems.

For an automotive VDC, some additional sensors beyond those used for ABS and TCS are usually required. Typically, a yaw rate sensor, a lateral acceleration sensor, and a steering wheel position sensor are included. The ABS systems usually already contain a longitudinal acceleration sensor. The wheel speed sensors together with the longitudinal accelerometer can be used to deduce the vehicle speed even when one or more of the wheels begins to lock. Although there are many refinements to VDC systems to account for special circumstances, it is fairly easy to see how to construct a strategy for using the brakes to improve the dynamics of a vehicle. The simple bicycle model used in Chapter 5 could be used as a reference model to predict the response to

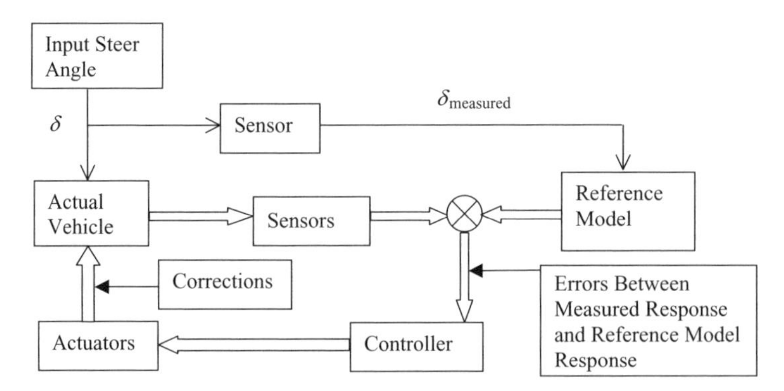

Figure 6.12 Model reference control scheme for influencing the steering dynamics of an automobile.

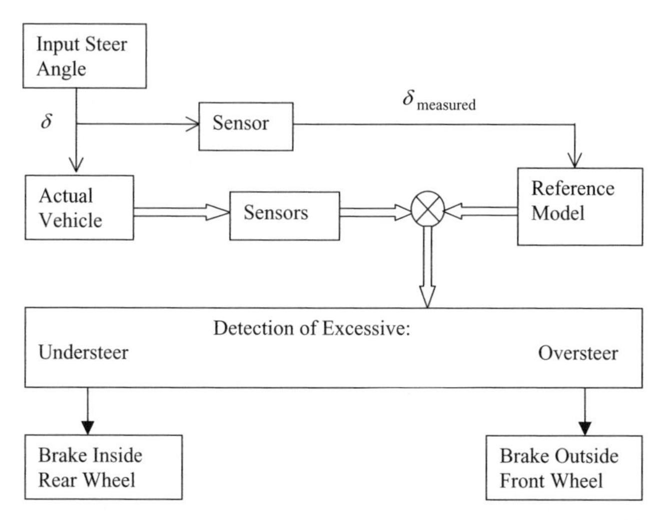

Figure 6.13 Basic version of vehicle stability control using ABS and TCS hardware and model reference control.

steering wheel inputs, for example. Then, knowing the speed and measuring the yaw rate and the lateral acceleration, it would be easy to see if the car was responding in a manner similar to the reference model response.

A basic scheme for using the brakes to correct an excessive oversteering or understeering condition is shown in Figure 6.13. Suppose, for example, that the yaw rate measured in a turn were more that the reference model yaw rate predicted by the computer with the same speed and steering angle. This would indicate an excessive oversteering condition, or in other words, that the rear slip angle was larger than normal. This could be caused by any number of conditions. It could be an underinflated rear tire, a slippery spot in the road being under the rear tires, or an unusually heavy load being concentrated at the rear of the vehicle. The control system could then apply braking to the outside front wheel. Since the rear slip angle is unusually large, the indication is that the front wheels have more traction than the rear and can safely generate a longitudinal braking force in addition to the necessary lateral force. The important point is that the braking force on the outside wheel will generate a yaw moment, tending to reduce the excess yaw rate.

On the other hand, if the measured yaw rate is less than the reference model yaw rate, the inside rear wheel brake would be applied. This would be an understeering condition in which the front slip angle was unusually large. The rear wheels then appear to have excessive traction compared to the front, and braking the inside rear wheel will produce a yaw moment, tending to increase the yaw rate and make the car turn more rapidly. As might be expected, this explanation of VDC is highly simplified and the several commercial versions differ in many details from one another. The point is that the stability properties of vehicles under active control may differ significantly from their inherent properties without control. Particularly in emergency situations, drivers are often not able to control their vehicles effectively. Heavy braking, for example, may turn a normally stable vehicle into an unstable one as the car pitches forward and the normal force at the front increases and the normal force at the rear decreases. This transient change in normal forces will increase the lateral force-generating capabilities at the front and reduce them at the rear. The tendency is to change a normally understeering and stable car into one that is at least temporarily oversteering and possibly unstable. The VDC system can restore stability to the car in a variety of situations and can make the car react in a manner much closer to the way the driver would expect under normal conditions.

One could certainly classify VDC systems as safety systems as well as stability enhancement systems. This is because the only actuators used are the brakes, and any action of VDC systems results in the vehicle slowing down. In most emergency situations, it is beneficial to slow the car down as well as to increase its stability. Even in this form of active stability control, there are differences in emphasis between maneuverability and safety or controllability. Sports car drivers often prefer a less active VDC system than drivers of family sedans because it leaves them with more flexibility to drive aggressively. Some cars offer the option of turning the VDC system off to let the driver control the car without electronic intervention.

Active Steering Systems

In principle, there are several other ways one could change the stability properties of an automobile besides selectively applying the brakes at individual wheels. For example, it is possible to steer the rear wheels of a car using an actuator and a controller responding to sensed deviations of motion from a reference vehicle response. A number of prototype vehicles have been produced with such schemes. Recently, the trend toward various types of drive-by-wire has resulted in the production of *active steering systems*, but most of these systems have acted on the front wheels.

In an active steering system, there need be no fixed relationship between the steering wheel angle (controlled by the driver) and the steering angle of the road wheels. Not only can the effective steering ratio be varied with speed, for example, but the road wheel angles can be controlled by a combination of driver and computer inputs. This means that the response of the vehicle to steering wheel inputs can be varied and an automobile can be made to respond to steering inputs much as a reference vehicle would. In such a case, the directional stability properties of a car could be made to be essentially constant even if there were physical changes in the vehicle.

Major changes in load, tire pressure, or surface traction pose a significant challenge to a driver because the steering response of the car can change drastically from the response under normal conditions. With active steering, a model reference type of controller can, within limits, make the car respond in a more normal fashion even when circumstances have changed the inherent stability properties of the vehicle. Historically, the introduction of power steering reduced the effort required at the steering wheel, but it did not disturb the fixed relationship between the steering wheel angle and the steering angle of the road wheels. Somewhat surprisingly, the steering ratios of cars offered with both manual steering and power steering are usually not very different, although power steering would permit much "faster" ratios. Historically, this was due partially to the desire to make the car steerable should the power steering system fail.

The possibility exists that active steering systems may not only improve the stability of a vehicle but may also be adapted to provide better controllability for particular maneuvers. A simple example of this idea that is now commercially in several forms is speed-dependent steering. At slow speed, the steering is made more responsive, so less motion of the steering wheel is required for parking, but at high speeds the steering ratio is made "slower" so that the driver does not feel that the car reacts too violently to steering wheel inputs. Steering the rear wheels in the direction opposite to the front wheels at low speeds and not steering them at high speeds can accomplish a similar effect. This has been done to improve the maneuverability of large trucks. The main reason for doing this is usually to reduce the minimum turning circle, but in effect the steering ratio is also changed. A more direct way to change the ratio is to steer the front wheels either "by-wire", using an actuator alone, or by using a differential in the steering column to allow the road wheels to be turned by a combination of driver input and actuator input. This way of accomplishing a variable steering ratio or changing the dynamic response to driver inputs is discussed briefly below.

The logic of these schemes has to do with the lateral acceleration and yaw rate gains that tend to increase with speed as discussed in Chapter 5, depending, of course, on the degree of understeering or oversteering the car exhibits. The idea is that as the acceleration and yaw rate gains increase with speed, it seems logical to reduce the ratio of steering wheel motion to road wheel motion. Of course, if the steering ratio were so high that the system was nearly pressure controlled rather than motion controlled, one might cast the argument in terms of a gain between pressure exerted on the wheel (or joystick) and the angular rate of change of the steer angle of the road wheels rather than in terms of a conventional steering ratio.

Stability Augmentation Using Front, Rear, or All-Wheel Steering

It is clear that an automobile can be steered from the front, rear, or both axles, and thus to improve the handling dynamics and stability of a car, one could consider a wide variety of active systems. For example, by extending the bicycle model developed in Chapter 5 to include rear wheel steering, one could use the idea of state-variable feedback shown in Figure 6.1 to change the basic dynamics of the car model by state variable feedback as shown in Figure 6.14. In the figure, $\delta_{f,\text{ref}}$ is the reference steer angle supplied by the driver using the steering wheel, δ_f is the front steer angle, and δ_r is the rear wheel steer angle. As indicated in Section 6.1, this type of scheme has the potential to change the steering dynamics of a vehicle significantly. Although many theoretical studies have been made concerning this type of system, few production systems have been realized.

Feedback Model Following Active Steering Control

The basic idea of model reference or model following control was introduced in Figure 6.12. The idea is to force an actual vehicle to respond nearly as an ideal reference

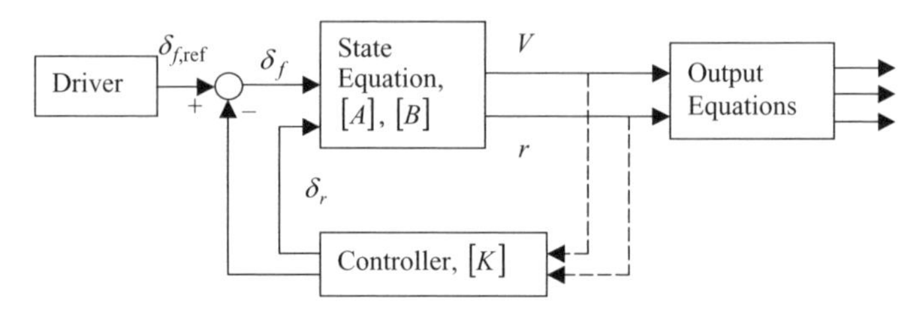

Figure 6.14 State-variable-feedback active steering.

vehicle would with the same command inputs. In the present example, we consider the simple automobile model developed in Chapter 5 and consider active front wheel steering only. The input will be the steering angle commanded by the driver, but the actual steering angle at the road wheels will be determined dynamically by the control system. If the controller is successful, the vehicle will respond much as the ideal vehicle would if given the same steer command.

To the extent that this control system functions as intended, it can solve many problems. For example, if the ideal model represents the real vehicle under normal conditions, the controller will attempt to maintain normal steering responses even when the real vehicle has become unstable. This could happen if a pickup truck were overloaded or if a rear tire were to lose air. Another possibility is that a vehicle may have poor steering characteristics because some other necessary aspects of its design preclude optimizing the dynamic responses. In this case the controller may render the vehicle much easier to drive than would otherwise be the case.

Furthermore, disturbance rejection is automatically a part of a model reference control. When the real vehicle is acted upon by disturbances such as wind gusts or sudden increases in drag on one side of the vehicle due to roadway conditions, it does not follow the model response, and the active system will act to bring the two responses closer together. Thus, it can be expected that a model reference active steering system has the potential to react less to disturbances than does the uncontrolled vehicle.

Finally, it is conceivable that the model parameters can be varied to suit either the task at hand or the driver's preference. During emergency maneuvers, the system could work with a VDC system as described above to do a better job of making the vehicle respond to the driver's commands. It is also conceivable that the driver could choose the response characteristics of the reference model either to make the controlled vehicle respond rapidly for negotiating a curvy road or to make it more stable for cruising a freeway. It should be obvious that the active system will have physical limitations. If the model response deviates too much from the actual vehicle response, it may be that the front wheels are simply not capable of generating enough force to bring the two responses together. Furthermore, for fail-safe operation, the authority of the controller actuator may be purposefully limited.

All of these considerations obviously mean that stability cannot be guaranteed in all situations. For example, on a very low friction surface, no possible active control can make a vehicle follow the responses of a model that presumes a normal friction surface. There will also always be some level of disturbance that will cause significant deviations between the model and the real vehicle. The designer of an active system must always perform computer or actual experiments to make certain that the controller does not make the situation worse under extreme conditions when its effectiveness may be much reduced.

Sliding Mode Control

In this section, a particular type of active steering system is discussed. Although the example to be presented will use linear models for the reference model as well as for the actual vehicle, and thus classical linear control techniques such as that shown in Figure 6.14 could be used, it is clear that these models are only linearized approximations to the nonlinear dynamics actually involved. Thus, it is reasonable to consider control techniques developed to be robust to variations in the linearized coefficients in the models as well as to disturbances. A robust control should continue to function even when the system is operating in the nonlinear region of some of its components. In the case of automotive vehicle lateral dynamics, the nonlinearities of the tire force generation are not severe until the car is maneuvering well beyond the normal range of acceleration in most cases. Therefore, there is a good chance that a robust control technique will yield useful improvements in most practical cases.

In this case we use the *sliding mode control* technique, [5, – 10]. The sliding mode control technique has been developed to be robust to changes in the system to be controlled. The sliding mode theory is presented only for the case of a single-input linear system. For simplicity, the active steering application will use the steer angle as the input, and linearized models as developed in Chapter 5 for the actual vehicle and the electronic reference model. As an introduction to the sliding mode concept, consider a dynamic system represented by the state equation

$$\dot{x} = Ax + bu + d \tag{6.57}$$

where x is an n-dimensional vector of states, u is a scalar input, and d is a bounded vector of disturbances. [The equations of motion, Eqs. (5.31) and (5.32), will be used in the active steering example.] Let $x_d(t)$ be a desired trajectory that the state $x(t)$ is intended to follow.

A function S is defined by the relation

$$S = c^{\mathrm{T}}(x - x_d) \tag{6.58}$$

where c^{T} is an n-dimensional row vector. Clearly, when $x = x_d$, $S = 0$. Then $S(x, x_d) = 0$ defines a surface in the state space that is called the *sliding surface*. This surface varies in time since $x_d = x_d(t)$. A controller is now constructed such that will tend to drive S toward zero or, equivalently, that will tend to drive x toward x_d. In the steering example, the desired trajectory x_d will be supplied by an electronic reference model that receives essentially the same steer angle input that the driver gives to the real vehicle. If S can be driven toward zero, the system state variables will approach the variables of the trajectory desired. For the active steering example, if S can be driven toward zero, the actual vehicle will react to the driver's steering command much as the reference model vehicle would if it had the identical steering inputs.

Consider the following behavior for $\dot{S}$:

$$\dot{S} = -\rho S - \alpha \operatorname{sgn} S = -\rho S - \alpha \frac{|S|}{S} \tag{6.59}$$

where ρ and α are positive constants. Then the derivative of S^2 is easily found:

$$\frac{d}{dt}\frac{S^2}{2} = S\dot{S} = -\rho S^2 - \alpha |S| \tag{6.60}$$

upon use of Eq. (6.59). Equation (6.60) states that the derivative of the square of S is always negative, no matter whether S is positive or negative. This is a sufficient condition to guarantee that S converges to zero [5].

A control strategy is derived by first differentiating Eq. (6.58) and substituting the result into Eq. (6.59):

$$\dot{S} = c^T(\dot{x} - \dot{x}_d) = -\rho S - \alpha \operatorname{sgn} S \tag{6.61}$$

Now Eq. (6.57) is used for $\dot{x}$ with the temporary assumption that the disturbance d is zero and Eq. (6.58) is used for S. When these relations are substituted into Eq. (6.61), the result is

$$c^{\mathrm{T}}Ax + c^{\mathrm{T}}bu - c^{\mathrm{T}}\dot{x}_d = -\rho c^{\mathrm{T}}(x - x_d) - \alpha \operatorname{sgn} c^{\mathrm{T}}(x - x_d) \tag{6.62}$$

When Eq. (6.62) is rearranged, a control law emerges:

$$u = -\frac{1}{c^{\mathrm{T}}b}\left[c^{\mathrm{T}}Ax - c^{\mathrm{T}}\dot{x}_d + \rho c^{\mathrm{T}}(x - x_d) + \alpha \operatorname{sgn} c^{\mathrm{T}}(x - x_d)\right] \tag{6.63}$$

In Eq. (6.63), it must of course be assumed that

$$c^{\mathrm{T}}b \neq 0 \tag{6.64}$$

The robustness of the control law, Eq. (6.63), can be illustrated by using it with the system equation, Eq. (6.57), but now including the disturbance $d(t)$. When this is done, instead of Eq. (6.61), a slightly different equation for S is found:

$$\dot{S} = -\rho S - \alpha \operatorname{sgn} S + c^{\mathrm{T}}d(t) \tag{6.65}$$

Now if the disturbance is bounded,

$$\left|c^{\mathrm{T}}d(t)\right| \leq \beta \tag{6.66}$$

Eq. (6.60) becomes

$$S\dot{S} \le -\rho S^2 - (\alpha - \beta)|S| \tag{6.67}$$

This implies that as long as

$$\alpha > \beta \tag{6.68}$$

the convergence to the sliding surface $S = 0$ is still guaranteed.

This means that for finite-sized disturbances, one can always make $x \to x_d$ by making α large enough. A similar proof shows that if α is large enough, convergence to the sliding surface $S = 0$ will be assured even when the parameters of the system to be controlled vary in a bounded way [5]. For the application to active steering, the conclusion is that even if the real vehicle responses vary because the vehicle responds as a nonlinear system depending on the situation in which it finds itself, the robust sliding mode controller should be able to force the system to respond much as the reference model does to the same steering inputs.

Although the control law, Eq. (6.63), is robust with respect to disturbances and parameter variations when α is sufficiently large, a chattering phenomenon may occur since the control law switches discontinuously whenever the sliding surface is crossed. (The signum function, $\text{sgn}\, S = S/|S|$, jumps between $+1$ and -1 whenever S changes sign.) In many cases this jerky behavior of the control variable is undesirable. In the case of steering control, it would not be desirable for the correction steer angle to chatter back and forth around some nominal value. The chattering problem is often solved while retaining the essential robustness of the controller by using a function that has a continuous transition between $+1$ and -1 for a small region near $S = 0$ in place of the signum function. An example of such a function would be $S/(|S| + \varepsilon)$, where ε is a small positive constant.

Finally, there is the task of choosing the components of c^{T} to assure reasonable behavior of the system when it is forced to remain on the surface $S = 0$ in the state space. One possibility is to consider only the linear part of the control law, Eq. (6.63), by setting $\alpha = 0$. With the assumptions that $x_d = 0$, $d = 0$, one can choose c^{T} such that the system has good stability properties. With these assumptions, when Eq. (6.63) is substituted into Eq. (6.57), the result is

$$\dot{x} = \left(A - \frac{bc^{\mathrm{T}}}{c^{\mathrm{T}}b} A - \rho \frac{bc^{\mathrm{T}}}{c^{\mathrm{T}}b} \right) x \tag{6.69}$$

One eigenvalue turns out always to be $s = -\rho$, since Eq. (6.59) becomes

$$\dot{S} = -\rho S \tag{6.70}$$

if only the linear part of the control law is used. The remaining eigenvalues depend on the choice of the components of c^{T} [9].

Another possibility is to choose c^{T} so that only easily measured components of x are involved in S and then to check to make sure that the system can have good stability properties with this choice of components. Although all the components of the state vector may be involved in the control law, Eq. (6.63), because of the term $c^{\mathrm{T}}Ax$, the other terms, $\rho c^{\mathrm{T}}(x - x_d)$ and $\alpha \operatorname{sgn} c^{\mathrm{T}}(x - x_d)$, would then not involve state variables that are hard to measure. In the example to follow, we will see that there are sometimes certain state variables that are not necessarily important in the control law. When this is the case, a simplified controller results.

State variables not directly measurable may, of course, be estimated using observers, but it is probably better to make as little use of observed variables as possible, for the sake of simplicity and reliability. The use of an observer to estimate a state variable that is difficult to measure directly is illustrated below.

Active Steering Applied to the Bicycle Model of an Automobile

The model of the lateral dynamics of an automobile with front wheel steering only developed in Chapter 5 is represented by Eqs. (5.31) and (5.32). The state variables are the lateral velocity of the center of mass V and the yaw rate r. The control input is the front steering angle δ. For present purposes, the steer angle will be assumed to be a combination of a driver-supplied steer angle, δ_f, and a correction steer angle supplied by the controller, δ_c:

$$\delta = \delta_f + \delta_c \tag{6.71}$$

When Eqs. (5.31) and (5.32) are put in the form of Eq. (6.57), the result is

$$\begin{bmatrix} \dot{V} \\ \dot{r} \end{bmatrix} = \begin{bmatrix} A_{11} & A_{12} \\ A_{21} & A_{22} \end{bmatrix} \begin{bmatrix} V \\ r \end{bmatrix} + \begin{bmatrix} \dfrac{C_f}{m} \\ \dfrac{aC_f}{I_z} \end{bmatrix} (\delta_f + \delta_c) + \begin{bmatrix} d_1 \\ d_2 \end{bmatrix} \tag{6.72}$$

where

$$\begin{aligned} A_{11} &= -\frac{C_f + C_r}{mU}, & A_{12} &= -\frac{aC_f - bC_r}{mU} - U \\ A_{21} &= -\frac{aC_f - bC_r}{I_z U}, & A_{22} &= -\frac{a^2C_f + b^2C_r}{I_z U}. \end{aligned} \tag{6.73}$$

The electronic reference model will be described by similar equations but with generally different parameters:

$$\begin{bmatrix} \dot{V}_d \\ \dot{r}_d \end{bmatrix} = \begin{bmatrix} A_{11d} & A_{12d} \\ A_{21d} & A_{22d} \end{bmatrix} \begin{bmatrix} V_d \\ r_d \end{bmatrix} + \begin{bmatrix} \dfrac{C_{fd}}{m} \\ \dfrac{aC_{fd}}{I_z} \end{bmatrix} \delta_f \tag{6.74}$$

in which it is assumed that the driver's steering angle input δ_f is measured and used as an input to the reference model.

The parameters of the model can be chosen freely so that the response of the model represents a desired response. For example, it might be logical to choose the parameters of the reference model to represent the actual vehicle when it is operating under design conditions. Thus, no corrections from the controller would be necessary unless the vehicle began to deviate from the type of response that the designers had in mind. This philosophy allows the model reference controller to function as a fault detector. It could give the driver a warning that the car was no longer responding in a normal fashion while at the same time correcting the steering response to keep the response closer to the design response than it would be without active control.

Another philosophy has to do with defining a desired response characteristic different from the response characteristic inherent to the uncontrolled vehicle. Often, there are practical reasons why a certain vehicle type will not have inherently desirable steering responses. In such a case, the model reference controller can improve the response. Since it has previously been discussed that a neutral steer vehicle is often thought to represent a sort of optimum vehicle in terms of steering, it might be logical to specialize the results for the neutral steer case and to consider these parameters for use in the reference model.

For neutral steering,

$$aC_f = bC_r \tag{6.75}$$

so C_r can be eliminated in favor of C_f, as was done in deriving the neutral steering transfer functions in Eqs. (5.58) – (5.60). A neutral steering reference model has the following simplified set of parameters to be used in Eq. (6.73):

$$\begin{aligned} A_{11d} &= -\frac{a+b}{bmU}C_f, & A_{12d} &= -U \\ A_{21d} &= 0, & A_{22d} &= \frac{a(a+b)}{I_zU}C_f \end{aligned} \tag{6.76}$$

In general, the sliding surface function S will involve all the state variables. In the case at hand, the yaw rate is easy to measure, and many production vehicles have a yaw rate sensor. In contrast, the lateral velocity is not simple to measure directly. The lateral acceleration is often measured in production vehicles, but this signal is not as straightforward to use in a control system, for a number of reasons (see, e.g., [6]). For this reason, and for the sake of simplicity in presentation, we develop a pure yaw rate controller.

Active Steering Yaw Rate Controller

The vehicle state variables and the reference model state variables for the bicycle model of Eqs. (5.31) and (5.32) are

$$x = \begin{bmatrix} V \\ r \end{bmatrix}, \qquad x_d = \begin{bmatrix} V_d \\ r_d \end{bmatrix} \tag{6.77}$$

To design a pure yaw rate controller, the sliding surface of Eq. (6.58) is made a function of the yaw rate alone:

$$S = c^{\mathrm{T}}(x - x_d) = r - r_d \tag{6.78}$$

which is achieved by choosing

$$c^{\mathrm{T}} = [\,0 \quad 1\,] \tag{6.79}$$

Then some of the terms in the control law, Eq. (6.63), are

$$c^{\mathrm{T}}b = \frac{aC_f}{I_z} \quad \text{and} \quad c^{\mathrm{T}}Ax = A_{21}V + A_{22}r \tag{6.80}$$

The final form of the yaw rate control law using Eqs. (6.78) – (6.80) is

$$u = \delta_f + \delta_c = -\frac{I_z}{aC_f}\left[(A_{21}V + A_{22}r) - \dot{r}_d + \rho(r - r_d) + \alpha\,\mathrm{sgn}(r - r_d)\right] \tag{6.81}$$

We can note that because of the choice of c^{T} in Eq. (6.79), the lateral velocity appears only once in the control law, whereas the rest of the feedback terms involve the measured yaw rate r.

Figure 6.15 is a block diagram of the yaw rate control together with the vehicle and the reference model. Note that although a model of the real vehicle is shown, the signals for lateral velocity and yaw rate that are shown flowing to the controller are actually measured signals coming from sensors on the actual vehicle. The mathematical model of the real vehicle is presented only to visualize the system and does not accurately represent the real vehicle in all cases, particularly when the vehicle is operating in a nonlinear region.

Since the stability of the system depends on the choice of the c^{T} coefficients in the definition of the sliding surface, it is worthwhile to use Eq. (6.69) to see whether the pure yaw rate control will result in a stable system when $S = 0$, considering only

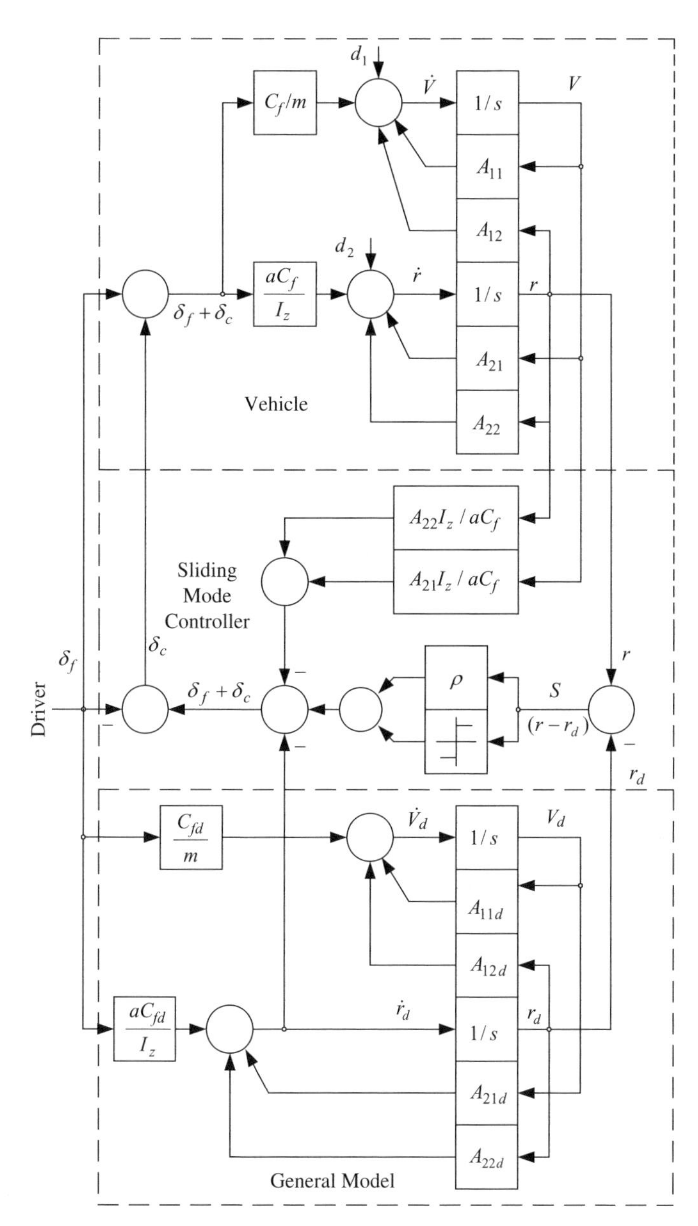

Figure 6.15 Yaw rate model following active steering control using the sliding mode technique (general case).

the linear part of the control law. From Eq. (6.72) the column vector b is

$$b = \begin{bmatrix} \dfrac{C_f}{m} \\ \dfrac{aC_f}{I_z} \end{bmatrix} \tag{6.82}$$

and using Eq. (6.79),

$$bc^T = \begin{bmatrix} 0 & \dfrac{C_f}{m} \\ 0 & \dfrac{aC_f}{I_z} \end{bmatrix}. \tag{6.83}$$

Using Eqs. (6.80) and (6.83), the terms in the closed-loop system matrix in Eq. (6.69) can be evaluated.

$$\begin{bmatrix} A_{11} & A_{12} \\ A_{21} & A_{22} \end{bmatrix} - \begin{bmatrix} 0 & \dfrac{I_z}{am} \\ 0 & 1 \end{bmatrix} \begin{bmatrix} A_{11} & A_{12} \\ A_{21} & A_{22} \end{bmatrix} - \rho \begin{bmatrix} 0 & \dfrac{I_z}{am} \\ 0 & 1 \end{bmatrix}$$

$$= \begin{bmatrix} A_{11} - \dfrac{A_{21} I_z}{am} & -A_{12} + \dfrac{A_{22} I_z}{am} - \dfrac{\rho I_z}{am} \\ 0 & -\rho \end{bmatrix} \tag{6.84}$$

The eigenvalues corresponding to the matrix in Eq. (6.84) are found from the characteristic equation,

$$\text{Det} \begin{bmatrix} s - A_{11} + \dfrac{A_{21} I_z}{am} & -A_{12} + \dfrac{A_{22} I_z}{am} + \rho I_z am \\ 0 & s + \rho \end{bmatrix} = 0 \tag{6.85}$$

When written out, Eq. (6.85) becomes

$$(s + \rho)\left(s - A_{11} + \frac{A_{12} I_z}{am}\right) = 0 \tag{6.86}$$

This yields the eigenvalues

$$s_1 = -\rho, \; s_2 = A_{11} - \frac{A_{21} I_z}{am} \tag{6.87}$$

As was predicted in Eq. (6.70), the first eigenvalue depends only on the control parameter ρ and has to do with the speed with which the system converges to the sliding surface $S = 0$. The second eigenvalue relates to the stability of the system when it is in the surface.

Using the vehicle parameters of Eq. (6.73), the eigenvalue s_2 can be evaluated:

$$s_2 = -\frac{C_f + C_r}{mU} - \frac{(aC_f - bC_r) I_z}{I_z U am} = -\frac{(a + b) C_r}{amU} \tag{6.88}$$

All the parameters in Eq. (6.88) are positive, so s_2 is negative. This negative eigenvalue shows that the controlled system is stable. As can be seen from Figure 6.15, because

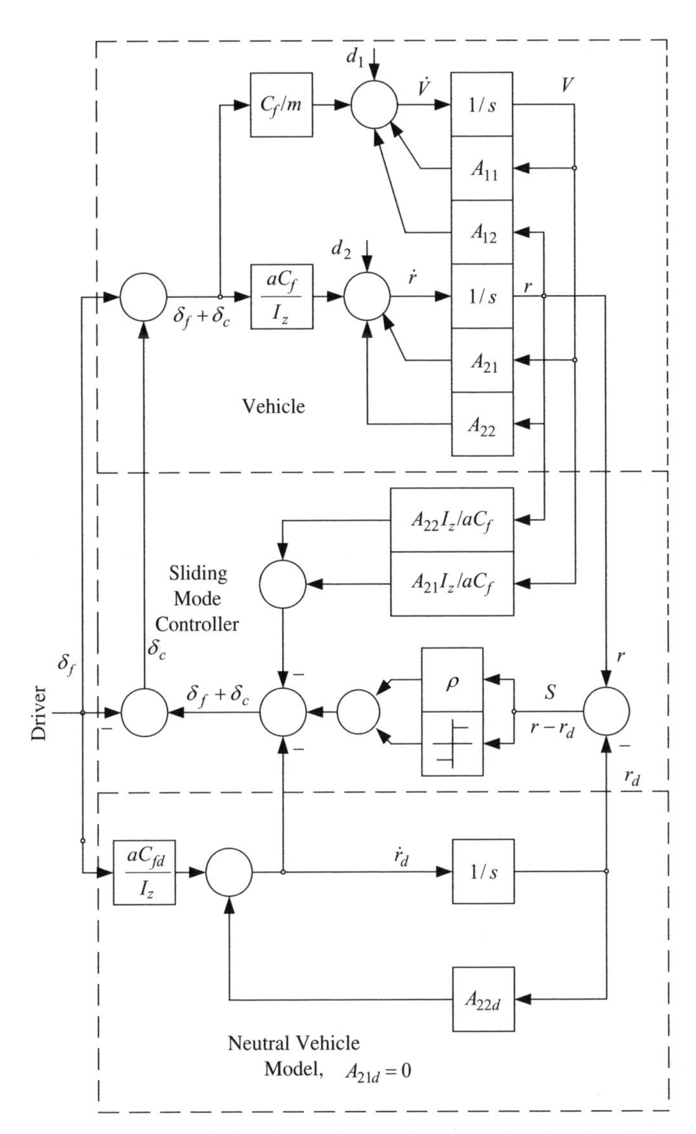

Figure 6.16 Yaw rate model following active steering control using the sliding mode technique (neutral steer reference model).

of the particular form chosen for c^{T} (a pure yaw rate controller), $\dot{V}_d$ is not needed for the controller, but V_d itself must be computed in order to find $\dot{r}_d$, which is needed for the controller. Figure 6.16 shows a further simplification when a neutral steer reference model is assumed. In this case only a simplified first-order yaw model will suffice because the parameter A_{21d} in Eq. (6.76) vanishes.

This robust sliding mode controller shown in Figure 6.16 can be related to model reference controllers studied previously, of the type shown in Figure 6.12 . One idea is to use only the yaw rate error, $r - r_d$, as a control signal and to compute the *steady* yaw rate corresponding to a given δ_f as the reference yaw rate, r_d [6]. Using Eqs. (6.74) and (6.76) for the neutral steer reference model, the steady yaw rate desired is found to be

$$r_{d,\text{ss}} = \frac{aC_f\delta_f}{I_z A_{22d}} = \frac{U\delta_f}{a+b} \tag{6.89}$$

As Figures 6.15 and 6.16 show, a model reference controller using only yaw rate error and a steady-state yaw rate for the desired yaw rate is essentially a simplification of the robust general sliding mode control discussed above. We have shown that it is feasible to construct a robust model following the steering correction controller using sliding mode techniques and yaw rate feedback. It is also possible to use a static reference model for a simplified model reference control system.

Active steering control systems can be realized in several ways. Yaw rate sensors are in common use and yaw accelerometers are available, although lateral velocity is harder to measure directly. The resulting systems can be stable and can make the vehicle respond much as the reference model would respond. The handling properties of the real vehicle can be changed by varying the parameters of the reference model and the vehicle controlled will automatically tend to reject external moment disturbances from wind gusts or braking on split friction coefficient surfaces. Of course, this presentation is highly simplified, and real control systems will have a number of special features necessary to assure good behavior under the many special circumstances that the vehicle and the driver will encounter. These examples are intended only to illustrate the philosophy of active control of handling dynamics.

Limitations of Active Stability Enhancement

It would be a mistake to think that the inherent dynamics of a vehicle are of little importance if active control is to be applied. One reason has to do with failure modes. An unstable vehicle may be stabilized actively, but should there be a failure in the control system, it would be preferable if the operator could still control the vehicle. It is certainly not a good idea to design a very dangerous vehicle and rely on an active control system to correct its bad behavior.

There also is a question of control authority. Any attempt to change the dynamic behavior of a vehicle with active means must consider the amount of control effort that is available. In theory, a system can be stabilized with virtually no control effort. In our analysis of vehicle stability, we considered only infinitesimal deviation from a basic motion. For an infinitesimal deviation from a basic motion, it takes only an infinitesimal force to bring the vehicle back to the basic motion. The control problem is to assure that the control force is properly related to the vehicle dynamics such that it will actually stabilize the system, but the amount of control effort is always proportional to the deviation. In reality, finite-size disturbances will affect the vehicle, and thus the controller must have enough authority to overcome destabilizing influences.

In some cases, large disturbances can be anticipated, and this means that large control forces will be required to stabilize the system. If these forces are limited, perhaps to assure fail-safe behavior, it may be that on occasion, the controller will be unable to stabilize the vehicle. In this case a drastically unstable vehicle that up to a point seemed stable would pose a severe problem for a human operator.

6.5 SUMMARY

In this chapter, a number of application areas were discussed in which mechanical dynamic systems were controlled by active means. Since electronics and computers or computer chips are almost always involved, one sometimes refers to such systems as *mechatronic systems* [2]. In modern machines, the reason for understanding and describing the dynamics of mechanical devices is increasingly to be able to design control systems that can improve the performance beyond what could be achieved without active control. The classical dynamics of the past aimed to understand how mechanical devices functioned dynamically, but today, the study of dynamics is often carried out to be able to use controllers to change the dynamics in positive ways.

REFERENCES

[1] D. D. Hoffman, and M. D. Rizzo, Chevrolet C5 Corvette vehicle dynamic control system, SAE Paper 980233, printed in *Vehicle Dynamics and Simulation*, SAE/SP-98/1361, society of Automotive Engineers, Warrendale, PA, 1998.

[2] D. C. Karnopp, D. L. Margolis, and R. C. Rosenberg, *System Dynamics: Modeling and Simulation of Mechatronic Systems*, 4th ed., Wiley, Hoboken, NJ, 2006.

[3] D. Karnopp, Active and passive isolation of random vibration, Section 3 of *Isolation of Mechanical Vibration Impact and Noise*, AMD Vol.1, No. 1, J. C. Snowdon and E. E. Ungar, Eds., ASME Monograph, American society of mechanical Engineers, New York, 1973, pp. 64–86.

[4] D. Karnopp, 1995, Active and semi-active vibration isolation, *Trans. ASME Special 50th Anniversary Design Issue*, Vol. 117(B), combined issue of *J. Mech. Des.* and *J. Vib. Acoust.*, pp. 177–185, 1995.

[5] P. Muraca, and G. Perone, Applications of variable structure systems to a flexible robot arm, *Proc. IMACS*, pp. 1415–1419, 1991.

[6] J. E. Nametz, R. E. Smith, and D. R. Sigma, *The design and testing of a microprocessor controlled four wheel steer concept car*, SAE Paper 885087, society of Automotive Engineers, Warrendale, PA, 1988.

[7] K. Ogata, *Modern Control Engineering*, Prentice-Hall, Englewood Cliffs, NJ, 1960.

[8] J.-J. Slotine, Sliding controller design for nonlinear systems, *Int. J. Control*, Vol. 40, No. 2, pp. 421–434, 1984.

[9] J.-J. E. Slotine, and S. S. Sasfry, Tracking control of nonlinear systems using sliding surfaces with applications to robot manipulators, *Int. J. Control*, Vol. 33, No. 2, pp. 465–492, 1983.

[10] J.-J. E. Slotine, and W. Li, *Applied Nonlinear Control*, 1991.

[11] W. T. Thomson, and M. D. Dahleh, *Theory of Vibrations with Applications*, Prentice Hall, Upper Saddle River, NJ, 1998.

[12] A. T. Van Zanten, R. Erhart, K. Landesfeind, and G. Pfaff, VDC *systems development*, SAE Paper 980235, printed in *ABS/Brake/VDC Technology*, SAE/SP-98/1339, Society of Automotive Engineers, Warrendale, PA, 1998.

PROBLEMS

6.1 Figure P6.1 illustrates the reduction of a negative feedback loop to an equivalent transfer function.

(a) Write the algebraic relation between y and x to demonstrate the truth of this block diagram equivalence.

(b) Apply this block diagram equivalence twice to the two loops in Figure 6.5b to find the transfer function between δ and F.

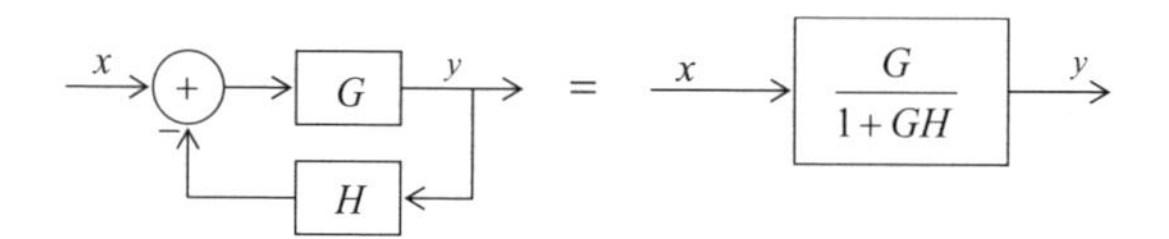

Figure P6.1

6.2 A standard way to write the normalized characteristic equation for a mass–spring–damper mechanical oscillator in terms of the undamped natural frequency ω_n and damping ratio ζ is $s^2 + 2\zeta\omega_n s + \omega_n^2 = 0$. Using physical parameters, the characteristic equation is $ms^2 + bs + k = 0$. By matching coefficients, derive formulas for ω_n and ζ as given in Eqs. (6.17) and (6.26).

6.3 Change the frequency response functions for the passive isolator, Eq. (6.16), and the active isololator, Eq. (6.25), into forms using the damping ratio ζ and the undamped natural frequency ω_n, as defined in Eqs. (6.17) and (6.26).

6.4 Using the results of Problem 6.3, show that for high forcing frequencies, $\omega > \omega_n$, the magnitude of the frequency response is proportional to the damping ratio and inversely proportional to the frequency, whereas for the active isolation system the response is inversely proportional to the frequency independent of the damping ratio. (This means that the active isolator can have a large amount of damping without compromising the isolation from high-frequency base motions.)

6.5 Figure 6.6 shows a scheme for applying a force to a mass that would mimic the optimum skyhook damping force but using an actuator in a practical location between the suspended mass and the moving base. Now consider a *semiactive damper* or controlled shock absorber as the actuator. (Semiactive dampers are quite common in automobile suspensions and in vibration isolation systems.) The damper reacts to the relative velocity $\dot{\delta}$, but it can be controlled by a rapidly reacting electrohydraulic valve or through the use of a magnetorheological fluid to produce a nearly arbitrarily controlled force in one direction. The limitation is that if the relative velocity is positive, the force on the mass can only pull the mass down, and if the relative velocity is negative, the force can only push the mass up. The relative velocity is affected by the instantaneous velocity of the base, $\dot{x}_0$. The limitation comes from the fact that a damper can only dissipate power; it cannot supply power as a general actuator could.

Consider that the desired force to be applied to the mass is the optimum linear force $-bv$, shown in Figure 6.6. The basic idea of a semiactive damper is that if the sign of the desired force is in the direction that the damper can produce, this force will be produced by a control system adjusting the damper characteristics. On the other hand, if the sign of the desired force is opposite in sign to the force that the damper could produce, the damper will be controlled to produce almost no force at all. Draw a space with the mass velocity $v = \dot{x}$ on the y-axis and the relative velocity $\dot{\delta} = \dot{x} - \dot{x}_0$ on the x-axis. Show regions in this space in which the desired force on the mass $-bv$ can be realized by a controlled damper and other regions in which the damper would be controlled to produce a nearly zero force.

6.6 Redraw Figure 6.7 in such a way that the steer angles are small and the turn radius is much larger than the wheelbase. Under these conditions it should be obvious that the angle between the two lines normal to the wheels, shown as dashed lines in Figure 6.7, is approximately $(a + b)/R$. Then show that this angle is also equal to δ, thus proving Eq. (6.28).

6.7 In relation to Figure 6.7, derive an expression for the lateral velocity V in terms of the steer angle δ and the parameters a, b, and U but without using the small-angle approximations. Check that when the steer angles are assumed to be small, the result becomes Eq. (6.30).

6.8 Derive Eq. (6.38) using Newton's laws by considering the vehicle to be an upside-down pendulum. The mass is m, the moment of inertia about the center of mass is I_1, and the center of mass is a distance h above the pivot point in Figure P6.8. The pivot acceleration to the right is given by $\dot{V} + rU$. (This formula comes from Chapter 5 or Appendix A.) You will need to write two equations involving linear accelerations and one involving angular

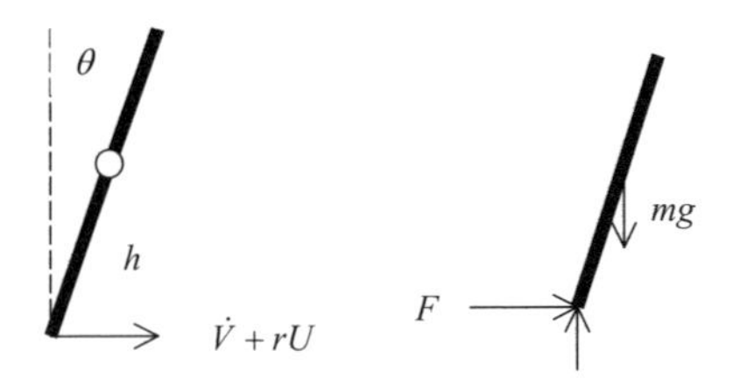

Figure P6.8

acceleration. Convince yourself that the vertical force at the pivot is almost equal to mg under the conditions that apply to Eq. (6.38) and that the acceleration of the center of mass is the acceleration of the pivot point plus some terms having to do with $\theta(t)$. (There are some other small terms in the center of mass acceleration having to do with the yaw rate and the lean angle, but these can be neglected.) Combine the equations to derive Eq. (6.38).

6.9 Equation (6.39) for the dynamics of a tilting vehicle was derived from Eq. (6.38), a linearized version of Eq. (6.37). What would Eq. (6.39) look like if Eq. (6.37) had been used without linearization? [This nonlinear equation could be used in computer simulation to find situations when the linear control scheme of Eq. (6.45) would no longer be effective in stabilizing the vehicle.]

6.10 Consider the state-variable feedback steering system shown in Figure 6.14, but with front wheel steering only. Write the open-loop equations in the form

$$\begin{bmatrix} \dot{V} \\ \dot{r} \end{bmatrix} = \begin{bmatrix} A_{11} & A_{12} \\ A_{21} & A_{22} \end{bmatrix} \begin{bmatrix} V \\ r \end{bmatrix} + \begin{bmatrix} B_1 \\ B_2 \end{bmatrix} \delta_f$$

where the A and B coefficients can be seen in Eqs. (6.72) and (6.73). Write the feedback law in the form

$$\delta_f = \delta_{f,\mathrm{ref}} - [\, K_1 \quad K_2 \,] \begin{bmatrix} V \\ R \end{bmatrix}$$

Combine these relations to find equations for the closed-loop system for the response of the lateral velocity and the yaw rate to the reference steer angle supplied by the driver. Do you think that by using the two gains K_1 and K_2, the closed-loop system can be given arbitrary eigenvalues (at least in theory)?

6.11 An argument in favor of rear wheel steering for an automobile is sometimes based on the response of the rear axle to a sudden steer input at the front axle. The concept is that only with rear wheel steering is there a direct control of the rear steer angle and thus the lateral force at the rear axle. Otherwise, the response at the rear axle to sudden steering inputs at the front is determined only by the basic dynamics of the vehicle, particularly by the mass distribution. This is a basic limitation to active front wheel steering. Consider the simple model of the lateral dynamics of an automobile used in Chapter 5 and illustrated in Figures 5.5 and 5.6. Suppose that a sudden *impulsive force* is applied at the front axle. Show that if $I_z = mab$, the immediate result would be no motion at all at the rear axle (i.e., in this case the back axle is located at the *center of percussion*). Consider the cases $I_z < mab$ and $I_z > mab$. In which case does the back axle have an initial velocity in the same direction as the front axle, and in which case does it have an initial velocity in the opposite direction? This will determine in which direction the force at the rear axle will be directed after sudden steering input at the front. Why is it not necessary to consider possible tire forces at the rear axle in these calculations?

6.12 Consider a model reference controller of the general type shown in Figure 6.12 using only yaw rate error, and the relation between steer angle and desired yaw rate given in Eq. (6.89). Construct a block diagram for this controller similar to Figure 6.15 but leaving the controller block unspecified, as in Figure P6.12. Compare this steering control to the sliding mode control in Figure 6.15.

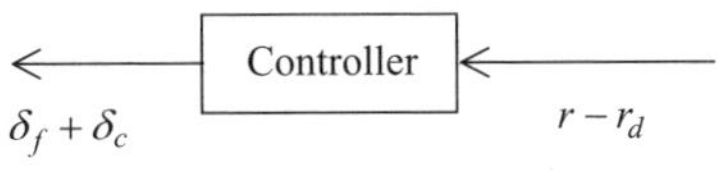

Controller

$\delta_f + \delta_c$ $r - r_d$

Figure P6.12

6.13 Consider a simple example of a sliding mode controller. Let the state variables be represented by $x = \begin{bmatrix} X \\ V \end{bmatrix}$ and the state equations be $\begin{bmatrix} \dot{X} \\ \dot{V} \end{bmatrix} = \begin{bmatrix} 0 & 1 \\ 0 & 0 \end{bmatrix} \begin{bmatrix} X \\ V \end{bmatrix} + \begin{bmatrix} 0 \\ 1 \end{bmatrix} U$. One can think of X as a position, V as a velocity, and U as a control force divided by a mass. Let the function S as defined in Eq. (6.58) be $S = [\, c_1 \quad c_2 \,] \begin{bmatrix} X \\ V \end{bmatrix} = c_1 X + c_2 V$. In this case, $x_d = 0$, so the controller will attempt to force both X and V to zero. Show that Eq. (6.61) yields the equation

$$\dot{S} = c_1 \dot{X} + c_2 \dot{V} = c_1 V + c_2 U = -\rho S - \alpha \operatorname{sgn} S$$

which gives a formula for the control variable as in Eq. (6.63):

$$U = -\frac{c_1}{c_2} V - \rho \frac{c_1 X + c_2 V}{c_2} - \frac{\alpha \operatorname{sgn}(c_1 X + c_2 V)}{c_2}$$

Plot the line $S = 0$ in the X–V plane, which is the state space for this system, assuming that c_1 and c_2 are positive quantities. Demonstrate that if the system is on the sliding surface, $S = 0$, $\dot{X} = -(c_1/c_2)X$, which represents and exponential motion toward the origin of the state space.

Equations (6.69) and (6.70) have to do with the controlled system, with the nonlinear part eliminated by setting $\alpha = 0$. Show that with $\alpha = 0$ the controlled system equations become

$$\begin{bmatrix} \dot{X} \\ \dot{V} \end{bmatrix} = \begin{bmatrix} 0 & 1 \\ -\rho \dfrac{c_1}{c_2} & -\dfrac{c_1}{c_2} - \rho \end{bmatrix} \begin{bmatrix} X \\ V \end{bmatrix}$$

Finally, show that the eigenvalues for the controlled system are $s = -\rho$, as predicted by Eq. (6.70) and $s = -(c_1/c_2)$, which relates to motion in the sliding surface:

$$\begin{bmatrix} \dot{X} \\ \dot{V} \end{bmatrix} = \begin{bmatrix} 0 & 1 \\ -\dfrac{\rho c_1}{c_2} & -\rho - \dfrac{c_1}{c_2} \end{bmatrix} \begin{bmatrix} X \\ V \end{bmatrix}$$

$$\text{Det} \begin{vmatrix} s & -1 \\ \dfrac{\rho c_1}{c_2} & s + \rho + \dfrac{c_1}{c_2} \end{vmatrix} = s^2 + s\left(\rho + \frac{c_1}{c_2}\right) + \frac{\rho c_1}{c_2} = (s+\rho)\left(s + \frac{c_1}{c_2}\right)$$

The eigenvalues are $s = -\rho$ and $s = -c_1/c_2$.

6.14 The elementary yaw rate controller shown in Figure P6.14 could be used to study how to choose a controller transfer function G such that the actual vehicle yaw rate would closely follow a desired yaw rate. In the figure, the transfer function between the front steer angle δ_f and the yaw rate r could be given for the bicycle model, Eq. (5.53). The controller transfer function G could be as simple as a proportional gain or could be a complicated compensator. Show that the closed-loop transfer function relating the desired yaw rate to the actual yaw rate is given by the expression

$$\frac{r}{r_d} = \frac{GH}{1 + GH}$$

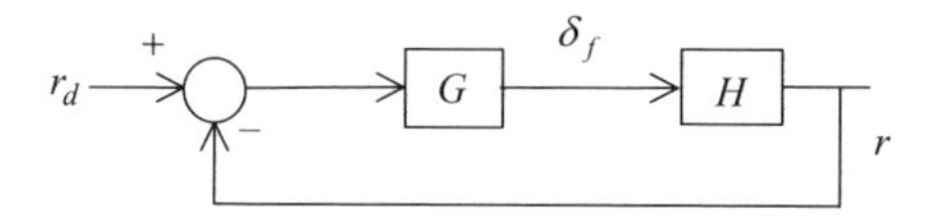

Figure P6.14

Assuming that G is a simple gain and that H is given by Eq. (5.53), find an expression for the characteristic equation for the closed-loop system. (You may not have the patience to write it out explicitly.)

6.15 Figure P6.15 depicts a yaw rate controller that computes the desired yaw rate r_d from the driver's steer input δ_f according to Eq. (6.89). The controller adds a compensating steer angle δ_c to the driver's steer angle by means of a differential element in the steering column. As in Problem 6.11, H is a vehicle transfer function such as Eq. (5.53), and G is a controller transfer function. Show that the transfer function relating the yaw rate to the driver's steer input is given by the expression

$$\frac{r}{\delta_f} = \frac{H[1 + GU/(a+b)]}{1 + GH}$$

Would you expect the characteristic equation for this controller to be different from the one derived in Problem 6.11?

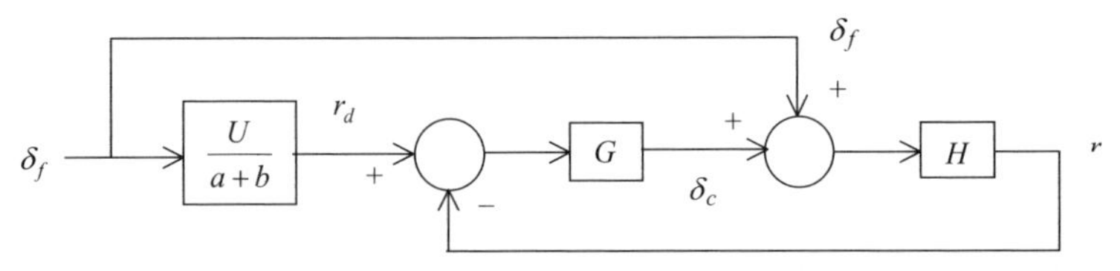

Figure P6.15

6.16 Figure P6.16 shows a mechanical system that is to be stabilized actively. The simple upside-down pendulum has mass m and length l and has negligible friction at its pivot. It is mounted on a movable cart that is controlled by an actuator. Assume that the actuator is strong and fast enough to move the cart in any manner called for by a control system; that is, there is no restriction on $X_0(t)$, $\dot{X}(t)$, or $\ddot{X}_0(t)$.

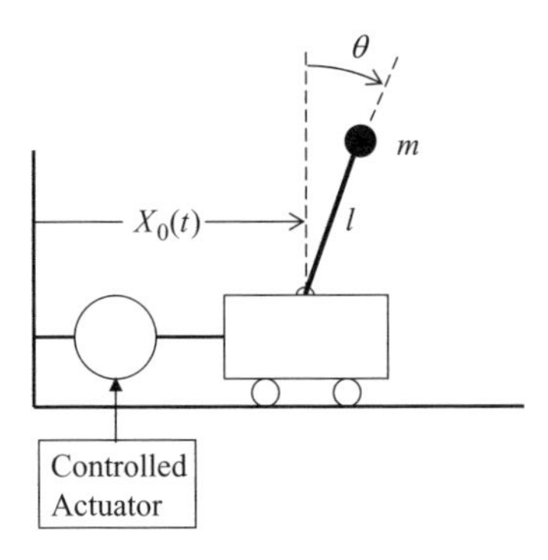

Figure P6.16

(a) Find the kinetic and potential energies of the system and derive a nonlinear equation of motion for the system.

(b) Show a block diagram for the nonlinear system using blocks containing the symbols sin(·) and cos(·) to produce $\sin\theta$ and $\cos\theta$ from the signal representing θ.

(c) Linearize the equation of motion for θ near zero and show a block diagram for the linearized equation.

(d) Show that state variable-feedback of $\dot{\theta}$ and θ using gains of $k_{\dot{\theta}}$ and k_θ with the proper signs can stabilize the linearized system so that θ will return to zero after a small diisturbance. Show a block diagram for the controlled system and give a criterion for the gains that will assure stability.

(e) Will the control scheme work for the nonlinear system after a large disturbance?

6.17 Figure P6.17*a* shows a simple single-degree-of-freedom model of a trailer being towed by a vehicle at a constant speed U. The trailer has mass m, centroidal moment of inertia I_c, and distances from the hitch point to the center of mass and from the center of mass to the axle of a and b, respectively. As in Chapter 5, we assume that the force on the axle can be expressed in terms of a cornering coefficient C_α and a slip angle α, $F = C_\alpha \alpha$. See Eq. (5.30) for similar expressions. The slip angle is approximately the lateral velocity of the wheels divided by the rolling velocity. (All angles will be considered to be small, although they are shown as relatively large in the figures for clarity.) For the first case, part (a), the slip angle is $\alpha = \theta + (a+b)\dot{\theta}/U$.

(a) Derive the equation of motion

$$\left(I_c + ma^2\right)\ddot{\theta} + C_\alpha \frac{(a+b)^2}{U}\dot{\theta} + C_\alpha \theta = 0$$

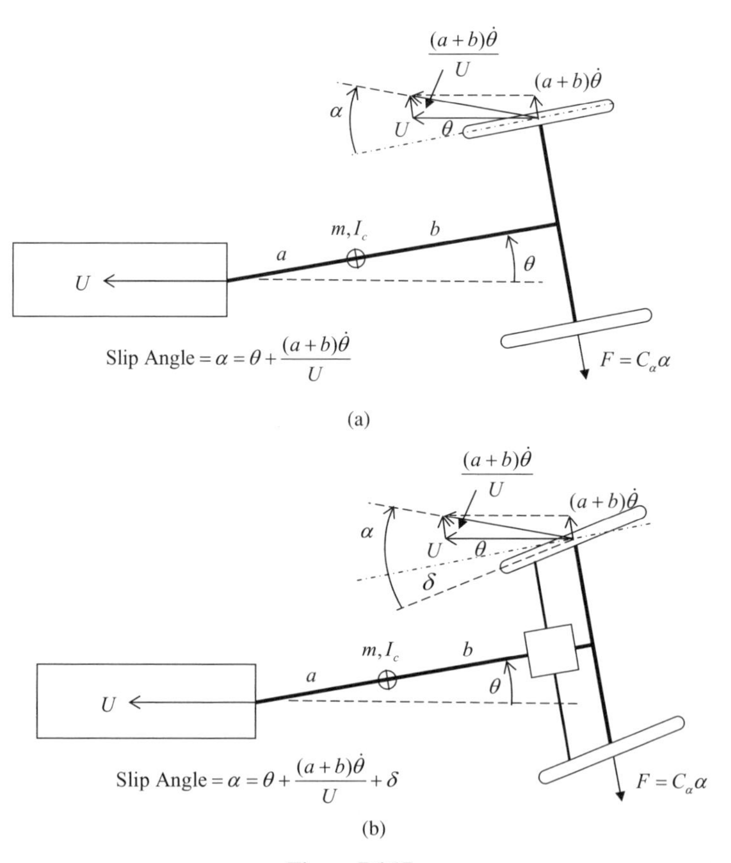

Figure P6.17

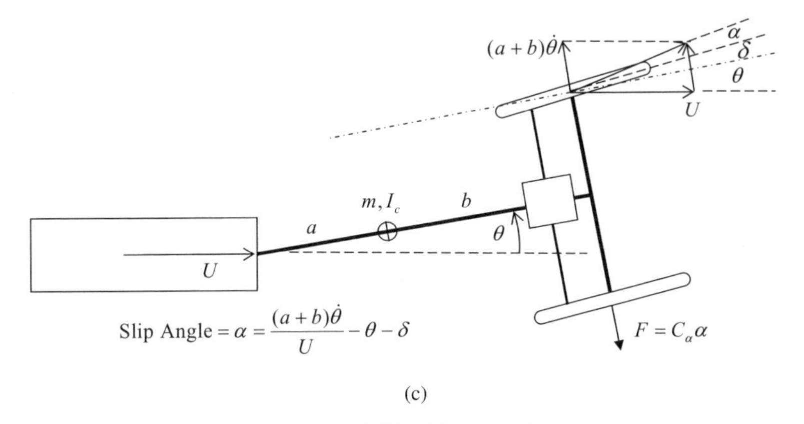

(c)

Figure P6.17 (Continued)

using either Lagrange's equation or basic dynamic principles. Note that the hitch can be considered a fixed point since it has no acceleration. You might need to use the parallel axis theorem from Appendix C.

(b) Noting that the equation of motion is of the form of an oscillator such as a mass–spring–damper system, what type of motion would you expect the trailer to make at very high speeds?

(c) Figure P6.17*b* shows an idea for improving the dynamics of the system by actively steering the trailer wheels. The steer angle δ is controlled by an actuator. The proposal is to use a sensor to measure the angular rate $\dot{\theta}$ and to use the control law $\delta = g\dot{\theta}$, where g is a gain, which could be constant or even speed dependent. Using the new slip angle formula shown in Figure P6.17*b*, derive a new equation of motion and comment on any possible improvement in the dynamics of the trailer.

(d) Figure P6.17*c* shows that if U is considered to be the speed of *backing up*, the slip angle formula changes quite a bit. It is common experience that backing up a trailer is a peculiarly unstable experience. Considering first that the steer angle δ is zero, derive a new equation of motion and show that the motion is unstable. Would the steering control law $\delta = g\dot{\theta}$ considered previously do any good to make the system stable?

(e) Would it do any good in helping to back up a trailer to base a steering control law on the angle θ? If you think a simple proportional control would work, how high should the gain be to guarantee stability?

6.18 There are sometimes sound reasons for designing an oversteering vehicle. For certain tasks, a three-wheeled vehicle makes sense. But a three-wheeled vehicle with a single wheel in front will almost inevitably be oversteering because the center of gravity must be toward the rear and all the roll stiffness must be at the rear. These are two factors that contribute to a tendency toward oversteering behavior and mean that such vehicles are normally restricted to low speeds. It has been suggested that yaw rate feedback active steering could be used to increase the critical speed of an oversteering vehicle from what it would be based only on the physical parameters of the vehicle. For example, the front steer angle δ could be based partly on the yaw rate r measured by a yaw rate sensor and partly by the driver input δ_d through the steering wheel. A simple version of the control law could be $\delta = -Gr + \delta_d$, with G being a proportional gain. In this problem the bicycle models developed in Chapter 5 are used to evaluate the proposal.

(a) Modify Eq. (5.34) to include this proposed active steering law.

(b) To study the stability aspects of the proposal, assume that $\delta_d = 0$ and then follow the derivation of Eq. (5.39) to arrive at a modified version of Eq. (5.40), now including the steering gain G.

(c) Equation (5.42) is the result of evaluation equation (5.40) after some algebraic simplification. You should be able to see that your modifications add only a few terms to Eq. (5.40), and without needing to evaluate the determinant from scratch, you can get a new version of the characteristic equation by adding only a few terms to Eq. (5.42).

(d) You should find that the new terms do not change the conclusion for the uncontrolled system that the coefficients of s^2 and s are inherently positive. The only term that could be negative (and cause instability) is the last term in the characteristic equation. Show that the modified system has a stability criterion that is a modified version of Eq. (5.45). After some algebraic manipulation, the stability criterion can be written as

$$(a+b)^2 C_f C_r \left[\frac{1}{U^2} + \frac{G}{(a+b)U} \right] + m(bC_r - aC_f) > 0$$

(e) In the oversteering case, $bC_r - aC_f$ is negative and there exists a critical speed above which the vehicle is unstable. Show that with yaw rate feedback, the critical speed can be increased beyond the value the critical speed would have without the active steering system.

7

RIGID-BODY MOTION IN THREE DIMENSIONS

The dynamics of a particle are fairly easy to describe, whether it moves in one, two, or three dimensions. The case for a rigid body is quite different because rotational motion must also be considered. Most of the cases considered in this book so far have involved the plane motion of a rigid body, which means that only a single rotation about an axis perpendicular to the motion plane needs to be considered. For a body-centered analysis of rigid body in plane motion, there are only two linear velocities of some point in the body, and a single angular velocity to be considered. If positions in inertial space have to be considered in addition to velocities, only three more variables are necessary: two position variables to locate the position of a point in the body (often, the center of mass is used) and a single angular position relating a line fixed in the body to a reference line in inertial space. One can say that the differential equations for a rigid body in plane motion will be sixth order if the positions as well as velocities are required. (Another way to put this is to say that a rigid body moving in plane motion has three degrees of freedom, yielding three second-order equations of motion or six first-order equations if the position of the body must be considered.)

As we have seen in Chapter 5, for some vehicle dynamic studies only the velocities are required, so the equations are only of third order in the case of plane motion. In fact, in the ground vehicle stability studies considered in Chapter 5, one velocity component was assumed to be essentially constant, so the equations of motion were only of second order. The general motion of a rigid body in three dimensions is quite complicated. There are three position variables necessary to locate a point in the body such as the center of mass in inertial space and three angular variables required to locate the body in rotation. In addition, there are three linear velocity components and three angular velocity components required to describe the motion, for a total of

12 variables. (A rigid body in three-dimensional motion is said to have six degrees of freedom, which correspond to a twelfth-order set of equations of motion.)

Although the differential equation set is of twelfth order in general, the situation is not quite as complex as this would suggest. First, one can basically separate translation and rotation into two separate sixth-order equation sets by using the center of mass as the point in the body used to describe the translation of the body. (The force-generating elements may eventually couple these equation sets.) Second, there are situations in which only a subset of the variables need to be considered. In this chapter we study several cases in which three-dimensional motion is involved but the order of the equation set required is much less than for the most general case.

Some facts about three-dimensional motion of rigid bodies are inescapable: (1) The description of the angular position and angular velocity of a body is more complicated than the description of the linear position and linear velocity of a point in the body, and (2) the inertial properties related to rotational motion of a rigid body are more complex than the inertial properties related to translational motion. These facts are not obvious if only plane motion is considered. Although most introductory texts do not treat three-dimensional motion of rigid bodies in much detail, more advanced treatments are readily available (see, e.g., [1–3]).

7.1 GENERAL EQUATIONS OF MOTION

In this section, basic principles of dynamics are reviewed as they apply to three-dimensional motion. Some of the principles have been described in previous chapters, but in most cases, only restricted versions of the equations were actually used. As far as translation is concerned, a rigid body behaves much as a particle. If $\mathbf{F}$ is the net vector force on the body and $\mathbf{a}$ is the vector acceleration *of the center of mass*, $\mathbf{v}$ is the velocity *of the center of mass* of the body, and m is the mass of the body, then

$$\mathbf{F} = m\frac{d\mathbf{v}}{dt} = m\mathbf{a} \tag{7.1}$$

just as if the body were a particle located at the center of mass. One can also define $\mathbf{P}$ to be the linear momentum with

$$\mathbf{P} = m\mathbf{v} \tag{7.2}$$

Then Eq. (7.1) can be expressed as

$$\mathbf{F} = \frac{d\mathbf{P}}{dt} \tag{7.3}$$

These relations are basically true for any coordinate frame but are most straightforward to evaluate for an inertial frame. Appendix A outlines how these equations appear in a body-fixed rotating frame in which the time derivatives have a special form.

In Chapter 5 we used these equations as they apply to the special case of plane motion.

If $\boldsymbol{\tau}_c$ represents the net moment or torque *about the center of mass* and $\mathbf{H}_c$ is the angular momentum of the body *evaluated about the center of mass*, a dynamic principle superficially similar to the linear momentum law Eq. (7.3) can be stated:

$$\boldsymbol{\tau}_c = \frac{d\mathbf{H}_c}{dt} \tag{7.4}$$

The advantage of using the center of mass for the moment center and the angular momentum is that the linear momentum law, Eq. (7.3), and the angular momentum law, Eq. (7.4), can then be written independently.

There is a special case in which a rigid body happens to be constrained in such a way that one point in the body is fixed in an inertial frame. Then, although Eq. (7.3) still applies, a torque–angular momentum relation similar to Eq. (7.4) can be written about the fixed point 0, and in many cases this is the only equation needed to describe the dynamics of the body:

$$\boldsymbol{\tau}_0 = \frac{d\mathbf{H}_0}{dt} \tag{7.4a}$$

In this special case, both the applied torque and the angular momentum are evaluated about the fixed point 0 rather than the center of mass c.

There is, however, a major difference between translational motion and rotational motion. Linear momentum is related to the center of mass velocity by the mass, a scalar quantity, as shown in Eq. (7.2). In contrast, the relation of the angular momentum to the angular velocity is a *tensor or matrix relation*. The relation can be expressed in matrix form if the vectors $\mathbf{H}_c$ and the angular velocity $\boldsymbol{\omega}$ are shown as column vectors:

$$[H_c] = [I_c][\omega], \qquad [H_c] = \begin{bmatrix} H_{cx} \\ H_{cy} \\ H_{cz} \end{bmatrix}, \qquad [\omega] = \begin{bmatrix} \omega_x \\ \omega_y \\ \omega_z \end{bmatrix} \tag{7.5}$$

The *centroidal* inertia matrix (discussed in Chapter 4 with regard to its role in computing kinetic energy) is

$$[I_c] = \begin{bmatrix} I_{xx} & I_{xy} & I_{xz} \\ I_{yx} & I_{yy} & I_{yz} \\ I_{zx} & I_{zy} & I_{zz} \end{bmatrix} \tag{7.6}$$

where the *moments of inertia* defined with respect to an orthogonal x–y–z coordinate system fixed in the body *at the center of mass* are

$$I_{xx} = \int (y^2 + z^2)\, dm, \qquad I_{yy} = \int (z^2 + x^2) dm, \qquad I_{zz} = \int (x^2 + y^2)\, dm \tag{7.7}$$

and the *products of inertia* are

$$I_{xy} = I_{yx} = -\int (xy)\,dm, \qquad I_{xz} = I_{zx} = -\int (xz)\,dm, \qquad I_{yz} = I_{zy} = -\int (yz)\,dm \tag{7.8}$$

These definitions are standard except possibly for the minus signs associated with the products of inertia. Some authors prefer to put the minus signs into the definitions as in Eq. (7.8), whereas others use plus signs in the definitions of the products of inertia, but then the minus signs appear in the off-diagonal terms in the matrix of Eq. (7.6). The relation between the angular momentum and the angular velocity, Eq. (7.5), is, of course, the same with either convention.

Again, in the special case in which one point in the body is fixed in inertial space, there is another inertia matrix $[I_0]$ just as in Eqs. (7.6)–(7.8), but with the origin of the coordinate system switched from the center of mass to the fixed point. The relationship between $[I_0]$ and $[I_c]$ involves the well-known *parallel axis theorem* given in Appendix C (see ref. [1] for a more complete discussion). It is another standard exercise to prove that the x–y–z coordinate system can always be rotated to a position at which all the products of inertia vanish (see, e.g., [1]). The coordinate directions are then called the *principal directions* and the moments of inertia are called the *principal moments of inertia*. See Appendix B for the principal moments of inertia of some common solid body shapes. If the principal direction are used for the x–y–z coordinate frame, the products of inertia vanish and the inertia matrix takes the form

$$[I_c] = \begin{bmatrix} I_1 & 0 & 0 \\ 0 & I_2 & 0 \\ 0 & 0 & I_3 \end{bmatrix} \tag{7.9}$$

The inertia matrix evaluated at a fixed point 0, $[I_0]$, also has principal axes for which the matrix has the form of Eq. (7.9), although the principal moments of inertia, I_{01}, I_{02}, and I_{03}, have different values from those for the center of mass in Eq. (7.9).

It is worth noting if a rigid body is constrained to move only in a plane, the complexities of the dynamics of rotation are reduced significantly. If the body moves only in the x–y plane, only a single component of the angular velocity, ω_z, exists and the z-component of the angular momentum from Eq. (7.5) is just

$$H_z = I_{zz}\omega_z \tag{7.10}$$

(Note that the z-axis does not necessarily have to be a principal axis, so I_{zz} is not necessarily the same as I_3.)

For plane motion, only the z-component of Eq. (7.4) is needed:

$$\tau_z = \frac{dH_z}{dt} = I_{zz}\frac{d\omega_z}{dt} = I_{zz}\alpha_z \tag{7.11}$$

Equation (7.11) has a close resemblance to components of Eq. (7.1), with the angular acceleration α_z playing a role analogous to the linear acceleration components, and the scalar I_{zz} playing a role similar to the scalar mass m. (Actually, there will be torque components required to keep the body in the x–y plane when the z-axis is not a principal axis, but these torque components have no influence on the dynamics of rotation about the z-axis.)

7.2 USE OF A BODY-FIXED COORDINATE FRAME

In Chapter 5, the use of a body-fixed coordinate frame was illustrated by the application to vehicle dynamics but only for cases in which the vehicles were restricted to plane motion. Appendix A gives the general equations of motion using a body-fixed frame for the general three-dimensional case, which can be seen to be quite complex. In this chapter we deal with three-dimensional motion but with some assumptions that reduce the complexity of the analysis. Figure 7.1 repeats the coordinate system used in Appendix A. Using the notation of Appendix A, we define the velocity components of the center of mass in the x–y–z coordinate system attached to the body by

$$\mathbf{v} = U\mathbf{i} + V\mathbf{j} + W\mathbf{k} \quad \text{or} \quad [v] = \begin{bmatrix} U \\ V \\ W \end{bmatrix} \tag{7.12}$$

and the angular velocity components of the body by

$$\boldsymbol{\omega} = p\mathbf{i} + q\mathbf{j} + r\mathbf{k} \quad \text{or} \quad [\omega] = \begin{bmatrix} p \\ q \\ r \end{bmatrix} \tag{7.13}$$

The net force components in the x–y–z coordinate directions are called X, Y, Z:

$$\mathbf{F} \equiv X\mathbf{i} + Y\mathbf{j} + Z\mathbf{k} \quad \text{or} \quad [F] = \begin{bmatrix} X \\ Y \\ Z \end{bmatrix} \tag{7.14}$$

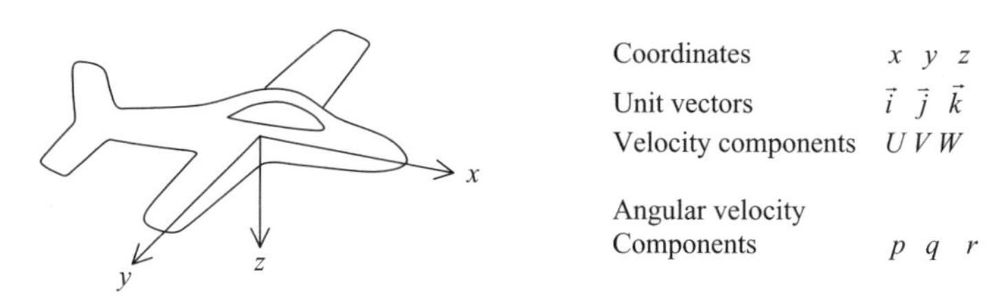

Figure 7.1 General notation used in Appendix A.

The net torque components evaluated about the center of mass in the x–y–x coordinate directions are called L, M, N:

$$\boldsymbol{\tau} \equiv L\mathbf{i} + Mb + N\mathbf{k} \quad \text{or} \quad [\tau] = \begin{bmatrix} L \\ M \\ N \end{bmatrix} \tag{7.15}$$

This particular notation is typically used for vehicle dynamic studies.

When Eq. (7.1) is evaluated in Appendix A using the formula for the time derivative in a rotating frame, Eq. (A.13), the result is

$$[F] = \begin{bmatrix} X \\ Y \\ Z \end{bmatrix} = m \begin{bmatrix} \dot{U} + qW - rV \\ \dot{V} + rU - pW \\ \dot{W} + pV - qU \end{bmatrix} \tag{7.16}$$

This equation, specialized for plane motion, has been used extensively in Chapter 5. The equivalent equation for rotation, Eq. (7.4), becomes much more complex in general:

$$[\tau] = \begin{bmatrix} L \\ M \\ N \end{bmatrix} = \begin{bmatrix} I_{xx}\dot{p} + I_{xy}\dot{q} + I_{xz}\dot{r} \\ I_{yx}\dot{p} + I_{yy}\dot{q} + I_{yz}\dot{r} \\ I_{zx}\dot{p} + I_{yz}\dot{q} + I_{zz}\dot{r} \end{bmatrix}$$
$$+ \begin{bmatrix} q(I_{zx}p + I_{zy}q + I_{zz}r) - r(I_{yx}p + I_{yy}q + I_{yz}r) \\ r(I_{xx}p + I_{xy}q + I_{xz}r) - p(I_{zx}p + I_{zy}q + I_{zz}r) \\ p(I_{yx}p + I_{yy}q + I_{yz}r) - q(I_{xx}p + I_{xy}q + I_{xz}r) \end{bmatrix} \tag{7.17}$$

In plane motion, this equation is simplified significantly to Eq. (7.11). Now we investigate some cases of three-dimensional motion for which a simplified version of Eq. (7.17) can be used.

Euler's Equations

There are many practical cases in which the principal axes of a body are known or can be computed. In some cases there is enough symmetry in the mass distribution of the body that the principal axes directions are obvious. In other cases, symmetry about a plane means that one of the principal directions is known and only the angle needs to be found on the other two perpendicular axes. A graphical way to find the remaining two principal directions is based on Mohr's circle [1]. Finally, standard computer programs are available to find the principal directions and moments of inertia given the inertia matrix for any arbitrary coordinate system. The problem of finding principal axes and moments of inertia is an example of a classical *eigenvalue problem*.

When the x–y–z axis system is a principal axis system with subscripts 1, 2, 3 corresponding to x, y, z, Eq. (7.17) simplifies quite a bit (see Appendix B):

$$[\tau] = \begin{bmatrix} L \\ M \\ N \end{bmatrix} = \begin{bmatrix} I_1\dot{p} \\ I_2\dot{q} \\ I_3\dot{r} \end{bmatrix} + \begin{bmatrix} (I_3 - I_2)qr \\ (I_1 - I_3)rp \\ (I_2 - I_1)pq \end{bmatrix} \tag{7.18}$$

These equations are often called *Euler's equations*. Note that in Chapter 5 when dealing with plane motion, two of the equations for translation, Eq. (7.17), were used with $W = 0$ and only one of the rotational equations, Eq. (7.18), was necessary since both p and q were zero. In fact, the third equation in Eq. (7.18) for $\dot{r}$ in plane motion is just the same as Eq. (7.11) if the z-axis is a principal axis. The complications of Eqs. (7.17) and (7.18) do not arise in plane motion because two of the three angular velocity components are zero.

Spin Stabilization of Satellites

As an application of the equations for three-dimensional motion using body-fixed coordinates, we consider the spin stabilization of a satellite in outer space. The satellite itself represents a nearly rigid body, and it has some interesting stability problems. Furthermore, Eqs. (7.16) and (7.17) or (7.18) are well suited to describing satellite motion. In outer space, the torques exerted on compact bodies are extremely small and for many purposes can be neglected. This means that according to Eq. (7.4), the angular momentum is constant, or nearly constant, in inertial space. If one only had experience with plane motion problems and thought about Eqs. (7.10) and (7.11), one might assume that a satellite would simply rotate at a constant rate about an axis fixed in space, depending only on its initial rotation rate. But for three-dimensional motion, the fact that the angular momentum is constant does not imply that the angular velocity is constant. The body can execute quite complicated motions even when the angular momentum remains fixed in direction and magnitude.

There is, of course, a significant gravity force that provides the acceleration associated with the orbit of the satellite, but if we assume a coordinate system attached to a spaceship moving in the same orbit with the satellite, then in this coordinate system the satellite appears "weightless" and the satellite would have no acceleration with respect to the spaceship. (In a uniform gravity field, the center of gravity, at which the gravity force acts, is also the center of mass, so the gravity force exerts no torque about the center of mass. In a nonuniform gravity field, this is not strictly the case, but for typical satellites the torque due to the small difference in location between the center of mass and the center of gravity is very small.) In the present analysis it is assumed that the situation for a satellite in outer space is equivalent to the case in which no forces and no torques act on the satellite.

With no force acting on the satellite, Eq. (7.1) shows that the velocity in absolute space would be constant. This means that a satellite traveling near a spaceship in the same basic orbit would have a constant velocity with respect to the spaceship. This fact is not quite so easy to appreciate in the body-fixed coordinate system except when the satellite is not rotating. If the forces X, Y, and Z are assumed to vanish, and if the angular velocity components p, q, and r also are assumed to be zero, Eq. (7.16) predicts that the linear velocity components U, V, and W will be constant. When the body is not rotating, the body-fixed coordinate system remains parallel to the inertial frame. However, when the satellite is rotating, with p, q, and r not zero, Eq. (7.16) shows that the components U, V, and W will change as the constant inertial velocity is projected along the rotating x–y–z axes attached to the satellite. As far as linear motion is concerned, the use of an inertial coordinate system shows immediately that the velocity in inertial space is constant, whereas the use of a body-fixed coordinate system requires more interpretation, since the relation between the inertial coordinate frame and the body-fixed frame is not simple.

The analysis of the linear motion of the center of mass of a body with no applied forces is quite straightforward with no surprises, but the study of rotation of a body with no applied torques is more interesting and complex than one might expect. The body-fixed coordinate system is well suited to this task. If one assumes that the moments L, M, and N are zero and that the angular rates p, q, and r are initially zero, then Eqs. (7.17) and (7.18) are satisfied when, $\dot{p}$, $\dot{q}$, and $\dot{r}$ are zero. That is, any torque-free nonrotating satellite remains nonrotating in space. On the other hand, if even a small moment or angular impulse from a particle hitting the satellite should occur, the satellite will acquire an angular velocity, and it will slowly change attitude. This may disturb the orientation of sensitive antennas. This change in orientation can be corrected if the satellite has thrusters, but the less the thrusters need to be used, the better, since the amount of rocket fuel on board is strictly limited.

One idea to reduce the influence of torque disturbances is *spin stabilization*. The idea is to spin the satellite around a principal axis. The satellite then acts somewhat like a gyroscope and is more resistant to changes in the direction of the spin axis due to small disturbance torques than it would be if it were not spinning. To analyze spin stabilization, let us first assume that the x–y–z axis system is a principal axis system. Then the simpler Euler equations of Eq. (7.18) can be used instead of the general equations (7.17). It is easy to see from Eq. (7.18) that the satellite can spin around any one of the three principal axes if the moments are zero and if the angular velocities about the other two axes also vanish. For example, suppose that the satellite is spinning around the x-axis with

$$q = r = 0 \tag{7.19}$$

Under this assumption, Eq. (7.18) then yields

$$\begin{bmatrix} I_1 \dot{p} \\ I_2 \dot{q} \\ I_3 \dot{r} \end{bmatrix} = \begin{bmatrix} 0 \\ 0 \\ 0 \end{bmatrix} \tag{7.20}$$

This result shows that if Eq. (7.19) is true at some initial time,

$$p = \Omega = \text{constant} \tag{7.21}$$

and Eq. (7.20) will remain true as time goes on. It is easy to show that similar results hold for spin about any one of the three principal axes.

We now consider spin about a principal axis as a *basic motion* and consider a *perturbed motion* in which angular velocities about the remaining two axes have small but nonzero values. This will lead to an analysis of the stability of spin stabilization. The assumption will be that the spin is about the x-axis,

$$p = \Omega, \qquad q = \Delta q, \qquad r = \Delta r \tag{7.22}$$

where Δq and Δr are supposed to be small with respect to the spin speed Ω. The first equation in Eq. (7.18) under the assumptions of Eq. (7.22) becomes

$$I_1 \dot{p} + \Delta q\, \Delta r(I_3 - I_2) = 0 \tag{7.23}$$

Under the assumption that the product of two small quantities $\Delta q\, \Delta r$ is negligible, we see that $\dot{p}$ is nearly zero and that Eq. (7.21) is approximately true for the perturbed as well as for the basic motion because $\dot{p}$ in Eq. (7.23) nearly vanishes.

The last two equations in Eq. (7.18) yield the following two equations for Δq and Δr after some minor manipulation:

$$\Delta \dot{q} + \frac{(I_1 - I_3)\Omega}{I_2} \Delta r = 0 \tag{7.24}$$

$$\Delta \dot{r} + \frac{(I_2 - I_1)\Omega}{I_3} \Delta q = 0 \tag{7.25}$$

An interesting pattern arises when these two equations are put in a vector–matrix form:

$$\begin{bmatrix} \Delta \dot{q} \\ \Delta \dot{r} \end{bmatrix} + \begin{bmatrix} 0 & \dfrac{(I_1 - I_3)\Omega}{I_2} \\ \dfrac{(I_2 - I_1)\Omega}{I_3} & 0 \end{bmatrix} \begin{bmatrix} \Delta q \\ \Delta r \end{bmatrix} = \begin{bmatrix} 0 \\ 0 \end{bmatrix} \tag{7.24a, 7.25a}$$

An eigenvalue analysis of these equations can proceed either by using the Laplace transform of the equations or by assuming exponential time dependence for the perturbation variables as has been done in previous chapters, $\Delta q = \Delta \bar{q} e^{st}$, $\Delta r = \Delta \bar{r} e^{st}$. In either case the algebraic equations to be solved are

$$\begin{bmatrix} s & \dfrac{(I_1 - I_3)\Omega}{I_2} \\ \dfrac{(I_2 - I_1)\Omega}{I_3} & s \end{bmatrix} \begin{bmatrix} \Delta \bar{q} \\ \Delta \bar{r} \end{bmatrix} = \begin{bmatrix} 0 \\ 0 \end{bmatrix} \tag{7.26}$$

The characteristic equation for Eq. (7.26) comes as usual from setting the determinant of the coefficient matrix to zero:

$$s^2 + \frac{(I_1 - I_3)(I_1 - I_2)\Omega^2}{I_2 I_3} = 0 \tag{7.27}$$

For simplicity, let

$$k = \frac{(I_1 - I_3)(I_1 - I_2)\Omega^2}{I_2 I_3} \tag{7.28}$$

Then Eq. (7.27) becomes

$$s^2 + k = 0 \tag{7.29}$$

and the eigenvalues are

$$s = \pm\sqrt{-k} \tag{7.30}$$

Now, if k is a positive number, the two eigenvalues will be complex-conjugate imaginary numbers $s = \pm j\sqrt{k}$. This means that the satellite will have angular velocity components Δq and Δr that vary in a sinusoidal manner at a radian frequency of $\sqrt{k}$, which, from Eq (7.28), is proportional to the spin speed Ω (see, e.g., [5]). Because the imaginary eigenvalues have a zero real part, the perturbations will neither grow nor decrease in magnitude as time goes on. In that case, one cannot say that the system is really stable, but at least it is not unstable. If the satellite is disturbed, the perturbation variables Δq and Δr will vary in time in a sinusoidal manner, but the amplitudes of the motion will neither grow nor decay. On the other hand, if k should be negative, then $-k$ is a positive number and Eq. (7.30) will have real eigenvalues. One will be positive and one will be negative. The real positive eigenvalue indicates that the system is unstable. It means that the perturbation variables will grow until the linearized equations of motion, Eqs. (7.24) and (7.25) , are no longer valid.

There are two situations for which the term k in Eq. (7.28) will be positive: The first is if I_1 is the largest of the three principal moments of inertia, and the second is if I_1 is the smallest. In the first case, the two terms in parentheses are both positive. In the second case, the terms are both negative but their product is positive. The conclusion is that if the satellite is spun around the axis with the maximum moment of inertia or about the axis with the minimum moment of inertia, the motion will be stable (or at least will not be unstable). In the case when I_1 is intermediate in magnitude between I_2 and I_3, one of the terms in parentheses is positive and the other term is negative. In this case, k in Eqs. (7.28) and (7.29) will be negative and the system is definitely unstable.

As we have seen, the satellite can theoretically spin about any principal axis if angular velocity is lined up *exactly* with the axis. If the angular velocity is only approximately lined up with the axis with the largest or the smallest moment of inertia, the angular velocity vector will wobble somewhat about the spin axis, but the magnitude of the wobble will neither grow nor decay with time. On the other hand, if the satellite is spin stabilized around the intermediate axis, the perturbation variables will act like the variables of a mass–spring system with a negative spring constant. A mass–spring system having a spring with a negative spring constant will deviate more and more from an equilibrium position if it is disturbed even slightly. From this analogy, one can see that if x is an intermediate principal axis, Δq and Δr will grow in time and the spin stabilization system is thus unstable. The satellite will execute a complicated tumbling motion.

In this analysis we have linearized nonlinear equations for the purpose of studying the stability of a basic motion. For an unstable system, the perturbation variables grow to a size such that the linearized equations no longer are accurate representations of the system. Practically, this means that while the perturbation spins do grow for a time, they really do not grow infinitely large in time. The satellite has only a finite amount of energy, so what ultimately happens is that the initial spin energy associated with the spin about the x-axis is distributed into spin energy associated with angular velocities about the y and z axes. If one attempts to spin-stabilize a satellite about an intermediate principal axis, the result is that the satellite will inevitably begin tumbling.

This result can even be demonstrated on Earth. A compact object (such as a blackboard eraser) thrown into the air behaves for awhile much like a torque-free rigid body. The only moments about the center of mass arise from the object's interaction with the air, and for awhile their effect is small. If the object has obvious principal axis directions and if it is clear which moment of inertia is the largest and which is the smallest, it is easy to throw the object up with a spin primarily around the principal axis with the largest or smallest moment of inertia. An attempt to spin the object about the principal axis with an intermediate moment of inertia will almost always result in an easily seen tumbling motion. There are rumors that an early satellite was planned to be spin-stabilized about the intermediate axis since the engineers knew that a rigid body can spin steadily about any principal axis. What they did not realize was that spinning about the intermediate axis is unstable (although applied mechanicians knew this long before satellites existed). The result was that the satellite tumbled uncontrollably and their spin stabilization would have been better called spin unstabilization.

7.3 USE OF AN INERTIAL COORDINATE FRAME

When inertial coordinates are used to describe the dynamics of rigid bodies, it is usual to start with the position of the body and then to describe velocities and accelerations in terms of time derivatives of position variables. This process is very straightforward for the position of the center of mass of a rigid body, and it readily leads to a description of the dynamics of translation. Figure 7.2 shows an inertial coordinate frame, X–Y–Z, with the position of the center of mass of a body described by

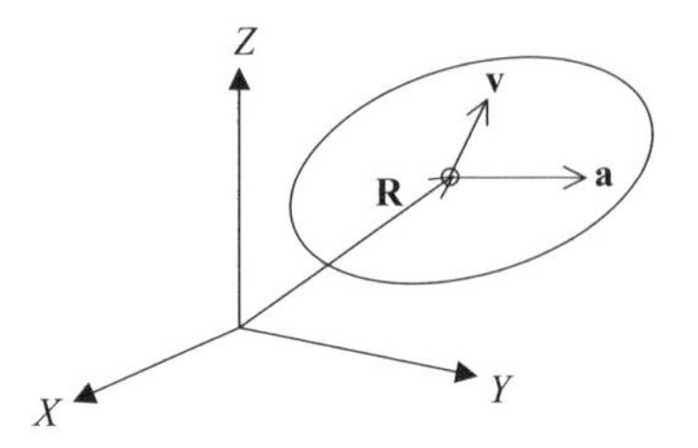

Figure 7.2 Position, velocity, and acceleration of the center of mass of a rigid body in an inertial coordinate frame.

$$\mathbf{R} = X\mathbf{i} + Y\mathbf{j} + Z\mathbf{k} \tag{7.31}$$

Since the coordinate frame is nonrotating and nonaccelerating, the velocity and acceleration are easily written.

$$\mathbf{v} = \dot{X}\mathbf{i} + \dot{Y}\mathbf{j} + \dot{Z}\mathbf{k} \tag{7.32}$$

$$\mathbf{a} = \ddot{X}\mathbf{i} + \ddot{Y}\mathbf{j} + \ddot{Z}\mathbf{k} \tag{7.33}$$

In this case, the position velocity and acceleration are all vectors and simply related to each other. Thus, the dynamics of translation, as represented by Eqs. (7.1)–(7.3), are as easy to use for a rigid body as for a particle in an inertial coordinate system.

Unfortunately, the description of the angular position of a rigid body is not as straightforward as the description of the linear position of a point in the body. There are several facts that should be known about the angular position of a rigid body. It is true that a rigid body can be taken from an initial angular position to any other by a finite angle of rotation about a single axis [1]. This is sometimes called *Euler's theorem* [2]. It takes two parameters to determine the direction of the rotation axis in three dimensions and another parameter to define how large the angle of rotation must be about the axis to achieve the final position. Thus, the angular position of a body relative to an inertial coordinate frame can be described by a line segment directed along the rotation axis with the length of the segment proportional to the angle of rotation needed to achieve the final position.

One might think that the directed line segment describing angular position relative to the inertial frame would be analogous to the position vector **R** used to describe the position of the center of mass. This is not the case, however. The usual demonstration that finite rotations cannot be represented as vectors involves two changes of position. For linear positions, the final position can be represented as the vector sum of two position vectors and it does not matter in which order the vectors are added or the positions changed. That is, for position vectors, $\mathbf{R} = \mathbf{R}_1 + \mathbf{R}_2 = \mathbf{R}_2 + \mathbf{R}_1$, but for two rotations, it makes a difference to the final angular position which rotation is done first. This fact is demonstrated in many introductory dynamics texts (e.g., [4]).

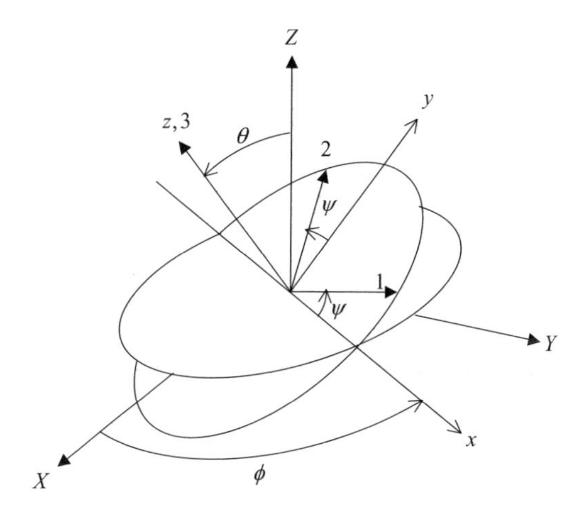

Figure 7.3 Definition of the three Euler angles θ, ϕ, and ψ.

There are a number of ways to describe the angular position of a rigid body, but for all of them, the connection between angular position, angular velocity, and angular acceleration is much more complicated than for the equivalent linear quantities in Eqs. (7.31)–(7.33) . This fact adds to the complexity of the three-dimensional angular motion already introduced by the inertia matrix, which relates the angular momentum to the angular velocity in Eq. (7.5).

Euler's Angles

One of the best known ways to describe the angular position of a rigid body is through the three Euler angles shown in Figure 7.3. The 1–2–3 coordinate is fixed in the body and rotates with it. The task is to represent the position of the 1–2–3 system with respect to the inertial X–Y–Z system. In addition, it is assumed that the 1–2–3 coordinate system is aligned with the principal directions in the body. In what follows, the principal moments of inertia I_1, I_2, and I_3 defined in Eq. (7.9) will refer to the 1–2–3 axis rather than the x–y–z coordinate system.

In Figure 7.3 a number of coordinate frames are shown all emanating from a point fixed in the body. In some applications this point is actually fixed in inertial space, and in others it may be the center of mass, so that Eqs. (7.3) and (7.4) can be used. In any case, the X–Y–Z system is nonrotating and is either an inertial system or at least remains parallel to an inertial system if the origin should be accelerating. One can think of the Euler angles as arising from a sequence of rotations that brings a frame from initial coincidence with the X–Y–Z frame to the angular position of the 1–2–3 frame. This is done with the aid of an auxiliary frame labeled x–y–z, which is initially lined up with X–Y–Z.

First, consider a rotation of x–y–z about the Z-axis with the angle of rotation called ϕ. This establishes the x-axis in the X–Y plane. Then consider a rotation of the z-axis about the x-axis of the angle θ. These two rotations bring the z-axis into coincidence with the 3-axis of the body. Finally, the angle ψ is the angle between the 1 and 2 axes and the x and y axes, respectively. You will have to stare at Figure 7.3 for awhile to appreciate the meaning of *Euler angles* θ, ϕ, and ψ. It should be clear from Figure 7.3 that the three generalized coordinates θ, ϕ, and ψ fix the position of the body with respect to the inertial frame. Given the Euler angles, one can find the body angular position or given the body position, one can find the angles that describe it. [There is, of course, the usual ambiguity associated with angles; one can add or subtract any number of full revolutions (2π radians) to the angles without changing the results.]

To use the dynamic equations (7.4) and (7.5), we need to find the angular velocity in terms of derivatives of the Euler angles. This is much more complicated than finding the linear velocity from the position in Eqs. (7.31) and (7.32). Since there are three separate coordinate systems in Figure 7.3, we use the notation **u** with a subscript to indicate unit vectors pointed along the coordinate directions in the three frames. The angular rates $\dot{\phi}$, $\dot{\theta}$, and $\dot{\psi}$ are components of the angular velocity relative to the X–Y–Z frame, but they are directed along the Z-, x-, and 3-axis directions. The angular velocity can be expressed thus:

$$\omega = \dot{\phi}\mathbf{u}_Z + \dot{\theta}\mathbf{u}_x + \dot{\psi}\mathbf{u}_3 \tag{7.34}$$

This mixed unit vector form is not very useful, but a number of relations between the unit vectors allows one to express ω in more useful ways. These relations simply represent unit vectors as projections of other unit vectors that lie in a common plane.

$$\begin{aligned} \mathbf{u}_Z &= \mathbf{u}_3 \cos\theta + \mathbf{u}_y \sin\theta \\ \mathbf{u}_x &= \mathbf{u}_1 \cos\psi - \mathbf{u}_2 \sin\psi \\ \mathbf{u}_x &= \mathbf{u}_X \cos\phi + \mathbf{u}_Y \sin\phi \\ \mathbf{u}_y &= \mathbf{u}_1 \sin\psi + \mathbf{u}_2 \cos\psi \end{aligned} \tag{7.35}$$

With the help of these relations, the angular velocity can be expressed in each of the three coordinate systems shown in Figure 7.3:

$$\boldsymbol{\omega} = (\dot{\phi}\sin\psi\sin\theta + \dot{\theta}\cos\psi)\mathbf{u}_1 + (\dot{\phi}\cos\psi\sin\theta - \dot{\theta}\sin\psi)\mathbf{u}_2 + (\dot{\phi}\cos\theta + \dot{\psi})\mathbf{u}_3 \tag{7.36}$$

$$\boldsymbol{\omega} = (\dot{\theta})\mathbf{u}_x + (\dot{\phi}\sin\theta)\mathbf{u}_y + (\dot{\phi}\cos\theta + \dot{\psi})\mathbf{u}_z \tag{7.37}$$

$$\boldsymbol{\omega} = (\dot{\theta}\cos\phi + \dot{\psi}\sin\theta\sin\phi)\mathbf{u}_X + (\dot{\theta}\sin\phi - \dot{\psi}\sin\theta\cos\phi)\mathbf{u}_Y + (\dot{\psi}\cos\theta + \dot{\phi})\mathbf{u}_Z \tag{7.38}$$

Kinetic Energy

The expressions in Eqs. (7.32) and (7.36)–(7.38) can be used to evaluate the kinetic energy as given in Eq. (4.4). If Eq. (7.32) represents the velocity of the center of mass and the angular velocity components in Eq. (7.36) along the principal directions are denoted ω_1, ω_2, and ω_3, the kinetic energy is

$$T = \tfrac{1}{2}m\left(\dot{X}^2 + \dot{Y}^2 + \dot{Z}^2\right) + \tfrac{1}{2}\left(I_1\omega_1^2 + I_2\omega_2^2 + I_3\omega_3^2\right) \tag{7.39}$$

where the principal moments of inertia I_1, I_2, and I_3 are evaluated about the center of mass. As noted in Chapter 4, for the special case in which the body has a point fixed in inertial space, not only can one use the special torque–angular momentum relation (7.4a), but the total kinetic energy can be expressed using the fixed-point inertia matrix $[I_0]$ without a term involving the center of mass velocity. If the matrix is evaluated in the principal directions, the expression corresponding to Eq. (7.39) is

$$T = \tfrac{1}{2}\left(I_{01}\omega_1^2 + I_{02}\omega_2^2 + I_{03}\omega_3^2\right) \tag{7.39a}$$

where the principal moments of inertia I_{01}, I_{02}, and I_{03} are now evaluated about the fixed point 0 instead of about the center of mass. When the coefficients ω_1, ω_2, and ω_3 from Eq. (7.36) are inserted into Eq. (7.39), the result is quite complex, but it can be used with Lagrange's equations to find equations of motion for rotation.

A special case of some interest is when the body has inertial symmetry about the 3-axis so that $I_1 = I_2$. This might be the case for a body with a bullet or disk shape. In this case the x and y axes are also principal axes, both with moment of inertia I_1. Then Eq. (7.37) can be used to evaluate Eq. (7.39) explicitly:

$$T = \tfrac{1}{2}m\left(\dot{X}^2 + \dot{Y}^2 + \dot{Z}^2\right) + \tfrac{1}{2}I_1\left(\dot{\phi}^2\sin^2\theta + \dot{\theta}^2\right) + \tfrac{1}{2}I_3\left(\dot{\phi}\cos\theta + \dot{\psi}\right)^2 \tag{7.40}$$

or if there happens to be a fixed point,

$$T = \tfrac{1}{2}I_{01}\left(\dot{\phi}^2\sin^2\theta + \dot{\theta}^2\right) + \tfrac{1}{2}I_{03}\left(\dot{\phi}\cos\theta + \dot{\psi}\right)^2 \tag{7.40a}$$

Steady Precession of Gyroscopes

A gyroscope typically consists of a wheel with bearings that allow it to spin about its axis of symmetry and mounted so that this axis of rotation can change in an arbitrary direction. Often, a system of *gimbals* holding the wheel keeps the center of mass stationary while allowing the axle to point in varying directions. Any rotating body will exhibit gyroscopic effects, but these effects are particularly evident in the case of a rapidly spinning wheel subjected to fairly small moments. Figure 7.4 shows a wheel (without the gimbal structure holding it) spinning rapidly about the z-axis but with a torque $\vec{\tau}$ about the x-axis. The wheel is assumed to have principal axes aligned with

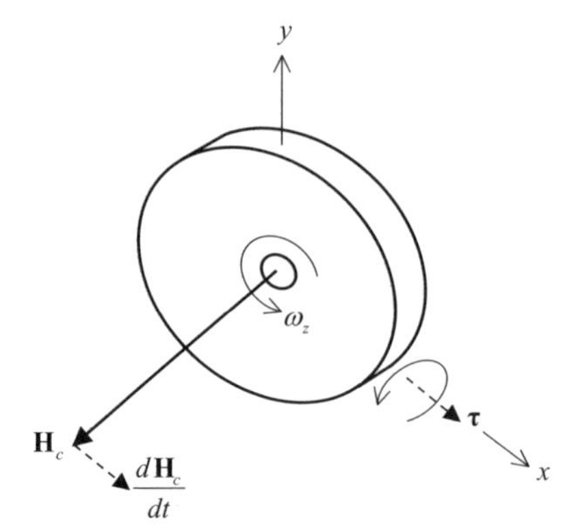

Figure 7.4 Wheel of a gyroscope with a large axial angular momentum.

the x–y–z axes with equal principal moments of inertia about any axes perpendicular to the axis of rotation. This is a special case because even though the x–y–z axis system is not rotating with the body, it can, none the less, be considered a principal axis coordinate frame.

Because of the large angular velocity component about the z-axis, the angular momentum according to Eqs. (7.5) and (7.9) is almost exactly along the z-axis, as shown in the figure. The magnitude of the angular momentum is essentially given by Eq. (7.10) with $I_{zz} = I_3$, even when there some small other components of angular velocity. The peculiar nature of gyroscopic motion is evident when we consider a torque acting along the x-axis. Common experience with nonspinning bodies would indicate that the torque about the x-axis should induce an angular acceleration about the x-axis. After a time, one would expect the body to be rotating about the x-axis, but in this case that is not what happens. Equation (7.4) states that the vector rate of change of the angular momentum is in the vector direction of the applied torque. In the case of the gyroscope, the applied torque does not cause the angular momentum to increase in magnitude but rather to change direction, as shown in Figure 7.4. The result is that the wheel is caused to rotate around the y-axis. The steady rotation caused by a constant torque is called *precession*. The speed of the precession is easy to compute. The magnitude of the angular momentum is

$$H_c \cong I_3\omega_z \tag{7.41}$$

and the magnitude of the rate of change of the angular momentum is just the length of the $\mathbf{H}_c$ vector times the rotational speed about the y-axis. (One can think of the rate of change of $\mathbf{H}_c$ to be the linear velocity of the tip of the vector.)

$$\frac{dH_c}{dt} = (I_3\omega_z)\omega_y = \tau_x \tag{7.42}$$

The precession is the rotation rate of the gyroscope about the y-axis is caused by torque about the x-axis:

$$\omega_y = \frac{\tau_x}{I_3 \omega_z} \tag{7.43}$$

The analysis just given is not exactly correct since it has neglected the component of the angular momentum associated with ω_y in Eq. (7.41), but for typical gyroscopes, the spin speed ω_z is very large, so for modest values of applied torque, ω_y is quite small. Also, this analysis assumes a steady motion in which the torque is constant. The dynamics of gyroscopes considering nonequilibrium initial conditions and time-varying torques is somewhat more complicated and is treated below.

The real surprise is that torque about the x-axis produces a rotation about the y-axis. This contradicts all experience with plane motion in which the familiar expression $\tau = I\alpha$ seems to be of the same nature as $F = ma$. In plane motion, though, the relation has to do with torque and angular acceleration about a single axis: the axis normal to the motion plane. Only in three dimensions is it clear that the seemingly analogous relations between net force and the rate of change of linear momentum, Eq (7.3), and net torque and the rate of change of angular momentum, Eq. (7.4), can lead to quite different types of motion response.

Dynamics of Gyroscopes

To study the dynamics of gyroscopes it is convenient to use Lagrange's equations and the use Euler angles as generalized coordinates. The equations of motion for the general case are quite complicated, but since practical gyroscopes have large values of axial spin and the dynamic deviations of the spin axis after a disturbance are usually small, we deal here with linearized versions of the equations. Referring to Figure 7.3, it should be clear that the direction of the spin axis is determined by θ and ϕ. The time history of these variables describes the dynamics of the gyroscope spin axis. (Because the wheel is spinning, the angle ψ itself is not of much interest since it keeps increasing rapidly, and indeed, $\dot{\psi}$ will turn out to be nearly constant.)

One might think that it would be possible to consider small deviations of θ and ϕ from *zero* steady values to describe the gyroscope dynamics, but when $\theta = 0$, the angles ϕ and ψ are actually measured about the same axis, so the three Euler angles do not, in this case, uniquely determine the position of the wheel. The position $\theta = 0$ is a singular position for Euler angles, and physically the phenomenon of *gimbal lock* occurs for gyroscopes if this position is reached. (Figure 7.6 shows a gimbal structure that would exhibit this phenomenon at $\theta = 0$.)

It is not possible to linearize the equations of motion successfully around the position $\theta = 0$ since the angle ϕ is not defined unambiguously at this value of θ. For this reason, we consider small angular deviations from another position shown in Figure 7.5. When θ and ϕ are both equal to $\pi/2$, the z, 3-axis is lined up with the X-axis and the x-axis is lined up with the Y-axis. Then the variables ξ_1 and ξ_2 are defined to be the small deviations from $\theta = \pi/2$ and $\phi = \pi/2$, respectively, as indicated in

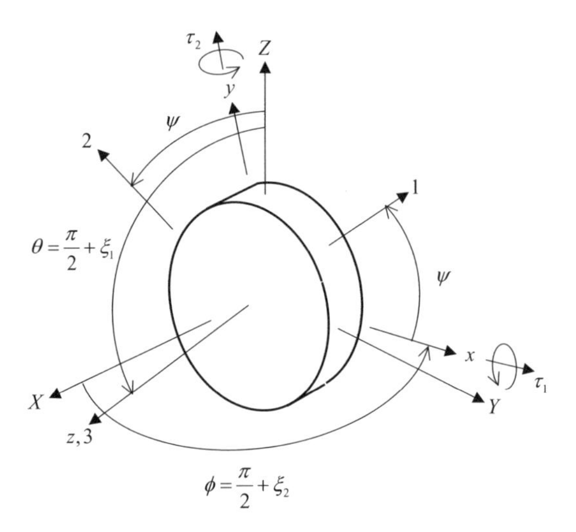

Figure 7.5 Angular deviations ξ_1 and ξ_2 away from steady Euler angles $\theta = \phi = \pi/2$.

Figure 7.5. This choice of the rest position will allow us to derive equations for the gyroscope in terms of the small time-varying deviation variables ξ_1 and ξ_2.

Assuming that the origin of the coordinate systems is at the center of mass of the wheel, and that the wheel has symmetry about the spin axis, the kinetic energy is given by Eq. (7.40) but without the linear velocity terms, or equivalently, by Eq. (7.40a) but with moments of inertia evaluated about the center of mass. The evaluation of the terms in Eq. (7.40a) is accomplished after noting the following relationships:

$$\begin{aligned} \theta &= \frac{\pi}{2} + \xi_1, \qquad \dot{\theta} = \dot{\xi}_1, \qquad \sin\left(\frac{\pi}{2} + \xi_1\right) = \cos\xi_1 \cong 1 \\ \phi &= \frac{\pi}{2} + \xi_2, \qquad \dot{\phi} = \dot{\xi}_2, \qquad \cos\left(\frac{\pi}{2} + \xi_1\right) = -\sin\xi_1 \cong -\xi_1 \end{aligned} \tag{7.44}$$

Using these relations in Eq. (7.40a) yields

$$T = \tfrac{1}{2} I_1 \left(\dot{\xi}_2^2 \cos^2 \xi_1 + \dot{\xi}_1^2\right) + \tfrac{1}{2} I_3 \left(-\dot{\xi}_2 \sin \xi_1 + \dot{\psi}\right)^2 \tag{7.45}$$

Although $\dot{\psi}$ is a large spin angular velocity, the variables ξ_1, $\dot{\xi}_1$, ξ_2, and $\dot{\xi}_2$ are all supposed to be small in magnitude. To find linearized equations of motion, one could use Eq. (7.45) as it is and then expand any nonlinear terms and retain only the *first-order* terms in the small quantities. Alternatively, the kinetic energy could be approximated up to the *second order* in the variables considered small: ξ_1, ξ_2, $\dot{\xi}_1$, and $\dot{\xi}_2$. This will automatically produce linear equations after the differentiation involved in Lagrange's equations. Using the second method, the approximate version of the

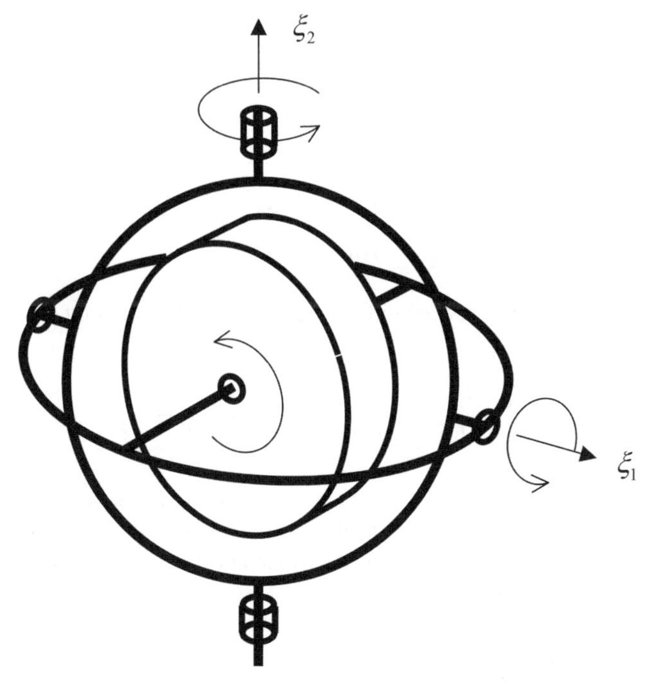

Figure 7.6 Gyroscope wheel supported by a gimbal structure.

kinetic energy is

$$T = \tfrac{1}{2}I_1\left(\dot{\xi}_2^2 + \dot{\xi}_1^2\right) + \tfrac{1}{2}I_3\left(-2\dot{\psi}\dot{\xi}_2\xi_1 + \dot{\psi}^2\right) \tag{7.46}$$

Now suppose that there are applied torques about the x and y axes but none about the z-axis. Calling the x-axis torque τ_1 and the y-axis torque τ_2 means that they are generalized forces for the ξ_1 and ξ_2 equations, as can be seen in Figure 7.5. There is no generalized force for the ψ equation. The Lagrange equations corresponding to ξ_1, ξ_2, and ψ are

$$\begin{aligned}
\frac{d}{dt} I_1\dot{\xi}_1 + I_3\dot{\psi}\dot{\xi}_2 &= \tau_1 \\
\frac{d}{dt}\left(I_1\dot{\xi}_2 - I_3\dot{\psi}\xi_1\right) &= \tau_2 \\
\frac{d}{dt}\left[I_3\left(\dot{\psi} - \dot{\xi}_2\xi_1\right)\right] &= 0
\end{aligned} \tag{7.47}$$

The first two equations are linear in the small variables, but the last one is actually nonlinear. However, the implication is that the quantity $\dot{\psi} - \dot{\xi}_2\xi_1$ is constant, and since we assume that $\dot{\psi}$ is a large spin and $\dot{\xi}_2\xi_1$ is the product of two small quantities, we can define

$$\omega_3 \cong \dot{\psi} \cong \text{constant} \tag{7.48}$$

to be the spin around the wheel axis. The two equations of motion then become

$$\begin{aligned} I_1 \ddot{\xi}_1 + I_3 \omega_3 \dot{\xi}_2 &= \tau_1 \\ I_1 \ddot{\xi}_2 - I_3 \omega_3 \dot{\xi}_1 &= \tau_2 \end{aligned} \tag{7.49}$$

In matrix form the equations exhibit *gyroscopic coupling* among the generalized velocities:

$$\begin{bmatrix} I_1 & 0 \\ 0 & I_1 \end{bmatrix} \begin{bmatrix} \ddot{\xi}_1 \\ \ddot{\xi}_2 \end{bmatrix} + \begin{bmatrix} 0 & I_3\omega_3 \\ -I_3\omega_3 & 0 \end{bmatrix} \begin{bmatrix} \dot{\xi}_1 \\ \dot{\xi}_2 \end{bmatrix} = \begin{bmatrix} \tau_1 \\ \tau_2 \end{bmatrix} \tag{7.50}$$

Note that in a steady motion in which the terms involving $\ddot{\xi}_1$ and $\ddot{\xi}_2$ can be neglected in Eqs. (7.49) and (7.50), steady torque about one axis creates a steady rotation about the other. For example, if τ_1 has a constant value and $\tau_2 = 0$, a steady-state solution of the equations is possible with

$$\dot{\xi}_2 = \frac{\tau_1}{I_3 \omega_3} \tag{7.51}$$

This is equivalent to Eq. (7.42), derived by considerating only the steady behavior. Thus, the dynamic equations also predict the steady precession behavior discussed previously. Of course, if Eq. (7.51) were to persist for a long time, ξ_2 would grow until the linearizing assumptions behind the dynamic equations we are using would no longer be valid.

The characteristic equation for the case with no torques is easily found from matrix equation (7.50):

$$\det \begin{vmatrix} I_1 s^2 & I_3\omega_3 s \\ -I_3\omega_3 s & I_1 s^2 \end{vmatrix} = I_1^2 s^4 + I_3^2 \omega_3^2 s^2 = 0 \tag{7.52}$$

so the eigenvalues are

$$s = 0, \qquad 0, \qquad +j\frac{I_3\omega_3}{I_1}, \qquad -j\frac{I_3\omega_3}{I_1} \tag{7.53}$$

The zero eigenvalues simply show that the spin axis can stay pointing in an arbitrary direction when there is no disturbing torque. The complex eigenvalues indicate that the axis can also oscillate sinusoidally with a natural frequency ω_n proportional to the spin speed.

$$\omega_n = \frac{I_3\omega_3}{I_1} \tag{7.54}$$

When the spin speed is large, as we have assumed, the frequency of the wobble of the spin axis occurs at a high frequency. In many cases, the slow precession described by Eqs. (7.42) and (7.51) is more easily observed than the high-frequency wobble that may also be present.

The eigenvalues found in Eq. (7.52) naturally lead to the conclusion that if a wobble is started, it will persist forever since there no energy dissipation is represented in the dynamic equations. An actual gyroscope has a number of bearings which have frictional torques and, when operating in air, has torques associated with the air drag. The assumption that the spin speed is constant in the presence of air drag and bearing friction is justified if a motor of some sort is incorporated to keep the rotor spinning. On the other hand, the friction torques at gimbal bearings will affect the gyrator dynamics and there is no simple way to eliminate their effect.

Figure 7.6 is a schematic diagram of a gimbal structure for a gyroscope in essentially the position shown in Figure 7.5. Note that parts of the gimbal structure rotate with the perturbation angles ξ_1 and ξ_2, so that the mass of the gimbal structure will affect the dynamic equations, Eq. (7.49) or (7.50) . Typically, the gyro wheel is massive and the gimbals are as light as possible, so for simplicity we will not take the effects of the gimbal rings on the moments of inertia into account. The bearing torques, on the other hand, do make the predicted undamped wobble actually dissipate.

It would be simple enough to use τ_1 and τ_2 to represent and type of bearing torque and then to use computer simulation to discover how the variables ξ_1 and ξ_2 behave after a disturbance. A comparison of Figures 7.5 and 7.6 shows that the torque τ_1 is associated with the bearings that allow the angular velocity $\dot{\xi}_1$, and ξ_1 has to do with the bearings that allow $\dot{\xi}_2$. Realistic bearing torques are functions of the bearing rotation rates, but they almost certainly are not linear functions. However, to proceed with the linearized analysis of gyroscope dynamics, we make the simple assumption that both torques are related to the appropriate angular rate by a single coefficient, b.

$$\tau_1 = -b\dot{\xi}_1, \qquad \tau_2 = -b\dot{\xi}_2 \tag{7.55}$$

The minus signs in Eq. (7.55) assure us that for $b > 0$, the friction torques resist the angular motion. After incorporating Eq. (7.55) into Eq. (7.50), the result is

$$\begin{bmatrix} I_1 & 0 \\ 0 & I_1 \end{bmatrix} \begin{bmatrix} \ddot{\xi}_1 \\ \ddot{\xi}_2 \end{bmatrix} + \begin{bmatrix} b & I_3\omega_3 \\ -I_3\omega_3 & b \end{bmatrix} \begin{bmatrix} \dot{\xi}_1 \\ \dot{\xi}_2 \end{bmatrix} = \begin{bmatrix} 0 \\ 0 \end{bmatrix} \tag{7.56}$$

The characteristic equation changes from Eq. (7.52) to

$$\det \begin{vmatrix} I_1 s^2 + bs & I_3\omega_3 s \\ -I_3\omega_3 s & I_1 s^2 + bs \end{vmatrix} = s^2 \left[(I_1 s + b)^2 + I_3^2 \omega_3^2 \right] = 0 \tag{7.57}$$

Finally, the eigenvalues change from those in Eq. (7.53) to

$$s = 0, \qquad 0, \qquad -\frac{b}{I_1} + j\frac{I_3\omega_3}{I_1}, \qquad -\frac{b}{I_1} - j\frac{I_3\omega_3}{I_1} \tag{7.58}$$

The zero eigenvalues again indicate that the spin axis can remain constant at any angular position. The complex-conjugate eigenvalues indicate that the spin axis can still wobble after a disturbance, but there will be a term in the solution due to the damping coefficient that makes the wobble die out in time.

The general form of the solution is

$$\begin{aligned} \xi_1 &= c_1 + e^{-bt/I_1}\left(c_3 \sin \frac{I_3}{I_1}\,\omega_3 t + c_4 \cos \frac{I_3}{I_1}\,\omega_3 t\right) \\ \xi_2 &= c_2 + e^{-bt/I_1}\left(-c_3 \cos \frac{I_3}{I_1}\,\omega_3 t + c_4 \sin \frac{I_3}{I_1}\,\omega_3 t\right) \end{aligned} \tag{7.59}$$

The four constants c_1, c_2, c_3, and c_4 are determined by the values of ξ_1, ξ_2, $\dot{\xi}_1$, and $\dot{\xi}_2$ at the initial time. Equation (7.59) shows that after an initial disturbance, the spin axis of the gyroscope will oscillate in a sinusoidal fashion at the frequency given by Eq. (7.54), but with exponentially decreasing amplitude. The time constant of the decay is just I_1/b. After a time, the values of the angles ξ_1 and ξ_2 will approach the values c_1 and c_2 respectively. The wobble described by Eq. (7.59) is not easily observed in toy gyroscopes when the spin speed is high, but as the spin slows down, the wobble frequency slows also and the wobbling motion is easier to see. At very low speeds, disturbances can cause large values of θ and ϕ which no longer obey the linearized equation for the small deviation variables ξ_1 and ξ_2. There may still be wobbling motion, but it will not be described accurately by Eq. (7.59).

7.4 SUMMARY

Although many practical problems involving rigid bodies can be analyzed under the assumption that the bodies move in a fixed plane, there are other problems that require a study of motion in three dimensions. While the dynamics of the translation of the center of mass of a rigid body in three dimensions is the same as for a particle, the rotation of a rigid body in three dimensions is considerably more complicated. Where the dynamic principles of translation involve vector operations and a scalar mass, the dynamics of rotation cannot rely only on vector operations (angular position cannot be described by a vector) and the inertial properties must be described by an inertia matrix or tensor. The result is that three-dimensional motion of a rigid body can become very complicated to analyze and the equations of motion can be complex and significantly nonlinear. Despite this, there are some applications of three-dimensional motion that can be described using either body-fixed or inertial coordinates that give insight into dynamic response without undue complication. In other cases, there is no alternative to computer simulation of the nonlinear dynamic equations in order to study the three-dimensional motions of a system.

REFERENCES

[1] S. H. Crandall, D. C. Karnopp, E. F. Kurtz, Jr., and D. C. Pridmore-Brown, *Dynamics of Mechanical and Electromechanical Systems*, McGraw-Hill, New York, 1968. (Reprint edition, Krieger Publishing, Melbourne, FL.)

[2] H. Goldstein, *Classical Dynamics*, Addison-Wesley, Reading, MA, 1959.

[3] D. T. Greenwood, *Principles of Dynamics*, Prentice Hall, Englewood Cliffs, NJ, 1988.

[4] R. C. Hibbler, *Engineering Mechanics Dynamics*, 11th ed., Pearson Prentice Hall, Upper Saddle River, NJ, 2007.

[5] K. Ogata, *Modern Control Engineering*, Prentice-Hall, Englewood Cliffs, NJ, 1970.

PROBLEMS

7.1 Figure P7.1 shows a uniform solid object in the form of a brick or blackboard eraser. Because of the symmetry of the object, the directions of the principal axes should be obvious. They are labeled 1–2–3 (see Appendix B).

(a) Which axis has the largest moment of inertia, which has the smallest, and which has the intermediate value for the moment of inertia?

(b) If such an object is made of a fairly dense material, when it is thrown into the air, it behaves for some time as if it were nearly torque-free. If you tried to throw it into the air with it spinning mainly around one of the three principal axes, about which axes do you think you would be successful and about which axes do you think you would be unlikely to be successful? You could answer this question experimentally.

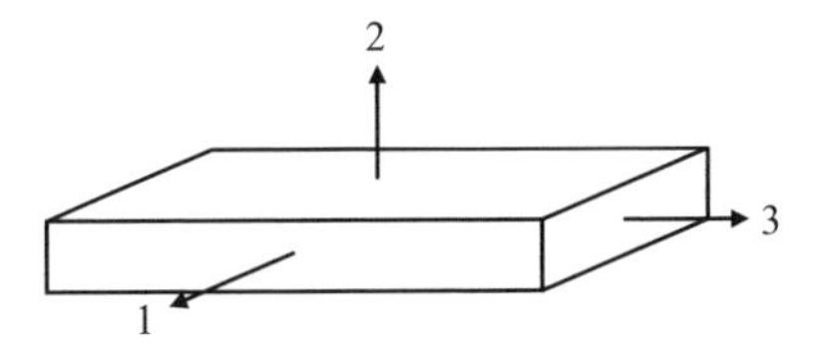

Figure P7.1

7.2 Figure P7.2 shows a satellite with principal axes x–y–z and the standard notation for the angular velocity. You are to see what advantage there might be to spin stabilization about the x-axis compared to the case without spin stabilization. Suppose that a small particle hits the satellite in such a way that a disturbance torque impulse is applied around the y-axis. The torque acts over a very short time period δt giving an impulse with the value

$$\text{imp} = \int_0^{\delta t} M\,dt$$

(a) First, assume that the satellite is not rotating at all initially, $p = q = r = 0$. Using Eq. (7.18), show that immediately after the particle hits, p and r are still zero but $q = \text{imp}/I_{yy}$. What type of motion is introduced by the particle in this case?

(b) Now assume that the satellite is spin stabilized about the x-axis with a large spin, and with $p = \Omega q = r = 0$ initially. Assume that the x-axis has either the maximum or the minimum principal moment of inertia. Note from Eq. (7.4) that the angular momentum is a constant vector in inertial space before the particle hits and later after it has hit the satellite, but it changes suddenly during the impulse. Show in a sketch the $\mathbf{H}_c$ vector before and after the impulsive torque.

(c) Estimate the angle between the x-axis and the angular momentum vector just after the impulse. Because the $\mathbf{H}_c$ vector remains constant in inertial space after the impulse, this means that the x-axis simply wobbles around its original position after the impulse, and the greater the spin, the less the amplitude of the wobble.

(d) By combining Eqs. (7.24) and (7.25), show that the equation for the perturbed system after the impulse is $\Delta \ddot{q} + k \, \Delta q = 0$, where k is defined in Eq. (7.28). Note that, similarly, $\Delta \ddot{r} + k \, \Delta r = 0$.

(e) If we consider time $t = 0$ to be just after the impulse, show that the solution for Δq is $\Delta q = (\text{imp}/I_{yy}) \cos k^{1/2} t$. (Note that Δr must also have a sinusoidal type of solution.)

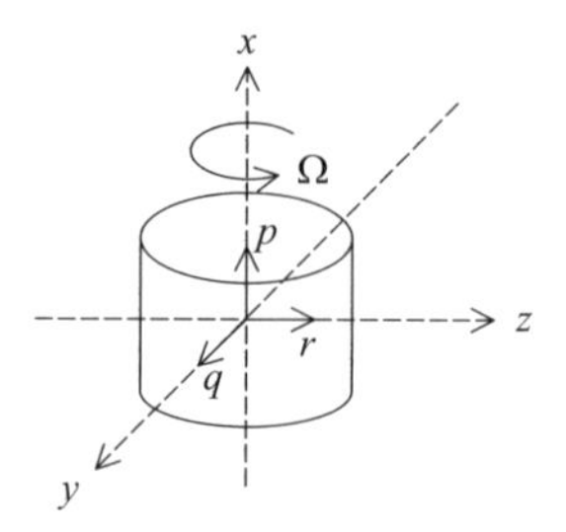

Figure P7.2

7.3 Suppose that a satellite were to be constructed such that $I_{yy} = I_{zz}$ and it was proposed to spin stabilize it by spinning it about the x-axis. Show that it does not matter for stability whether I_{xx} is larger or smaller than I_{yy}.

7.4 Write Euler's equations, Eqs. (7.18), for the case that $I_{yy} = I_{zz}$ and for no applied torque about the x-axis, $L = 0$.

(a) Show that p can be constant with a value Ω.

(b) Compare the differential equations for q and r to the linearized equations for the gyroscope. Is there gyroscopic coupling?

(c) Why are Euler's equations not exactly the same as the linearized gyroscope equations?

7.5 Use Eq. (7.45) and Lagrange's equations to find the equations of motion of a gyroscope when the small-angle approximations are not made. (You may not want to carry the operations out to the final result. I don't either.)

7.6 Figure P7.6 shows the view from above of the left front wheel of a car making a right-hand turn. The wheel has a moment of inertia about its axle of I and a rolling radius r. The forward speed of the wheel is U and the radius of the turn is R. Using the results presented in the section on steady precession, answer the following questions:

(a) Assuming that the turn radius is much greater that the wheel rolling radius, what is the approximate magnitude of the angular momentum of the wheel?

(b) What is the rate of precession of the wheel?

(c) What is the magnitude of the moment that the axle must be exerting on the wheel to account for the forced precession?

(d) We know that there is a tendency of cars to roll toward the outside of a turn and even to overturn if turns are taken too fast. Does the gyroscopic effect of the wheel tend to increase this effect or to decrease it? (Remember, the moment from the wheel onto the car is the opposite of the moment from the car onto the wheel.)

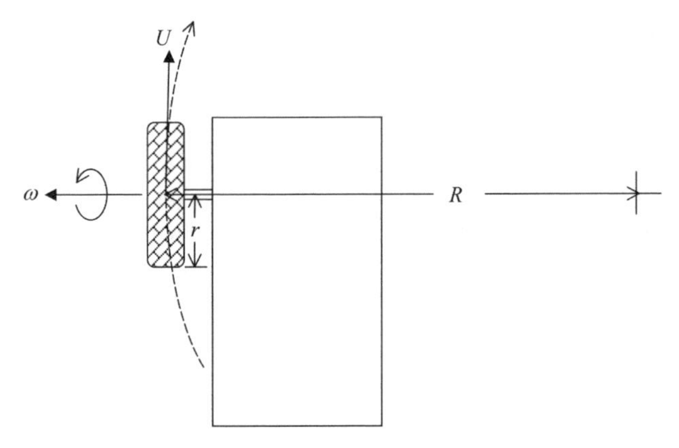

Figure P7.6

7.7 A wheel with mass m and axial moment of inertia I is spinning rapidly with speed ω and hanging form the ceiling on a string (Figure P7.7). The distance from the wheel center of mass to the string attachment point of the light axle is l and the acceleration of gravity is g.

(a) Assuming that the wheel precesses steadily, what is the precession rate Ω, and seen from above, is it clockwise or counterclockwise?

(b) Estimate the small angle that the string makes with the vertical because of the precession. Assume that the distance the string attachment point on the axle moves away from the vertical is small compared to l.

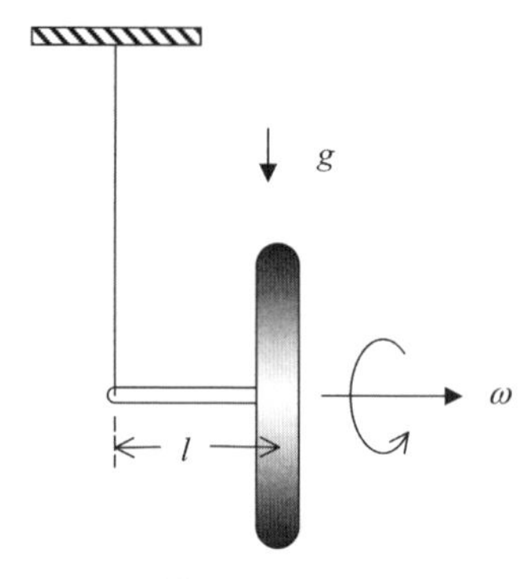

Figure P7.7

7.8 A rigid cone rolls without slip on a horizontal surface X–Y such that its constant rotation rate about the Z-axis is Ω (Figure P7.8). As shown in the side view of the cone, it has a half-angle of α. The origin of the coordinate system, 0, is a point fixed in inertial space and is also a point in the cone. A basic principle of mechanics is that the relation between applied torque and the rate of change of angular momentum applies when one point of a rigid body is fixed in inertial space, as is the tip of the cone in this case. The equation $\boldsymbol{\tau}_0 = d\mathbf{H}_0/dt$ is similar to Eq. (7.4) except that the torque and the angular momentum are evaluated about the fixed point instead of the center of mass. Furthermore, the angular momentum is related by a relationship similar to Eq. (7.5), $[\mathbf{H}_0] = [I_0][\omega]$, but now the moments of inertia are evaluated about point 0 instead of the center of mass. (The angular velocity $\boldsymbol{\omega}$ of the body is the same whether a fixed point or the center of mass is being considered.) Consider that the cone has moments of inertia of I_1, $I_1$and I_3 evaluated with respect to point 0, where I_3 is the moment of inertia for the axis of symmetry and I_1 applies to any axis through the tip of the cone perpendicular to the axis of symmetry (see Appendix B).

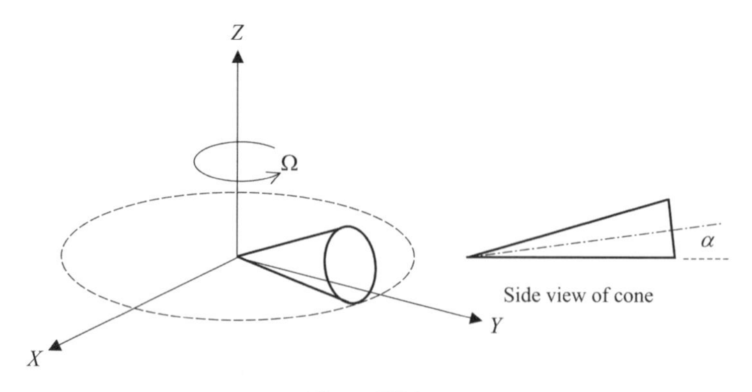

Figure P7.8

(a) What is the angular velocity of the cone? Show the components in a sketch. *Hint:* Can you find a line on the cone about which it is instantaneously rotating? The direction of this line is the direction of the $\boldsymbol{\omega}$ vector.
(b) Find the angular momentum of the cone and show it on a sketch.
(c) Find $d\mathbf{H}_0/dt$.
(d) Describe the forces and torques that must exist on the cone to maintain this motion.

7.9 Consider the Euler equations (7.18) for a body for a special case: (1) There is no torque at all on the body, and (2) the body is symmetrical about the z-axis so that $I_{xx} = I_{yy} = I_1$ and $I_{zz} = I_3$.
(a) Show that the angular rate r is constant.
(b) Show that the equations for the angular rates p and q have a particularly simple form if the constant quantity $s = (I_3 - I_1)r/I_1$ is introduced.
(c) Verify that the sinusoidal oscillations of p and q can be represented by $p = \omega_0 \sin(st + \varphi)$, $q = \omega_0 \cos(st + \varphi)$, where ω_0 is an arbitrary amplitude and φ is an arbitrary phase angle.

7.10 The motion of a spinning top shown in Figure P7.10 can be well described by the Euler angles defined in Figure 7.3 . The reason that tops are of interest to children is that when they are spinning fast enough, instead of falling over under the influence of gravity as expected, they can execute complicated motions. These motions include *precession* of the 3-axis around the vertical direction, and *nutation*, which is an oscillation of the tilt angle θ around an average value. The equations of motion for a top are fairly easy to derive if several simplifying assumptions are made:

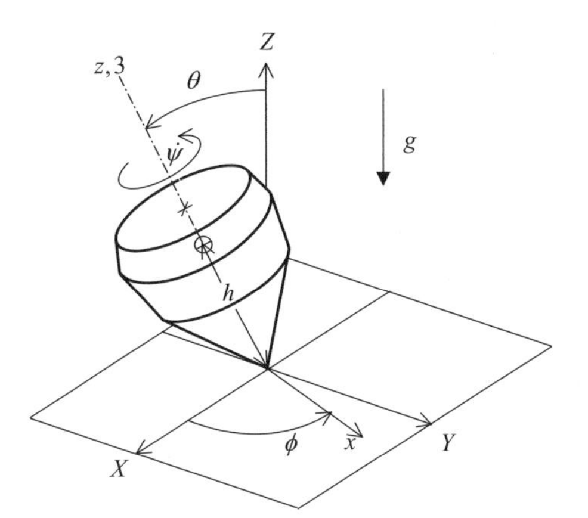

Figure P7.10

1. Suppose that the point of the top is a fixed in the X–Y–Z inertial space. This means that Eq. (7.4a) could be used to express the relation between the torque applied around the fixed point and the rate of change of the angular

momentum again evaluated about the fixed point. Also, the kinetic energy can use the special form of Eq. (7.39a) using principal moments of inertia evaluated about the fixed point.

2. Tops are normally symmetrical, so the principal moments of inertia about the 1 and 2 axes can be assumed equal. This means that the special form of Eq. (7.40a) can be used for the kinetic energy.
3. Tops typically spin for quite a long time before slowing down significantly, so it is logical to neglect friction torques at the point and from air drag in describing the dynamic motions. The most significant torque is then due to the gravity force mg acting a distance h from the pivot point.
   (a) Use Lagrange's equations to find the equations of motion for θ, ϕ and ψ.
   (b) Show that ω_3 given in Eq. (7.36) is constant under the assumptions above.
   (c) Find the condition for steady procession when $\theta =$ constant ($\dot{\theta} = \ddot{\theta} = 0$) and $\dot{\phi} =$ constant ($\ddot{\phi} = 0$).
   (d) Does the result in part (c) match the result in Eq. (7.43) when $\theta = \pi/2$?

7.11 This problem is related to the problem of *critical speeds of rotating shafts.* Many shaft or rotor systems can rotate at low speeds with very small deflections, but when the speed exceeds a critical value, the system becomes unstable and the shaft suddenly experiences a large and potentially catastrophic deflection. Figure P7.11 shows a thin uniform rod of length l and mass m freely pivoted at point o on a shaft forced to spin at a constant rate ω_0 and acted upon by gravity. The rod is constrained to remain in a plane defined by the disk, which is welded to the rod and rotates with angular speed ω_0. The moment of inertia for an axis directed along the rod can be assumed to be zero, and for any axis through o perpendicular to the rod the moment of inertia is $ml^2/3$. It is obvious that the rod could spin with $\theta = 0$ at any speed, but at a sufficiently high spin speed, there is another possible constant angle which could spin. The speed at which the second equilibrium angle first appears is the critical speed.

(a) Find the kinetic energy using Eq. (7.39a) and the potential energy due to gravity, and use these with a Lagrange's equation to find the equation of motion for the angle θ.

(b) Using the equation of motion, find the equilibrium angles, θ_0, the rod could assume as a function of the spin speed ω_0. (These are angles at which $\ddot{\theta} = 0$ and the equation of motion is still satisfied.) You will find that below a critical speed only one equilibrium angle is possible, but above the critical speed, another equilibrium is possible.

(c) Analyze the stability of motion for small deviations from the possible equilibrium angles by linearizing the nonlinear equation of motion. The procedure is to replace the nonlinear equation of the form $\ddot{\theta} + f(\theta) = 0$ with the linearized form $\Delta\ddot{\theta} + [df(\theta)/d\theta|_{\theta_0}]\Delta\theta = 0$. Then if the derivative of f evaluated at θ_0, $[df(\theta)/d\theta|_{\theta_0}]$, is positive, the system is stable; if not, it is unstable.

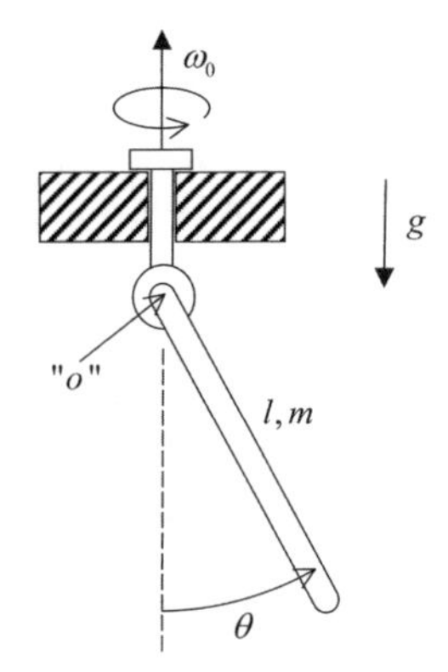

Figure P7.11

7.12 A spherical pendulum with mass m and length l hangs in a gravity field g. The position of the pendulum is described by the variables θ and ϕ as shown in Figure P7.12. The three unit vectors in the order $\mathbf{u}_\theta$, $\mathbf{u}_r$, $\mathbf{u}_\phi$ form a right-handed orthogonal set of vectors attached to the body and moving with it.

(a) Assume first that the pendulum bob is a point mass. Note that the mass has velocity components only in the $\mathbf{u}_\theta$ and $\mathbf{u}_\phi$ directions and that the magnitudes of these components can be found by a study of the figure. Find the kinetic and potential energies.

(b) Write the Lagrange equations for θ and ϕ.

(c) Interpret the ϕ equation in terms of angular momentum.

(d) Now suppose that the pendulum bob is a sphere of radius a made of a material with uniform mass density and that it has a centroidal moment of inertia $I = \frac{2}{5}ma^2$ about any axis. Convince yourself that the

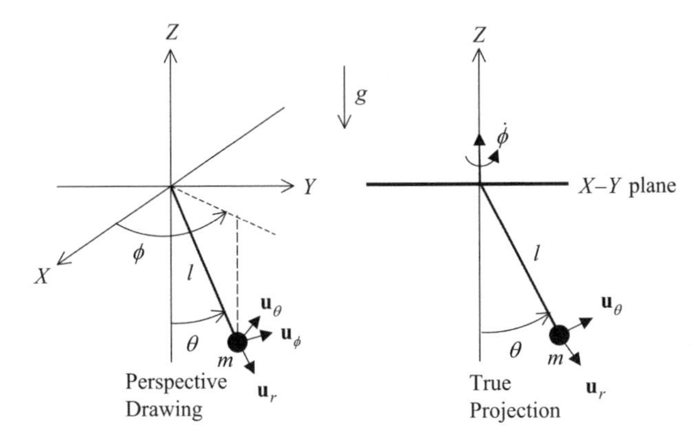

Figure P7.12

angular velocity of the $\mathbf{u}_\theta$, $\mathbf{u}_r$, $\mathbf{u}_\phi$ system (and the body) is given by $\boldsymbol{\omega} = (\dot{\phi}\sin\theta)\mathbf{u}_\theta + (-\dot{\phi}\cos\theta)\mathbf{u}_r + (-\dot{\theta})\mathbf{u}_\phi$.

(e) Write a new expression for the kinetic energy, including the effect of the finite-sized spherical pendulum bob. Interpret this result in light of the parallel axis theorem for moments of inertia.

(f) Write new Lagrange equations for the system.

7.13 A horizontal disk with axial moment of inertia I can rotate with negligible friction in fixed bearings (Figure P7.13). On it is mounted a lightweight frame carrying a spinning second disk with mass m and principal centroidal moments of inertia I_1, I_1, and I_3. (Note that because two of the principal moments of inertia are equal, any diametral axis is a principal axis. For the calculation of kinetic energy diametral axes not spinning with the body can be used.) The spin of the second disk measured with respect to a horizontal plane is $\dot{\psi}$ and the angular velocity of the horizontal disk is $\dot{\phi}$. The second disk swings on the frame as a pendulum with length l in the gravity field with the angle θ. (The three angles used to describe the motion of the system have some similarity to the Euler angles of Figure 7.3, but they are not Euler angles.)

(a) Find the kinetic and potential energy of the system.

(b) Find the equations of motion.

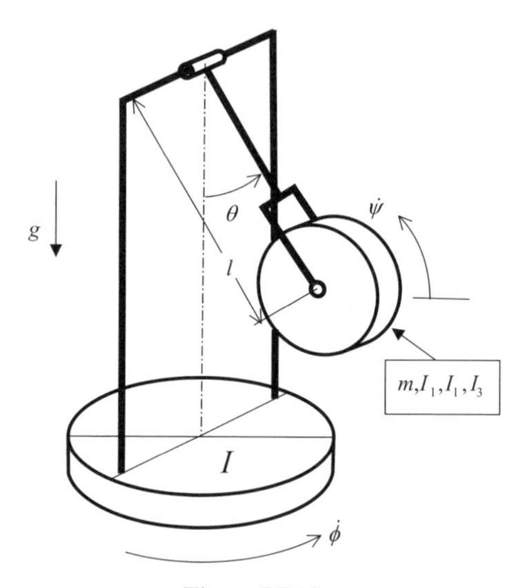

Figure P7.13

7.14 Figure P7.14 shows the system of Problem 7.12 but with the spinning disk rotating about a different axis. Now, $\dot{\psi}$ represents the angular velocity of the disk with respect to the frame rod. Note that the frame rod itself has angular velocity components due to the rates $\dot{\phi}$ and $\dot{\theta}$. The parameters of the elements are the same as in Problem 7.12.

(a) Find new expressions for the kinetic and potential energy.
(b) Derive the equations of motion.
(c) Show that the axial component of the angular velocity of the spinning disk ω_3 is constant, and use this fact to simplify the equations of motion.

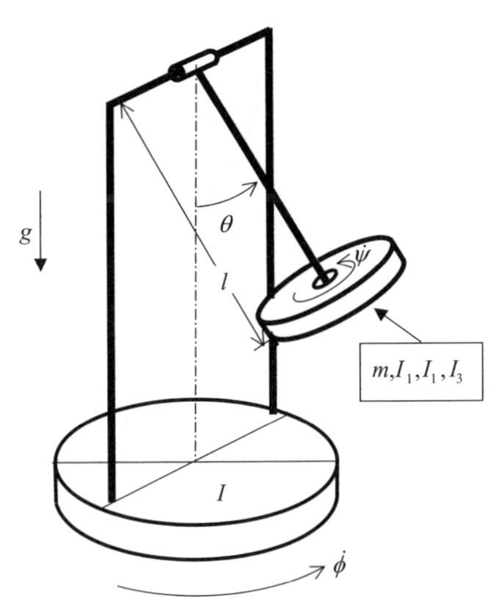

Figure P7.14

7.15 Figure P7.15*a* shows a disk rotating at constant angular velocity Ω. A ball with mass m, radius a, and centroidal moment of inertia I (about any axis) rolls on the disk without slipping (see Appendix B). The inertial coordinates X and Y describe not only the position of the center of the ball as shown in part *b*, but also the point of contact between the ball and the disk. Part *c* shows the velocity of the contact point when the the point is a distance $r = \sqrt{X^2 + Y^2}$ from the axis of rotation. (In the position shown, the component v_X is in the negative X-direction.) Finally, part *d* shows the velocity components of the center of the ball, $\dot{X}$ and $\dot{Y}$, as well as the components of the angular velocity of the ball, ω_X and ω_Y. There is no geometric relationship between X and Y but the constraint that the ball rolls without slipping must be enforced. This *nonholonomic constraint* means that we cannot use Lagrange's equations for this problem, but, of course, we can still use the direct method of applying the laws of mechanics to study this problem.

(a) Using part *c* of the figure, convince yourself that a point *on the disk* at the location X, Y (the contact point) has the velocity components $v_Y = \Omega X$, $v_X = -\Omega Y$.
(b) Using part *d* of the figure, find the X and Y components of the velocity of the bottom of the ball in terms of $\dot{X}$, $\dot{Y}$, ω_X, and ω_Y. Write the nonholonomic constraints by making the velocity components at the bottom of the ball match the velocity components in part (a). (These constraints are clearly

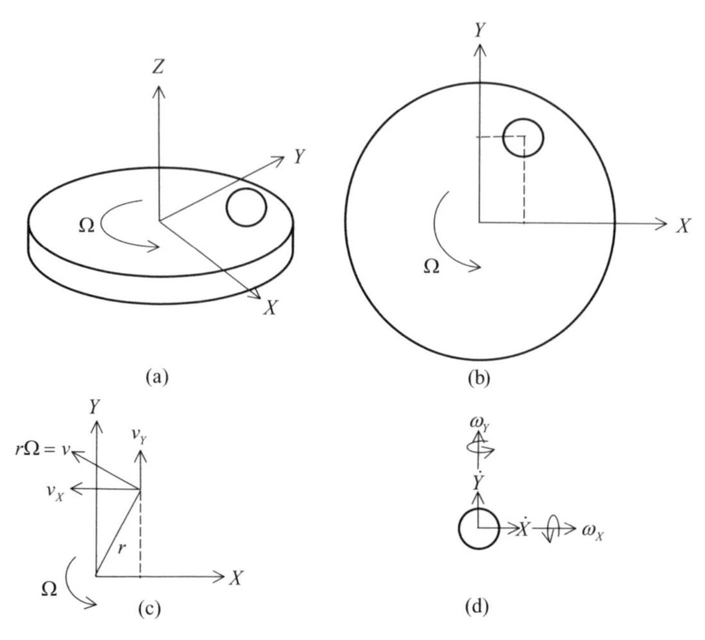

Figure P7.15

not just among position coordinates as is the case in the *holonomic* systems studied in Chapter 4.)

(c) Define the force components from the disk on the ball contact point as F_X and F_Y. The dynamic equations are $F_X = m\ddot{X}$, $F_Y = m\ddot{Y}$, $aF_Y = I\dot{\omega}_X$, $-aF_X = I\dot{\omega}_Y$. Using these four equations and the constraint equations, eliminate the forces and the angular accelerations to find two equations involving only the derivatives of X and Y.
The final versions of the equations should be in the form

$$\dot{X} + kY = c_1$$
$$\dot{Y} - kX = c_2$$

where k, c_1, and c_2 are constants.

(d) Show that the ball will actually move in a circle such as $(X - X_0)^2 + (Y - Y_0)^2 = R^2$, where the center is at X_0, Y_0 and the radius is R. The parameters of the circle depend on how the ball got started rolling initially.

7.16 A disk with uniform mass density and total mass M rolls without slip on a horizontal circle of radius L (Figure P7.16). The axle from point 0 to A through the center of the disk has negligible mass and rotates about the post with angular speed ω_0. Note that the coordinate frame $0xyz$ rotates with the axle around the post but does not participate in the spin of the disk around the axle.

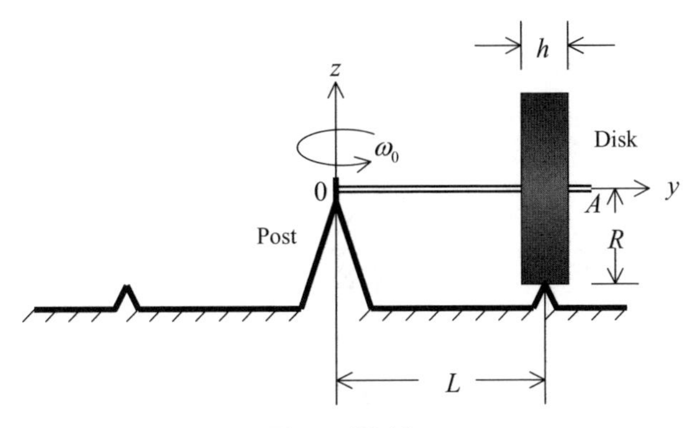

Figure P7.16

(a) Using the results in Appendix B and the parallel axis theorem, Appendix C, verify that the frame $0xyz$ can be considered to be a principal axis coordinate system for the disk with moments of inertia

$$I_{xx} = I_{zz} = M\left(\frac{R^2}{4} + \frac{h^2}{12} + L^2\right), \; I_{yy} = M\frac{R^2}{2}$$

(b) Using the no-slip condition, find the angular velocity *of the disk* ω_{disk} and express it in terms of the unit vectors $\mathbf{u}_x$, $\mathbf{u}_y$, $\mathbf{u}_z$ associated with $0xyz$.

(c) Find the kinetic energy of the disk.

(d) Now find the angular momentum of the disk about point 0, $\mathbf{H}_0$, and express the components again in the rotating coordinate system using $\mathbf{u}_x$, $\mathbf{u}_y$, $\mathbf{u}_z$.

(e) Find the angular velocity of $0xyz$, $\boldsymbol{\omega}_{xyz}$.

(f) Using the torque–angular momentum law $\boldsymbol{\tau}_0 = d\mathbf{H}_0/dt = (\partial\mathbf{H}_0/dt)_{\text{rel}} + \boldsymbol{\omega}_{xyz} \times \mathbf{H}_0$, find the torque exerted on the disk for the case that ω_0 is constant.

(g) Verify your answer to part (f) by making a sketch showing how the $\mathbf{H}_0$ vector rotates with the xyz coordinate system and showing that $\dot{\mathbf{H}}_0$ is the velocity of the tip of the $\mathbf{H}_0$ vector.

(h) Evaluate the term $(\partial\mathbf{H}_0/\partial t)_{\text{rel}}$ in the torque expression.

7.17 A long thin rod of mass m and length l is pivoted with negligible friction at its center on a shaft that is forced to spin around a vertical axis with angular speed $\dot{\phi}$ by the applied torque $\tau(t)$ (Figure P7.17). Use the principal axes 1, 2, and 3 as shown.

(a) What are the moments of inertia I_1, I_2, and I_3?

(b) Using θ and ϕ as generalized coordinates, find the three components ω_1, ω_2, and ω_3 of the angular velocity of the rod.

(c) Find the kinetic energy T.

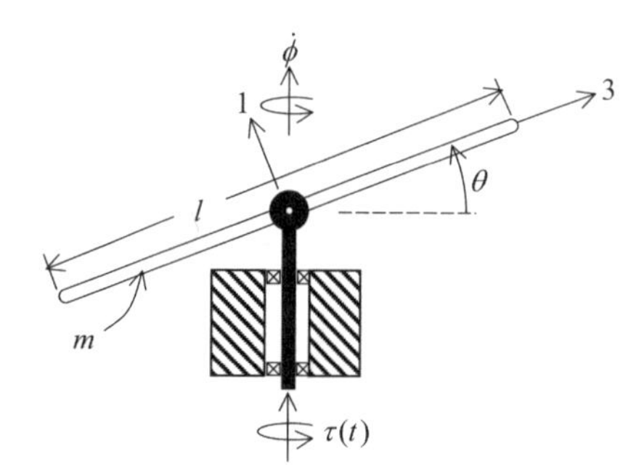

Figure P7.17

(d) Write the Lagrange equations for θ and ϕ, including any necessary generalized forces.

(e) For the special case in which $\tau = 0$, show that the generalized momentum associated with ϕ is constant, and using this fact, determine the way that $\dot{\phi}$ varies with θ.

7.18 Figure P7.18 shows a rotor mounted in a boat on bearings (not shown) and spinning rapidly in the direction indicated by the arrow. If the boat makes a steady right turn, show in a sketch the bearing forces due to the gyroscopic effect of the rotor, and determine whether the boat tends to pitch nose up or nose down.

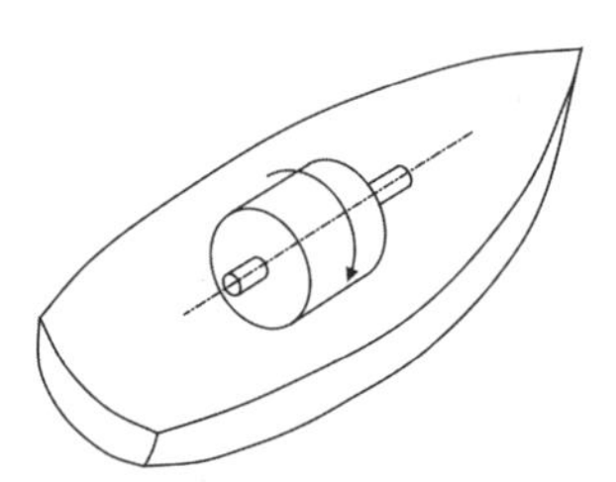

Figure P7.18

8

VIBRATION OF MULTIPLE-DEGREE-OF-FREEDOM SYSTEMS

When inertial elements such as mass particles and rigid bodies interact with compliance elements such as torsional springs, pneumatic springs, and belvel washer springs, the interaction is oscillatory at a natural frequency dependent on the inertial and springlike properties of the elements. If a system containing these interacting elements has forcing input near the natural frequency, the response builds to a resonance and can cause destruction of the system, noise, and other unpleasantness. Systems with many interacting elements can have many natural frequencies, corresponding to many potential resonances. Vibrations is a classic engineering topic that first develops the concepts of natural frequency and resonance and then develops methods to control or limit vibration response.

In a complete presentation of vibration analysis we would first determine the effectiveness of passive approaches to vibration control. These passive approaches would include passive isolation using shock absorbers and springs to separate the offending system from the structure to which it is attached. We would further develop active approaches to vibration control, where sensors, actuators, and signal-processing devices are used to keep vibrations at an arbitrarily small level. From a dynamics point of view, the topic of vibrations is one of the few application areas where linear analysis is justified. Vibrational motion tends to be associated with rather small motions of structures. Helicopters present a terrible vibration environment, and even though they are constructed from exotic materials, linear vibration concepts can lead to very technically advanced approaches to vibration control. The shaking of buildings, dams, and bridges can all benefit from linear vibration concepts. Suspensions for ground vehicles and their drive trains and engine mounts are all excellent applications of linear vibration considerations.

In this chapter some fundamental concepts in vibrations are first developed using the modeling and derivation approaches that have been used throughout this book. This development is followed by engineering uses for this general development in prescribing measures to control vibrations.

8.1 NATURAL FREQUENCY AND RESONANCE OF A SINGLE-DEGREE-OF-FREEDOM OSCILLATOR

Free Response

Figure 8.1*a* shows a mass particle attached to ground through a spring and damper and acted upon by an external force. This system is called a force-excited single-degree-of-freedom oscillator. The figure also shows velocity, acceleration, and force diagrams. This oscillator is a classic starting point for initial vibration studies. The position of the mass x is measured from equilibrium in a gravity field and as shown in Section 2.5, there is no need to include the gravity force in the force diagram when using such coordinate definitions. For this force-excited system the spring and damper are considered to be positive in tension. This is indicated in the force diagram of Figure 8.1*d*, where the spring and damper forces are shown acting downward on the mass particle. When the spring and damper are in tension and assumed positive, and x and $\dot{x}$ are positive as indicated, the spring and damper forces are as shown in the force diagram. The equation of motion is then

$$-kx - b\dot{x} + F(t) = m\ddot{x} \tag{8.1}$$

or

$$\ddot{x} + \frac{b}{m}\dot{x} + \frac{k}{m}x = \frac{F(t)}{m} \tag{8.2}$$

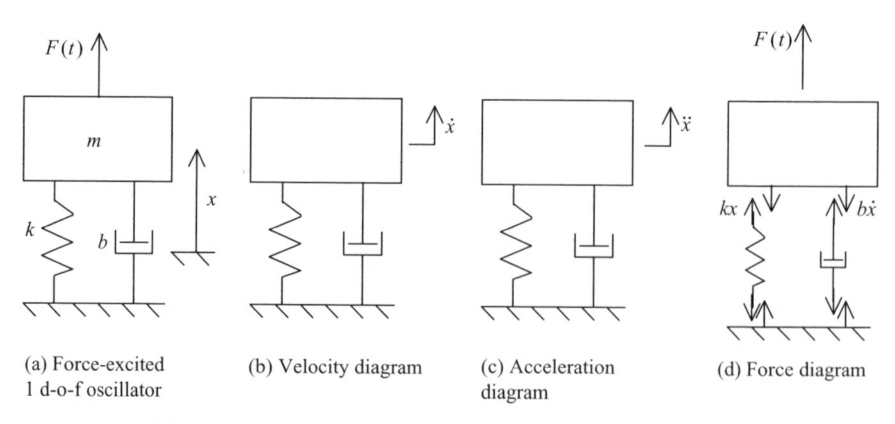

Figure 8.1 Force-excited single-degree-of-freedom oscillator.

We first pursue the eigenvalue problem as described in Section 3.2. We let the forcing be zero and the solution be

$$x(t) = Xe^{st} \tag{8.3}$$

Substituting into (8.2) yields

$$\left(s^2 + \frac{b}{m}s + \frac{k}{m}\right) Xe^{st} = 0 \tag{8.4}$$

We cannot let $X = 0$, or we produce the uninteresting solution that the system remains at rest. Instead, we require the characteristic equation to be zero,

$$s^2 + \frac{b}{m}s + \frac{k}{m} = 0 \tag{8.5}$$

This has the solution

$$s_{1,2} = -\frac{b}{2m} \pm \sqrt{\left(\frac{b}{2m}\right)^2 - \frac{k}{m}} \tag{8.6}$$

First consider these eigenvalues when there is no damping (i.e., $b = 0$). Then the eigenvalues are only imaginary:

$$s_{1,2} = \pm j\sqrt{\frac{k}{m}} \tag{8.7}$$

This means that the free response would have the appearance

$$x(t) = X_1 e^{j\sqrt{kt/m}} + X_2 e^{-j\sqrt{kt/m}} \tag{8.8}$$

where you should recall that X_1 and X_2 are constants but because they are associated with complex-conjugate eigenvalues, they are complex-conjugate constants. The free response from (8.8) reduces to the real form

$$x(t) = A\cos\sqrt{\frac{k}{m}}t + B\sin\sqrt{\frac{k}{m}}t \tag{8.9}$$

Thus, the undamped free response will oscillate at a frequency called the *undamped natural frequency*,

$$\omega_n = \sqrt{\frac{k}{m}} \tag{8.10}$$

Returning to Eq. (8.6) and reinstalling the damping, let's look at the case when

$$\left(\frac{b}{2m}\right)^2 < \frac{k}{m}$$

and rewrite (8.6) as

$$s_{1,2} = -\frac{b}{2m\omega_n}\,\omega_n \pm j\omega_n\sqrt{1 - \left(\frac{b}{2m\omega_n}\right)^2} \tag{8.11}$$

The square root of -1, j, has been pulled from the radical, and the first term on the right has been multiplied and divided by ω_n. It is customary to define a damping ratio ζ such that

$$\zeta = \frac{b}{2m\omega_n} \tag{8.12}$$

which reduces (8.11) to

$$s_{1,2} = -\zeta\omega_n \pm j\omega_n\sqrt{1 - \zeta^2} = -\zeta\omega_n \pm j\omega_d \tag{8.13}$$

where ω_d is called the *damped natural frequency*. Notice that the two eigenvalues have negative real parts and complex-conjugate imaginary parts. This means that the free response will oscillate at ω_d (rad/s) and will decay at a rate depending on the damping ratio ζ. After some simplifications, the free response can be written as the real-time function

$$x(t) = e^{-\zeta\omega_n t}(A\cos\omega_d t + B\sin\omega_d t) \tag{8.14}$$

In a more complete treatment of vibrations, we would return to the eigenvalues of Eq. (8.6) and discuss the case when

$$\left(\frac{b}{2m}\right)^2 \geq \frac{k}{m}$$

but in this case the radical would be positive and the two eigenvalues would be negative real numbers. If this were then used in the solution from Eq. (8.3), there would be no oscillation, only exponential decay of the solution. This is not an interesting case when studying the topic of vibrations. It should be noted, however, that as ζ increases, the oscillations decay quicker. Thus, finding ways to increase the damping might be a good approach to vibration control.

Forced Response

The concept of a frequency response, presented in Section 3.2 and used elsewhere in this book, is the primary analysis tool for vibration studies. In the Laplace domain the forced equation [8.2] would be

$$(s^2 + 2\zeta\omega_n s + \omega_n^2)X(s) = \frac{F(s)}{m} \tag{8.15}$$

and the input/output transfer function would be

$$\frac{X(s)}{F(s)} = \frac{1/m}{s^2 + 2\zeta\omega_n s + \omega_n^2} \tag{8.16}$$

The complex frequency response function comes from letting $s = j\omega$ in the transfer function, with the result

$$\frac{X(j\omega)}{F(j\omega)} = \frac{1/m}{(j\omega)^2 + 2\zeta\omega_n j\omega + \omega_n^2} = \frac{1/m}{\omega_n^2 - \omega^2 + j2\zeta\omega_n\omega} \tag{8.17}$$

The magnitude of (8.17) is the response amplitude, and the angle of (8.17) is the response phase angle. To present results in a nice nondimensional form, both sides of (8.17) are multiplied by the stiffness k and some manipulations are done to produce

$$\begin{aligned} \left|\frac{kX}{F}\right| &= \frac{1}{[(1 - \omega^2/\omega_n^2)^2 + (2\zeta(\omega/\omega_n))^2]^{1/2}} \\ \frac{kX}{F} &= 0 - \tan^{-1}\frac{2\zeta(\omega/\omega_n)}{1 - \omega^2/\omega_n^2} \end{aligned} \tag{8.18}$$

The frequency response can be sketched or plotted using a computer. The magnitude response is shown in Figure 8.2 for different damping ratios, ζ. The response near the natural frequency $\omega/\omega_n = 1$ exhibits what is called *resonance*. When a lightly damped system is excited near its natural frequency, there is amplification of the response. This resonance of a dynamic system can be quite destructive and noisy, and a large part of vibration studies is devoted to methods of controlling resonance in structures. As can be seen in Figure 8.2, increasing the damping tends to reduce the resonance. In some actual applications it is possible to increase the damping by adding passive elements such as shock absorbers or energy-absorbing materials. However, many structures are built from inherently lightly damped materials such as metal beams in bridges and buildings. For such structures it is quite difficult to add passive damping and alternative approaches are needed to prevent resonance.

Comparison of Two Suspension Geometries

Figure 8.3 shows two similar single-degree-of-freedom oscillators with excitation at their base. If we use our imaginations, the system in Figure 8.3*a* can be thought of as

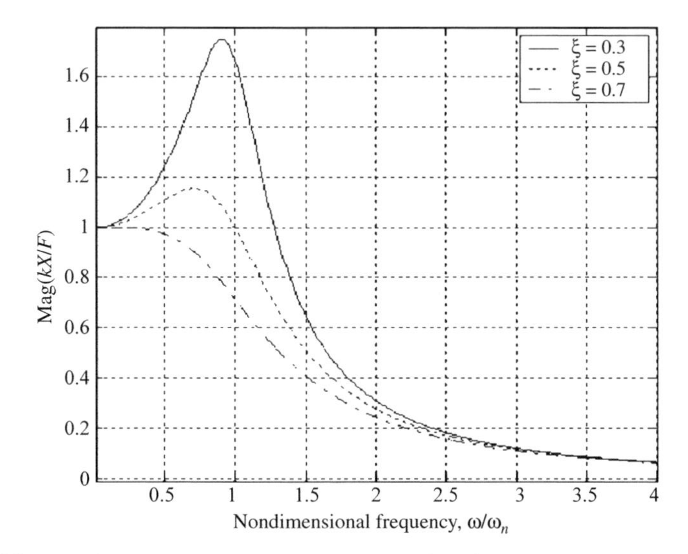

Figure 8.2 Nondimensional frequency response of a forced single-degree-of-freedom oscillator.

a ground vehicle with a conventional passive suspension. The mass m represents the sprung mass, and the spring and damper k and b represent the suspension elements. The roadway unevenness is represented by the input $x_{\text{in}}(t)$. In part e of the figure the damper has been moved from the conventional location between the wheel and body to a location where it provides inertial damping of the mass. In this configuration one end of the damper is attached to inertial space, and on a vehicle this would require attachment to the sky. This configuration is called *skyhook damping* [1]. Figure 8.3 also shows acceleration and force diagrams for the two systems. In both systems the spring and damper are assumed positive in compression, as is indicated by the positive configurations shown in the force diagrams. The equations of motion are

$$k(x_{\text{in}} - x) + b(\dot{x}_{\text{in}} - \dot{x}) = m\ddot{x} \tag{8.19}$$

for the conventional system and

$$k(x_{\text{in}} - x) - b\dot{x} = m\ddot{x} \tag{8.20}$$

for the skyhook system.

These equations can be straightforwardly cast into the Laplace domain and the transfer functions derived relating the response displacement to the input displacement:

$$\frac{X(s)}{X_{\text{in}}(s)} = \frac{(b/m)s + k/m}{s^2 + (b/m)s + k/m} \tag{8.21}$$

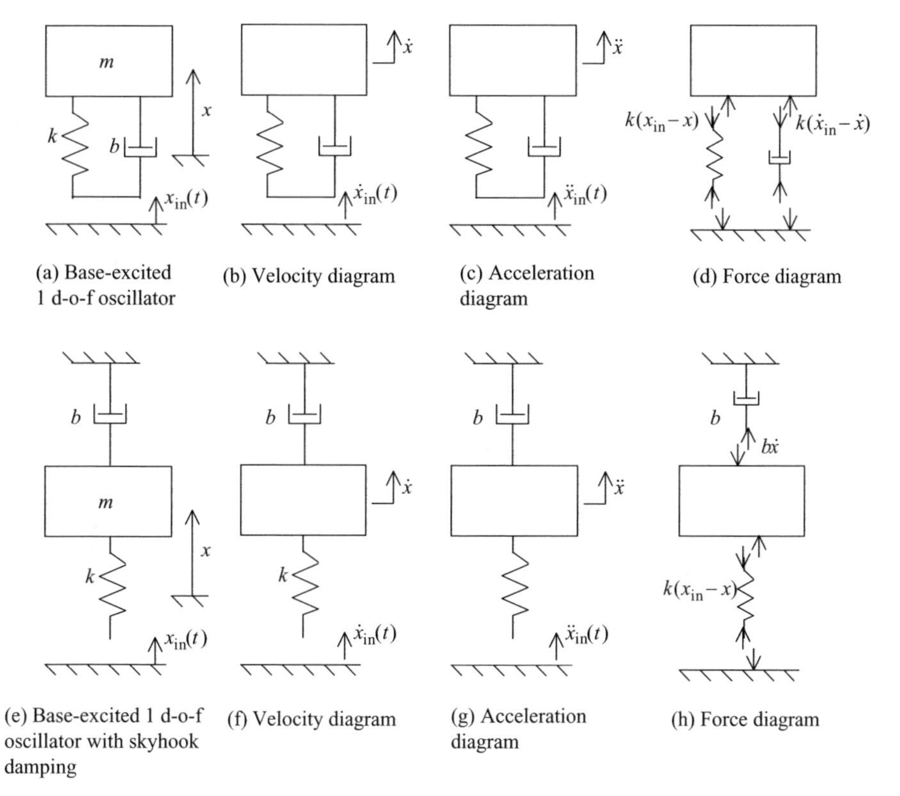

Figure 8.3 Conventional and skyhook isolation systems.

for the conventional system and

$$\frac{X(s)}{X_{\text{in}}(s)} = \frac{k/m}{s^2 + (b/m)s + k/m} \tag{8.22}$$

for the skyhook system.

Notice that both configurations have the same denominator, and this is identical to the denominator for the force-excited system from Eq. (8.16). Thus, all these oscillators have the same eigenvalues. The two base-excited systems differ only in the numerators of their transfer functions. The frequency response of each system is obtained by first letting $s = j\omega$ in the transfer functions to generate the respective complex frequency response functions followed by taking the magnitude of the result to produce the magnitude of the frequency response. These steps were taken, producing the result

$$\left|\frac{X}{X_{\text{in}}}\right| = \frac{\left[1 + (2\zeta(\omega/\omega_n))^2\right]^{1/2}}{[(1 - \omega^2/\omega_n^2)^2 + (2\zeta(\omega/\omega_n))^2]^{1/2}} \tag{8.23}$$

for the conventional system and

$$\left|\frac{X}{X_{\text{in}}}\right| = \frac{1}{\left[(1-\omega^2/\omega_n^2)^2 + (2\zeta(\omega/\omega_n))^2\right]^{1/2}} \tag{8.24}$$

for the skyhook system. Comparing these two configurations yields pretty good insight into the trade-offs required of a conventional passive suspension. The Matlab program [2] was used to produce the results shown in Figure 8.4. In part *a* of the figure for the conventional configuration with low damping, there is substantial resonance near the natural frequency $\omega/\omega_n = 1$, followed by a steep decrease in response as the forcing frequency is raised. A softly sprung vehicle exhibits this type of behavior. There tends to be substantial motion or wallowing as the vehicle turns or stops, but small bumps traversed at high speeds are hardly felt at all. The wallowing motion tends to be near the natural frequency; the small bumps are higher-frequency inputs. As the damping increases, the response near the natural frequency is reduced, but the response at higher frequencies is increased. A stiffly sprung vehicle does not exhibit much motion due to low-speed maneuvers, but higher-frequency roadway bumps tend to come booming through to the body. We say that such vehicles allow one to "feel the road," as opposed to a lightly sprung vehicle, which feels "vague" in its handling. The vibration response of passive suspensions composed of springs and shock absorbers is reasonably represented by the simple single-degree-of-freedom model of Figure 8.3*a*.

The base-excited system of Figure 8.3*e* with frequency response from Eq. (8.24) has the magnitude response shown in Figure 8.4*b*. For low damping we get the same resonant behavior as in a passive conventional system. However, this time, as damping is increased, the resonance is controlled and there is no increase in the high-frequency response. As mentioned above, the damper to ground is called skyhook damping and is not very practical for vehicles driving on roadways. Nevertheless, the response of the skyhook system is so good that many studies, inventions, and commercialization have taken place that attempt to duplicate the response of the skyhook system.

Consider the single-degree-of-freedom oscillator of Figure 8.5 with control actuator located between the ground input and the sprung mass. The acceleration and force diagrams are shown in the figure and the equations of motion are

$$k(x_{\text{in}} - x) + F_c = m\ddot{x} \tag{8.25}$$

where F_c is the control force. The actual actuator that could generate the control force is not shown. Imagine that the velocity of the sprung mass is measured and fed back to a computer that commands the control force to be

$$F_c = -b_c\dot{x} \tag{8.26}$$

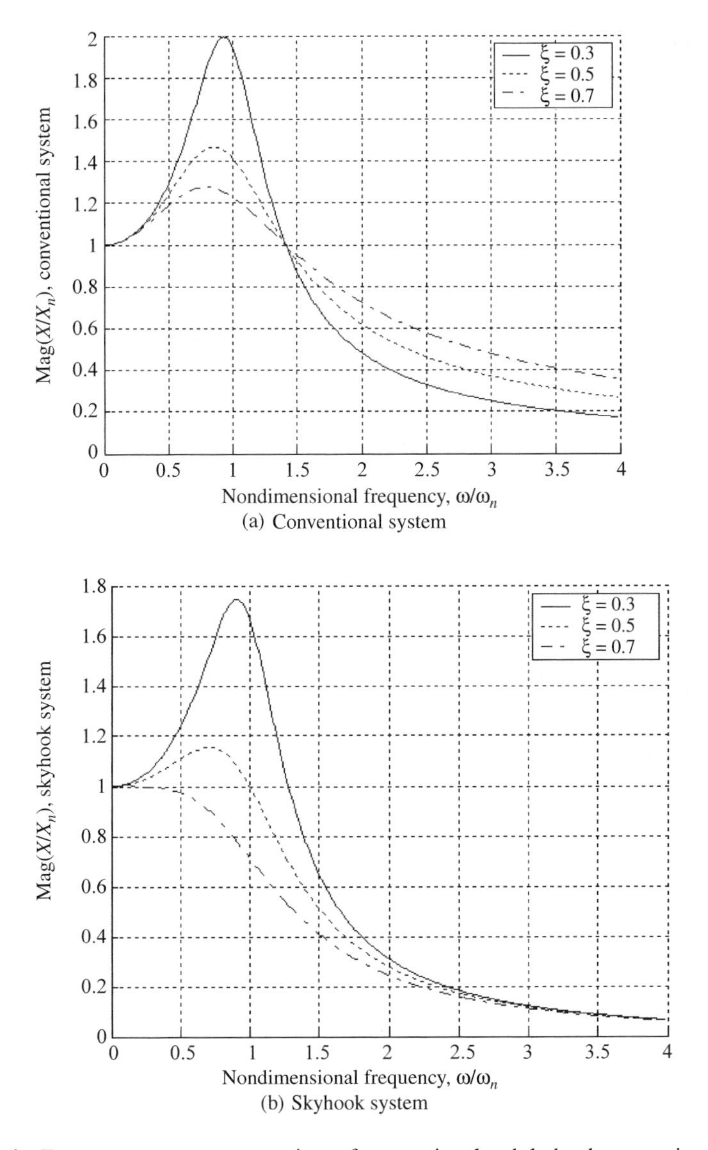

Figure 8.4 Frequency response comparison of conventional and skyhook suspension systems.

If this is possible, using (8.26) in (8.25) results in

$$m\ddot{x} + b_c\dot{x} + kx = kx_{\text{in}} \tag{8.27}$$

This is identical to the equation of motion (8.20) for the skyhook system, indicating that this controlled system will perform identically to the skyhook system. Thus, by coupling automatic control and vibration studies we are able to realize an outstanding isolation strategy. The interested reader is directed to refs. [3–5].

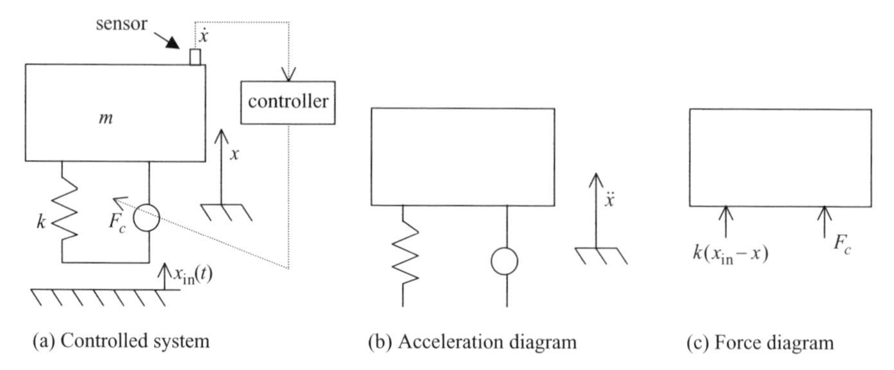

Figure 8.5 Skyhook control system.

8.2 TWO-DEGREE-OF-FREEDOM SYSTEMS

The single-degree-of-freedom oscillator yields excellent insight into the concepts of eigenvalues, natural frequency, frequency response, and resonance. But a two degree-of-freedom system is required to introduce the next important concept in vibrations: modal dynamics. Consider the two-mass three-spring system shown in Figure 8.6. It has two degrees of freedom in the two position coordinates x_1 and x_2. These are inertial coordinates measured from the relaxed position of the springs. The positive velocity and acceleration components are indicated in the figure. Also shown in Figure 8.6 is the force diagram assuming that the springs k_1 and k_2 are positive in tension

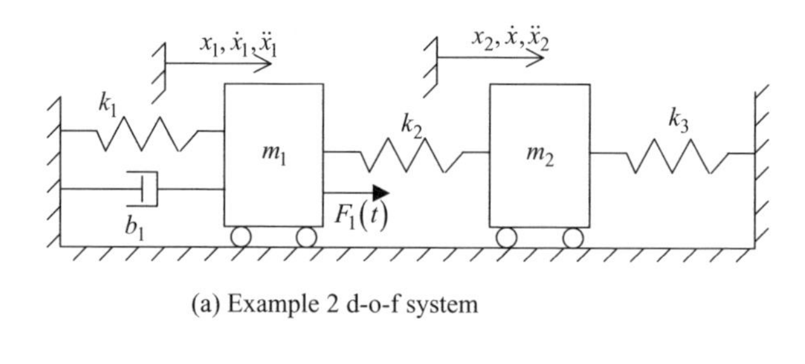

(a) Example 2 d-o-f system

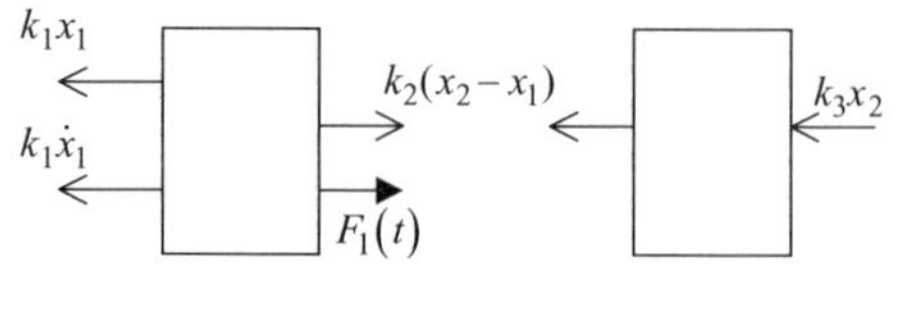

(b) Force diagram

Figure 8.6 Example two-degree-of-freedom system.

and the spring k_3 is positive in compression. The equations of motion are derived straightforwardly as

$$\begin{aligned} -k_1x_1 - b_1\dot{x}_1 + k_2(x_2 - x_1) + F_1(t) &= m_1\ddot{x}_1 \\ -k_2(x_2 - x_1) - k_3x_2 &= m_2\ddot{x}_2 \end{aligned} \tag{8.28}$$

These can be put into the matrix form

$$\begin{bmatrix} m_1 & 0 \\ 0 & m_2 \end{bmatrix}\begin{bmatrix} \ddot{x}_1 \\ \ddot{x}_2 \end{bmatrix} + \begin{bmatrix} b_1 & 0 \\ 0 & 0 \end{bmatrix}\begin{bmatrix} \dot{x}_1 \\ \dot{x}_2 \end{bmatrix} + \begin{bmatrix} k_1 + k_2 & -k_2 \\ -k_2 & k_2 + k_3 \end{bmatrix}\begin{bmatrix} x_1 \\ x_2 \end{bmatrix} = \begin{bmatrix} 1 \\ 0 \end{bmatrix} F_1 \tag{8.29}$$

This form of the equations was pointed out in Section 2.5 to be of the general form of Eq. (2.77), repeated here:

$$\mathbf{A}\ddot{\mathbf{x}} + \mathbf{B}\dot{\mathbf{x}} + \mathbf{C}\mathbf{x} = \mathbf{F} \tag{8.30}$$

which is an excellent starting point for vibration analysis.

Free, Undamped Response

We start the analysis by letting the forcing function $F_1(t)$ and the damping b_1 both be zero. The resulting equations are of the form

$$\mathbf{A}\ddot{\mathbf{x}} + \mathbf{C}\mathbf{x} = 0 \tag{8.31}$$

For an undamped free response, after release with arbitrary initial conditions we expect that the system will oscillate in some fashion and the two masses will move with some phase relationship relative to each other. We could guess that the response is of the form

$$\begin{aligned} x_1 &= A_1 \cos(\omega_n t + \phi_1) \\ x_2 &= A_2 \cos(\omega_n t + \phi_2) \end{aligned} \tag{8.32}$$

and substitute into the governing equations and attempt to solve for the amplitudes and phase angles. Another approach is to use the complex notation and guess the solution is

$$\begin{aligned} x_1 &= X_1 e^{j\omega_n t} \\ x_2 &= X_2 e^{j\omega_n t} \end{aligned} \tag{8.33}$$

where X_1 and X_2 are constants, or more generally,

$$\mathbf{x} = \mathbf{X}e^{j\omega_n t} \tag{8.34}$$

where **X** is a vector of the constants:

$$\mathbf{X} = \begin{bmatrix} X_1 \\ X_2 \end{bmatrix} \tag{8.35}$$

Substituting (8.33) into (8.29) yields

$$\begin{bmatrix} -m_1\omega_n^2 + k_1 + k_2 & -k_2 \\ -k_2 & -m_2\omega_n^2 + k_2 + k_3 \end{bmatrix} \begin{bmatrix} X_1 \\ X_2 \end{bmatrix} = 0 \tag{8.36}$$

or more generally, using (8.34) in (8.31) yields

$$\left[-\mathbf{A}\omega_n^2 + \mathbf{C}\right] \mathbf{X} = 0 \tag{8.37}$$

Equations (8.36) and (8.37) are eigenvalue problems similar to that described in Section 3.2, but in this formulation the eigenvalue is ω_n^2 rather than the s-variable used in the first-order form of linear equations.

The condition for nonzero values of **X** is that

$$\text{Det}\left[-\mathbf{A}\omega_n^2 + \mathbf{C}\right] = 0 \tag{8.38}$$

for the general problem, and in particular for our example, taking the determinant of (8.36) will result in

$$\omega_n^4 - \left(\frac{k_1 + k_3}{m_2} + \frac{k_1 + k_2}{m_1}\right)\omega_n^2 + \frac{k_1k_2 + k_1k_3 + k_2k_3}{m_1m_2} = 0 \tag{8.39}$$

The resulting equation from (8.38) for the general problem and the resulting fourth-order polynomial of (8.39) for our example are called *frequency equations* and their solution yields the natural frequencies of the system. The frequency equation is guaranteed to be factorable into

$$\left(\omega_n^2 - \omega_1^2\right)\left(\omega_n^2 - \omega_2^2\right)\cdots\left(\omega_n^2 - \omega_m^2\right) = 0 \tag{8.40}$$

where m equals the number of degrees of freedom.

For our two-degree-of-freedom system there will be two solutions for ω_n^2,

$$\omega_n^2 = \omega_1^2$$

and

$$\omega_n^2 = \omega_2^2 \tag{8.41}$$

and four solutions for ω_n,

$$\begin{aligned}\omega_n &= +\omega_1\\ \omega_n &= -\omega_1\end{aligned}$$

and (8.42)

$$\begin{aligned}\omega_n &= +\omega_2\\ \omega_n &= -\omega_2\end{aligned}$$

and the solution from Eq. (8.33) gets expanded to

$$\begin{aligned}x_1 &= X_{11}e^{j\omega_1 t} + X_{12}e^{-j\omega_1 t} + X_{13}e^{j\omega_2 t} + X_{14}e^{-j\omega_2 t}\\ x_2 &= X_{21}e^{j\omega_1 t} + X_{22}e^{-j\omega_1 t} + X_{23}e^{j\omega_2 t} + X_{24}e^{-j\omega_2 t}\end{aligned} \tag{8.43}$$

It appears that there are eight unknown constants but only four initial conditions for x_1 and x_2: the initial position and velocity of each mass.

Returning to Eq. (8.36), we should remember that the condition for the existence of nonzero values of the constants X_1 and X_2 was that the determinant of the coefficient matrix be zero. We enforced this requirement, and this led to the frequency equation and the natural frequencies. Now we must determine the associated nonzero values of X_1 and X_2. This is accomplished by substituting the solution frequencies back into Eq. (8.36) and solving the resulting equation,

$$\begin{bmatrix} -m_1\omega_i^2 + k_1 + k_2 & -k_2 \\ -k_2 & -m_2\omega_i^2 + k_2 + k_3 \end{bmatrix} \begin{bmatrix} X_{1i} \\ X_{2i} \end{bmatrix} = 0 \tag{8.44}$$

where the sub i refers to the ith frequency and the associated constants. It appears now that (8.44) generates two equations for each of the pairs of constants, and thus each constant can be determined explicitly and the initial conditions are not needed. This is not the case, in that the original requirement that the determinant of the coefficient matrix be zero also eliminates the independence of the resulting equations. Thus, only one independent equation comes from (8.44) and the two implied equations carry exactly the same information. For example, if ω_1 is used in (8.44), the two resulting equations are

$$\begin{aligned}(-m_1\omega_1^2 + k_1 + k_2)X_{11} - k_2X_{21} &= 0\\ -k_2X_{11} + (-m_2\omega_1^2 + k_2 + k_3)X_{21} &= 0\end{aligned} \tag{8.45}$$

Although the two equations do not appear to be the same, in fact they are the same, and using either the first or second of (8.45) yields one relationship between the two constants. Using the first of (8.45), for example, yields

$$X_{21} = \frac{-m_1\omega_1^2 + k_1 + k_2}{k_2}X_{11} \tag{8.46}$$

We can carry out this step for each of the natural frequencies, with the result

$$X_{22} = \frac{-m_1\omega_1^2 + k_1 + k_2}{k_2} X_{12}$$
$$X_{23} = \frac{-m_1\omega_2^2 + k_1 + k_2}{k_2} X_{13} \tag{8.47}$$
$$X_{24} = \frac{-m_1\omega_2^2 + k_1 + k_2}{k_2} X_{14}$$

Notice that substituting the frequency $-\omega_1$ produces the same result as substituting $+\omega_1$; thus,

$$\frac{X_{22}}{X_{12}} = \frac{X_{21}}{X_{11}}$$
$$\frac{X_{24}}{X_{14}} = \frac{X_{23}}{X_{13}} \tag{8.48}$$

Equations (8.46) and (8.47) are four additional equations which when combined with the four initial conditions allows determination of the eight constants in the solution (8.43).

We are not so interested in carrying out all the steps outlined here to come up with the free response of a system. However, the relationship of the natural frequencies and associated constants yield great insight into the behavior of multidegree-of-freedom vibratory systems. Equation (8.43) can be put into vector form as

$$\begin{bmatrix} x_1 \\ x_2 \end{bmatrix} = \begin{bmatrix} X_{11} \\ X_{21} \end{bmatrix} e^{j\omega_1 t} + \begin{bmatrix} X_{12} \\ X_{22} \end{bmatrix} e^{-j\omega_1 t} + \begin{bmatrix} X_{13} \\ X_{23} \end{bmatrix} e^{j\omega_2 t} + \begin{bmatrix} X_{14} \\ X_{24} \end{bmatrix} e^{-j\omega_2 t} \tag{8.49}$$

If the first constant in each vector is factored and the remaining entry divided by the factored constant, the result is

$$\begin{bmatrix} x_1 \\ x_2 \end{bmatrix} = X_{11} \begin{bmatrix} 1 \\ \mu_1 \end{bmatrix} e^{j\omega_1 t} + X_{12} \begin{bmatrix} 1 \\ \mu_1 \end{bmatrix} e^{-j\omega_1 t} + X_{13} \begin{bmatrix} 1 \\ \mu_2 \end{bmatrix} e^{j\omega_2 t} + X_{14} \begin{bmatrix} 1 \\ \mu_2 \end{bmatrix} e^{-j\omega_2 t} \tag{8.50}$$

where μ_i the come from (8.46) and (8.47). The vectors that are factored in this way are called *mode vectors* or *modal vectors* or *eigenvectors*. Before interpreting the meaning of the mode vectors, Eq. (8.50) can be reduced further by

$$\begin{bmatrix} x_1 \\ x_2 \end{bmatrix} = \begin{bmatrix} 1 \\ \mu_1 \end{bmatrix} \left(X_{11}e^{j\omega_1 t} + X_{12}e^{-j\omega_1 t}\right) + \begin{bmatrix} 1 \\ \mu_2 \end{bmatrix} \left(X_{13}e^{j\omega_2 t} + X_{14}e^{-j\omega_2 t}\right) \tag{8.51}$$

The constants that are associated with complex-conjugate exponentials are themselves complex-conjugate constants, and the parenthetical terms allow reduction to the final form

$$\begin{bmatrix} x_1 \\ x_2 \end{bmatrix} = \begin{bmatrix} 1 \\ \mu_1 \end{bmatrix} (A_{11} \cos \omega_1 t + B_{11} \sin \omega_1 t) + \begin{bmatrix} 1 \\ \mu_2 \end{bmatrix} (A_{22} \cos \omega_2 t + B_{22} \sin \omega_2 t) \tag{8.52}$$

In this final form the four remaining real constants are exposed and could be determined from initial conditions. We are now in a position to interpret the meaning of the natural frequencies and the associated mode vectors.

First notice that this two-degree-of-freedom system can only oscillate at the two natural frequencies ω_1 and ω_2, and starting from any arbitrary initial conditions, regardless of how complex the motion might appear, it is composed of oscillations at only the two natural frequencies. Now imagine that the initial conditions were such that

$$x_1(0) = 1 \text{ units}$$

and

$$x_2(0) = \mu_1 \text{ units}$$

and the initial velocities are both zero. In other words, the initial displacements are scaled according to the first mode vector. The constants would then come from applying these initial conditions such that

$$\begin{aligned} 1 &= A_{11} + A_{22} \\ \mu_1 &= \mu_1 A_{11} + \mu_2 A_{22} \\ 0 &= B_{11}\omega_1 + B_{22}\omega_2 \\ 0 &= \mu_1 B_{11}\omega_1 + \mu_2 B_{22}\omega_2 \end{aligned} \tag{8.53}$$

Solving (8.53) results in

$$\begin{aligned} A_{11} &= 1 \\ A_{22} &= B_{11} = B_{22} = 0 \end{aligned}$$

or

$$\begin{aligned} x_1(t) &= 1 \cos \omega_1 t \\ x_2(t) &= \mu_1 \cos \omega_1 t \end{aligned} \tag{8.54}$$

Equation (8.54) shows that if the initial displacements are scaled according to a mode vector, the system responds only at the associated natural frequency and remains forever in the mode configuration. This would be true if the initial conditions were scaled according to the second mode, but the response would be at the second natural frequency. A final conclusion is that the free response of a lightly damped two-degree-of-freedom system is nothing more than a linear combination of modal configurations oscillating at the respective natural (or modal) frequencies. There are special initial conditions where only one mode is excited.

Returning to the general representation of Eqs. (8.37) and (8.38), there are some conclusions that can be made about multi-degree-of-freedom vibratory systems. From the solution of the frequency equation (8.37), there will result m natural frequencies squared $\omega_n^2 = \omega_1^2,\ \omega_2^2,\ \omega_3^2, \ldots,\ \omega_m^2$, and corresponding to each of these will be a modal vector $\mathbf{r}_1,\ \mathbf{r}_2,\ \mathbf{r}_3, \ldots,\ \mathbf{r}_m$ such that the free response would have the general appearance

$$\begin{aligned}\mathbf{x} = \mathbf{r}_1(A_{11}\cos\omega_1 t + B_{11}\sin\omega_1 t) + \mathbf{r}_2(A_{22}\cos\omega_2 t + B_{22}\sin\omega_2 t) + \cdots \\ + \mathbf{r}_m(A_{mm}\cos\omega_m t + B_{mm}\sin\omega_m t)\end{aligned} \tag{8.55}$$

and the free response is nothing more than a linear combination of modal configurations oscillating at their respective natural frequencies.

A more physical interpretation can be gained by considering the example system from Figure 8.6 with two equal masses and three equal springs: thus,

$$m_1 = m_2 = m$$
$$k_1 = k_2 = k_3 = k$$

The frequency equation from (8.39) becomes

$$\omega_n^4 - \left(4\frac{k}{m}\right)\omega_n^2 + \left(3\frac{k^2}{m^2}\right) = 0 \tag{8.56}$$

The solution of this is

$$\omega_n^2 = 2\frac{k}{m} \pm \frac{k}{m} \tag{8.57}$$

or

$$\begin{aligned}\omega_1^2 &= \frac{k}{m} \\ \omega_2^2 &= 3\,\frac{k}{m}\end{aligned} \tag{8.58}$$

Using Eqs. (8.46) and (8.47), the associated mode vectors come from

$$\begin{aligned}X_{21} &= \frac{-m\omega_1^2 + 2k}{k}X_{11} = 1X_{11} \\ X_{23} &= \frac{-m\omega_2^2 + 2k}{k}X_{13} = -1X_{13}\end{aligned} \tag{8.59}$$

By letting $X_{11} = X_{13} = 1$, the mode vectors are scaled such that

$$\mathbf{r}_1 = \begin{bmatrix} 1 \\ 1 \end{bmatrix}$$
$$\mathbf{r}_2 = \begin{bmatrix} 1 \\ -1 \end{bmatrix} \tag{8.60}$$

The interpretation of the mode frequencies and mode vectors are shown in Figure 8.7. The first mode corresponds to the two masses moving in the same direction with the same amplitude while oscillating at the first natural frequency, and the second mode corresponds to the two masses moving in opposite directions while oscillating at the second natural frequency. Any general motion of the system will be some combination of these two simple motions, as indicated by Eq. (8.52). This probably seems like a lot of work to obtain modal frequencies and modal vectors, but computers can do much of the work. The job of the reader is pretty much always the same. You must formulate a reasonable model in terms of reasonable coordinates and apply the dynamic principles to obtain the equations of motion. If the system is modeled as linear or linearized after equation derivation, the system of equations can

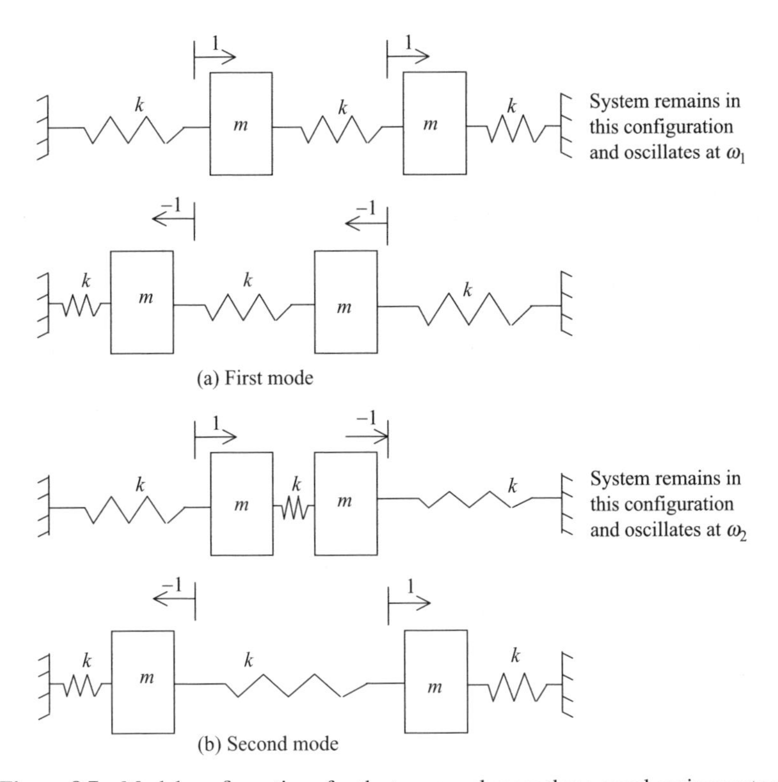

Figure 8.7 Modal configurations for the two-equal-mass three-equal-spring system.

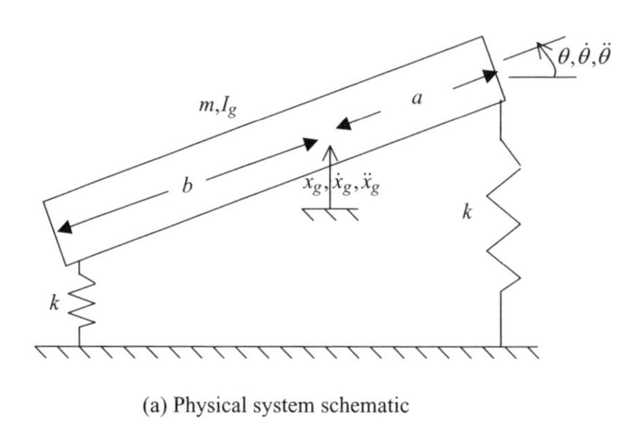

(a) Physical system schematic

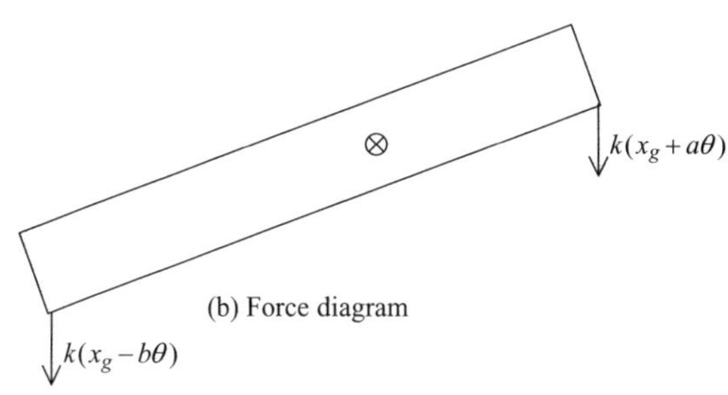

(b) Force diagram

Figure 8.8 Half-car model.

be put into the matrix form of Eq. (8.37). At this point, once parameters have been defined, a computer can output the modal frequencies and associated mode shapes.

As an example, consider the half-car model shown in Figure 8.8. This is a side view of the vehicle, and the front axle is on the right. The center of gravity is located distance a from the front axle and b from the rear axle. This is a much simplified version of the model presented in Figure 2.8. The rigid body has mass m and c.g. moment of inertia I_g. The springs are the same at the left and right.

Figure 8.8*b* shows a force diagram when both springs are considered positive in extension. The motion is constrained to be small to justify using a linear model, and with this assumption the spring forces are as shown in the figure. The equations of motion are derived as

$$\begin{aligned} -k(x_g + a\theta) - k(x_g - b\theta) &= m\ddot{x}_g \\ k(x_g - b\theta)b - k(x_g + a\theta)a &= I_g\ddot{\theta} \end{aligned} \tag{8.61}$$

These can be put into the standard matrix format as

$$\begin{bmatrix} m & 0 \\ 0 & I_g \end{bmatrix} \begin{bmatrix} \ddot{x}_g \\ \ddot{\theta} \end{bmatrix} + \begin{bmatrix} 2k & k(a-b) \\ k(a-b) & k(a^2+b^2) \end{bmatrix} \begin{bmatrix} x_g \\ \theta \end{bmatrix} = 0 \tag{8.62}$$

TABLE 8.1. Parameters for the Half-Car Vehicle Model

Mass of the body,
$m = 2500/2.2\,\text{kg}$
Length of the vehicle, $L = 8(0.3058)\,\text{m}$
$\frac{a}{b} = 0.9$
Calculate:
$b = \frac{L}{1 + a/b}$

$a = L - b$
Moment of inertia,

$I_g = \frac{mL^2}{12}$
Define a frequency $f_n = 1.0\,\text{Hz}$

Calculate:

$k = \frac{1}{2}m(2\pi f_n)^2$

Table 8.1 defines the parameters for the system.
Using the Matlab program [2], the matrices **A** and **C** are first defined from (8.62) as

$$\mathbf{A} = \begin{bmatrix} m & 0 \\ 0 & I_g \end{bmatrix}$$
$$\mathbf{C} = \begin{bmatrix} 2k & k(a-b) \\ k(a-b) & k(a^2+b^2) \end{bmatrix} \tag{8.63}$$

The matrix

$$\mathbf{F} = \mathbf{A}^{-1}\mathbf{C} \tag{8.64}$$

is then formulated, which yields the eigenvalue problem

$$\mathbf{I}\omega^2 = \mathbf{F} \tag{8.65}$$

Matlab then computes the eigenvalues and eigenvectors by the command

```
[eigvec,omega_sq]=eig(F)
```

The result of this is the output

```
eig_vec =
        -0.9979  0.0319
        -0.0642  -0.9995
```

and (8.66)

```
omega_sq =
        39.3154  0
        0        118.9264
```

The interpretation of this result is that the first natural frequency is

$$\omega_1^2 = 39.3154 \,\mathrm{rad}^2/\mathrm{s}^2$$

or

$$\omega_1 = 6.27 \,\mathrm{rad/s}$$

or

$$f_1 = \frac{\omega_1}{2\pi} = 1.0 \,\mathrm{Hz}$$

and the first mode vector comes from the first column of `eig_vec` and after scaling is

$$\mathbf{r}_1 = \begin{bmatrix} 1 \\ 0.06 \end{bmatrix}$$

The second natural frequency is

$$\omega_2^2 = 118.92 \,\mathrm{rad}^2/\mathrm{s}^2$$

or

$$\omega_2 = 10.91 \,\mathrm{rad/s}$$

or

$$f_2 = \frac{\omega_2}{2\pi} = 1.74 \,\mathrm{Hz}$$

with the corresponding mode vector coming from the second column of `eig_vec`, which after scaling becomes

$$\mathbf{r}_2 = \begin{bmatrix} 1 \\ -31.33 \end{bmatrix}$$

The interpretation of the mode vectors is a little complicated because our coordinates are in mixed units (i.e., displacement and angular rotation). Figure 8.9 shows the general configuration for the two modes of this system.

The first mode oscillates at $f_1 = 1.0\,\mathrm{Hz}$, and if the center of gravity is displaced upward by one unit, there is virtually no corresponding rotation. This is called the *heave mode* of a vehicle. The second mode oscillates at $f_2 = 1.74\,\mathrm{Hz}$ and for a unit displacement of the center of gravity there is a large negative rotation of the body.

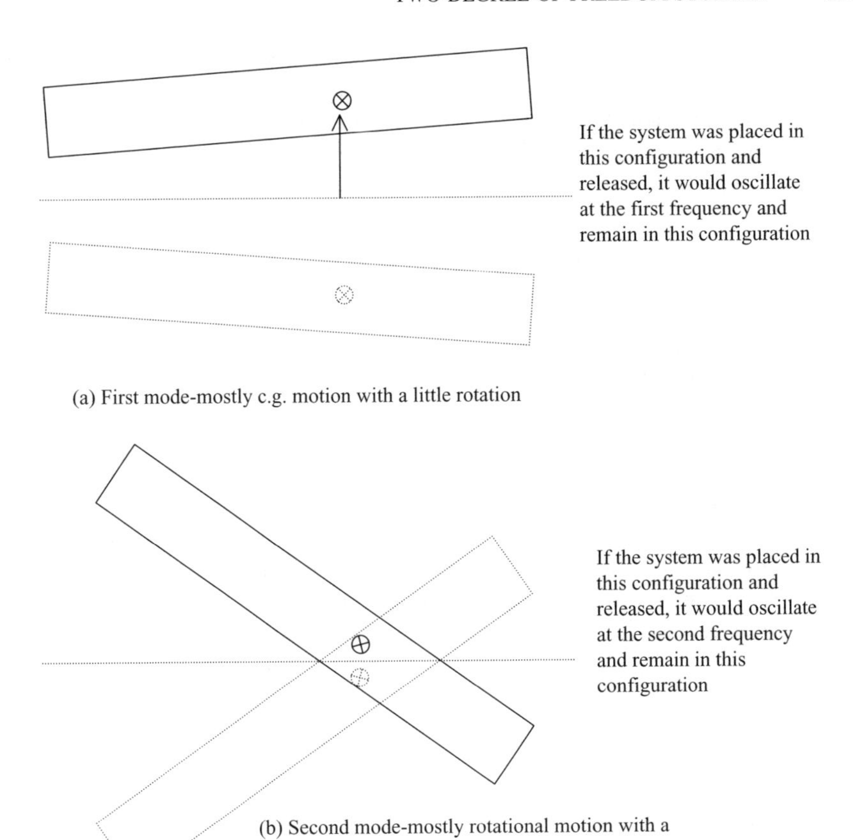

Figure 8.9 Modes of the half-car system.

This is called the *pitch mode* of the vehicle. Depending on the location of the center of gravity and the stiffness distribution between the front and rear, the heave mode will exhibit some amount of rotation, and the pitch mode will include some amount of c.g. motion.

If the vehicle is perfectly symmetrical, the modes are totally decoupled. We can see this by letting $a/b = 1$ and running the Matlab program again. The result is

```
eig_vec  =
   1     0
   0     1
```

and (8.67)

```
omega_sq  =
39.4784       0
   0      118.4353
```

This time the first mode vector indicates c.g. displacement with no associated rotation, and the second mode vector indicates rotation with no associated c.g. displacement. The modes are decoupled. It turns out that it is easier to control resonance in systems where the modes are decoupled. Thus, a strategy for vibration control is to design an isolation system that tends to decouple the system modes.

Forced Response of Two-Degree-of-Freedom Systems

The frequency response is the most important forced response to consider for any vibrational system. It is interesting to see what differs between the single-degree-of-freedom frequency response covered in Figure 8.4 and that for the two-degree-of-freedom system. We use a computational procedure to evaluate the frequency response, but first it is instructive to derive the transfer functions and complex frequency response functions for the two-degree-of-freedom system of Figure 8.6. The equations of motion were derived previously as Eqs. (8.28) and in matrix form in Eq. (8.29). As developed in Section 3.2, the easiest way to derive the frequency response is to start with appropriate input/output transfer functions and then generate the complex frequency response functions by letting $s = j\omega$.
The matrix equations (8.29) are cast into the Laplace s-domain with the result

$$\begin{bmatrix} m_1s^2 + b_1s + k_1 + k_2 & -k_2 \\ -k_2 & m_2s^2 + k_2 + k_3 \end{bmatrix} \begin{bmatrix} X_1 \\ X_2 \end{bmatrix} = \begin{bmatrix} 1 \\ 0 \end{bmatrix} F_1 \tag{8.68}$$

The denominator of the transfer functions is the determinant of the matrix on the left-hand side of (8.68). Thus,

$$D = m_1m_2\left(s^4 + \frac{b_1}{m_1}s^3 + \frac{k_2+k_3}{m_2} + \frac{k_1+k_2}{m_1}s^2 + \frac{b_1}{m_1}\frac{k_2+k_3}{m_2}s + \frac{k_1k_2+k_1k_3+k_2k_3}{m_1m_2}\right) \tag{8.69}$$

The transfer function for the displacement of the first mass comes from substituting the column vector on the right of (8.68) for the first column in the matrix on the left and taking the determinant, with the result

$$\frac{X_1(s)}{F_1(s)} = \frac{m_2(s^2 + (k_2+k_3)/m_2)}{D(s)} \tag{8.70}$$

and the transfer function for the displacement of the second mass comes from the determinant of the matrix on the left of (8.68), having first substituted the column vector on the right for the second column, thus,

$$\frac{X_2(s)}{F_1(s)} = \frac{k_2}{D(s)} \tag{8.71}$$

It is convenient to collect the parameters into groups,

$$\omega_{n1}^2 = \frac{k_1}{m_1}, \qquad \frac{b_1}{m_1} = 2\zeta\omega_{n1}$$

and write the transfer functions in nondimensional form,

$$\frac{k_1 X_1(s)}{F_1(s)} = \frac{s^2/\omega_{n1}^2 + (k_2/k_1 + k_3/k_1)/(m_2/m_1)}{\frac{s^4}{\omega_{n1}^4} + 2\zeta\frac{s^3}{\omega_{n1}^3} + \left[\frac{k_2/k_1 + k_3/k_1}{m_2/m_1} + \left(1 + \frac{k_2}{k_1}\right)\right]\frac{s^2}{\omega_{n1}^2} + 2\zeta\frac{k_2/k_1 + k_3/k_1}{m_2/m_1}\frac{s}{\omega_{n1}} + \frac{k_2/k_1 + k_3/k_1 + (k_2/k_1)(k_3/k_1)}{m_2/m_1}} \tag{8.72}$$

$$\frac{k_1 X_2(s)}{F_1(s)} = \frac{\frac{(k_2/k_1)}{(m_2/m_1)}}{\frac{s^4}{\omega_{n1}^4} + 2\zeta\frac{s^3}{\omega_{n1}^3} + \left[\frac{k_2/k_1 + k_3/k_1}{m_2/m_1} + \left(1 + \frac{k_2}{k_1}\right)\right]\frac{s^2}{\omega_{n1}^2} + 2\zeta\frac{k_2/k_1 + k_3/k_1}{m_2/m_1}\frac{s}{\omega_{n1}} + \frac{k_2/k_1 + k_3/k_1 + (k_2/k_1)(k_3/k_1)}{m_2/m_1}} \tag{8.73}$$

To generate the complex frequency response functions, we would next let $s = j\omega$ in the transfer functions and then generate the response magnitude by evaluating the magnitude of the resulting complex numbers. It is probably not worthwhile to write this down since the frequency response will ultimately be evaluated computationally. But if we imagine letting $s = j\omega$ in (8.72) and (8.73), there are several interesting observations to be made. If the damping ζ is set to zero, the denominator of the transfer functions is the frequency equation derived earlier when discussing the free response. In nondimensional form the frequency equation for this case is

$$\frac{\omega^4}{\omega_{n1}^4} - \frac{k_2/k_1 + k_3/k_1}{m_2/m_1} + \left(1 + \frac{k_2}{k_1}\right)\frac{\omega^2}{\omega_{n1}^2} + \frac{k_2/k_1 + k_3/k_1 + (k_2/k_1)(k_3/k_1)}{m_2/m_1} = 0 \tag{8.74}$$

The two solutions to (8.74) will be the nondimensional frequencies, where the frequency response will exhibit resonant behavior.

A very interesting observation comes from the numerator of (8.72). If $s = j\omega$, the numerator of the frequency response magnitude becomes

$$\text{numerator of (8.72)} = \left[\left(\frac{k_2/k_1 + k_3/k_1}{m_2/m_1} - \frac{\omega^2}{\omega_{n1}^2}\right)^2\right]^{1/2} \tag{8.75}$$

which goes to zero for the special nondimensional frequency

$$\frac{\omega_0^2}{\omega_{n1}^2} = \frac{k_2/k_1 + k_3/k_1}{m_2/m_1} \tag{8.76}$$

This implies that if the input force was a harmonic input at this special frequency, the response of the first mass would be zero. Such an occurrence happens often with multi-degree-of-freedom systems. It is very common to take advantage of such dynamics in the design of isolation systems. This concept is developed further below.

For the case

$$\frac{k_2}{k_1} = \frac{k_3}{k_1} = \frac{m_2}{m_1} = 1$$

the Matlab program [2] was used to evaluate the frequency responses from Eqs. (8.72) and (8.73). The results are shown in Figure 8.10 for a few values of the damping ratio ζ. For light damping the two resonances are very apparent and they are reduced as the damping is increased. As indicated by the transfer function from Eq. (8.72), there is a special frequency where the response is zero independent of the damping. It is interesting that the response of the second mass at this special frequency is really not noteworthy. It does not resonate at this special frequency, and in fact the response appears to be independent of the damping at this frequency.

As mentioned above, the appearance of a zero response in frequency responses of higher-order systems is quite common, and the next topic deals with taking advantage of this interesting dynamic condition.

8.3 TUNED VIBRATION ABSORBERS

If a lightly damped oscillator is attached to a structure that is vibrating at a fixed frequency, with proper tuning of the oscillator, the structure can be made to have virtually no response at the tuned frequency of the oscillator. The simplest system to demonstrate this dynamic effect is the system shown in Figure 8.11. The structure is a mass m_s that is acted upon by a disturbance force $F_d(t)$. This structure could be a wall of a building where unknown forces are creating motion in the wall, a helicopter fuselage that is shaking due to the action of the rotor system, or motorcycle handlebars that "buzz" due to engine vibration. Attached to the structure mass in Figure 8.11 is the lightly damped spring–mass–damper tuned vibration absorber (TVA) system of mass m_a, stiffness k_a, and damping b_a. The coordinate x_s locates the structure mass, and the coordinate x_a locates the TVA mass. The coordinates are inertial and are measured from equilibrium in a gravity field. The force diagram is shown in Figure 8.11, where the spring and damper are assumed positive in compression.

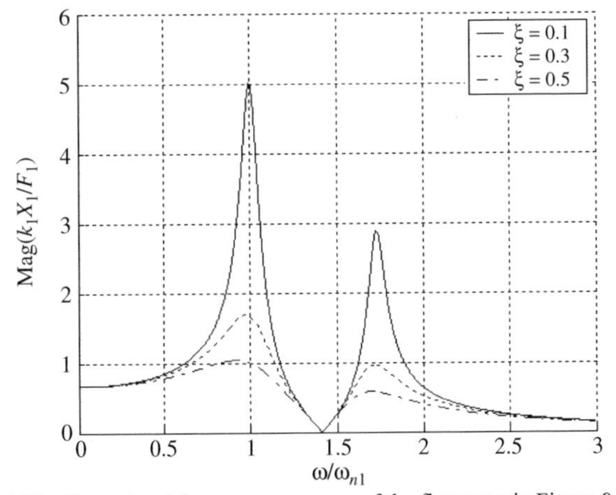

(a) Nondimensional frequency response of the first mass in Figure 8.6

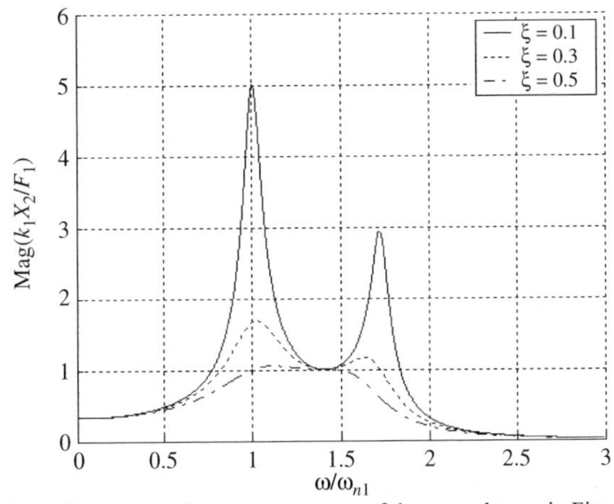

(b) Nondimensional frequency response of the second mass in Figure 8.6

Figure 8.10 Nondimensional frequency response of a two-degree-of-freedom system.

Application of Newton's laws yields

$$\begin{aligned} F_d - b_a(\dot{x}_s - \dot{x}_a) - k_a(x_s - x_a) &= m_s\ddot{x}_s \\ b_a(\dot{x}_s - \dot{x}_a) + k_a(x_s - x_a) &= m_a\ddot{x}_a \end{aligned} \tag{8.77}$$

or in matrix form,

$$\begin{bmatrix} m_s & 0 \\ 0 & m_a \end{bmatrix}\begin{bmatrix} \ddot{x}_s \\ \ddot{x}_a \end{bmatrix} + \begin{bmatrix} b_a & -b_a \\ -b_a & b_a \end{bmatrix}\begin{bmatrix} \dot{x}_s \\ \dot{x}_a \end{bmatrix} + \begin{bmatrix} k_a & -k_a \\ -k_a & k_a \end{bmatrix}\begin{bmatrix} x_s \\ x_a \end{bmatrix} = \begin{bmatrix} 1 \\ 0 \end{bmatrix} F_d \tag{8.78}$$

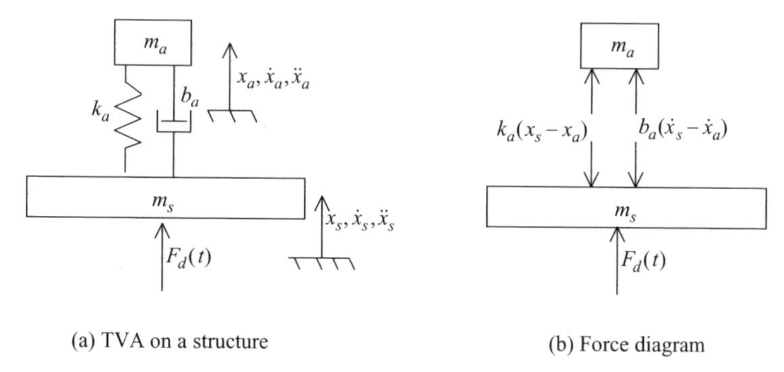

Figure 8.11 Simple structure to demonstrate a TVA.

It is instructive to derive the transfer functions relating the structure mass response to the input force and the TVA mass response to the input force. In the Laplace domain Eq. (8.78) becomes

$$\begin{bmatrix} m_s s^2 + b_a s + k_a & -(b_a s + k_a) \\ -(b_a s + k_a) & m_a s^2 + b_a s + k_a \end{bmatrix} \begin{bmatrix} X_s \\ X_a \end{bmatrix} = \begin{bmatrix} 1 \\ 0 \end{bmatrix} F_d \tag{8.79}$$

Cramer's rule is used to derive the transfer functions,

$$\frac{X_s(s)}{F_d(s)} = \frac{m_a s^2 + b_a s + k_a}{m_s m_a s^2[s^2 + (b_a/m_a + b_a/m_s)s + (k_a/m_a + k_a/m_s)]} \tag{8.80}$$

and

$$\frac{X_a(s)}{F_d(s)} = \frac{b_a s + k_a}{m_s m_a s^2[s^2 + (b_a/m_a + b_a/m_s)s + (k_a/m_a + k_a/m_s)]} \tag{8.81}$$

Define the parameter groups $k_a/m_a = \omega_0^2$ and $b_a/m_a = 2\zeta\omega_0$ and Eqs. (8.80) and (8.81) can be written nondimensionally as

$$\frac{k_a X_s(s)}{F_d(s)} = \frac{(m_a/m_s)\left[(s^2/\omega_0^2) + (2\zeta(s/\omega_0) + 1)\right]}{\left(s^2/\omega_0^2\right)\left[(s^2/\omega_0^2) + 2\zeta(1 + m_a/m_s)(s/\omega_0) + (1 + m_a/m_s)\right]} \tag{8.82}$$

$$\frac{k_a X_a(s)}{F_d(s)} = \frac{(m_a/m_s)\,[2\zeta(s/\omega_0) + 1]}{\left(s^2/\omega_0^2\right)\left[s^2/\omega_0^2 + 2\zeta(1 + m_a/m_s)(s/\omega_0) + (1 + m_a/m_s)\right]} \tag{8.83}$$

To determine the frequency response, let $s = j\omega$ in the transfer functions and compute the magnitudes of the resulting complex numbers. Rather than show the entire frequency response, let's look at the numerator for the response of the structure,

Eq. (8.82), when $s = j\omega$. The magnitude of the parenthetical term is

$$\left| 1 - \frac{\omega^2}{\omega_0^2} + j2\varsigma\frac{\omega}{\omega_0} \right| = \left[\left(1 - \frac{\omega^2}{\omega_0^2}\right)^2 + \left(2\varsigma\frac{\omega}{\omega_0}\right)^2 \right]^{1/2} \tag{8.84}$$

For $\varsigma \to 0$, this magnitude is very small for a forcing frequency equal to the tuned frequency of the oscillator, ω_0. This indicates that the structure response X_s is virtually zero for a forcing frequency at the TVA tuned frequency, and the tuned frequency of the TVA can be set by adjustment of its own parameters, k_a and m_a, independent of any structure parameters. Thus, for systems subject to fixed-frequency inputs, TVAs are an inexpensive approach to minimizing the response of the system at the fixed input frequency.

Obviously, not all systems that have vibration problems are operating at fixed frequencies, and TVAs are not effective for such systems. However, there are a large number of very practical everyday systems that do operate at nearly fixed frequencies. Once at cruise altitude, aircraft operate at nearly fixed rpm. Helicopters operate at nearly constant rotor speed. Air-conditioning units on the roofs of buildings operate at a fixed motor speed. Many manufacturing machines operate at fixed speeds. There really are countless applications for TVAs in vibration control. Armed with your knowledge of how lightly damped systems resonate for inputs near their natural frequencies, you might imagine that lightly damped systems exhibit a large response near their natural frequencies even if the disturbances to the system are not harmonic at a fixed frequency. For example, if the system in Figure 8.8 was dropped onto the ground from some height and the springs then stuck to the ground, the system would tend to heave and pitch at its natural frequencies. Thus if TVAs are placed on structures and tuned to the structure natural frequencies, some level of vibration control will be accomplished, as the structure is subject to various disturbances from the environment even though these disturbances are not harmonic at fixed frequencies.

A very simple and typical TVA is shown in Figure 8.12. It consists of cylindrical shaft of length L and diameter D with an attached mass m_a. The shaft, which is the "spring" of the TVA, is made from a material with Young's modulus E. If the shaft were rigidly connected to a structure at its end, it would behave like a cantilever beam

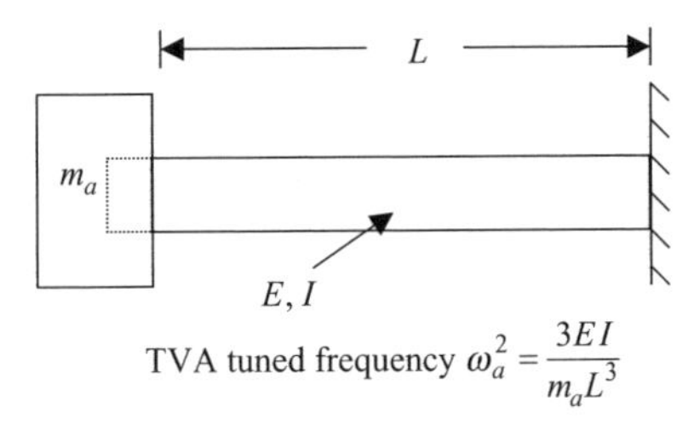

Figure 8.12 Simple TVA configuration.

of effective stiffness

$$k_a = \frac{3EI}{L^3} \tag{8.85}$$

where I is the area moment of inertia. The TVA frequency would be

$$\omega_a^2 = \frac{k_a}{m_a} = \frac{3EI}{m_a L^3} \tag{8.86}$$

A nice feature of using a circular shaft for the TVA spring is that it will allow motion in any direction. In many applications the primary disturbance direction is not known, so it is nice that the TVA will "seek" the proper direction for its response. Another nice feature of this simple design is that the stiffness is dependent on the length L. If the mass is threaded onto the shaft, the TVA frequency is adjustable by moving the mass in or out. This is often done for fine tuning the TVA. Now if we could just put a little motor in the mass so that it could rotate itself in or out on the shaft and then use a sensor to determine vibration levels, we could have the TVA tune itself automatically. Before getting too excited about this new invention, it already exists and is called an *adaptive TVA*. Such devices come in many different designs and configurations.

Some Configurations for TVAs

The system in Figure 8.13 consists of a mass suspended on a spring with a base excitation. Attached to the base is a pinned disk of moment of inertia I_g. The mass has an extension arm that moves without slip on the circumference of the disk. Part *b* of the figure shows the force diagram for the system with the internal force F_I exposed. From Newton's law the equation of motion for the mass is

$$F_I + k(x_i - x) = m\ddot{x} \tag{8.87}$$

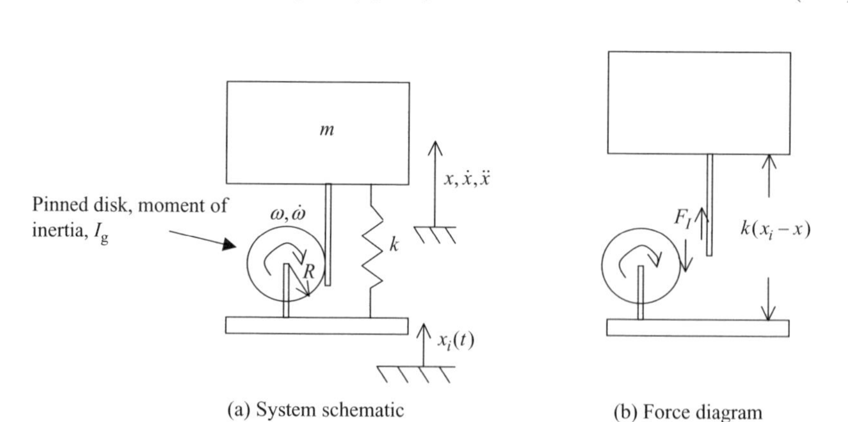

(a) System schematic (b) Force diagram

Figure 8.13 TVA using a rotational inertia.

and the rotational equation for the disk is

$$F_I R = I_g \dot{\omega} \tag{8.88}$$

There is a kinematic relationship due to the no-slip assumption between the extension arm and the disk. For the positive directions indicated in the figure, the absolute vertical velocity at the disk edge is

$$v_1 = \dot{x}_i - R\omega$$

using the base as the reference, and this must equal the absolute velocity of the same point starting from the mass; thus,

$$\dot{x}_i - R\omega = \dot{x} \tag{8.89}$$

or

$$\omega = \frac{\dot{x}_i - \dot{x}}{R} \tag{8.90}$$

Differentiating (8.90) and substituting into (8.88) allows substitution for F_{I} in (8.87), with the result

$$\left(1 + \frac{I_g}{mR^2}\right) m\ddot{x} + kx = \frac{I_g}{mR^2} m\ddot{x}_i + kx_i \tag{8.91}$$

To obtain the frequency response, we first deriver the transfer function,

$$\frac{X(s)}{X_i(s)} = \frac{(I_g/mR^2)ms^2 + k}{(1 + I_g/mR^2)ms^2 + k} \tag{8.92}$$

This can be written as

$$\frac{X(s)}{X_i(s)} = \frac{I_g/mR^2}{(1 + I_g/mR^2)} \frac{s^2 + \omega_0^2}{s^2 + \omega_n^2} \tag{8.93}$$

where

$$\begin{aligned} \omega_0^2 &= \frac{1}{I_g/mR^2} \frac{k}{m} \\ \omega_n^2 &= \frac{1}{(1 + I_g/mR^2)} \frac{k}{m} \end{aligned} \tag{8.94}$$

If we let $s = j\omega$ in the transfer functions and solved for the frequency response magnitude, we would see that the mass is motionless for an input frequency at ω_0, showing that this configuration is a TVA. We would further determine that there is an undamped resonance at the natural frequency ω_n and that the resonance occurs at a lower frequency than the TVA frequency. This could cause some problems if the system started from rest and needed to pass through a frequency range before reaching the operating frequency corresponding to the TVA tuned frequency.

Another TVA configuration is shown in Figure 8.14. The mass to be isolated is connected to the base excitation through the lever mechanism shown. The lever is massless but has the TVA mass m_a at its end. The TVA mass is connected to the base through the spring k_a. The lever lengths are indicated in the figure. Inertial coordinates are used, measured from equilibrium in a gravity field. We will see if there is a special frequency where a fixed-frequency harmonic input of $x_i(t)$ results in no motion of the suspended mass.

Figure 8.14*b* shows the force diagram with forces shown in their assumed positive directions. The equations of motion are

$$\begin{aligned} F_2 &= m\ddot{x}_m \\ k_a(x_i - x_a) + F_m &= m_a\ddot{x}_a \end{aligned} \tag{8.95}$$

for the two mass elements in the system. The lever arm is assumed massless with no rotational inertia; therefore, it is constrained to be in moment and force equilibrium. Taking moments about the center pivot yields the constraint

$$F_2 l_1 = F_m l_2 \tag{8.96}$$

There is also a kinematic constraint among the variables that is most easily seen by first recognizing that for small clockwise rotations, the lever angle θ can be expressed as

$$\theta = \frac{x_i - x_m}{l_1} \tag{8.97}$$

and then the displacement of the TVA mass becomes

$$x_a = x_i + l_2\theta = x_i + \frac{l_2}{l_1}(x_i - x_m) = \left(1 + \frac{l_2}{l_1}\right)x_i - \frac{l_2}{l_1}x_m \tag{8.98}$$

Using (8.96) in the first of (8.95) yields

$$\frac{l_2}{l_1}F_m = m\ddot{x}_m \tag{8.99}$$

If we solve (8.99) for F_m and substitute into the second of (8.95), we get

$$k_a(x_i - x_a) + \frac{m\ddot{x}_m}{l_2/l_1} = m_a\ddot{x}_a \tag{8.100}$$

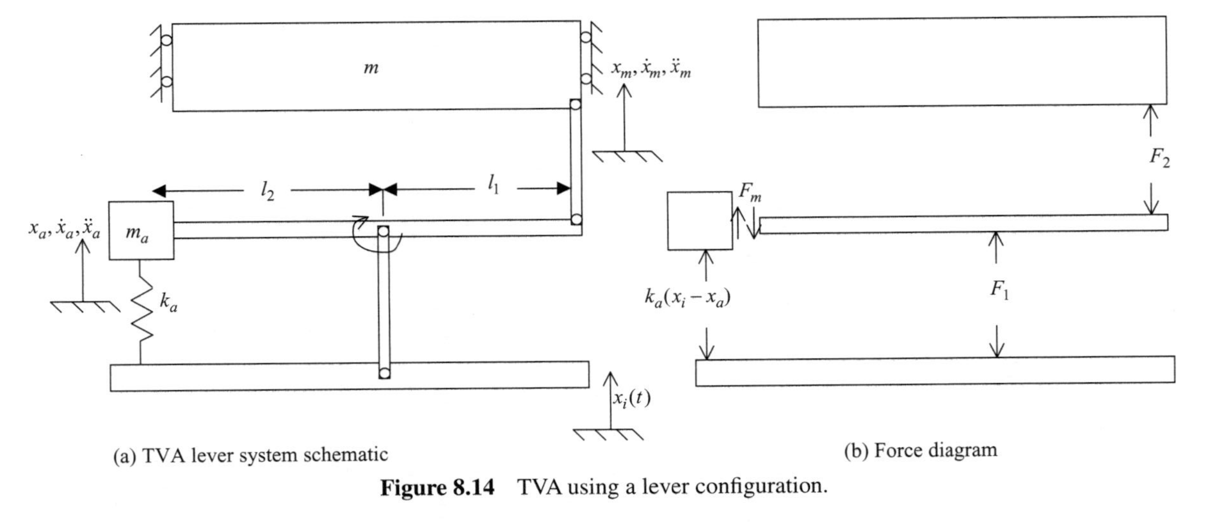

(a) TVA lever system schematic

(b) Force diagram

Figure 8.14 TVA using a lever configuration.

Finally, using Eq. (8.98) on the left of (8.100) and differentiating (8.98) twice and using the result on the right of (8.100) yields, after some rearrangement of terms,

$$\left[m + m_a\left(\frac{l_2}{l_1}\right)^2\right]\ddot{x}_m + k_a\left(\frac{l_2}{l_1}\right)^2 x_m = m_a\frac{l_2}{l_1}\left(1+\frac{l_2}{l_1}\right)\ddot{x}_i + k_a\left(\frac{l_2}{l_1}\right)^2 x_i \tag{8.101}$$

For frequency response we first derive the transfer function by casting (8.101) into the Laplace domain, with the result

$$\frac{X_m(s)}{X_i(s)} = \frac{(l_2/l_1)(1+l_2/l_1)}{(m/m_a)+(l_2/l_1)^2}\,\frac{s^2 + \dfrac{(l_2/l_1)}{1+l_2/l_1}\dfrac{k_a}{m_a}}{s^2 + \dfrac{(l_2/l_1)^2}{(m/m_a)+(l_2/l_1)^2}\dfrac{k_a}{m_a}} \tag{8.102}$$

If we let $s = j\omega$ in the transfer function, it is easily seen that a frequency ω_0 exists where the response of the mass is zero; that is,

$$\omega_0^2 = \frac{l_2/l_1}{1+l_2/l_1}\,\frac{k_a}{m_a} \tag{8.103}$$

Notice that the TVA frequency is dictated by parameters associated with the TVA and does not depend on parameters of the system of which the TVA is a part. It turns out that this configuration for a TVA is used often in helicopters between the rotor system and the fuselage. It is tuned to the blade pass frequency at the steady-state operating speeds of the helicopter.

As a final example of a two-degree-of-freedom system exhibiting useful dynamic characteristics, consider the interesting centrifugal pendulum shown in Figure 8.15. The system consists of a pinned rotating disk of centroidal moment of inertia I_g with a pinned pendulum at the circumference of the disk. We use the inertial coordinate θ and its derivatives to describe the disk motion, and we will use the relative angle ϕ and its derivatives to describe the motion of the pendulum. The reader should verify that these two coordinates allow placement of the system in any of its possible configurations. A torque $\tau_i(t)$ acts on the disk.

Imagine that the disk is rotating at a pretty high steady-state rotational speed ω_s, and under this steady-state condition the pendulum is staying in the position where $\phi = 0$. Now imagine that the input torque acts at a fixed frequency such that the disk speeds up and slows down just a little from its steady-state speed and the pendulum moves back and forth just a little from its steady-state position. We desire equations of motion for this scenario.

Figure 8.15 shows velocity, acceleration, and force diagrams for the system. All components are written with respect to a coordinate frame attached to the disk at its center and rotating with the disk. The transfer formulas for velocity and acceleration from Chapter 1 were used to write down the components shown in the figure.

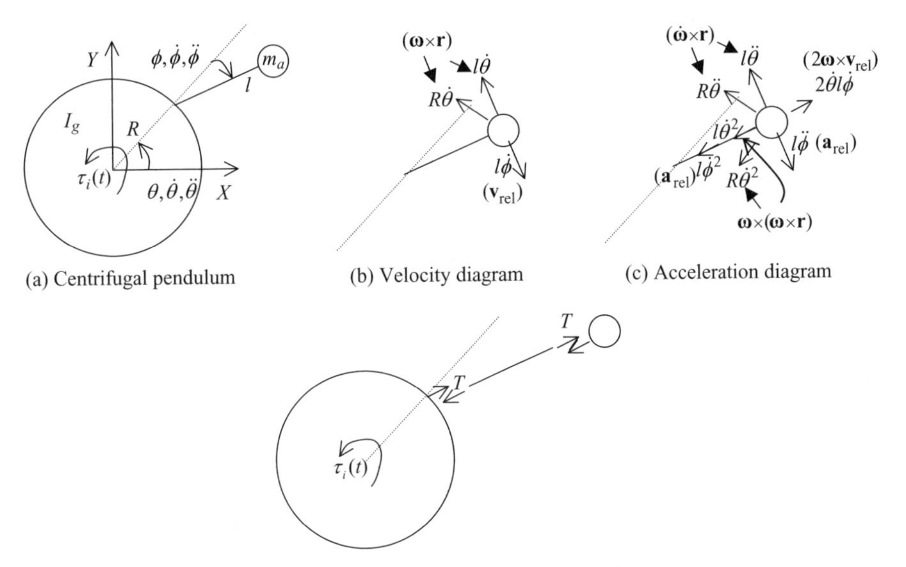

Figure 8.15 Centifugal pendulum TVA.

The vector expressions for these transfer relationships are

$$
\begin{aligned}
\mathbf{v}_p &= \mathbf{v}_0 + \boldsymbol{\omega} \times \mathbf{r} + \mathbf{v}_{\mathrm{rel}} \\
\mathbf{a}_p &= \mathbf{a}_0 + \dot{\boldsymbol{\omega}} \times \mathbf{r} + \boldsymbol{\omega} \times (\boldsymbol{\omega} \times \mathbf{r}) + 2\boldsymbol{\omega} \times \mathbf{v}_{\mathrm{rel}} + \mathbf{a}_{\mathrm{rel}}
\end{aligned}
\tag{8.104}
$$

The frame angular velocity ω has magnitude $\dot{\theta}$ with positive direction out of the page. The **r** vector is officially the vector from the base of the frame (disk center) to the point being described (pendulum mass). For this example the **r** vector is composed of the vector sum of the component of length R from the disk center to the pendulum pin joint plus the component of length l from the pendulum pin joint to the pendulum mass.

Application of Newton's laws for the pendulum mass using the orthogonal directions along the pendulum length and perpendicular to the pendulum length yields

$$
\begin{aligned}
-T &= m_a(2l\dot{\theta}\dot{\phi} - l\dot{\theta}^2 - l\dot{\phi}^2 - R\dot{\theta}^2\cos\phi - R\ddot{\theta}\sin\phi) \\
0 &= m_a(l\ddot{\phi} + R\dot{\theta}^2\sin\phi - l\ddot{\theta} - R\ddot{\theta}\cos\phi)
\end{aligned}
\tag{8.105}
$$

and for the disk the moment equation is

$$
-TR\sin\phi + \tau_i(t) = I_g\ddot{\theta} \tag{8.106}
$$

Equations (8.105) and (8.106) would have to be solved computationally to predict all possible dynamic responses of this system. For vibration considerations we linearize these equations about the steady-state operating point described above.

Using the first of (8.105) in (8.106) yields a good starting point for the linearization procedure:

$$\begin{aligned} &m_a(2l\dot{\theta}\dot{\phi} - l\dot{\theta}^2 - l\dot{\phi}^2 - R\dot{\theta}^2\cos\phi - R\ddot{\theta}\sin\phi)R\sin\phi + \tau_i(t) = I_g\ddot{\theta} \\ &0 = m_a(l\ddot{\phi} + R\dot{\theta}^2\sin\phi - l\ddot{\theta} - R\ddot{\theta}\cos\phi) \end{aligned} \tag{8.107}$$

For small motions of ϕ, $\cos\phi \approx 1$ and $\sin\phi \approx \phi$ but remember that θ is not small and $\dot{\theta}$ is almost constant at $\dot{\theta} = \omega_s$. For linearization we neglect terms that contain products of variables. In the first of (8.107) the first term is neglected, as there is a product of $\dot{\phi}\phi$. The second term remains, but the third term is neglected. The fourth term remains and the fifth term is neglected. In the second of (8.107), all the terms remain, with the final linearized result,

$$\begin{aligned} -m_a R(R+l)\omega_s^2\phi + \tau_i(t) &= I_g\ddot{\theta} \\ l\ddot{\phi} + R\omega_s^2\phi - (R+l)\ddot{\theta} &= 0 \end{aligned} \tag{8.108}$$

Equations (8.108) are next put into the Laplace domain, with the result

$$\begin{bmatrix} I_g s^2 & m_a R(R+l)\omega_s^2 \\ -(R+l)s^2 & ls^2 + R\omega_s^2 \end{bmatrix} \begin{bmatrix} \Theta(s) \\ \Phi(s) \end{bmatrix} = \begin{bmatrix} 1 \\ 0 \end{bmatrix} T_i(s) \tag{8.109}$$

Finally, we can derive the transfer function relating the response of the disk Θ to the input torque T_i,

$$\frac{\Theta(s)}{T_i(s)} = \frac{(1/I_g)\left(s^2 + (R/l)\omega_s^2\right)}{s^2\left\{s^2 + [1 + m_a l^2/I_g(1 + R/l)^2](R/l)\omega_s^2\right\}} \tag{8.110}$$

For frequency response we let $s = j\omega$ to derive the complex frequency response function, and the response amplitude is calculated as the magnitude of the resulting complex number. If this is done, the numerator of (8.10) is zero at the special frequency

$$\omega_0^2 = \frac{R}{l}\omega_s^2 \tag{8.111}$$

This shows that the centrifugal pendulum configuration is a TVA where the TVA tuned frequency depends on the operating angular velocity of the disk.

An interesting aspect of this application is that the torque input frequency is typically some multiple of the operating speed. If the disk is the flywheel of a six-cylinder four-stroke engine, there will be three torque impulses per revolution of the engine; thus, $\omega_0 = 3\omega_s$, and $R/l = 9$. Thus $l = 1/9R$ would substantially cancel the unsteady vibration of the flywheel for all operating speeds. If the disk is the four-bladed rotor system of a helicopter, there will be four torque impulses per revolution of the rotor system, and the tuned TVA frequency would be $\omega_0 = 4\omega_s$ and $R/l = 16$. Thus, $l = \frac{1}{16}R$ would produce substantial reductions in vibrations.

It turns out that the centrifugal pendulum is used on heavy truck flywheels and is very typical in helicopter rotor systems.

8.4 SUMMARY

This chapter has been a brief introduction to the very broad topic of vibrations. It is one of the engineering topics that can be addressed using linear analysis, and the results are directly applicable to real engineering systems. A single-degree-of-freedom system was used to introduce the concept of natural frequency, frequency response, and resonance. A two-degree-of-freedom system was needed to present the concept of modal dynamics. Some general discussion described how the two-degree-of-freedom concepts could be expanded to multiple degrees of freedom. The two-degree-of-freedom system allowed introduction to the concept of a tuned vibration absorbers and several configurations were presented. Interested readers are directed to refs. [6] and [7] for treatment of the topic of vibrations in much more depth.

Hopefully, the reader recognizes a certain consistency in the treatment of all the engineering applications of dynamics. In all cases a model is postulated that contains the dynamic elements and kinematic motions that the modeler deems important for the application at hand. This step is among the most difficult in the process and requires some experience to postulate reasonable models. Appropriate coordinates are then chosen, and velocity, acceleration, and force diagrams are constructed if Newton's laws are to be used to derive the equations of motion. Acceleration and force diagrams are not needed if Lagrange's equations are to be used to formulate equations of motion. Equations of motion are the goal of the formulation process. Once equations are derived, the next step is to obtain some information from the model. This generally requires simulation when the equations of motion are nonlinear and for some special cases, such as the vibrations presented in this chapter, the equations can be linearized and analytical results pursued.

In Chapter 9 we expand on the topic of vibrations to include distributed systems in which mass and stiffness are not unique elements but instead, are distributed in some continuous fashion throughout the system. The resulting equations of motion are partial differential equations rather than total differential equations. Do not think that this occurrence overwhelmingly complicates the topic of dynamics. As will be seen, there are some fundamental similarities between lumped system vibrations and distributed system vibrations. These are developed in Chapter 9.

REFERENCES

[1] D. C. Karnopp, M. J. Crosby, and R. Hardwood, Vibration control using semi-active force generators, ASME Paper 73-DET-122, *Trans. ASME J. Eng. Ind.*, Vol. 96, Ser. B, No. 2, pp. 619–626, May 1974.

[2] Matlab and Simulink software, Mathworks Co., Natick, MA.

[3] D. L. Margolis, Semi-active control of wheel hop in group vehicles, *Veh. Sys. Dyn.*, Vol. 12, No. 6, pp. 317–330, 1983.

[4] R. C. Redfield, and D. C. Karnopp, Performance sensitivity of an actively damped vehicle suspension to feedback variation, *J. Dyn. Syst., Meas. Control*, Vol. 111, No. 1, pp. 51–60, 1989.

[5] C. Yue, T. Butsuen, and J. K. Heddrick, Alternative control laws for automotive active suspensions, *J. Dyn. Syst., Meas., Control,* Vol. 111, No. 2, pp. 286–291, 1989.

[6] L. Meirovitch, *Elements of Vibration Analysis*, McGraw-Hill, New York, 1975.

[7] D. J. Inman, *Engineering Vibration*, Prentice Hall, Upper Saddle River, NJ, 1996.

PROBLEMS

8.1 Shown in Figure P8.1 are two very similar single-degree-of-freedom systems. System a has a force excitation $F_{\text{in}}(t)$ acting on the mass and there is a *transmitted* force F acting on the ground. In system b the base of the system is excited and the mass position is the output. For both systems the coordinate for the mass is measured from equilibrium in a gravity field.

(a) Derive the equations of motion for these two systems. Derive the transfer functions relating $F(s)$ to $F_{\text{in}}(s)$ in system a and $x(s)$ to $x_{\text{in}}(s)$ for system b. Note that the force transmitted in system a is the sum of the spring and damper forces. Comment on the similarities and differences of the two transfer functions.

(b) Let $s = j\omega$ and derive the complex frequency response functions for the two systems. Sketch the magnitude of the frequency response versus the forcing frequency.

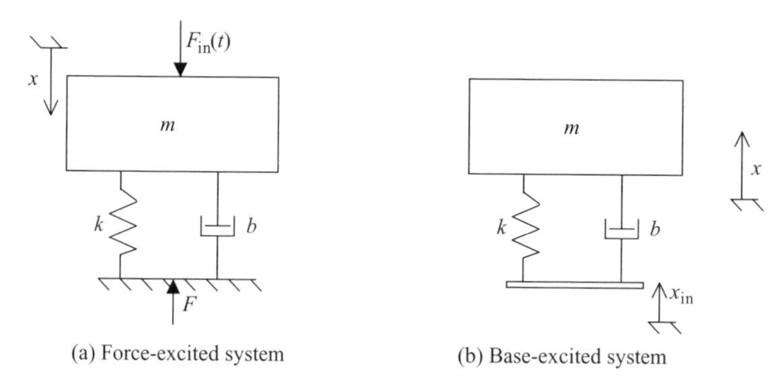

Figure P8.1

8.2 **(a)** This system consists of a massless lever with various components attached as shown in Figure P8.2. It is a single-degree-of-freedom system and the coordinate to use is the angle θ. For small motions of the angle, derive the linear equations of motion for this system. When $\theta = 0$ there is no force in the spring.

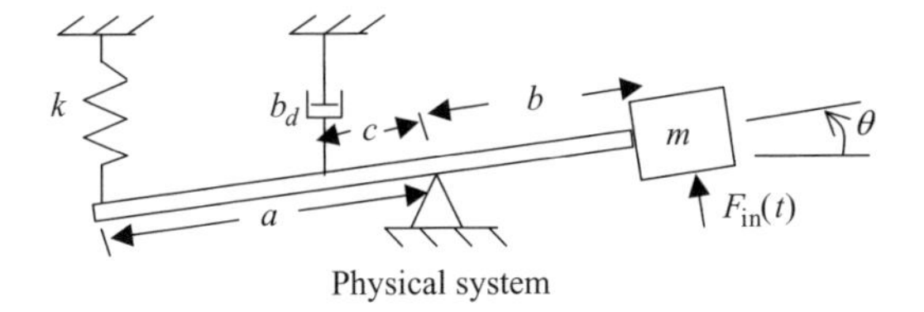

Physical system

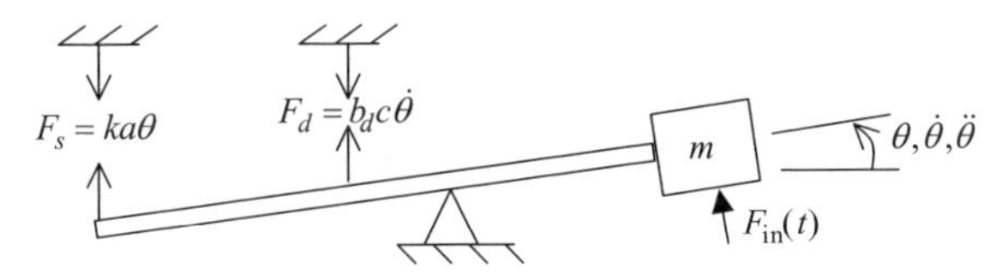

Velocity, acceleration, and force diagrams combined

Figure P8.2

(b) Let the forcing $F_{\text{in}}(t) = 0$ determine the system eigenvalues. Interpret the meaning of these eigenvalues. Write down the general form of the free response. Derive expressions for the time constant τ, the undamped natural frequency ω_n, and the period of oscillation T.

(c) Derive the transfer function relating $\theta(s)$ to $F_{\text{in}}(s)$, and from this result derive the complex frequency response function. Sketch the magnitude of the frequency response for low damping.

8.3 This problem is identical to Problem 8.2 except that the excitation is the prescribed motion at the end of the spring (Figure P8.3).

(a) Derive the small-angle equations of motion for this system.

(b) Derive the transfer function relating $\theta(s)$ to $x_{\text{in}}(s)$. From this result derive the complex frequency response function, and from this result determine the magnitude $|\theta(j\omega)/x_{in}(j\omega)|$ of the frequency response. Sketch the magnitude versus the forcing frequency.

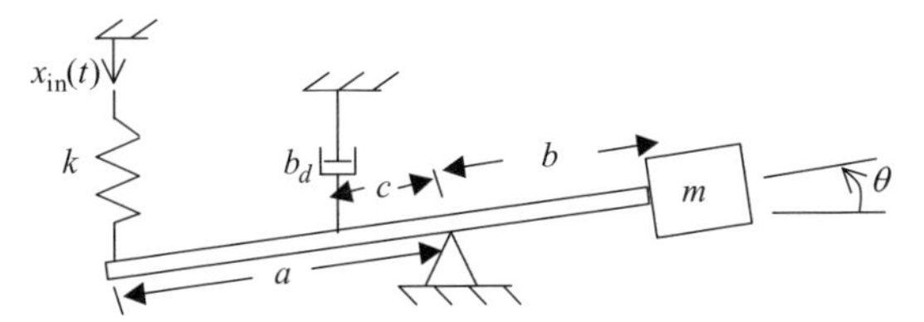

Figure P8.3

8.4 In this system the disk of rotational inertia I_g is attached to inertial ground through the rotational damper and spring as shown in Figure P8.4. The torsional damper generates a retarding torque proportional to the relative angular velocity across it and the torsional spring generates a retarding torque proportional to the twist in the spring. An input torque $\tau_{\text{in}}(t)$ is acting on the disk.

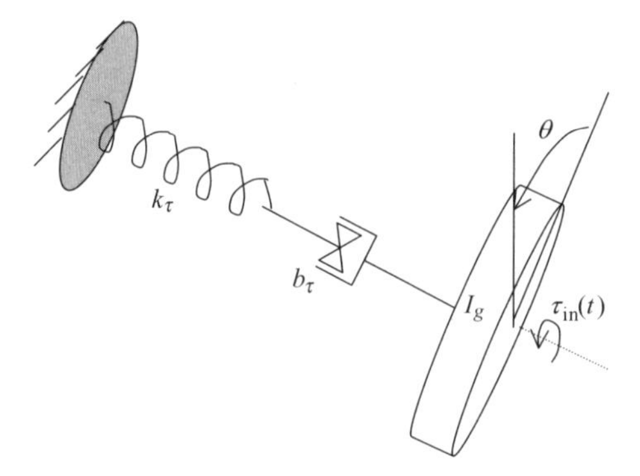

Figure P8.4

(a) Derive the equations of motion suitable for small angular motions of the disk.

(b) Derive the transfer function relating $\theta(s)$ to the input torque $\tau_{in}(s)$. Next, derive the complex frequency response function and sketch the magnitude of the response versus the forcing frequency.

8.5 Figure P8.5 shows two base-excited single-degree-of-freedom systems. In part *a* the damper is between the base and the mass, and in part *b* the damper is between the mass and inertial ground.

(a) Derive the equations of motion for these two systems.

(b) Transform the equations into the Laplace s-domain and derive the transfer functions relating the output $x_m(s)$ to the input $x_{in}(s)$. Use the definitions of natural frequency $\omega_n^2 = k/m$ and damping ratio $\zeta = b/2m\omega_n$ and express the transfer functions in terms of ω_n and ζ.

(c) Derive the complex frequency response function for each system and express the magnitude of the frequency response. Sketch the magnitude response versus forcing frequency for a range of damping ratio ς. Comment on similarities and differences.

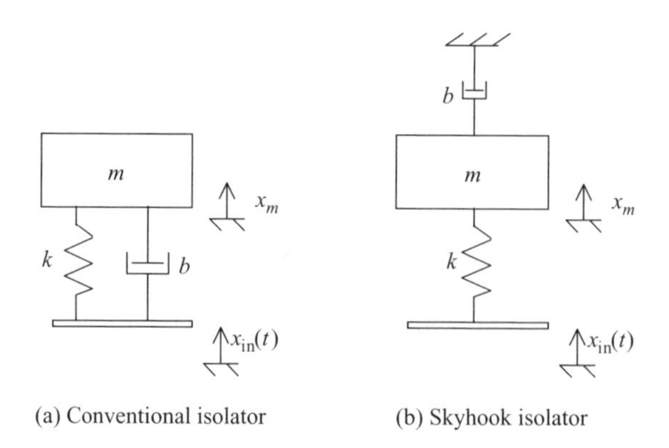

(a) Conventional isolator (b) Skyhook isolator

Figure P8.5

8.6 Shown in Figure P8.6 is a base-excited single-degree-of-freedom system with the spring and damper in a mechanical series. A temporary variable x' is introduced to locate the connection between the spring and damper.

(a) Derive the equations of motion for this system. A good approach is to leave x' in the formulation and start with the first-order form by using the definition $\dot{x}_m = v_m$.

(b) Derive the transfer function relating $x_m(s)$ to the input $x_{\text{in}}(s)$. Use the definitions for natural frequency $\omega_n^2 = k/m$ and damping ratio $\zeta = k/2b\omega_n$ and express the transfer function in terms of ω_n and ζ. Derive the complex frequency response function.

(c) Sketch the magnitude of the frequency response versus forcing frequency for a range of damping ratio ζ. Do you find it interesting how the damping constant affects the damping ratio for this system configuration?

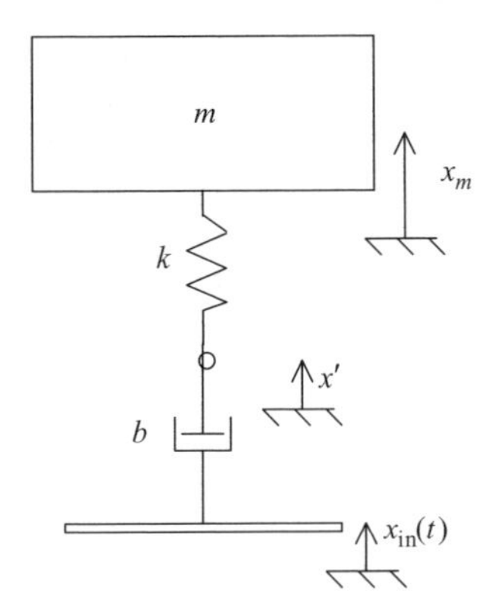

Figure P8.6

8.7 A system consists of two identical pendulums connected at their pivots by a torsional spring (Figure P8.7). The spring generates a torque or moment proportional to its twist angle. The coordinates are the angles θ_1 and θ_2 of the pendulums.

(a) Derive the equations of motion for small angular rotations of the pendulums. Put the equations in the matrix form of Eq. (8.31).

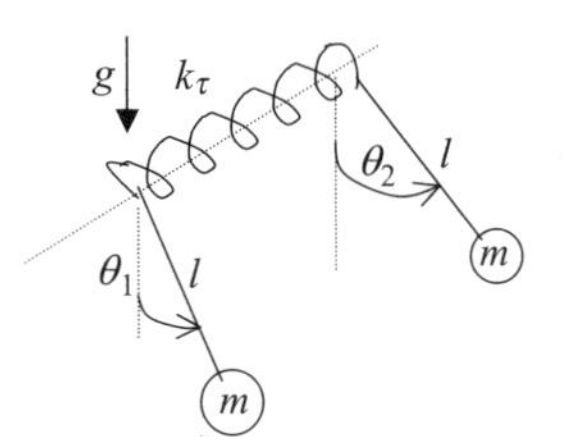

Figure P8.7

(b) Assume that the solution of the free response is $\theta_1 = \Theta_1 e^{j\omega_n t}$ and $\theta_2 = \Theta_2 e^{j\omega_n t}$ and represent the result as an eigenvalue problem of the form of Eq. (8.37). Derive the frequency equation. Solve for the natural frequencies ω_{n1}^2 and ω_{n2}^2.

(c) Solve for the modal vectors, letting the first entry in each vector equal unity. Sketch the mode vectors and interpret their meaning.

8.8 The system shown in Figure P8.8 is the two-degree-of-freedom, half-car model used as an example in the chapter 8 text. In the chapter example, in Figure P8.8 the coordinates used were x_g and θ, and the analysis was performed computationally. This time use the coordinates x_1 and x_2 and perform the analysis analytically.

(a) Derive the equations of motion for this system restricted to small motions using the coordinates x_1 and x_2, and put into second-order matrix form similar to Eq. (8.31).

(b) Assume that the free response is governed by $x_1 = X_1 e^{j\omega_n t}$ and $x_2 = X_2 e^{j\omega_n t}$, and represent the result as an eigenvalue problem of the form of Eq. (8.37).

(c) Derive the frequency equation and solve for the natural frequencies and modal vectors. Interpret the meaning of the mode vectors. To make the manipulations easier, let $k_2/k_1 = 2$ and $I_g = ml^2/12$.

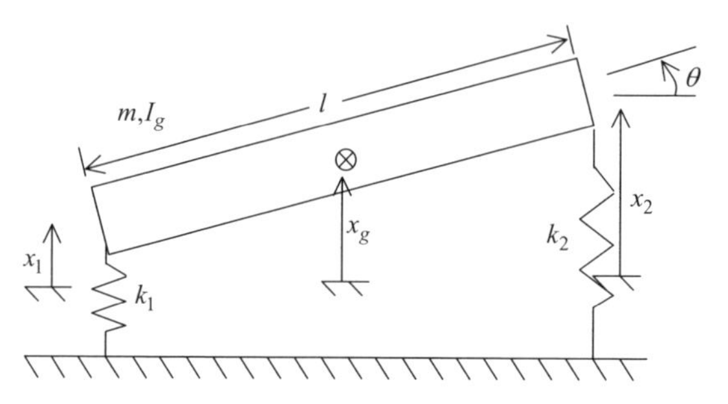

Figure P8.8

8.9 Two equal mass elements are attached to the ends of a massless rod (Figure P8.9). A spring is located at the center of the rod. This is a two-degree-of-freedom system with coordinates x_1 and x_2, which locate the respective mass elements with respect to an inertial frame. When the coordinates are zero, there is no force in the spring.

(a) Derive the equations of motion for this system restricted to small motions and put into the second-order matrix form of Eq. (8.31).

(b) Assume that the free response is governed by $x_1 = X_1 e^{j\omega_n t}$ and $x_2 = X_2 e^{j\omega_n t}$ and represent the result as an eigenvalue problem of the form of Eq. (8.37).

(c) Derive the frequency equation and solve for the natural frequencies and modal vectors. Interpret the meaning of the mode vectors.

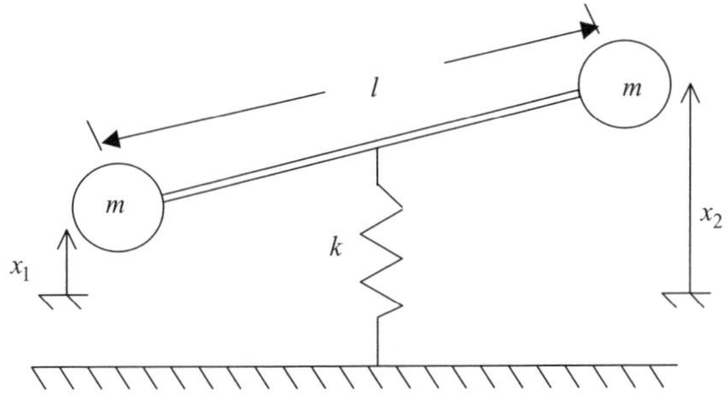

Figure P8.9

8.10 This is the same system as Problem 8.9, part this time a torsional spring of stiffness K_τ has been added at the center of the connecting rod (Figure P8.10). This torsional spring generates an opposing torque or moment proportional to the twist of the spring. The coordinates are still x_1 and x_2, which locate the respective mass elements. If it is helpful, you can use $k_\tau = 2kl^2$.

(a) Derive the equations of motion for this system restricted to small motions and put into the second-order matrix form of Eq. (8.31).

(b) Assume that the free response is governed by $x_1 = X_1 e^{j\omega_n t}$ and $x_2 = X_2 e^{j\omega_n t}$ and represent the result as an eigenvalue problem of the form of Eq. (8.37).

(c) Derive the frequency equation and solve for the natural frequencies and modal vectors. Interpret the meaning of the mode vectors.

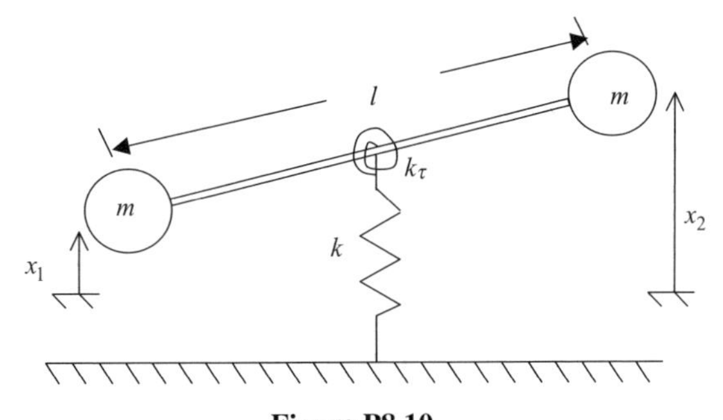

Figure P8.10

8.11 This is the same problem as Problem 8.10 except that the coordinates have been changed to x_g, which locates the center of the connecting rod and θ (Figure P8.11). If it is helpful, you can use $k_\tau = 2kl^2$.

(a) Derive the equations of motion for this system restricted to small motions and put into the second-order matrix form of Eq. (8.31).

(b) Assume that the free response is governed by $x_g = X_g e^{j\omega_n t}$ and $\theta = \Theta e^{j\omega_n t}$ and represent the result as an eigenvalue problem of the form of Eq. (8.37).

(c) Derive the frequency equation and solve for the natural frequencies and modal vectors. Interpret the meaning of the mode vectors.

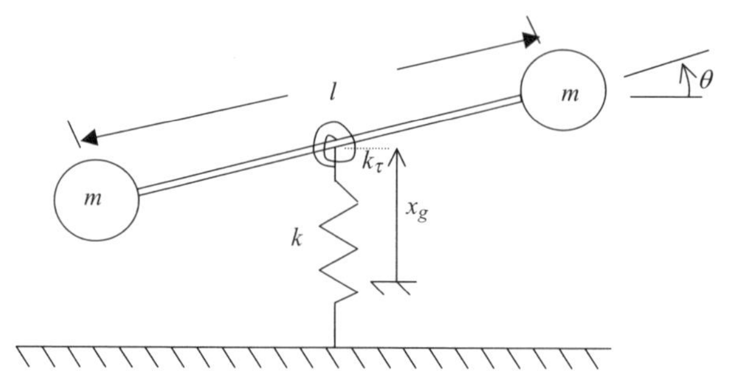

Figure P8.11

8.12 This problem is similar to Problem 8.11, but this time the spring k and torsional spring k_τ have been moved to the left end (Figure P8.12). The coordinates are x, which locates the left end from some equilibrium position, and the rotational angle θ. If it is helpful, you can use $k_\tau = kl^2$.

(a) Derive the equations of motion for this system restricted to small motions and put into the second-order matrix form of Eq. (8.31).

(b) Assume that the free response is governed by $x = Xe^{j\omega_n t}$ and $\theta = \Theta e^{j\omega_n t}$ and represent the result as an eigenvalue problem of the form of Eq. (8.37).

(c) Derive the frequency equation and solve for the natural frequencies and modal vectors. Interpret the meaning of the mode vectors.

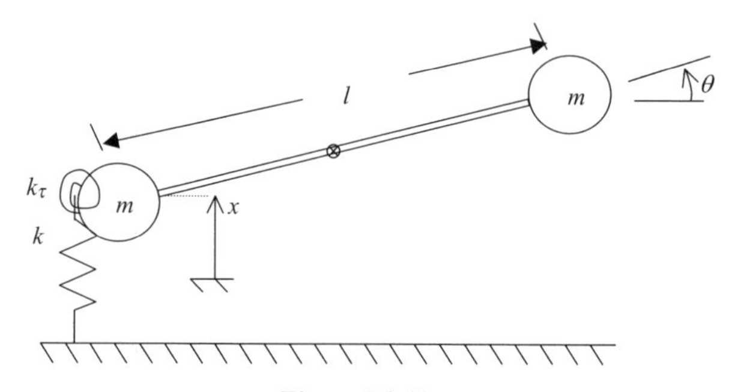

Figure P8.12

8.13 A rigid body of mass m and c.g. moment of inertia I_g is suspended at its left end by a vertical spring k and a torsional spring k_τ (Figure P8.13). The coordinate x locates the left end from some equilibrium position, and the coordinate θ indicates the angular rotation of the body. If it is helpful, you can use $k_\tau = kl^2$ and $I_g = ml^2/12$.

(a) Derive the equations of motion for this system restricted to small motions and put into the second-order matrix form of Eq. (8.31).

(b) Assume that the free response is governed by $x = Xe^{j\omega_n t}$ and $\theta = \Theta e^{j\omega_n t}$ and represent the result as an eigenvalue problem of the form of Eq. (8.37).

(c) Derive the frequency equation and solve for the natural frequencies and modal vectors. Interpret the meaning of the mode vectors.

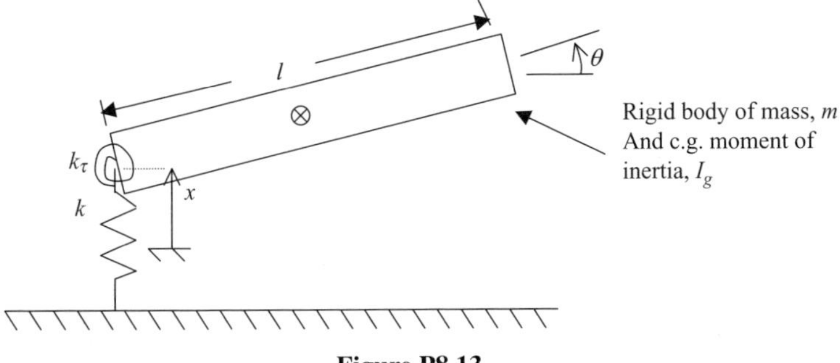

Figure P8.13

8.14 This two-degree-of-freedom system consists of three equal springs and two equal masses with the right-hand mass being forced by the input $F(t)$ (Figure P8.14).

(a) Derive the equations of motion and put them in the second-order matrix form of Eq. (8.30).

(b) Derive the transfer functions relating $x_1(s)$ and $x_2(s)$ to the input $F(s)$. If the damping is set to zero, derive expressions for the natural frequencies of the system.

(c) Reinstall the damping, derive the complex frequency, response functions, and determine the magnitude of the frequency responses for the two outputs. Is there a frequency where either frequency response has zero response?

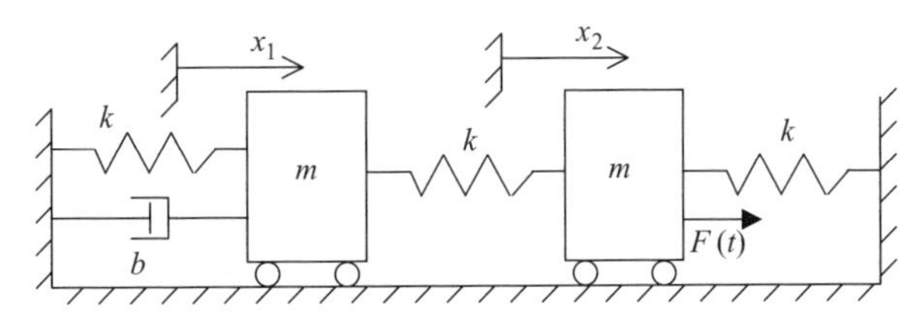

Figure P8.14

8.15 This is the same system as Problem 8.14, except that this time the left end of the system has a prescribed displacement input $x_{in}(t)$ (Figure P8.15).

(a) Derive the equations of motion and put them in the second-order matrix form of Eq. (8.30).

(b) Derive the transfer functions relating $x_1(s)$ and $x_2(s)$ to the input $x_{in}(s)$. If the damping is set to zero, derive expressions for the natural frequencies of the system.

(c) Reinstall the damping, derive the complex frequency response functions, and determine the magnitude of the frequency responses for the two outputs. Is there a frequency where either frequency response has zero response?

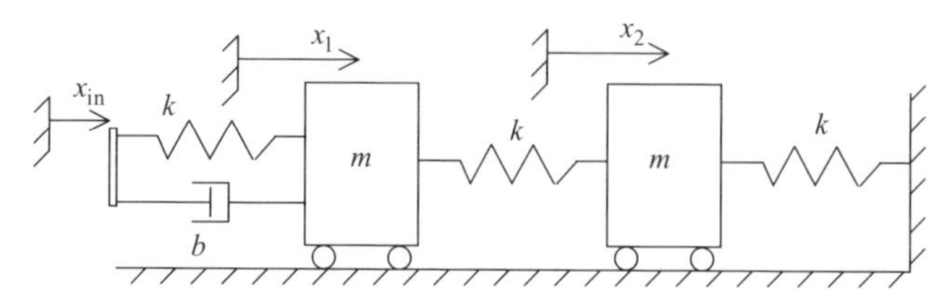

Figure P8.15

8.16 A structure m is acted upon by a force $F(t)$. (Figure P8.16). Attached to the structure are two spring–mass systems, each with different mass and stiffness. This is a three-degree-of-freedom system with a coordinate needed to locate each of the three mass elements. The system is constrained to move only vertically. We know from the section on TVAs that the addition of one spring mass device is a TVA that drives the response x_m to zero at a special input frequency. What happens when two TVA devices are attached to the same structure?

(a) Derive the equations of motion for this system and express in the second-order matrix form of Eq. (8.30).

(b) Derive the transfer function relating the structure mass motion $x_m(s)$ to the input $F(s)$. Of particular interest is the numerator of the transfer function.

(c) Let $s = j\omega$ and determine if there is a frequency or frequencies where the structure mass response goes to zero. If there is, derive expressions for these frequencies.

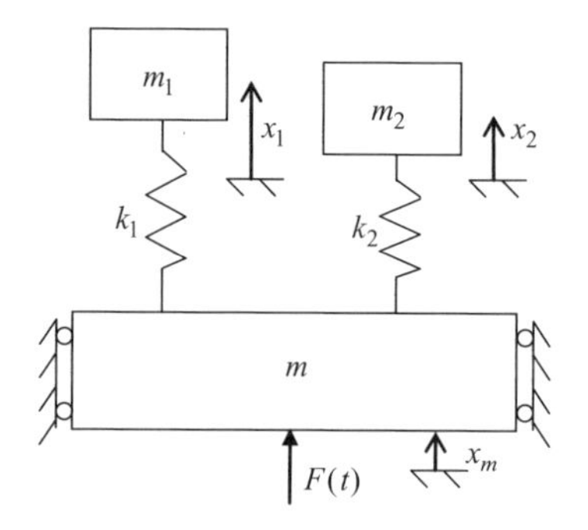

Figure P8.16

8.17 This system is similar to that shown in Problem 8.16. The two single-TVA systems have been replaced by a rigid body sitting atop a structure mass m. (Figure P8.17). We wonder if this two-degree-of-freedom system will provide two adjustable frequencies where the structure mass is motionless under harmonic forcing from the input force $F(t)$.

(a) Derive the equations of motion for this system restricted to small motions and express in the second-order matrix form of Eq. (8.30).

(b) Derive the transfer function relating the structure mass motion $x_m(s)$ to the input $F(s)$. Of particular interest is the numerator of the transfer function.

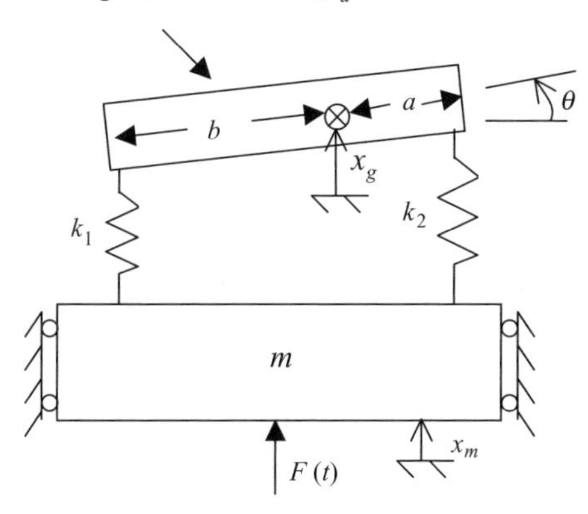

Figure P8.17

(c) Let $s = j\omega$ and determine if there is a frequency or frequencies where the structure mass response goes to zero. If there is, derive expressions for these frequencies.

8.18 A cart of mass m_c is acted upon by an external force $F(t)$ and has an attached pendulum of length l and mass m_p (Figure P8.18). This is a two-degree-of-freedom system and the coordinates are the position of the cart x and the angular position of the pendulum θ.

(a) Derive the equations of motion for this system restricted to small motions and put into the second-order matrix form of Eq. (8.30).

(b) Let the external forcing be zero and solve the eigenvalue problem for the natural frequencies and the associated mode shapes. Interpret the mode shapes using sketches and words.

(c) Reinstall the forcing and derive the transfer function between $x(s)$ and $F(s)$. Derive the complex frequency response function. Derive the magnitude of the frequency response and sketch the magnitude versus forcing frequency. Does the pendulum behave as a TVA?

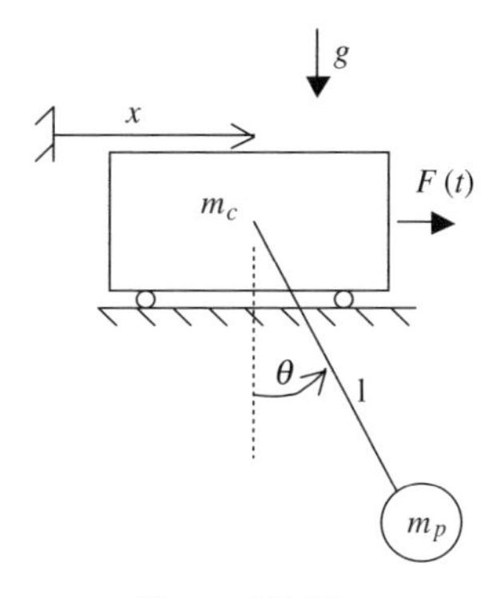

Figure P8.18

9

DISTRIBUTED SYSTEM VIBRATIONS

In Chapter 8, the dynamic principles were applied to systems that exhibit oscillatory motion. When compliance-like elements interact with inertial elements, potential resonances can occur whenever external excitation is supplied. The study of vibrations is generally concerned with understanding and calming vibration problems through passive isolation such as rubber pads with damping, dynamic isolation such as TVAs, and through the use of automatic control as described in Chapter 6. The vibrational systems described in Chapter 8 are composed of individual elements representing springs, dampers, masses, inertias, and so on. When such elements are configured into a system, it is called a *lumped representation* of the system.

In reality, springs have mass, rigid bodies can deflect, and mass elements can deform. Sometimes it is important to recognize these effects in system models in order to portray reality more accurately. When this is done, the mass and stiffness effects become distributed over the system and are no longer individual elements at distinct locations. When Newton's laws are applied to such systems, the resulting equations of motion are partial differential equations rather than the total differential equations we saw in earlier chapters. We will see that the vibration concepts of natural frequency, mode shapes, resonance, and frequency response described in Chapter 8 apply to distributed systems. Furthermore, it is shown that the vibration calming concepts developed in Chapter 8 remain applicable to distributed systems.

9.1 STRESS WAVES IN A ROD

Figure 9.1 shows a circular rod of mass density ρ, length L, and cross-sectional area A, constructed from a material of Young's modulus E with an input force $F(t)$ acting at its left end. How does this device behave dynamically? For some applications it

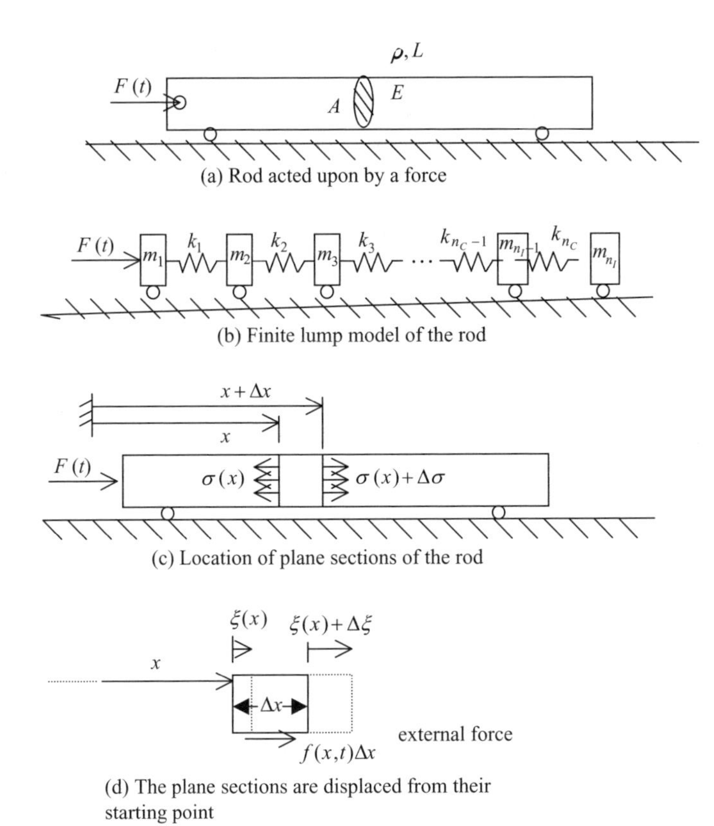

Figure 9.1 Rod with an external force at its end.

may be all right to assume that the rod is rigid and treat it as a point mass that simply accelerates horizontally under action of the external force. In reality, when force is applied at the left end, the right end is unaware that anything has happened before a wave of energy travels the length of the rod and arrives at the right end. At this time the right end begins to move. The wave then reflects off the right end and travels back toward the left all the way, affecting the motion of each cross-sectional plane of the rod. The rod may appear to move as a rigid body under the action of the force, but the reality is that there is much vibration of the rod as it accelerates. If the vibration of the rod is unimportant, the rod can be treated as a rigid body. Otherwise, an alternative description must be developed.

The determining factor in whether distributed effects must be considered is typically the frequency range of the input. As mentioned above, the physical reality is that the wave travels along the rod after application of the input force. For low-frequency input, the wave will travel back and forth many times over one period of input. Under such conditions, the difference between the motion at the left and right ends will be unnoticed, and the rod can be treated as a rigid body. As the frequency of the input increases, there may be many cycles of input before the first wave reaches the right end. Under these conditions, the rod must be treated as a distributed system.

There exists a commercial product consisting of a long slender rod that is purposely excited at a resonant frequency. This device is used to drill through hard rock. An engineer who did not understand distributed system vibrations could have had no input into the design of such a device.

It is always possible to think of a distributed system as being composed of many lumps of conventional elements. Figure 9.1*b* shows a rod reticulated into a finite lump model. The mass of the rod is

$$m = \rho A L \tag{9.1}$$

so if n_I lumps of mass are included in the model, a reasonable choice for each mass element might be

$$m_i = \frac{\rho A L}{n_I} \tag{9.2}$$

If the rod was built in at the left end and a force was applied at the right end, the rod would stretch a bit. We know that the uniform stress σ in the rod is related to the strain by

$$\sigma = E\frac{\Delta L}{L} \tag{9.3}$$

and that the force at the right end would be σA. Thus, at the right end of our imaginary built-in rod,

$$F = \frac{EA}{L}\Delta L \tag{9.4}$$

or the stiffness k of the rod is

$$k = \frac{EA}{L} \tag{9.5}$$

The compliance of the rod is the inverse of the stiffness. If we use n_c compliance elements in the finite lump model, a reasonable choice for each compliance element is

$$C_i = \frac{1/k}{n_c} \tag{9.6}$$

or the stiffness of each spring element would be

$$k_i = n_c k \tag{9.7}$$

In Figure 9.1*b*, if a force is applied at the left end, the first mass will move and compress the first spring, which will move the second mass, and so on. A "wave" will move to the right and the actual physics of the distributed system will be approximated. The problem with this approach to approximating distributed system dynamics is that it requires a large number of lumps to attain reasonable accuracy in the predictions. In Chapter 8 it was shown that as the number of degrees of freedom increases, so does the number of system natural frequencies. It turns out that as more and more lumps are used in the finite lump approach of Figure 9.1*b*, the lower natural frequencies become more accurate but the additional frequencies from the increased degrees of freedom are very inaccurate. It is true that as the number of lumps approaches infinity the system converges to the actual distributed system, but the number of lumps required to approximate the real system becomes prohibitive and we must resort to another approach.

In Figure 9.1*c* a plane section of the rod is located by the position x, and another plane section is identified at the position just to the right of the first section at $x + \Delta x$. A stress wave has just passed that section and the stress on the right-side plane is indicated as slightly different from that on the left. When dealing with small changes in quantities over small distances, we use the Taylor series expansion and retain only the first-order terms. For the stress variable we say that

$$\sigma(x + \Delta x) = \sigma(x) + \frac{\partial \sigma}{\partial x}\Delta x + \frac{\partial^2 \sigma}{\partial x^2}\frac{(\Delta x)^2}{2} + \cdots \tag{9.8}$$

and retain only up to the first-order term; thus,

$$\sigma(x + \Delta x) = \sigma(x) + \frac{\partial \sigma}{\partial x}\Delta x \tag{9.9}$$

and

$$\Delta\sigma = \frac{\partial \sigma}{\partial x}\Delta x \tag{9.10}$$

The argument for this is that we will ultimately let the distances between plane sections shrink to zero, $\Delta x \to 0$, and the higher order terms will become vanishingly small.

In Figure 9.1*d* the plane sections are shown slightly displaced from their original positions. The left-side plane has moved to the right a distance ξ and the right-side plane section has moved a distance $\xi + \Delta\xi$. For uniform stress across the plane, the stress–strain relationship from Eq. (9.3) for this very small section would be

$$\sigma = E\frac{\Delta\xi}{\Delta x} = E\frac{(\partial\xi/\partial x)\Delta x}{\Delta x} = E\frac{\partial \xi}{\partial x} \tag{9.11}$$

It is typical when dealing with distributed systems to characterize any external forcing as a force per unit length acting over the length of the system. This is shown as $f(x,t)$

in Figure 9.1*d*. For our example system there is only external forcing acting at the left end of the rod. It will be shown below how to handle "point" forces as distributed forces per unit length.

We can now write the equations of motion for this differential element. Newton's law states that

$$\left(\sigma + \frac{\partial \sigma}{\partial x}\Delta x\right) A - \sigma A + f(x,t)\,\Delta x = \rho A\,\Delta x \frac{\partial^2 \xi}{\partial t^2} \tag{9.12}$$

where the second-order partial derivative on the right-hand side is the acceleration of the plane section of width Δx. Substituting (9.11) in (9.12) yields

$$EA\frac{\partial^2 \xi}{\partial x^2} + f(x,t) = \rho A \frac{\partial^2 \xi}{\partial t^2} \tag{9.13}$$

When the external forcing is set zero, to Eq. (9.13), called the *simple wave equation*, describes the distributed dynamics of quite a few engineering systems. For lumped systems the mass and stiffness elements are discretely located and their displacement is dependent only on time t. When the mass and stiffness are distributed, the displacement has to be identified by both the position and the time; thus, we say that ξ is a function of both x and t or $\xi = \xi(x,t)$. This spatial dependence of the variable ξ requires the additional complication of specifying boundary conditions, a step that is not needed for lumped system models.

For our example system in Figure 9.1*a*, the external force is purposely shown as acting just to the right of the left end. Thus, at both the left and right ends the system is force- or stress-free. From Eq. (9.11) the boundary conditions can be written as

$$\frac{\partial \xi(0,t)}{\partial x} = \frac{\partial \xi(L,t)}{\partial x} = 0 \tag{9.14}$$

It must be admitted that deriving the equations of motion for distributed systems is an advanced concept and can be challenging. Fortunately, these equations are derived for us for many different systems and can be found documented in such books as ref. [1]. The inclusion of distributed systems in engineering systems is the subject of this book, so let's continue toward a solution that we can use.

Free Response: Separation of Variables

Returning to Eq. (9.13), we set the forcing zero to temporarily and assume that the solution to the unforced equation can be separated into the product of a function of x times a function of t. We ask if

$$\xi(x,t) = Y(x)g(t) \tag{9.15}$$

is a possible solution to the unforced equation. (9.13). This attempt at a solution is cleverly named *separation of variables*. Using (9.15) in the unforced (9.13) results in

$$E\frac{d^2Y}{dx^2}g = \rho Y\frac{d^2g}{dt^2} \tag{9.16}$$

The next step is to divide both sides by ρYg, with the result

$$\frac{E}{\rho}\frac{1}{Y}\frac{d^2Y}{dx^2} = \frac{1}{g}\frac{d^2g}{dt^2} \tag{9.17}$$

Notice now that regardless of what $Y(x)$ turns out to be, the left side of (9.17) is a function of x only, and similarly the right side is a function of t only. It is unlikely that we can find a $Y(x)$ and a $g(t)$ that will produce equal left and right sides of (9.17) for all x between 0 and L and for all time t. Thus, we must conclude that separation of variables has led us nowhere, or notice that if each side of (9.17) equals the same constant, we can continue. It is customary to let the right-hand side of (9.17) equal the constant $-\omega^2$, resulting in the two total differential equations,

$$\begin{aligned}\frac{d^2g}{dt^2} + \omega^2 g &= 0\\ \frac{d^2Y}{dx^2} + \beta^2 Y &= 0\end{aligned} \tag{9.18}$$

where $\beta^2 = \rho/E\,\omega^2$. At this point we do not know the specific values of ω^2 but we do know that the first of (9.19) is the simple harmonic equation and has the solution of sine and cosine of ωt. For the unforced response, we are more interested in the second of Eq. (9.18). It also has the form of the simple harmonic equation but has dependence on space rather than time. Its general solution is

$$Y(x) = A\cos\beta x + B\sin\beta x \tag{9.19}$$

To solve the spatial equation we must make use of the boundary conditions from Eq. (9.14). Using (9.15), it can be stated that

$$\begin{aligned}\frac{\partial\xi}{\partial x}(0,t) &= \frac{dY(0)}{dx}g(t) = 0\\ \frac{\partial\xi}{\partial x}(L,t) &= \frac{dY(L)}{dx}g(t) = 0\end{aligned} \tag{9.20}$$

Since $g(t)$ can have any value, the boundary conditions reduce to the requirement that

$$\frac{dY(0)}{dx} = \frac{dY(L)}{dx} = 0 \tag{9.21}$$

Using (9.19) and carrying out the operations from (9.21) results in

$$\begin{aligned} 0 &= -A\beta \sin 0 + B\beta \cos 0 = B\beta \\ 0 &= -A\beta \sin \beta L + B\beta \cos \beta L \end{aligned} \tag{9.22}$$

For now we make the restriction that $\beta \neq 0$. For this case, $B = 0$ and the second of (9.23) requires that either $A = 0$ or

$$\sin \beta L = 0 \tag{9.23}$$

If $A = 0$, then $Y = 0$ and then $\xi = 0$, which means that one solution to the unforced system is that it not move. A more interesting solution comes from Eq. (9.23), which indicates that there are an infinite number of solutions such that $\sin \beta_n L = 0$; therefore,

$$\beta_n L = n\pi \tag{9.24}$$

or

$$\beta_n = \frac{n\pi}{L} \tag{9.25}$$

Equation (9.23), called the *frequency equation*, is the distributed system counterpart to the frequency equation from Chapter 8 for multi-degree-of-freedom systems. The solution yields the natural frequencies from Eqs. (9.18):

$$\omega_n^2 = \frac{E}{\rho} \left(\frac{n\pi}{L} \right)^2 \tag{9.26}$$

Associated with each natural frequency is a mode shape that is the counterpart of an eigenvector for lumped systems. For distributed systems it is called an *eigenfunction* and comes from Eq. (9.19), where we are still using $B = 0$; thus,

$$Y_n(x) = A_n \cos \beta_n x = A_n \cos n\pi \frac{x}{L} \tag{9.27}$$

The constants A_n are arbitrary and are set to unity for the presentation here. This is similar to what was done in Chapter 8, where the first entry of an eigenvector was set to unity and the rest of the vector was scaled appropriately. In a manner analogous to lumped systems, we say that for each natural frequency from Eq. (9.26), there is a corresponding mode shape given by Eq. (9.27). We could combine the mode shapes with their corresponding time functions from the solution of the first of (9.18) and show that the free response of this distributed system is a linear combination of mode shapes oscillating at their respective natural frequencies. We could show further that if it was possible to initialize the displacements in the rod according to a mode shape configuration and then release the rod, the rod would remain in that modal configuration and oscillate at its respective natural frequency.

Figure 9.2 shows the first few modes of an unforced rod. The interpretation is, for the first mode, for example, that if the left end of the rod were displaced one unit to the right and the right end of the rod one unit to the left and the rest of the rod was displaced according to the half-cosine shape and then released in this configuration, the rod would oscillate at ω_1 from Eq. (9.26) and remain in this modal configuration. The interpretation is similar for all the mode shapes. The extreme positions of the oscillation are shown by the solid and dashed lines in Figure 9.2.

We are far more interested in the forced response of distributed systems, so we pursue that next.

Forced Response

The expansion theorem [1] states basically that any instantaneous displacement state of a distributed system can be reconstructed by a linear combination of the mode shape functions. This is true even if the instantaneous displacement state was acquired due to external forcing. This interesting result indicates how important the unforced eigenfunctions are for obtaining the forced response, and it allows for the forced response to be expressed as

$$\xi(x, t) = \sum_n Y_n(x)\eta_n(t) \tag{9.28}$$

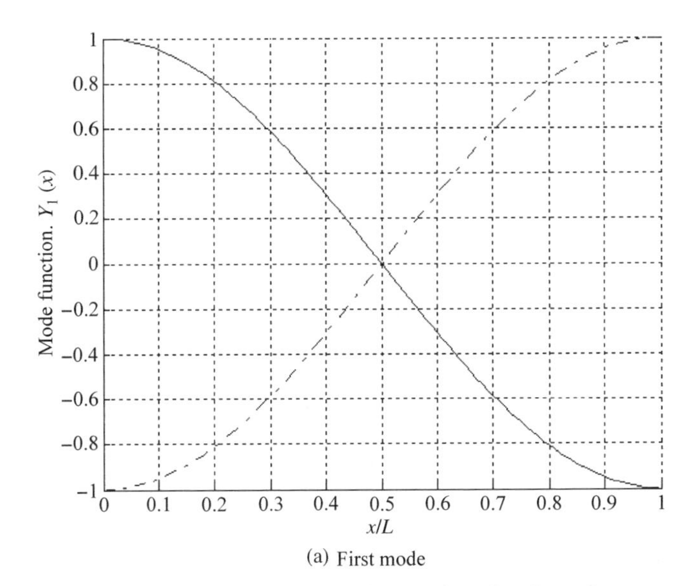

(a) First mode

Figure 9.2 Mode shapes of rod unforced at the end.

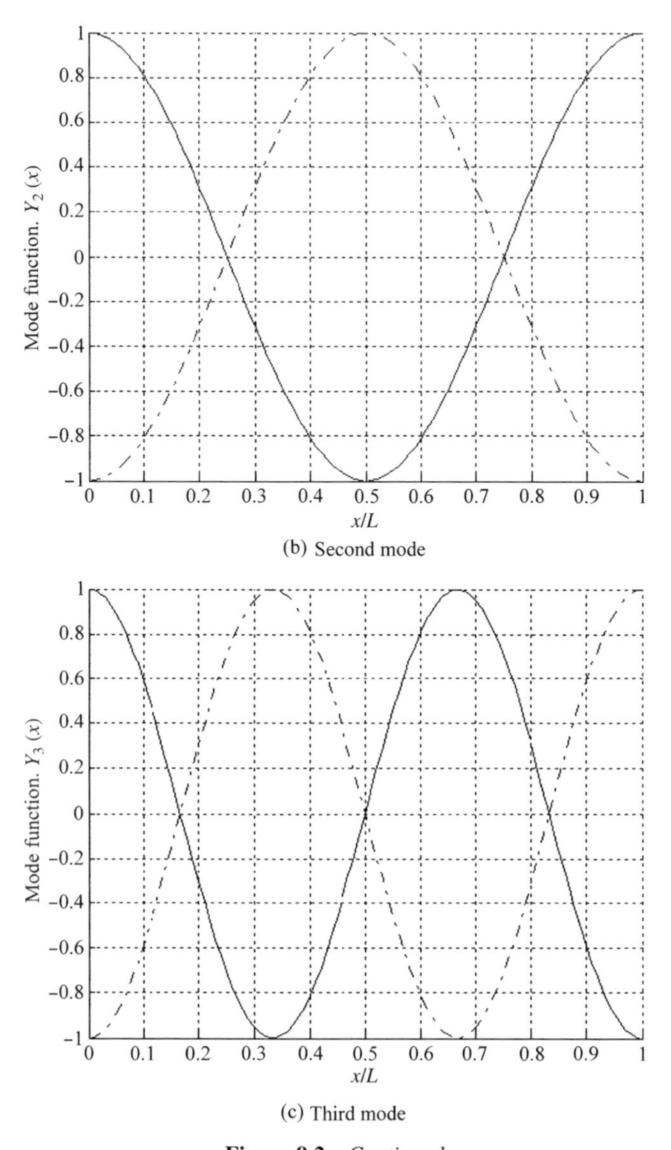

(b) Second mode

(c) Third mode

Figure 9.2 *Continued*

where the $\eta_n(t)$ are unknown time functions, not to be confused with the time functions used previously for the unforced response. At any instant of time, the $\eta_n(t)$ are just a collection of constants and (9.28) simply states that the instantaneous displacement at any position along the rod is composed of a summation of constants times the mode shapes evaluated at the position of interest. The summation would officially be over the infinity of modes. In practice, the summation is over a finite number of modes.

Orthogonality of Mode Functions

Before developing the forced response of the rod, it is important to recognize a special property of the eigenfunctions or mode shapes that were obtained for the unforced system. The property, called *orthogonality*, is expressed as

$$\int_0^L Y_n(x)Y_m(x)dx = 0 \qquad \text{if } n \neq m \tag{9.29}$$

For the rod example,

$$\begin{aligned} Y_n(x) &= \cos n\pi \frac{x}{L} \\ Y_m(x) &= \cos m\pi \frac{x}{L} \end{aligned} \tag{9.30}$$

and the orthogonality property becomes

$$\int_0^L \cos n\pi \frac{x}{L} \cos m\pi \frac{x}{L} dx = 0 \qquad \text{if } n \neq m \tag{9.31}$$

The interested reader should carry out this integration or look it up in a handbook and determine that it is in fact true. It can be shown [1] that in general, subject to some conditions, mode shape functions obey this orthogonality property for distributed vibrational systems. We make use of orthogonality to develop the forced response.

Representation of Point Forces

When developing the equations of motion it was stated that external forcing is typically represented as a distributed force per unit length such that the total force on the system is calculated through integration. For the rod we used $f(x, t)$ as the external force per unit length to derive the forced equation of motion (9.13). The total time-varying force on the rod, $F_T(t)$, would then be

$$F_T(t) = \int_0^L f(x,t)\,dx \tag{9.32}$$

We are interested in *point forces*, forces that are time varying but act at only one point on the distributed system. Thus, we need to represent a point force as a force per unit length in order to pursue the forced response.

The *delta function* $\delta(x)$ is perfect for representing point forces in distributed systems. The delta function is defined such that it is extremely tall and extremely thin, exists only when its argument is zero, and has a finite area of unity. Symbolically, we say that

$$\begin{aligned}
&\delta(x - x_0) = 0, && x \neq x_0 \\
&\delta(x - x_0) = \infty, && x = x_0 \\
&\int_0^L \delta(x - x_0)dx = 1
\end{aligned} \tag{9.33}$$

The integral is nondimensional and dx has the dimension of length. Therefore, the delta function actually has the dimension of inverse length or 1 divided by length. Now imagine that the rod of Figure 9.1 has external point forces $F_1(t)$ acting at the location x_1 and $F_2(t)$ acting at the location x_2. Consider representing the force per unit length as

$$f(x, t) = F_1(t)\delta(x - x_1) + F_2(t)\delta(x - x_2) \tag{9.34}$$

With this representation the external forces exist only at the locations $x = x_1$ and $x = x_2$, and from Eq. (9.32) the total force acting is

$$F_T(t) = \int_0^L F_1(t)\delta(x - x_1)\,dx + \int_0^L F_2(t)\delta(x - x_2)\,dx$$

or

$$F_T(t) = F_1(t) + F_2(t) \tag{9.35}$$

Thus, Eq. (9.34) allows point forces to be represented as forces per unit length, and Eq. (9.35) shows that the total force is properly accounted for through integration. This meets the criteria for using the delta function to represent point forces in distributed systems. For the rod example there is only one external force and it is acting just inside the left boundary. The force per unit length for the example system is then

$$f(x, t) = F(t)\delta(x - \varepsilon) \tag{9.36}$$

where $\varepsilon = 0$ for all practical purposes.

With this point force representation the equation of motion from (9.13) becomes

$$\rho A \frac{\partial^2 \xi}{\partial t^2} - EA \frac{\partial^2 \xi}{\partial x^2} = F(t)\delta(x) \tag{9.37}$$

Substituting Eq. (9.28) as a “guess” for the solution yields

$$\sum_n \rho A Y_n \ddot{\eta}_n - \sum_n EA \eta_n \frac{d^2 Y_n}{dx^2} = F(t)\delta(x) \tag{9.38}$$

From the free response development Eq. (9.18) we know that

$$\frac{d^2 Y_n}{dx^2} = -\beta_n^2 Y_n = -\frac{\rho}{E}\omega_n^2 Y_n \tag{9.39}$$

Substituting into (9.38) results in

$$\sum_n \rho A Y_n \ddot{\eta}_n + \sum_n \rho A \omega_n^2 \eta_n Y_n = F(t)\delta(x) \tag{9.40}$$

The next step is to multiply through by $Y_m(x)$ and integrate from 0 to L. Thus

$$\sum_n \rho A \ddot{\eta}_n \int_0^L Y_n Y_m \, dx + \sum_n \rho A \omega_n^2 \eta_n \int_0^L Y_n Y_m \, dx = F(t) \int_0^L Y_m \delta(x) \, dx \tag{9.41}$$

where some liberties have been taken in interchanging the operations of summation and integration. We now make use of the orthogonality property of the mode functions and use Eq. (9.29). As the summation over n is carried out, the integral terms on the left side of (9.41) are all zero until $n = m$, at which point only one equation remains:

$$m_m \ddot{\eta}_m + k_m \eta_m = F(t) \int_0^L Y_m \delta(x) \, dx \tag{9.42}$$

where

$$m_m = \int_0^L \rho A Y_m^2 \, dx \quad \text{and} \quad k_m = m_m \omega_m^2 \tag{9.43}$$

The first of Eqs. (9.43) is called the *modal mass* and the second is called the *modal stiffness*. Calculation of the modal mass requires knowledge of the mode shape functions and may require numerical integration to actually calculate values for each of the modes included. The modal stiffness requires knowledge of the mode frequencies.

The right-hand side of (9.42) simplifies nicely. The argument is that when an integrand is composed of any continuous function such as a mode shape, and this continuous function is multiplying a delta function, over the spatial duration of the delta function the continuous function is virtually constant and equal to its value evaluated at the location of the external force: at $x = 0$ for this example. After pulling the virtually constant $Y_m(0)$ from the integral, the remaining integral is equal to unity by definition of the delta function. Thus, Eq. (9.42) reduces to its final form:

$$m_m \ddot{\eta}_m + k_m \eta_m = F(t) Y_m(0) \tag{9.44}$$

This result shows that the apparently very complicated task of obtaining the forced response of the distributed rod is reduced to solving the single Eq. (9.44) for as many time functions as are deemed necessary, $\eta_1(t)$, $\eta_2(t)$, $\eta_3(t)$, ..., and then reconstructing the forced response from Eq. (9.28) as

$$\xi(x,t) = Y_1\eta_1 + Y_2\eta_2 + Y_3\eta_3 + \cdots \tag{9.45}$$

For the rod example, where the mode shape functions are cosine functions, the modal mass and stiffnesses can be assessed analytically. From (9.43) we need the integral of the mode shape function squared. For the rod example,

$$m_m = \rho A \int_0^L \cos^2 m\pi \frac{x}{L}\, dx = \frac{\rho AL}{2} = \frac{m_{\text{rod}}}{2} \tag{9.46}$$

and the modal stiffnesses are

$$k_n = \frac{m_{\text{rod}}}{2}\omega_n^2 \tag{9.47}$$

Rigid-Body Mode

The rod example being demonstrated here has one more interesting aspect that requires attention before a solution can be finalized. You may recall when pursuing the free response that the possibility of $\beta = 0$ was not allowed in Eqs. (9.22). This means that the natural frequency $\omega_n = 0$ was not allowed at that point in the development. It turns out that the boundary conditions determine whether or not a *zero-frequency eigenvalue* is permitted. When the physical system is not attached to a physical boundary that constrains its motion, a zero-frequency eigenvalue will exist. Such is the case for the rod example. The zero-frequency case is treated separately from the dynamic modes and we return to Eq. (9.18) with $\beta = 0$ and write

$$\frac{d^2Y_0}{dx^2} = 0 \tag{9.48}$$

for the zero-frequency mode. The mode then becomes

$$Y_0(x) = C_1x + C_2 \tag{9.49}$$

and is subject to the same boundary conditions as those of all the modes. The boundary conditions come from Eq. (9.21), with the result

$$C_1 = 0$$

and therefore,

$$Y_0(x) = C_2 \tag{9.50}$$

The constant C_2 is arbitrary and is set to unity for this example. The zero-frequency mode is orthogonal in the same manner as all the other modes, meaning that

$$\int_0^L 1 \cos n\pi \frac{x}{L}\, dx = 0, \qquad n = 1, 2, 3, \ldots \tag{9.51}$$

and the modal mass for the zero frequency mode m_0 is

$$m_0 = \int_0^L \rho A (1)^2\, dx = \rho A L = m_{\text{rod}} \tag{9.52}$$

so the modal mass for the zero-frequency mode equals the mass of the rod.

The contribution of the zero-frequency mode to the forced response comes from solving

$$m_0 \ddot{\eta}_0 = F(t) \tag{9.53}$$

and the total forced response will have the general appearance

$$\xi(x, t) = Y_0 \eta_0 + Y_1 \eta_1 + Y_2 \eta_2 + Y_3 \eta_3 + \cdots \tag{9.54}$$

The astute reader will notice that Eq. (9.53) is nothing more than Newton's law applied to the rod as a rigid body. Thus, the distributed dynamic response is the rigid-body response plus the contribution from the modal dynamics.

Back to the Forced Response

Let's retain three dynamic modes plus the rigid-body mode and formulate the complete forced response of the rod. The mode frequencies and mode shapes come from Eqs. (9.26) and (9.27), the modal masses and stiffnesses come from Eqs. (9.46) and (9.47), and the equations to be solved come from Eqs. (9.44) and (9.53). For a three-dynamic-mode model we need

$$\begin{aligned} m_0 \ddot{\eta}_0 &= F(t) \\ m_1 \ddot{\eta}_1 + k_1 \eta_1 &= F(t) Y_1(0) \\ m_2 \ddot{\eta}_2 + k_2 \eta_2 &= F(t) Y_2(0) \\ m_3 \ddot{\eta}_3 + k_3 \eta_3 &= F(t) Y_3(0) \end{aligned} \tag{9.55}$$

and the response of the rod is expressed as

$$\xi(x, t) = (1)\eta_0(t) + Y_1(x)\eta_1(t) + Y_2(x)\eta_2(t) + Y_3(x)\eta_3(t) \tag{9.56}$$

For computational solution it is convenient to put the equations into first-order form; thus,

$$
\begin{aligned}
\dot{\eta}_0 &= v_0 \\
\dot{v}_0 &= \frac{F(t)}{m_0} \\
\dot{\eta}_1 &= v_1 \\
\dot{v}_1 &= \frac{F(t)}{m_1} Y_1(0) - \frac{k_1}{m_1} \eta_1 \\
\dot{\eta}_2 &= v_2 \\
\dot{v}_2 &= \frac{F(t)}{m_2} Y_2(0) - \frac{k_2}{m_2} \eta_2 \\
\dot{\eta}_3 &= v_3 \\
\dot{v}_3 &= \frac{F(t)}{m_3} Y_3(0) - \frac{k_3}{m_3} \eta_3
\end{aligned}
\tag{9.57}
$$

These equations were programmed using Matlab [2] and the parameters are shown in Table 9.1. Using the table we can calculate the natural frequencies from Eq. (9.26) as

$$
\begin{aligned}
\omega_1 &= \sqrt{\frac{E}{\rho}}\frac{\pi}{L} \\
\omega_2 &= \sqrt{\frac{E}{\rho}}\left(2\frac{\pi}{L}\right) \\
\omega_3 &= \sqrt{\frac{E}{\rho}}\left(3\frac{\pi}{L}\right)
\end{aligned}
\tag{9.58}
$$

TABLE 9.1. Parameters for the Rod Example

Density $\rho = 7850\,\text{kg/m}^3$
Length, 10.0 m
Diameter $D = 0.0425\,\text{m}$
Calculate area:
$A = \frac{\pi D^2}{4}\,\text{m}^2$
Young's modulus $E = 2.1 \times 11\,\text{N/m}^2$
Force at left end is suddenly applied with magnitude $F = 500\,\text{N}$

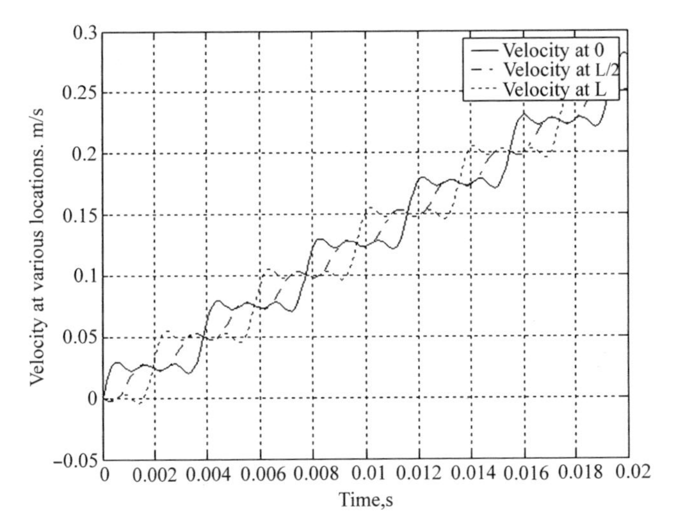

Figure 9.3 Velocity response of the rod for a slip input at the left end.

or the cyclic frequencies are

$$\begin{aligned} f_1 &= 259\,\text{Hz} \\ f_2 &= 517\,\text{Hz} \\ f_3 &= 776\,\text{Hz} \end{aligned} \tag{9.59}$$

Figure 9.3 shows the velocity response at the left end, middle, and right end of the rod for a step input force at the left end. It is shown for just a brief amount of time. The rigid-body response is obvious, as the general response exhibits an increasing velocity. The effect of the distributed dynamics can be seen as the various cross sections respond at different instants as the "wave" travels down the rod and reflects off the boundaries. In Figure 9.4 the rigid-body mode has been subtracted from the response so that the modal response is prominent. Keep in mind that this is only a three-mode model. If the number of modes is increased, the accuracy will improve.

A topic that is not addressed fully in this book is the question of how many modes to retain for a given dynamic system. Reference [3] covers this topic quite thoroughly. In an engineering application we usually have some idea of the frequency range of interest or the frequency content of the system inputs. The distributed system analysis presented here requires acquisition of the system natural frequencies. It is suggested that the number of modes retained exceed by a factor of 2 and perhaps 5 the highest frequency of interest for the application. This rule becomes modified somewhat as more subsystems are attached to the distributed system, which is typically the case in an engineering application.

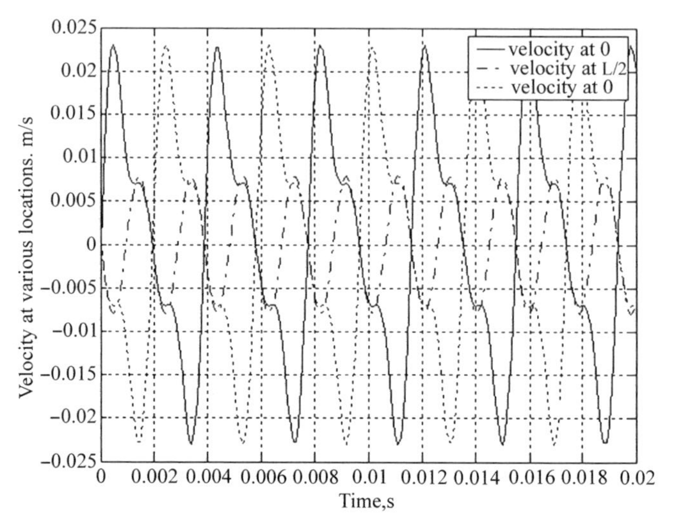

Figure 9.4 Velocity response of the rod with the rigid-body mode removed.

9.2 ATTACHING THE DISTRIBUTED SYSTEM TO EXTERNAL DYNAMIC COMPONENTS

Consider the distributed rod in a system where it is attached to additional external dynamics as shown in Figure 9.5. A cart of mass m_c rolls without friction on the surface of the rod and is attached to the rod through the spring and damper k_{s1}, and b_{d1} shown. The spring–damper is attached to the rod at the midpoint. The spring k_{s2} is attached to inertial ground at the right end. Figure 9.5*b* is a force diagram for the

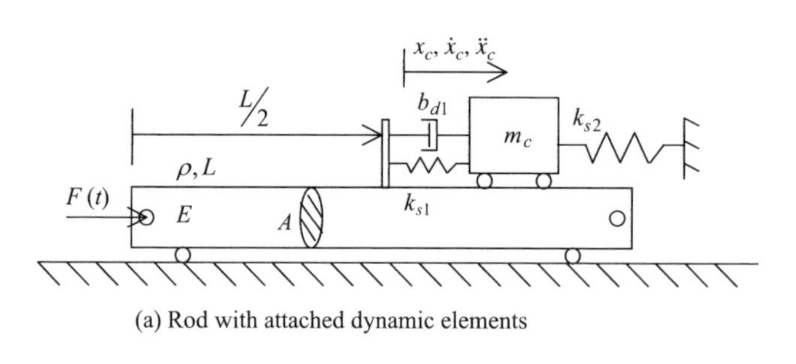

(a) Rod with attached dynamic elements

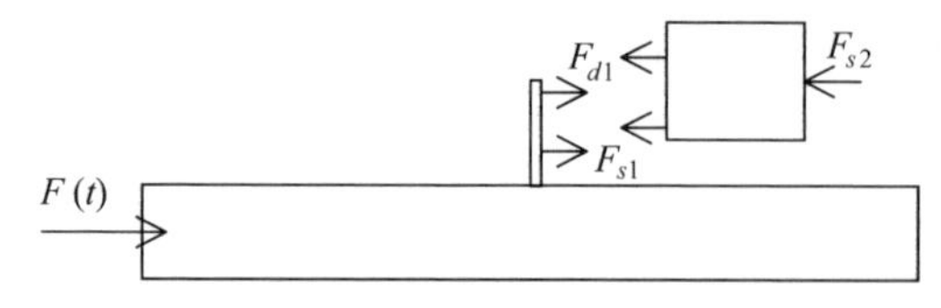

(b) Force diagram with coordinate to locate the cart

Figure 9.5 Distributed rod attached to external dynamic elements.

system. An additional coordinate x_c and its derivatives are used to locate the cart with respect to inertial ground. For the rod, the equations of motion change only because there is an additional force at the midpoint of the rod. Equations (9.55) change to

$$\begin{aligned} m_0\ddot{\eta}_0 &= F(t) + F_{d1} + F_{s1} \\ m_1\ddot{\eta}_1 + k_1\eta_1 &= F(t)Y_1(0) + (F_{d1} + F_{s1})Y_1\left(\frac{L}{2}\right) \\ m_2\ddot{\eta}_2 + k_2\eta_2 &= F(t)Y_2(0) + (F_{d1} + F_{s1})Y_2\left(\frac{L}{2}\right) \\ m_3\ddot{\eta}_3 + k_3\eta_3 &= F(t)Y_3(0) + (F_{d1} + F_{s1})Y_3\left(\frac{L}{2}\right) \end{aligned} \tag{9.60}$$

where the additional external force, composed of the sum of the spring and damper forces is multiplied by the appropriate mode shape function evaluated at the location of the force as required by Eq. (9.44).

For the cart the equation of motion is

$$-F_{d1} - F_{s1} - F_{s2} = m_c\ddot{x}_c \tag{9.61}$$

The additional springs and damper forces come from

$$\begin{aligned} F_{s1} &= k_{s1}\left[x_c - \xi\left(\frac{L}{2}, t\right)\right] \\ F_{d1} &= b_{d1}\left[x_c - \frac{\partial\xi}{\partial t}\left(\frac{L}{2}, t\right)\right] \\ F_{s2} &= k_{s2}x_c \end{aligned} \tag{9.62}$$

and the rod displacements and velocities come from Eq. (9.56) as

$$\begin{aligned} \xi\left(\frac{L}{2}, t\right) &= (1)\eta_0(t) + Y_1\left(\frac{L}{2}\right)\eta_1(t) + Y_2\left(\frac{L}{2}\right)\eta_2(t) + Y_3\left(\frac{L}{2}\right)\eta_3(t) \\ \frac{\partial\xi}{\partial t}\left(\frac{L}{2}, t\right) &= (1)\dot{\eta}_0(t) + Y_1\left(\frac{L}{2}\right)\dot{\eta}_1(t) + Y_2\left(\frac{L}{2}\right)\dot{\eta}_2(t) + Y_3\left(\frac{L}{2}\right)\dot{\eta}_3(t) \end{aligned} \tag{9.63}$$

Equations (9.60) through (9.63) are a complete representation of the dynamic system of Figure 9.5. There are some commercial equation solvers that would allow the equations to remain in this form, and the software will sort the equations for proper computational solution. If we do not use such a program, we have some additional work to do to predict the motion–time history of this system.

Equations. (9.63) are used in (9.62) and the result substituted into the first-order form of (9.61) and of (9.60). The final result can be represented as

$$\dot{\mathbf{x}} = \boldsymbol{A}\mathbf{x} + \mathbf{b}F \tag{9.64}$$

where the state vector is

$$\mathbf{x} = \begin{bmatrix} \eta_0 \\ v_0 \\ \eta_1 \\ v_1 \\ \eta_2 \\ v_2 \\ \eta_3 \\ v_3 \\ x_c \\ v_c \end{bmatrix} \tag{9.65}$$

and the **A**-matrix components are

$$a(1,1) = 0,\ a(1,2) = 1.0,\ a(1,3) = 0,\ a(1,4) = 0,\ a(1,5) = 0$$

$$a(1,6) = 0,\ a(1,7) = 0,\ a(1,8) = 0,\ a(1,9) = 0,\ a(1,10) = 0$$

$$a(2,1) = -\frac{k_{s1}}{m_0},\ a(2,2) = -\frac{b_{d1}}{m_0},\ a(2,3) = -\frac{k_{s1}}{m_0} Y_1\left(\frac{L}{2}\right)$$

$$a(2,4) = -\frac{b_{d1}}{m_0} Y_1\left(\frac{L}{2}\right),\ a(2,5) = -\frac{k_{s1}}{m_0} Y_2\left(\frac{L}{2}\right)$$

$$a(2,6) = -\frac{b_{d1}}{m_0} Y_2\left(\frac{L}{2}\right),\ a(2,7) = -\frac{k_{s1}}{m_0} Y_3\left(\frac{L}{2}\right)$$

$$a(2,8) = -\frac{b_{d1}}{m_0} Y_3\left(\frac{L}{2}\right),\ a(2,9) = \frac{k_{s1}}{m_0},\ a(2,10) = \frac{b_{d1}}{m_0}$$

$$a(3,1) = 0,\ a(3,2) = 0,\ a(3,3) = 0,\ a(3,4) = 1.0,\ a(3,5) = 0$$

$$a(3,6) = 0,\ a(3,7) = 0,\ a(3,8) = 0,\ a(3,9) = 0,\ a(3,10) = 0$$

$$a(4,1) = -\frac{k_{s1}}{m_1} Y_1\left(\frac{L}{2}\right),\ a(4,2) = -\frac{b_{d1}}{m_1} Y_1\left(\frac{L}{2}\right),\ a(4,3) = -\frac{k_{s1}}{m_1} Y_1^2\left(\frac{L}{2}\right) - \frac{k_1}{m_1}$$

$$a(4,4) = -\frac{b_{d1}}{m_1} Y_1^2\left(\frac{L}{2}\right),\ a(4,5) = -\frac{k_{s1}}{m_1} Y_1\left(\frac{L}{2}\right) Y_2\left(\frac{L}{2}\right)$$

$$a(4,6) = -\frac{b_{d1}}{m_1} Y_1\left(\frac{L}{2}\right) Y_2\left(\frac{L}{2}\right),\ a(4,7) = -\frac{k_{s1}}{m_1} Y_1\left(\frac{L}{2}\right) Y_3\left(\frac{L}{2}\right)$$

$$a(4,8) = -\frac{b_{d1}}{m_1} Y_1\left(\frac{L}{2}\right) Y_3\left(\frac{L}{2}\right),\ a(4,9) = \frac{k_{s1}}{m_1} Y_1\left(\frac{L}{2}\right),\ a(4,10) = \frac{b_{d1}}{m_1} Y_1\left(\frac{L}{2}\right)$$

$$a(5, 1) = 0,\ a(5, 2) = 0,\ a(5, 3) = 0,\ a(5, 4) = 0,\ a(5, 5) = 0$$

$$a(5, 6) = 1.0,\ a(5, 7) = 0,\ a(5, 8) = 0,\ a(5, 9) = 0,\ a(5, 10) = 0$$

$$a(6, 1) = -\frac{k_{s1}}{m_2} Y_2\left(\frac{L}{2}\right),\ a(6, 2) = -\frac{b_{d1}}{m_2} Y_2\left(\frac{L}{2}\right),\ a(6, 3) = -\frac{k_{s1}}{m_2} Y_2\left(\frac{L}{2}\right) Y_1\left(\frac{L}{2}\right)$$

$$a(6, 4) = -\frac{b_{d1}}{m_2} Y_2\left(\frac{L}{2}\right) Y_1\left(\frac{L}{2}\right),\ a(6, 5) = -\frac{k_{s1}}{m_2} Y_2^2\left(\frac{L}{2}\right) - \frac{k_2}{m_2}$$

$$a(6, 6) = -\frac{b_{d1}}{m_2} Y_2^2\left(\frac{L}{2}\right),\ a(6, 7) = -\frac{k_{s1}}{m_2} Y_2\left(\frac{L}{2}\right) Y_3\left(\frac{L}{2}\right)$$

$$a(6, 8) = -\frac{b_{d1}}{m_2} Y_2\left(\frac{L}{2}\right) Y_3\left(\frac{L}{2}\right),\ a(6, 9) = \frac{k_{s1}}{m_2} Y_2\left(\frac{L}{2}\right),\ a(6, 10) = \frac{b_{d1}}{m_2} Y_2\left(\frac{L}{2}\right)$$

$$a(7, 1) = 0,\ a(7, 2) = 0,\ a(7, 3) = 0,\ a(7, 4) = 0,\ a(7, 5) = 0$$

$$a(7, 6) = 0,\ a(7, 7) = 0,\ a(7, 8) = 1.0,\ a(7, 9) = 0,\ a(7, 10) = 0$$

$$a(8, 1) = -\frac{k_{s1}}{m_3} Y_3\left(\frac{L}{2}\right),\ a(8, 2) = -\frac{b_{d1}}{m_3} Y_3\left(\frac{L}{2}\right),\ a(8, 3) = -\frac{k_{s1}}{m_3} Y_3\left(\frac{L}{2}\right) Y_1\left(\frac{L}{2}\right)$$

$$a(8, 4) = -\frac{b_{d1}}{m_3} Y_3\left(\frac{L}{2}\right) Y_1\left(\frac{L}{2}\right),\ a(8, 5) = -\frac{k_{s1}}{m_3} Y_3\left(\frac{L}{2}\right) Y_2\left(\frac{L}{2}\right)$$

$$a(8, 6) = -\frac{b_{d1}}{m_3} Y_3\left(\frac{L}{2}\right) Y_2\left(\frac{L}{2}\right),\ a(8, 7) = -\frac{k_{s1}}{m_3} Y_3^2\left(\frac{L}{2}\right) - \frac{k_3}{m_3}$$

$$a(8, 8) = -\frac{b_{d1}}{m_3} Y_3^2\left(\frac{L}{2}\right),\ a(8, 9) = \frac{k_{s1}}{m_3} Y_3\left(\frac{L}{2}\right),\ a(8, 10) = \frac{b_{d1}}{m_3} Y_3\left(\frac{L}{2}\right)$$

$$a(9, 1) = 0,\ a(9, 2) = 0,\ a(9, 3) = 0,\ a(9, 4) = 0,\ a(9, 5) = 0$$

$$a(9, 6) = 0,\ a(9, 7) = 0,\ a(9, 8) = 0,\ a(9, 9) = 0,\ a(9, 10) = 1.0$$

$$a(10, 1) = -\frac{k_{s1}}{m_c},\ a(10, 2) = -\frac{b_{d1}}{m_c},\ a(10, 3) = -\frac{k_{s1}}{m_c} Y_1\left(\frac{L}{2}\right)$$

$$a(10, 4) = -\frac{b_{d1}}{m_c} Y_1\left(\frac{L}{2}\right),\ a(10, 5) = -\frac{k_{s1}}{m_c} Y_2\left(\frac{L}{2}\right)$$

$$a(10, 6) = -\frac{b_{d1}}{m_c} Y_2\left(\frac{L}{2}\right),\ a(10, 7) = -\frac{k_{s1}}{m_c} Y_3\left(\frac{L}{2}\right)$$

$$a(10, 8) = -\frac{b_{d1}}{m_c} Y_3\left(\frac{L}{2}\right),\ a(10, 9) = \frac{k_{s1}}{m_c},\ a(10, 10) = \frac{b_{d1}}{m_c}$$

The accompanying forcing vector is

$$\mathbf{b} = \begin{bmatrix} 0 \\ \dfrac{1}{m_0} \\ 0 \\ \dfrac{1}{m_1} \\ 0 \\ \dfrac{1}{m_2} \\ 0 \\ \dfrac{1}{m_3} \\ 0 \\ 0 \end{bmatrix} \tag{9.66}$$

These linear equations were programmed in Matlab [2] and the response for the same step input as used previously is shown in Figures 9.6 and 9.7. The parameters for the additional dynamic elements are given in Table 9.2.

Figure 9.6 shows the velocity responses for the same output points as Figure 9.3, but this time for the modified system of Figure 9.5. Because the cart is attached to inertial ground through the second spring, the rod is no longer capable of moving horizontally unconstrained. This is apparent from the velocity response at any of the cross sections. The modal representation of the rod is unchanged and still contains the rigid-body mode plus the three included dynamic modes. The modal representation

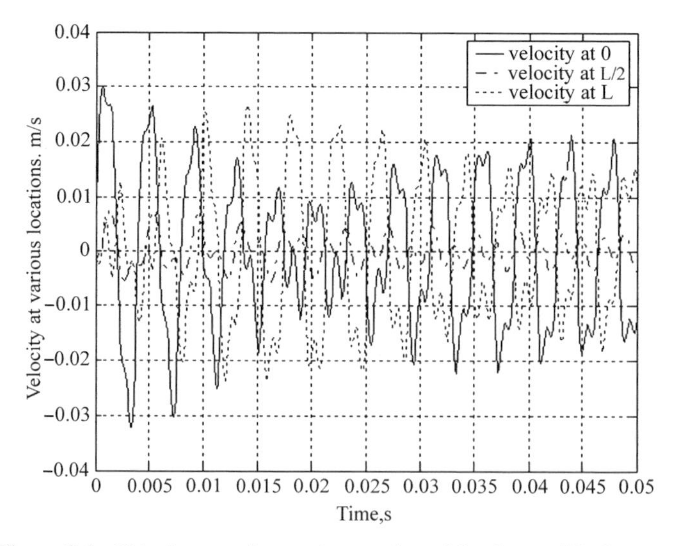

Figure 9.6 Velocity at various points on the rod for the modified system.

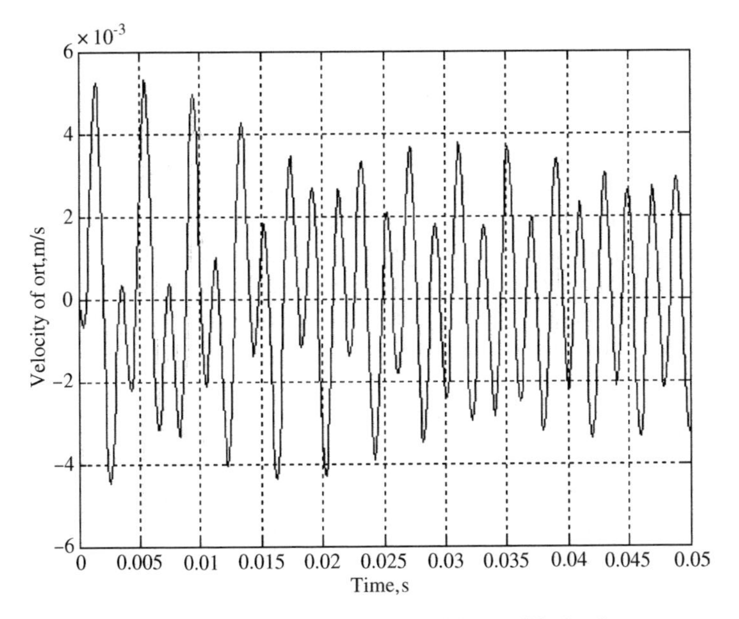

Figure 9.7 Velocity of the cart is part of the modified rod system.

is quite capable of simulating the constrained system with no changes in the modal model. Figure 9.7 shows the velocity response of the cart. Because there is some damping in the model, all the responses show some signs of decay, even over the brief period displayed in the figures.

When an engineering system requires modeling some components as distributed dynamic elements, the finite mode approach developed here is the most useful. In fact, the finite mode model of a distributed system is the most accurate low-order model that can be defined. If one decides to include five modes in a model, the five mode frequencies are exactly correct. There are no convergence issues such as those associated with using many distinct lumps to approximate the distributed dynamics. Of course, if the finite mode representation is to be used, the mode functions and mode frequencies must be obtained.

TABLE 9.2. Parameters for the Rod Shown in Figure 9.5

Mass of cart $m_c = m_{\text{rod}}$
Define a frequency $f = 500\,\text{Hz}$
Calculate:

$k_{s1} = m_c(2\pi\, f)^2$
$k_{s2} = k_{s1}$

Define a damping ratio $\zeta = 0.3$
Calculate:

$b_{d1} = 2\zeta(2\pi\, f)m_c$

The distributed rod was used here to demonstrate the analytical procedure for obtaining mode functions and mode frequencies. Quite often, the distributed system physics are so complex that the analytical procedure cannot be used and separation of variables cannot be carried out by hand calculations. For these cases there are finite element and finite difference software packages [4] that can deliver numerical solutions for mode functions and associated frequencies. We then follow the finite mode procedure for including external dynamics to construct overall models. There are also experimental methods for determining modes and frequencies. The conclusion is that the reader should spend some time becoming familiar with the finite mode approach for including distributed dynamics in an overall engineering model. To help with this education, the next examples are presented.

9.3 TIGHTLY STRETCHED CABLE

Figure 9.8 shows a tightly stretched cable of length L and mass per unit length ρ. The ends of the cable are attached to massless carts that can move vertically due to the motion of the cable. The cable is stretched such that it has an internal tension T and this is assumed constant even though the cable stretches if it is deflected from its original horizontal position. Two point forces, $F_1(t)$ and $F_2(t)$, are acting at positions x_1 and x_2 measured from the left end. We model the stretched cable as a distributed system, but Figure 9.8*b* shows the cable represented by several lumped mass elements connected by springs. If enough masses and springs are included in the finite lump representation, we would expect the model to behave more or less like a true distributed system. However, as with the rod example, it requires many lumped elements to get accuracy in the first few modal frequencies, and each additional mass and spring introduces additional frequencies that are inaccurate. As more and more lumps are included, there will come a point where the finite lump approach will be abandoned in favor of finite modes.

In Figure 9.8*c* the force on the *i*th lumped mass element is shown. It is assumed that the springs are tightly stretched and each spring has tension force T. For small vertical motion, the equation of motion for this mass element is

$$-T\theta_i + T\theta_{i+1} = m_i \ddot{w}_i \tag{9.67}$$

But the small angles of the springs are approximately

$$\begin{aligned} \theta_i &= \frac{w_i - w_{i-1}}{\Delta x} \\ \theta_{i+1} &= \frac{w_{i+1} - w_i}{\Delta x} \end{aligned} \tag{9.68}$$

where Δx is the distance between the mass elements. Thus, substituting into (9.67), we have

$$-\frac{T}{\Delta x}(w_i - w_{i-1}) + \frac{T}{\Delta x}(w_{i+1} - w_i) = m_i \ddot{w}_i \tag{9.69}$$

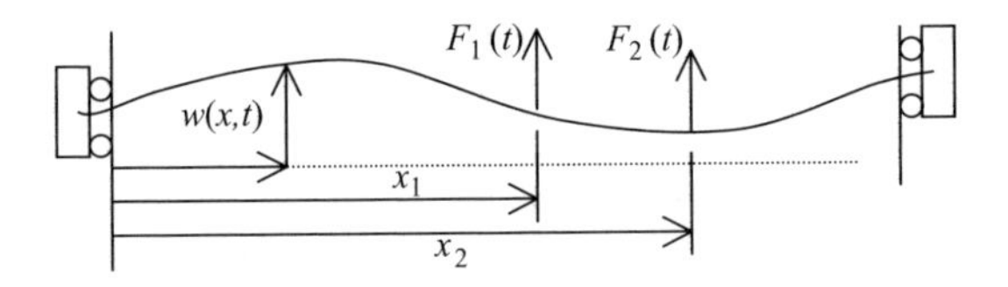

(a) Tightly stretched cable of length L and mass per unit length ρ

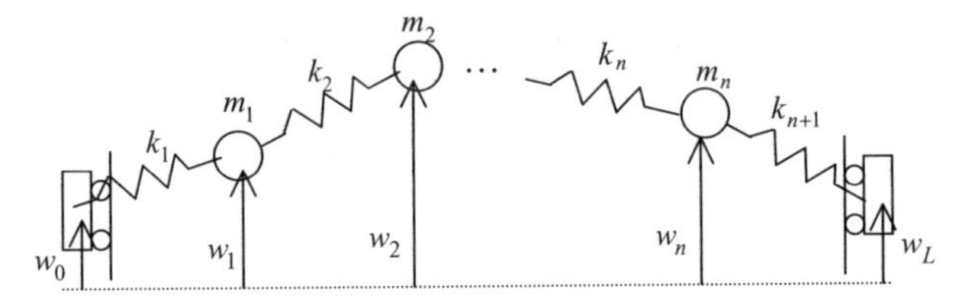

(b) Finite lump model of the stretched cable

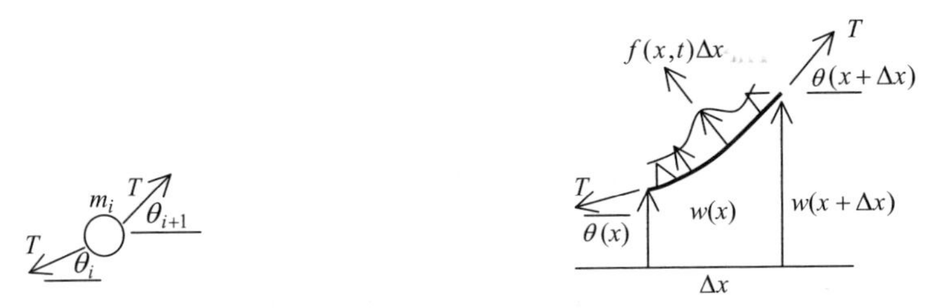

(c) Force on the ith lumped mass element (d) Differential element of the stretched cable

Figure 9.8 Tightly stretched cable

If we think of the parenthetical terms as the relative displacement across a spring, the spring constants in the finite lump model become

$$k_i = \frac{T}{\Delta x} \tag{9.70}$$

If the model contains n_I mass elements, then

$$(n_I + 1)\Delta x = L, \quad \text{or} \quad \Delta x = \frac{L}{n_I + 1}, \quad \text{or} \quad k_i = \frac{T}{L}(n_I + 1) \tag{9.71}$$

The mass of the cable is, $m_c = \rho L$, so for n_I lumped mass elements each individual mass would be

$$m_i = \frac{m_c}{n_I} \tag{9.72}$$

If the intent of the overall model is to have a very approximate representation of the distributed dynamics of the stretched cable, a "few"-lump model is all right and the lumped element parameters can be calculated from Eqs. (9.70) and (9.72).

The true distributed dynamics come from considering the differential element shown in Figure 9.8*d*. The tension in the cable is assumed constant. The small element of length Δx is displaced from its original horizontal position, and each variable is shown at the right of the element slightly changed from its value at the left end. The external force per unit length is also acting on the element. Newton's law applied to the cable element in the vertical direction yields, for small displacements,

$$-T\theta(x) + T\theta(x+\Delta x) + f(x,t)\,\Delta x = \rho\Delta x \frac{\partial^2 w}{\partial t^2} \tag{9.73}$$

Using the Taylor series expansion and retaining only first-order terms yields

$$\theta(x+\Delta x) = \theta(x) + \frac{\partial \theta}{\partial x}\Delta x \tag{9.74}$$

Substituting into (9.73) yields

$$T\frac{\partial \theta}{\partial x} + f(x,t) = \rho\frac{\partial^2 w}{\partial t^2} \tag{9.75}$$

Keeping in mind that the length of the element in Figure 9.8*d* is very small, the angle at the left end is very nearly

$$\theta(x) \approx \frac{w(x+\Delta x) - w(x)}{\Delta x}$$

which becomes

$$\theta(x) = \frac{\partial w}{\partial x} \tag{9.76}$$

as $\Delta x \to 0$. Equation (9.75) then becomes

$$T\frac{\partial^2 w}{\partial x^2} + f(x,t) = \rho\frac{\partial^2 w}{\partial t^2} \tag{9.77}$$

This equation of motion is identical to Eq. (9.14) for the rod. Thus, the stretched cable vertical dynamics is identical to the longitudinal dynamics of the rod. When the forcing is set to zero temporarily, Eq. (9.77) is the simple wave equation.

If we use the delta function to represent the point forces shown in Figure 9.8*a* the equation that needs to be solved is

$$\rho\frac{\partial^2 w}{\partial t^2} - T\frac{\partial^2 w}{\partial x^2} = F_1\delta(x-x_1) + F_2\delta(x-x_2) \tag{9.78}$$

Since the carts that are attached at each boundary are assumed massless, there can be no net vertical force acting on them or there will be infinite acceleration. Since the vertical force is the cable tension multiplied by the angle the cable makes relative to horizontal, the boundary condition must be that there is no slope of the cable at the two ends. Thus, the boundary conditions become

$$\frac{\partial w}{\partial x}(0, t) = \frac{\partial w}{\partial x}(L, t) = 0 \tag{9.79}$$

Free Response: Separation of Variables

Setting the forcing to zero in Eq. (9.78), we assume a separated solution,

$$w(x, t) = Y(x)g(t) \tag{9.80}$$

and substitute into (9.78), with the result

$$\rho Y\ddot{g} - Tg\frac{d^2Y}{dx^2} = 0 \tag{9.81}$$

Dividing both sides by ρYg yields

$$\frac{1}{g}\ddot{g} - \frac{T}{\rho}\frac{1}{Y}\frac{d^2Y}{dx^2} = 0 \tag{9.82}$$

The first term on the left is a function of time only, and the second term on the left is a function of displacement only. For this to be true for all time and all displacements, each term must be equal to the same constant. As was done before for the rod example, it is conventional to let the first term on the left side of (9.82) equal $-\omega^2$, with the result

$$\begin{aligned} \ddot{g} + \omega^2 g &= 0 \\ \frac{d^2Y}{dx^2} + \beta^2 Y &= 0 \\ \beta^2 &= \frac{\rho}{T}\omega^2 \end{aligned} \tag{9.83}$$

The solution to the second of Eqs. (9.83) is

$$Y(x) = A\cos\beta x + B\sin\beta x \tag{9.84}$$

subject to the conditions from Eq. (9.79),

$$\frac{dY}{dx}(0) = \frac{dY}{dx}(L) = 0 \tag{9.85}$$

Using (9.85) in (9.84) yields $B = 0$ from the first condition if we restrict our attention to the case when $\beta \neq 0$, and from the second condition,

$$A\beta \sin \beta L = 0 \tag{9.86}$$

As before for the rod example, Eq. (9.86) is called the *frequency equation* and is the continuous system counterpart to the discrete frequency equation for lumped systems.

One solution to (9.86) is that $A = 0$. But combined with $B = 0$, this yields an uninteresting response. We are currently not allowing $\beta = 0$, so the only solution to (9.86) is

$$\sin \beta L = 0 \tag{9.87}$$

or

$$\beta_n L = n\pi \tag{9.88}$$

or, using the last of (9.83),

$$\omega_n^2 = \frac{T}{\rho}\left(\frac{n\pi}{L}\right)^2 \tag{9.89}$$

These are the natural frequencies of the system, and the corresponding mode functions come from Eq. (9.84) as

$$Y_n(x) = \cos n\pi \frac{x}{L} \tag{9.90}$$

where the arbitrary constant has been set to unity. These mode shapes are identical to those for the rod example, except that this time the modal displacements are vertical rather than longitudinal.

The mode shapes are shown in Figure 9.2; only their interpretation is different than was the case for the rod. For the first mode of the tightly stretched cable, if the left end is displaced one unit vertically upward and the right end is displaced one unit vertically downward and the rest of the cable was displaced into the half-cosine shape shown in Figure 9.2*a*, when released the cable will oscillate at the first natural frequency,

$$\omega_1^2 = \frac{T}{\rho}\left(\frac{\pi}{L}\right)^2$$

and remain in this first modal configuration. The interpretation of the higher-frequency modes is similar.

The boundary conditions for this example allow for unconstrained motion of the cable; thus, a zero-frequency or rigid-body mode exists. As done previously with the rod example, if $\omega = 0$ is a permissible frequency, then $\beta = 0$ from Eq. (9.83) and

$$\frac{d^2 Y_0}{dx^2} = 0 \tag{9.91}$$

The solution is

$$Y_0(x) = C_1 x + C_2 \tag{9.92}$$

The rigid-body mode must satisfy the boundary conditions from Eq. (9.85), with the result

$$Y_0(x) = C_2 = 1 \tag{9.93}$$

after the arbitrary constant is set to unity. The mode shapes are orthogonal, in that

$$\int_0^L Y_n(x) Y_m(x)\, dx = 0, \qquad n \neq m \tag{9.94}$$

and this includes the rigid-body mode $Y_0(x)$.

Forced Response

The forced equation of motion is given in Eq. (9.78). Following the identical procedure as that used for rod, we assume the forced response for the cable deflection follows

$$w(x,t) = \sum_n Y_n(x)\eta_n(t), \qquad n = 0, 1, 2, \ldots, \tag{9.95}$$

Substituting this into (9.78), multiplying through by $Y_m(x)$, making use of (9.83), and integrating over the length of the cable results in equations for the time functions:

$$\begin{aligned} m_0\ddot{\eta}_0 &= F_1(t)Y_0(x_1) + F_2(t)Y_0(x_2) && \text{for the } \textit{zero}\text{th or rigid body mode} \\ m_n\ddot{\eta}_n + k_n\eta_n &= F_1(t)Y_n(x_1) + F_2(t)Y_n(x_2) && \text{for all the dynamic modes} \end{aligned} \tag{9.96}$$

Once the external forces F_1 and F_2 are specified, the time functions can be determined and the forced response constructed from Eq. (9.95) using a finite number of modes.

If $F_1(t)$ and $F_2(t)$ are actually the result of attached external systems such as shown in Figure 9.9, the equations of motion for the combined lumped and distributed system can be derived using Newton's laws. The spring at the left end would be

$$F_1(t) = -kw(0,t) = -k\sum_n Y_n(0)\eta_n(t) \tag{9.97}$$

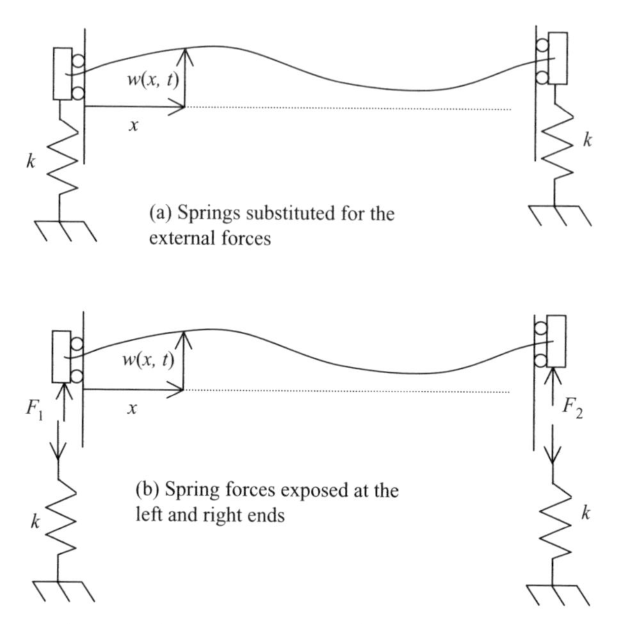

Figure 9.9 Tightly stretched cable with a spring at the boundaries.

and the one at the right end, would be

$$F_2(t) = -kw(L, t) = -k \sum_n Y_n(L)\eta_n(t) \tag{9.98}$$

Once the number of retained modes is selected, Eqs. (9.97) and (9.98) can be used in (9.96) and the time functions $\eta_n(t)$ can be computed. Then the deflection of the cable can be constructed from (9.95). The details of this are pursued in the problems for this chapter.

The sign convention used for the attached components to distributed systems is sometimes a source of confusion. In Eqs. (9.97) and (9.98) there is a minus sign that the reader might be wondering about. When dealing with distributed systems it is always a good idea to maintain the sign convention used when deriving the governing partial differential equations. For the cable example the deflections are positive upward and the external forces are positive upward. This convention resulted in Eqs. (9.96), where the external forces are needed to proceed.

When the external forces were actually from the attached springs rather than just specified as step inputs or harmonic inputs or whatever, it is important to maintain the positive upward convention. This means that the springs must be defined positive in compression in order that their positive force pushes upward. Since the spring deflections at their attachment points are positive upward, a tension force would naturally occur in the springs. Thus, for a positive-upward deflection, the minus sign is required to define tension forces as negative to go along with compression forces being positive.

An interesting observation for the cable with the springs at the ends is that as the spring stiffness is made larger and larger, the ends of the cable are constrained to move less and less. In the limit as the stiffness becomes infinite, the ends of the cable cannot move at all and the distributed system becomes one with fixed boundary conditions where instead of the boundary conditions from Eq. (9.79), the boundary conditions are

$$w(0,t) = w(L,t) = 0 \tag{9.99}$$

A good question is whether a cable with free ends will converge to a cable with fixed ends if the springs are made increasingly stiff. This question will be answered in the problems for this chapter. In Problem 9.4 the separated solution for the stretched cable is repeated using the fixed boundary conditions from (9.99). Mode shapes and associated modal frequencies are derived. In Problem 9.10, using only a couple of retained modes, it is asked to show that the original unconstrained modes do, in fact, approach the fixed-end modes as the spring stiffness approaches infinity.

9.4 BERNOULLI–EULER BEAM

The final example of distributed system vibrations is the Bernoulli–Euler beam shown in Figure 9.10. A beam is a very typical engineering component where its vibration behavior is important in structures such as buildings, truck frames, and bridges. In Figure 9.10*b* a plane section of the beam is isolated and all the forces and moments acting on the section are shown. The Bernoulli–Euler beam assumes that the shear stiffness is infinite such that sections of the beam perpendicular to the neutral axis when the beam is at rest remain perpendicular to the neutral axis when the beam is deflected. Another assumption is that the rotational inertia of any plane section of the beam is negligible, requiring any plane section to be in moment equilibrium.

Figure 9.11 shows a lumped version of the beam, treating sections of the beam as mass elements attached to each other through springs that resist bending and springs that resist shear. In the Bernoulli–Euler model the sections of the beam have mass, but the rotational inertia is assumed to be zero. The shear springs k_s are infinite in this model such that the mass elements remain perpendicular to the links that represent the neutral axis. Several of these lumps could be put together to represent beam dynamics, however, as with all distributed systems, a large number of lumps must be included to obtain an accurate representation of the low-frequency behavior of the system.

From the beam element of Figure 9.10, Newton's law for the element results in

$$-V(x) + \left[V(x) + \frac{\partial V}{\partial x}\,\Delta x\right] + f(x,t)\,\Delta x = \rho A \Delta x\,\frac{\partial^2 w}{\partial t^2}$$

or (9.100)

$$\frac{\partial V}{\partial x} + f(x,t) = \rho A\,\frac{\partial^2 w}{\partial t^2}$$

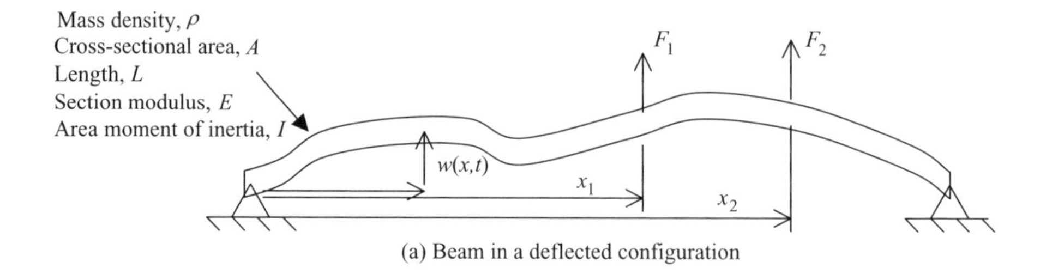

(a) Beam in a deflected configuration

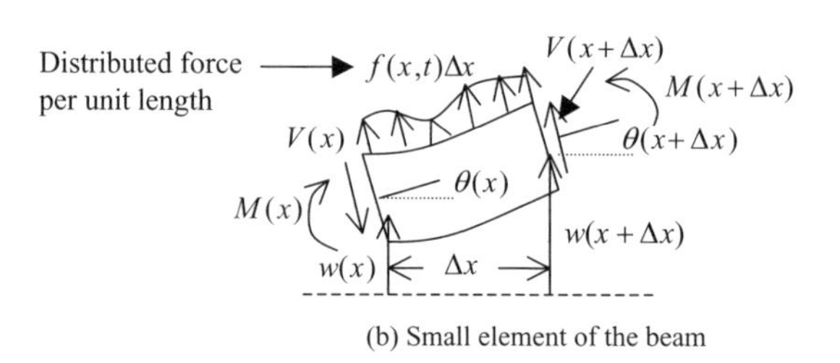

(b) Small element of the beam

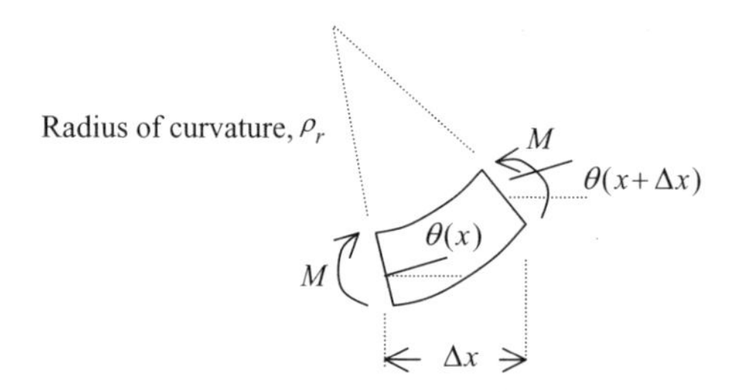

(c) Beam element in pure bending

Figure 9.10 Bernoulli–Euler beam and beam elements.

Taking moments about the left end of the element, moment equilibrium requires that

$$-M(x)+\left[M(x)+\frac{\partial M}{\partial x}\Delta x\right]+\left[V(x)+\frac{\partial V}{\partial x}\Delta x\right]\,\Delta x\,+f(x,t)\,\Delta x\frac{\Delta x}{2}=0 \tag{9.101}$$

Simplifying by neglecting higher-order terms such as $\Delta x \rightarrow 0$ results in

$$V=-\frac{\partial M}{\partial x} \tag{9.102}$$

Using (9.102) in (9.100) yields the intermediate result,

$$-\frac{\partial^2 M}{\partial x^2}+f(x,t)\,=\,\rho A\frac{\partial^2 w}{\partial t^2} \tag{9.103}$$

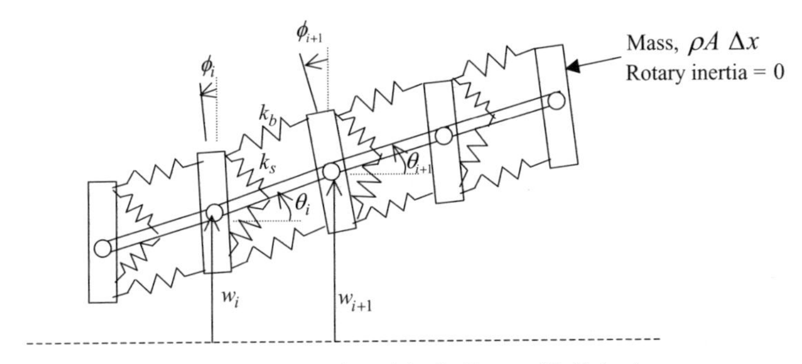

Figure 9.11 Lumped model of a Bernoulli–Euler beam.

Figure 9.10*c* shows a small plane section subject to pure bending. In an actual beam, pure bending does not occur, but it is a reasonable assumption. Reference [1] has a more complete treatment of this topic. For pure bending,

$$M = \frac{EI}{\rho_r} \tag{9.104}$$

But the radius of curvature times $\Delta\theta$ is the arc length of the section, which is approximately equal to the length Δx; thus,

$$\rho_r \frac{\partial\theta}{\partial x}\Delta x \approx \Delta x \quad \text{or} \quad \frac{1}{\rho_r} = \frac{\partial\theta}{\partial x} \tag{9.105}$$

Combining with (9.104) yields

$$M = EI\frac{\partial\theta}{\partial x} \tag{9.106}$$

The derivation is almost complete when it is recognized that

$$\theta \approx \frac{w(x+\Delta x) - w(x)}{\Delta x} = \frac{\partial w}{\partial x} \tag{9.107}$$

when $\Delta x \to 0$. Finally, we use (9.107) in (9.106), with the result

$$M = EI\frac{\partial^2 w}{\partial x^2} \tag{9.108}$$

and use this in (9.103) to yield

$$-EI\frac{\partial^4 w}{\partial x^4} + f(x,t) = \rho A\frac{\partial^2 w}{\partial t^2} \tag{9.109}$$

For the point forcing shown in Figure 9.10*a*, the final forced equation becomes

$$EI\frac{\partial^4 w}{\partial x^4} + \rho A\frac{\partial^2 w}{\partial t^2} = F_1\delta(x - x_1) + F_2\delta(x - x_2) \tag{9.110}$$

Equation (9.110) is the forced Bernoulli–Euler beam equation for the case of uniform properties along the length of the beam. This is the starting point for determining modal frequencies and associated mode shapes.

Free Response: Separation of Variables

The boundary conditions for the example system are pinned ends as shown in Figure 9.10*a*. These boundary conditions require that there be no deflection or moment at the two ends. Using Eq. (9.108), the boundary conditions are

$$w(0, t) = w(L, t) = \frac{\partial^2 w}{\partial x^2}(0, t) = \frac{\partial^2 w}{\partial x^2}(L, t) = 0 \tag{9.111}$$

We assume a separated solution

$$w(x, t) = Y(x)g(t) \tag{9.112}$$

and substitute into the unforced equation (9.110), with the result

$$EI\frac{d^4 Y}{dx^4}g + \rho AY\frac{d^2 g}{dt^2} = 0 \tag{9.113}$$

This result is divided through by ρAYg, resulting in

$$\frac{EI}{\rho A}\frac{1}{Y}\frac{d^4 Y}{dx^4} + \frac{1}{g}\frac{d^2 g}{dt^2} = 0 \tag{9.114}$$

The first term on the left is a function of x only, and the second term on the left is a function of t only. This can be true for all x and all t only if both terms are equal to the same constant. As was done for the rod and cable examples, it is conventional to let the second term in (9.114) equal $-\omega^2$, with the result

$$\begin{aligned} \frac{d^2 g}{dt^2} + \omega^2 g &= 0 \\ \frac{d^4 Y}{dx^4} - \beta^4 Y &= 0 \end{aligned} \tag{9.115}$$

where

$$\beta^4 = \frac{\rho A}{EI}\omega^2 \tag{9.116}$$

The second of Eq. (9.115) has an analytical solution of

$$Y(x) = A\cosh\beta x + B\sinh\beta x + C\cos\beta x + D\sin\beta x \tag{9.117}$$

Interested readers should try (9.117) in the governing equation and assure themselves that this is a solution. Although this result is admittedly more complex than the counterparts for the rod and cable, nevertheless Eq. (9.117) is simply a function of displacement and there are four constants that must be determined from the boundary conditions, rather than only two for the earlier examples.

It turns out that most boundary conditions will result in nonzero values for all the constants, and the resulting mode shape functions are a bit cumbersome. The pinned boundary conditions used here were chosen for demonstration purposes, as they produce the simplest possible mode shape functions. From Eqs. (9.111) and (9.112) we can conclude that the boundary conditions apply to the mode shape functions as

$$Y(0) = Y(L) = \frac{d^2Y}{dx^2}(0) = \frac{d^2Y}{dx^2}(L) = 0 \tag{9.118}$$

Using these in Eq. (9.117) results in

$$\begin{aligned} 0 &= A + C \\ 0 &= A\beta^2 - C\beta^2 \end{aligned} \tag{9.119}$$

from the conditions at $x = 0$. For $\beta \neq 0$ the solutions is

$$A = C = 0 \tag{9.120}$$

Using the conditions at $x = L$ results in

$$\begin{aligned} 0 &= B\sinh\beta L + D\sin\beta L \\ 0 &= B\beta^2\sinh\beta L - D\beta^2\sin\beta L \end{aligned} \tag{9.121}$$

Again, if $\beta \neq 0$, then since $\sinh\beta L$ cannot be zero, we must conclude that $B = 0$ and

$$D\sin\beta L = 0 \tag{9.122}$$

If we let $D = 0$, all the constants are 0 and $Y(x) = 0$. The more interesting conclusion is that

$$\sin\beta_n L = 0 \quad \text{and} \quad \beta_n L = n\pi, \qquad n = 1, 2, 3, \ldots \tag{9.123}$$

and from Eq. (9.116) the natural frequencies are

$$\omega_n^2 = \frac{EI}{\rho A}\left(\frac{n\pi}{L}\right)^4 \tag{9.124}$$

The associated mode shape functions are

$$Y_n(x) = D_n \sin n\pi \frac{x}{L} \tag{9.125}$$

where D_n are arbitrary constants and are set to unity for the rest of the presentation.

The interpretation of the mode shapes and frequencies is that if the beam were displaced in a modal configuration and released, the beam would oscillate at the associated frequency and remain in the respective modal configuration. Figure 9.12 shows the first few mode shapes of the beam.

The mode shape functions are orthogonal in the sense that

$$\int_0^L Y_n(x)Y_m(x)\,dx = 0, \quad n \neq m \tag{9.126}$$

and this is by now pretty obvious for mode shapes given by (9.125). It is interesting that this orthogonality condition applies to general mode shapes given by Eq. (9.117). From ref. [1], for a beam that is clamped at the left end and free at the right end, separation of variables leads to the frequency equation,

$$\cos\beta_n L \cosh\beta_n L = -1 \tag{9.127}$$

and corresponding mode shapes,

$$\begin{aligned} Y_n(x) &= (\sin\beta_n L - \sinh\beta_n L)(\sin\beta_n x - \sinh\beta_n x) \\ &\quad + (\cos\beta_n L + \cosh\beta_n L)(\cos\beta_n x - \cosh\beta_n x), \quad n = 1, 2, 3, \ldots \end{aligned} \tag{9.128}$$

The frequency equation is a bit more complicated than (9.123), for the pinned beam, and (9.127) must be solved numerically for the values of $\beta_n L$. The mode shape functions are much more complicated than those for the pinned beam, Eq. (9.125), but these are still orthogonal and obey the condition from Eq. (9.126).

Forced Response

Returning to Figure 9.10 and the forced equation of motion (9.110), we carry out the same steps as was done previously to arrive at the forced response of rod and cable. We start with the assumed solution,

$$w(x, t) = \sum_n Y_n(x)\eta_n(t) \tag{9.129}$$

and substitute this into the forced equation, with the result

$$\sum_n EI\frac{d^4Y_n}{dx^4}\eta_n + \sum_n \rho A Y_n \frac{d^2\eta_n}{dt^2} = F_1\delta(x - x_1) + F_2\delta(x - x_2) \tag{9.130}$$

Using the second of Eqs. (9.115) in the first term on the left of (9.130) results in

$$\sum_n \rho A\omega_n^2 Y_n\eta_n + \sum_n \rho A Y_n \frac{d^2\eta_n}{dt^2} = F_1\delta(x - x_1) + F_2\delta(x - x_2) \qquad (9.131)$$

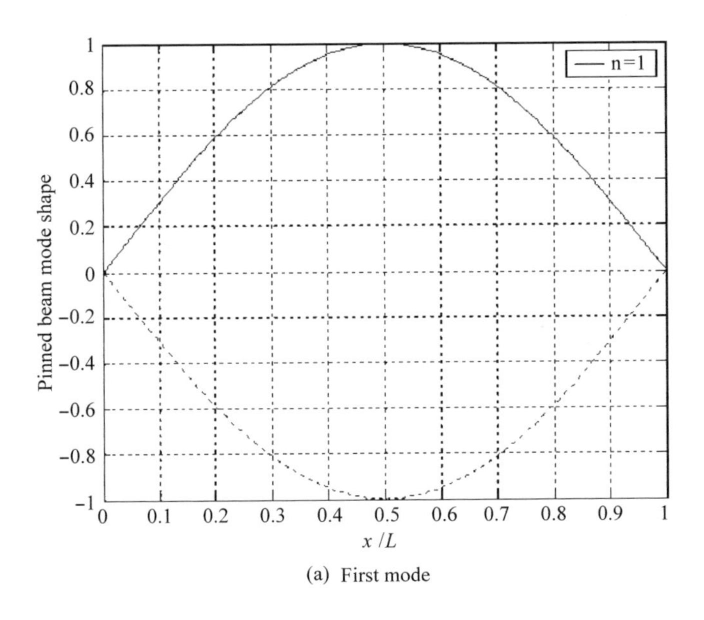

(a) First mode

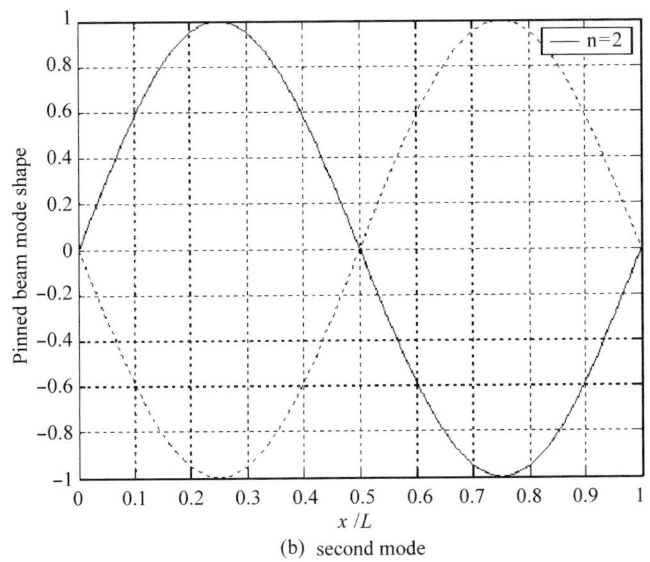

(b) second mode

Figure 9.12 The First three mode shapes of a pinned Bernoulli–Euler beam.

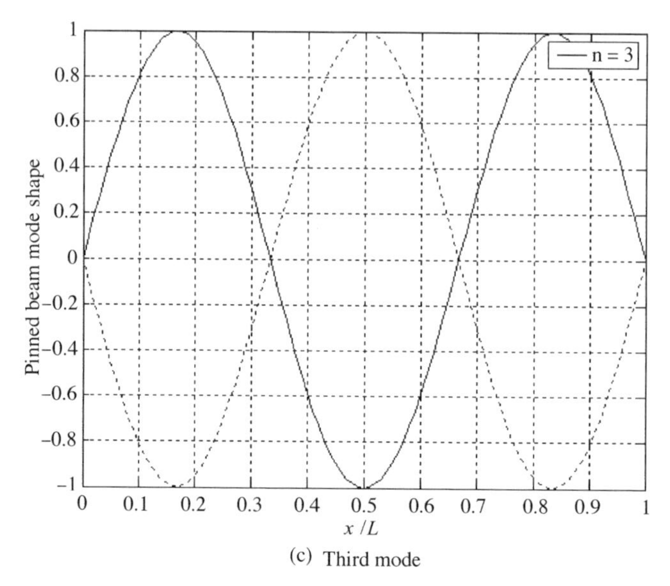

(c) Third mode

Figure 9.12 *continued*

We next multiply through by $Y_m(x)$ and integrate over the length of the beam. Using the orthogonal property of the move shapes, we get

$$m_m \ddot{\eta}_m + k_m \eta_m = F_1 Y_m(x_1) + F_2 Y_m(x_2) \tag{9.132}$$

where the modal mass comes from

$$m_m = \int_0^L \rho A Y_m^2 \, dx = \frac{\rho A L}{2} = \frac{m_{\text{beam}}}{2} \tag{9.133}$$

and the modal stiffnesses are

$$k_m = m_m \omega_m^2 = m_m \frac{EI}{\rho A} \left(\frac{m\pi}{L} \right)^4 \tag{9.134}$$

Once the external forces are specified, (9.132) can be solved for the time functions and the beam displacement constructed from Eq. (9.129).

Now consider the system in Figure 9.13, where the external forces on the beam are replaced by a system consisting of a rigid body of mass m_r and c.g. moment of inertia I_r attached to the beam at locations x_1 and x_2. The coordinate y_g locates the center of gravity of the rigid body, and the angle θ, its counter clockwise orientation. We are interested in the equations of motion for this system, retaining three modes for the beam representation.

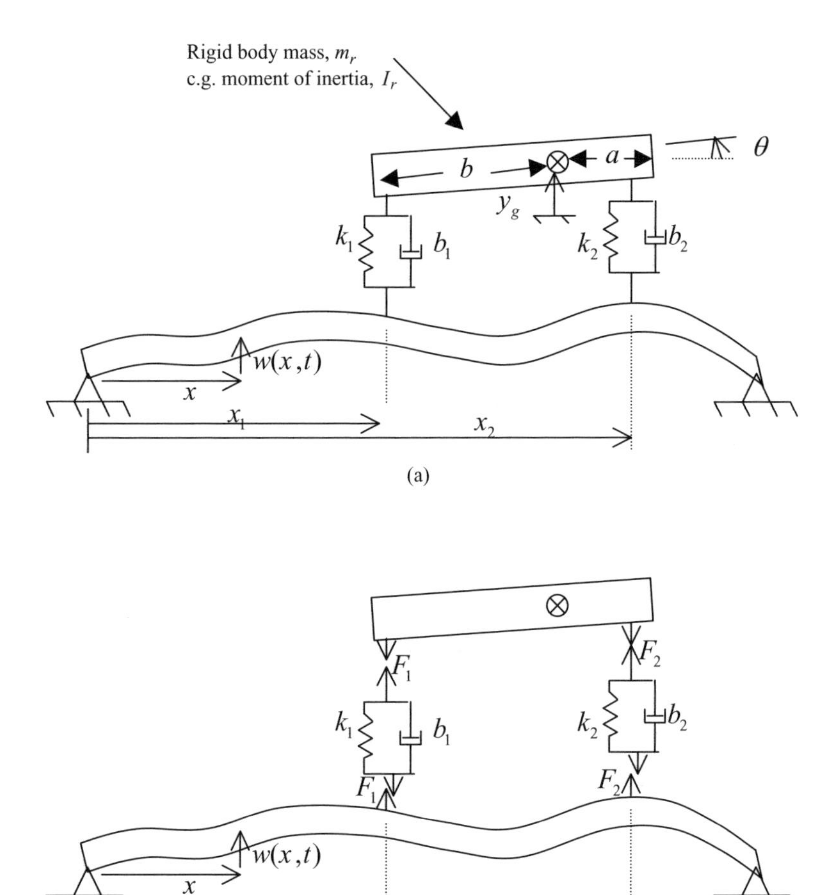

(b) Force diagram

Figure 9.13 Beam with rigid body connected.

For forces to be positive upward acting on the beam, the spring and damper at the two ends of the rigid body must be positive in tension. The displacement of the right end of the rigid body is $y_g + a\theta$, and the displacement at the left end is $y_g - b\theta$ assuming small angular displacements. Thus, the tension forces in the spring–damper modules are

$$
\begin{aligned}
F_1 &= k_1[(y_g - b\theta) - w(x_1, t)] + b_1 \left[(\dot{y}_g - b\dot{\theta}) - \frac{\partial w}{\partial t}(x_1, t) \right] \\
F_2 &= k_2[(y_g + a\theta) - w(x_2, t)] + b_2 \left[(\dot{y}_g + a\dot{\theta}) - \frac{\partial w}{\partial t}(x_2, t) \right]
\end{aligned}
\tag{9.135}
$$

where

$$
\begin{aligned}
w(x_1, t) &= \sum_{n=1}^{3} Y_n(x_1)\eta_n(t) \\
w(x_2, t) &= \sum_{n=1}^{3} Y_n(x_2)\eta_n(t) \\
\frac{\partial w}{\partial t}(x_1, t) &= \sum_{n=1}^{3} Y_n(x_1)\dot{\eta}_n(t) \\
\frac{\partial w}{\partial t}(x_2, t) &= \sum_{n=1}^{3} Y_n(x_2)\dot{\eta}_n(t)
\end{aligned}
\tag{9.136}
$$

The equations of motion for the attached rigid body are

$$
\begin{aligned}
-F_1 - F_2 &= m_r \ddot{y}_g \\
F_1 b - F_2 a &= I_r \ddot{\theta}
\end{aligned}
\tag{9.137}
$$

The forces from Eqs. (9.135) can now be used in the modal equations (9.132) and the rigid-body equations (9.137) along with substitutions from Eqs. (9.136) to form a complete set of equations of motion allowing the three mode beam model to interact with the attached rigid-body system. To solve the system of equation we would need to manipulate the final set into first-order form ready for computer solution.

9.5 SUMMARY

When modeling physical engineering systems it is sometimes necessary to include the distributed dynamic behavior in order to get a reasonable representation of the system. When representing the dynamics of large tractor/trailer rigs, the fuselage of a helicopter, or the movement of a bridge due to wind, if the distributed dynamics are not included, the modeling exercise is a waste of time. In this chapter the concept of separation of variables was presented, leading to a finite mode representation of the distributed parts of a system. It was shown that the modes can be made to interact with external components attached to the distributed system in order to represent an overall system in which the distributed dynamics are only one part.

The problems that follow will assist the reader in gaining experience in working with finite mode models and connecting these modes to external systems to build overall system models. The challenge is ultimately to come up with a set of computable equations that are ready for computer solution. This was demonstrated for one of the examples in this chapter and will be asked of the reader in the problems section.

REFERENCES

[1] L. Meirovitch, *Analytical Methods in Vibrations*, Macmillan, New York, 1967.

[2] Matlab and Simulink software, Mathworks Co., Natick, MA.

[3] D. C. Karnopp, D. L. Margolis, and R. C. Rosenberg, *System Dynamics: Modeling and Simulation of Mechatronic Systems*, 4th ed., Wiley, Hoboken, NJ, 2006.

[4] Nastran finite element analysis software, Noran Engineering, Westminster, CA.

PROBLEMS

9.1 The distributed longitudinal vibrations of a rod with force-free boundaries resulted in the orthogonal modes and frequencies,

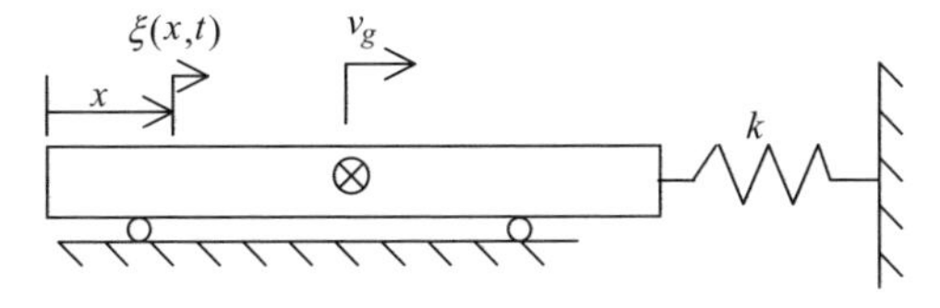

Rod of length, L
Cross-sectional area, A
Mass density, ρ
Young's modulus, E

Figure P9.1

$$\begin{aligned} \omega_0 &= 0 \\ Y_0 &= 1 \\ \omega_n^2 &= \frac{E}{\rho}\left(\frac{n\pi}{L}\right)^2 \\ Y_n(x) &= \cos n\pi\,\frac{x}{L}, \qquad n = 1, 2, 3, \ldots \end{aligned} \tag{P9.1-1}$$

The forced response of the rod is given by

$$\xi(x, t) = \sum_{0}^{n} Y_n(x)\eta_n(t) \tag{P9.1-2}$$

In Figure P9.1 the rod is moving to the right and has just struck the spring at the right end. Include the rigid-body mode and two dynamic modes and derive the equations of motion for $\eta_n(t)$. Put your result in first-order state equation format.

9.2 In this problem the rod from Problem 9.1 is interacting with a spring and mass at its right end (Figure P9.2). An external force is acting on the mass m. The force-free modes and frequencies of the rod are

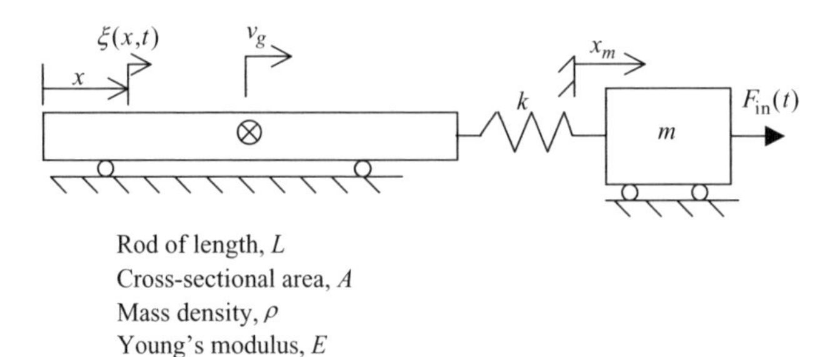

Figure P9.2

$$\begin{aligned}
\omega_0 &= 0 \\
Y_0 &= 1 \\
\omega_n^2 &= \frac{E}{\rho}\left(\frac{n\pi}{L}\right)^2 \\
Y_n(x) &= \cos n\pi \frac{x}{L}, \qquad n = 1, 2, 3, \ldots
\end{aligned} \tag{P9.2-1}$$

The forced response for the rod is

$$\xi(x, t) = \sum_0^n Y_n(x)\eta_n(t) \tag{P9.2-2}$$

Retaining the rigid body mode plus two dynamic modes, derive the equations of motion for the $\eta_n(t)$ and reduce to first-order state equations. Put your the matrix form

$$\dot{\mathbf{x}} = \mathbf{A}\mathbf{x} + \mathbf{b}F_{\text{in}} \tag{P9.2-3}$$

9.3 This problem expands on Problem 9.2, where the mass element at the right end of the rod has been replaced by a second rod that exhibits distributed dynamics (Figure P9.3). The rod on the left is now referred to as rod 1 and the rod on the right is rod 2. The displacement of a cross section of rod 2 is located by the position z. An external force acts on the right end of rod 2. The mode shapes and associated natural frequencies were derived for force-free boundary conditions and resulted in the following mode shapes and frequencies for rod 1,

$$\begin{aligned}
\omega_{1_0} &= 0 \\
Y_{1_0} &= 1 \\
\omega_{1_n}^2 &= \frac{E_1}{\rho_1}\left(\frac{n\pi}{L_1}\right)^2 \\
Y_{1_n}(x) &= \cos n\pi \frac{x}{L_1}, \qquad n = 1, 2, 3, \ldots
\end{aligned} \tag{P9.3-1}$$

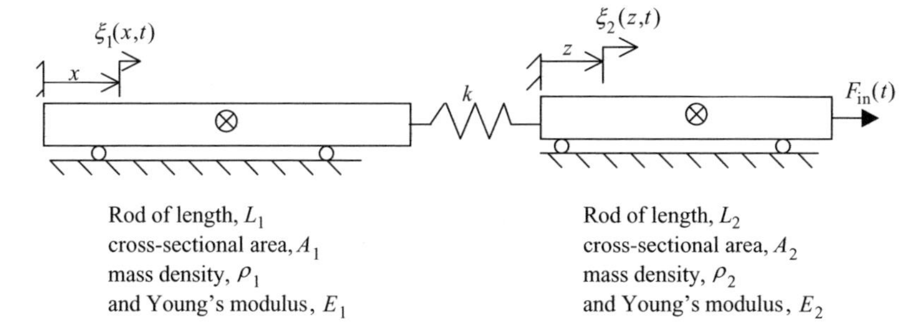

Figure P9.3

and the mode shapes and frequencies for rod 2,

$$\begin{aligned}
\omega_{2_0} &= 0 \\
Y_{2_0} &= 1 \\
\omega_{2_n}^2 &= \frac{E_2}{\rho_2}\left(\frac{n\pi}{L_2}\right)^2 \\
Y_{2_n}(z) &= \cos n\pi\,\frac{z}{L_2}, \qquad n = 1, 2, 3, \ldots
\end{aligned} \tag{P9.3-2}$$

The forced response of rod 1 is

$$\xi_1(x, t) = \sum_0^n Y_{1_n}(x)\eta_{1_n}(t) \tag{P9.3-3}$$

and the forced response for rod 2 is

$$\xi_2(z, t) = \sum_0^n Y_{2_n}(z)\eta_{2_n}(t) \tag{P9.3-4}$$

Retain the rigid body mode and two dynamic modes for each rod and derive the equations of motion for the $\eta_{1_n}(t)$ and $\eta_{2_n}(t)$. Explain how you would put the final equations into first-order form.

9.4 Figure P9.4 shows a tightly stretched cable with fixed boundary conditions

$$w(0, t) = w(L, t) = 0 \tag{P9.4-1}$$

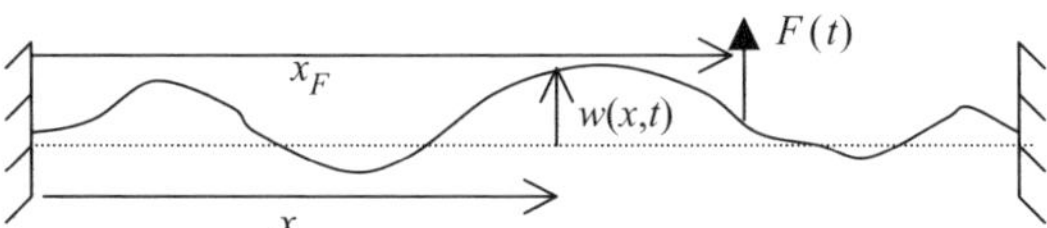

Figure P9.4

There is one external force acting at the location x_F. The governing equation of motion is

$$T\frac{\partial^2 w}{\partial x^2} + F\delta(x - x_F) = \rho\frac{\partial^2 w}{\partial t^2} \qquad \text{(P9.4-2)}$$

(a) Use separation of variables for the unforced response where it is assumed that

$$w(x, t) = Y(x)g(t) \qquad \text{(P9.4-3)}$$

and derive the mode shapes and associated natural frequencies for this system. Show that the mode shapes and frequencies are

$$\begin{aligned} \omega_n^2 &= \frac{T}{\rho}\left(\frac{n\pi}{L}\right)^2 \\ Y_n(x) &= \sin n\pi\,\frac{x}{L} \end{aligned} \qquad \text{(P9.4-4)}$$

(b) For the forced response, the solution is

$$w(x, t) = \sum_n Y_n(x)\eta_n(t) \qquad \text{(P9.4-5)}$$

Derive the equations of motion for the unknown time functions $\eta_n(t)$. Derive expressions for the modal mass and modal stiffness.

9.5 Figure P9.5 shows a tightly stretched cable that is fixed at the left end and free at the right end. An external force $F(t)$ is acting at the location x_F along the cable. The boundary conditions for this system are

$$w(0, t) = \frac{\partial}{\partial x}w(L, t) = 0 \qquad \text{(P9.5-1)}$$

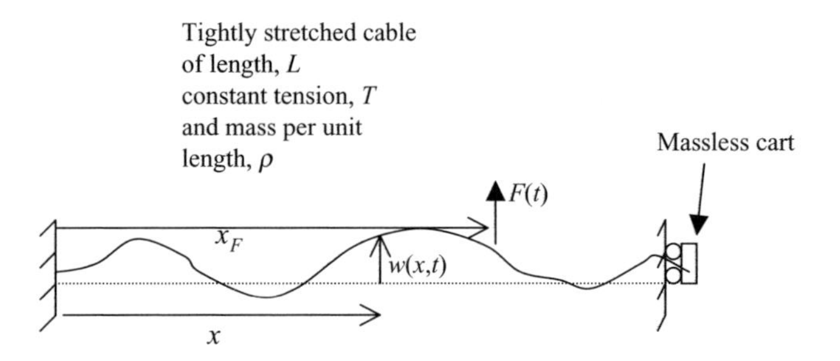

Figure P9.5

The governing equation of motion is

$$T\frac{\partial^2 w}{\partial x^2} + F\delta(x - x_F) = \rho\frac{\partial^2 w}{\partial t^2} \tag{P9.5-2}$$

(a) Use separation of variables for the unforced response where it is assumed that

$$w(x, t) = Y(x)g(t) \tag{P9.5-3}$$

and derive the mode shapes and associated natural frequencies for this system.

(b) For the forced response the solution is

$$w(x, t) = \sum_n Y_n(x)\eta_n(t) \tag{P9.5-4}$$

Derive the equations of motion for the unknown time functions $\eta_n(t)$. Derive expressions for the modal mass and modal stiffness.

9.6 Figure P9.6 shows a tightly stretched cable with fixed boundary conditions. At location x_m, a system consisting of a spring, damper, and mass is attached to the cable. The mass is located by the coordinate y_m measured from equilibrium. For fixed boundaries the mode shapes and frequencies are

$$\begin{aligned} \omega_n^2 &= \frac{T}{\rho}\left(\frac{n\pi}{L}\right)^2 \\ Y_n(x) &= \sin n\pi\,\frac{x}{L} \end{aligned} \tag{P9.6-1}$$

The forced response of the distributed part of the system is given by

$$w(x, t) = \sum_n Y_n(x)\eta_n(t) \tag{P9.6-2}$$

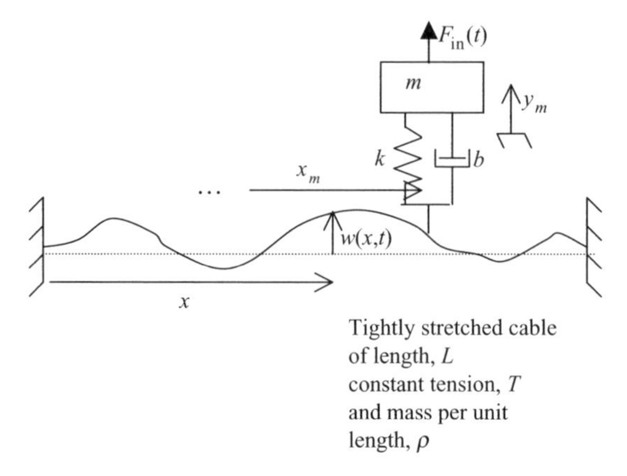

Figure P9.6

where $\eta_n(t)$ are unknown time functions. Derive the equations of motion for the unknown time functions as they interact with the attached system. Retain two modes for the cable and write down a complete first-order form representation of this system.

9.7 A tightly stretched cable has an external force $F(t)$ located at the position x_F and an additional point mass m located at the position x_m (Figure P9.7). The forced response of the cable is given by

$$w(x, t) = \sum_{n} Y_n(x)\eta_n(t) \tag{P9.7-1}$$

where $Y_n(x)$ are the mode shapes for the fixed–fixed cable and $\eta_n(t)$ are unknown time functions. Show the interaction force between the cable and the attached mass element and derive the equations of motion for the $\eta_n(t)$. Retain two modes for the cable and write a complete first-order representation for this system.

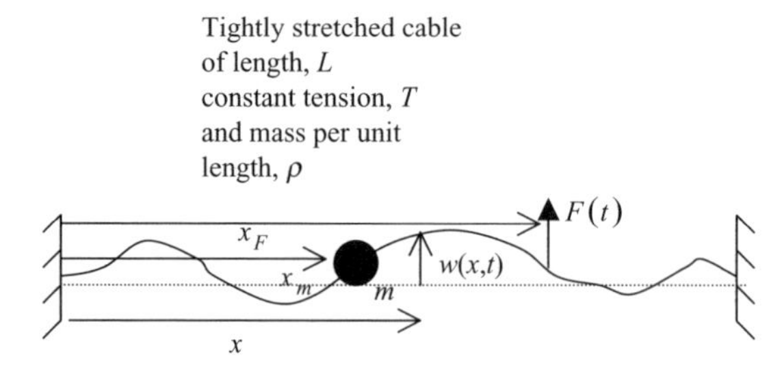

Figure P9.7

9.8 Figure P9.8 shows a two-spring two-mass lumped model of a distributed rod with a force acting on the free end. For the distributed rod with a fixed left end

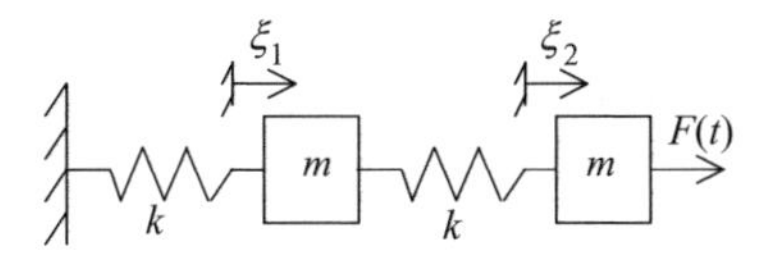

Figure P9.8

and a free right end, the mode shapes and frequencies are

$$\omega_n^2 = \frac{E}{\rho}\left[(2n-1)\frac{\pi}{2L}\right]^2 \qquad \text{(P9.8-1)}$$

$$Y_n(x) = \sin(2n-1)\frac{\pi}{2}\frac{x}{L}$$

It was shown in the chapter text that for a two-mass representation of the distributed rod, the mass elements would be

$$m = \frac{\rho AL}{2} = \frac{m_{\text{rod}}}{2} \qquad \text{(P9.8-2)}$$

and the stiffness elements would be reasonably represented by

$$k = 2\frac{EA}{L} \qquad \text{(P9.8-3)}$$

where E is Young's modulus and A is the rod cross-sectional area. For the lumped model here, derive the equations of motion, perform a vibration analysis, and derive expressions for the two natural frequencies of this lumped representation. Compare your results to the first two natural frequencies of the actual distributed rod. Any comments about the accuracy of the lumped modeling approach for distributed systems?

9.9 Shown in Figure P9.9 is a two-mass model of a tightly stretched cable with fixed boundary conditions at both ends. For the distributed version of this model, the natural frequencies and associated mode shapes are

$$\omega_n^2 = \frac{T}{\rho}\left(\frac{n\pi}{L}\right)^2 \qquad \text{(P9.9-1)}$$

$$Y_n(x) = \sin n\pi\frac{x}{L}$$

In Section 9.2 it was shown that reasonable parameters for a two-mass model of a distributed cable are

$$m = \frac{\rho L}{2}$$
$$k = 3\frac{T}{L} \qquad \text{(P9.9-2)}$$

where ρ is the mass per unit length of the cable, L is the length of the cable, and T is the constant tension. For the lumped model shown here, derive the equations of motion and perform a vibration analysis to determine the two natural frequencies for this system. Compare your results to the first two natural frequencies for the distributed version of this system, and comment on the accuracy of the finite lump model.

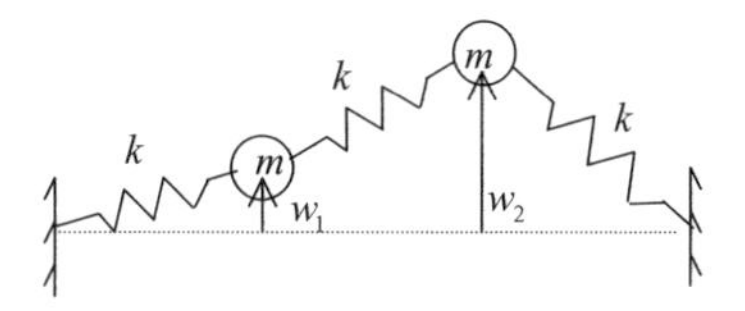

Figure P9.9

9.10 This problem will demonstrate whether the natural frequencies associated with a force-free boundary condition can converge to the frequencies associated with a fixed boundary condition if the free end is constrained not to move. Figure P9.10*a* below shows a tightly stretched cable with a fixed boundary at the left and a free boundary at the right. There is an external force $F(t)$ acting at the right end.

(a) For the fixed–free boundary conditions, propose a separated solution $w(x,t) = Y(x)g(t)$ and show that the natural frequencies and mode shapes are

$$\omega_n^2 = \frac{T}{\rho}\left[(2n-1)\frac{\pi}{2L}\right]^2$$
$$Y_n(x) = \sin(2n-1)\frac{\pi}{2}\frac{x}{L} \qquad \text{(P9.10-1)}$$

(b) The forced solution for the system has the form $w(x,t) = \sum_n Y_n(x)\eta_n(t)$. Use this in the forced equations of motion and derive the equations for the unknown time functions $\eta_n(t)$. Your result should be

$$m_n\ddot{\eta}_n + k_n\eta_n = FY_n(L) \qquad \text{(P9.10-2)}$$

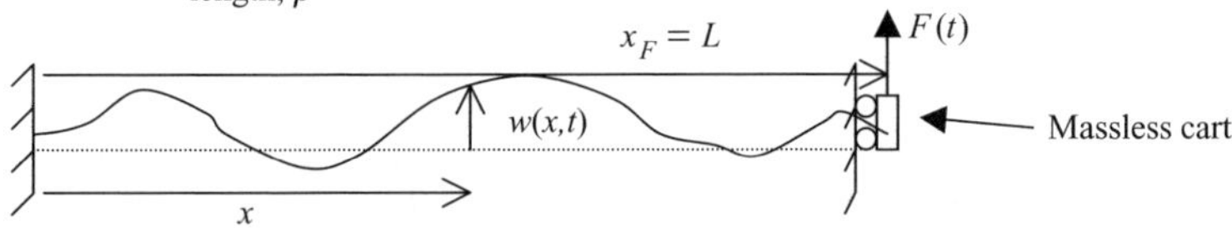

(a) Tightly stretched cable with an external force at the right end

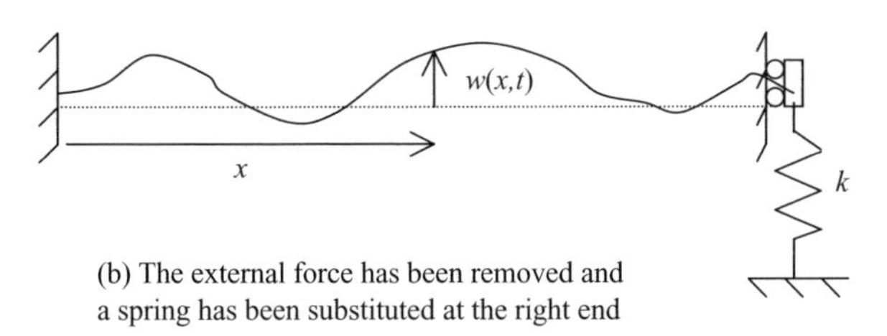

(b) The external force has been removed and a spring has been substituted at the right end

Figure P9.10

where

$$m_n = \int_0^L \rho Y_n^2 \, dx \quad \text{and} \quad k_n = m_n \omega_n^2 \tag{P9.10-3}$$

(c) If the spring k is substituted for the external force at the right end, the force F in Eq. (P9.10-2) would be

$$F = -kw(L, t) = -k \sum_n Y_n(L)\eta_n(t) \tag{P9.10-4}$$

where, remember, the mode shape functions are for the fixed–free cable. Retain two modes and write down the equations for $\eta_1(t)$ and $\eta_2(t)$. Perform a vibration analysis and derive the frequency equation for the system.

(d) Let $k \rightarrow \infty$ in the frequency equation and retain only those terms that contain k. Write down the remaining natural frequency in terms of the mode shapes and frequencies from Eqs. (P9.10-1). Substitute the appropriate values for the modes and frequencies and derive a final expression for the system in Figure P9.10*b*.

The natural frequencies for a fixed–fixed cable were requested in Problem 9.4. The natural frequencies are given by

$$\omega_n^2 = \frac{T}{\rho}\left(\frac{n\pi}{L}\right)^2 \qquad \text{(P9.10-5)}$$

and the first frequency is $\omega_1 = \sqrt{T/\rho}(\pi/L)$. Compare this to your result for the very stiff spring at the right end and comment on the possibility of modes associated with free boundaries converging to modes associated with fixed boundaries as the free boundary is more and more constrained.

9.11 Figure P9.11 shows a pinned Bernoulli–Euler beam with an attached spring–mass–damper system at position x_m. The coordinate z locates the mass element measured from equilibrium. This system is similar to that shown in Figure 9.13. The free response mode shapes and frequencies were derived in the chapter and are

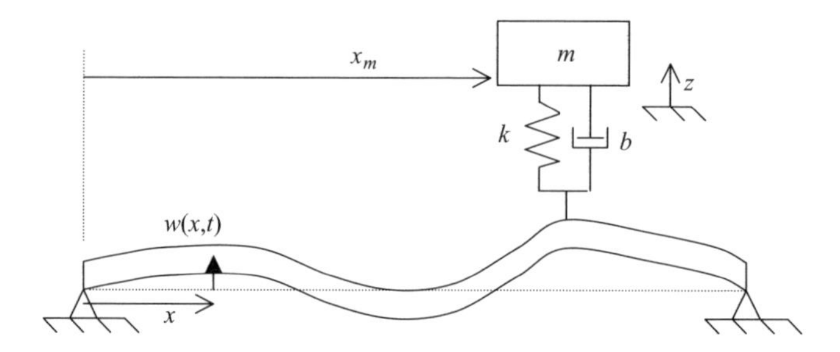

Bernoulli-Euler beam of mass density, ρ, Young's modulus, E, cross-sectional area, A, area moment of inertia, I

Figure P9.11

$$\omega_n^2 = \frac{EI}{\rho A}\left(\frac{n\pi}{L}\right)^4 \qquad \text{(P9.11-1)}$$

$$Y_n(x) = \sin n\pi\,\frac{x}{L} \qquad \text{(P9.11-2)}$$

If an external force was located at position x_m rather than the system shown, the forced response and the equations for the unknown time functions in the forced response would be

$$w(x,t) = \sum_n Y_n(x)\eta_n(t) \qquad \text{(P9.11-3)}$$

and

$$m_n \ddot{\eta}_n + k_n \eta_n = F(t) Y_n(x_m) \qquad \text{(P9.11-4)}$$

(a) Replace the external force with the system shown and derive the equations of motion for the attached system and an expression that will be substituted for the external force in Eq. (P9.11-4).

(b) Retain two modes in the beam representation and write down a complete set of equations for this system.

9.12 Figure P9.12 is the same as that for Problem 9.11 except now we allow the attached system to move at constant velocity U across the beam such that the location of the attached system changes with time. This system might be thought of as a simple vehicle model traversing a span of elevated roadway. In fact, such models are used to determine the vehicle–roadway interactions for high-speed trains traversing bridges.

For the pinned Bernoulli–Euler beam the mode shapes and frequencies are

$$\omega_n^2 = \frac{EI}{\rho A} \left(\frac{n\pi}{L} \right)^4 \qquad \text{(P9.12-1)}$$

and

$$Y_n(x) = \sin n\pi \frac{x}{L} \qquad \text{(P9.12-2)}$$

If the attached vehicle system was removed and an external force was in its place, the forced response would be

$$w(x, t) = \sum_n Y_n(x) \dot{\eta}_n(t) \qquad \text{(P9.12-3)}$$

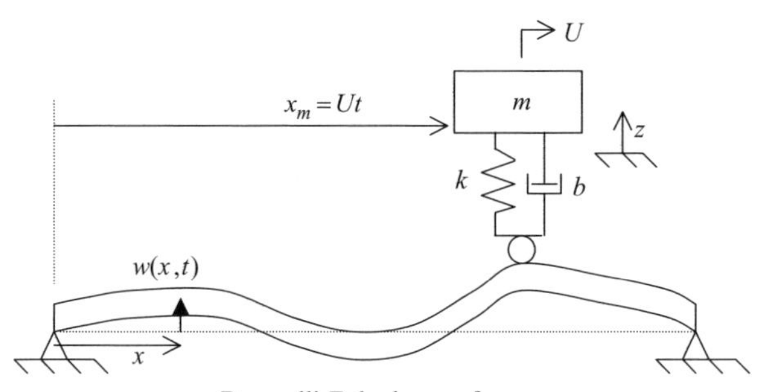

Figure P9.12

and the unknown time functions would come from the solution of

$$m_n \ddot{\eta}_n + k_n \eta_n = F(t) Y_n(x_m) \tag{P9.12-4}$$

(a) For the combined vehicle system and distributed beam, what is your best guess as to what would change in the equations of motion between a fixed vehicle location and a moving vehicle location?

If you guessed that you need only change the location where the mode shapes are evaluated to be varying with $x_m = Ut$, this would be an excellent guess. This would mean that the displacement at the base of the vehicle would be

$$w(x_m, t) = \sum_n Y_n(Ut) \eta_n(t) \tag{P9.12-5}$$

which would be needed as an input to the spring, and the velocity at the base of the vehicle would be

$$\frac{\partial}{\partial t} w(x_m, t) = \sum_n Y_n(Ut)\, pt \dot{\eta}_n(t) \tag{P9.12-6}$$

which is needed as an input to the damper.

Unfortunately, your guess is not quite complete. If a vehicle is traversing a fixed roadway with an uneven surface, there is still a vertical velocity at the base of the vehicle equal to the forward speed multiplied by the slope of the roadway under the wheel, and the integral of this vertical velocity is the displacement at the base of the vehicle. The slope of our "roadway" is given by

$$\frac{\partial}{\partial x} w(x, t) = \sum_n \frac{d}{dx} Y_n(x) \eta_n(t) \tag{P9.12-7}$$

where we simply need the derivatives of the already derived mode functions. Thus, the input velocity at the base of the vehicle $v_v(x, t)$ is given by

$$v_v(x_m, t) = \frac{\partial}{\partial t} w(x_m, t) + U \frac{\partial}{\partial x} w(x_m, t) \tag{P9.12-8}$$

where the first term on the right side accounts for the vertical velocity of the beam at the point of contact with the vehicle and the second term accounts for the "convected" component of vertical velocity at the point of contact of the vehicle. In terms of the mode shapes and unknown time functions,

$$v_v(x_m, t) = \sum_n Y_n(Ut)\dot{\eta}_n + U \sum_n \frac{dY_n(Ut)}{dx} \eta_n \tag{P9.12-9}$$

and the vertical displacement at the base of the vehicle is

$$q_v(x_m, t) = \int \left[\sum_n Y_n(Ut)pt\dot{\eta}_n + U \sum_n \frac{dY_n(Ut)}{dx} \eta_n \right] dt \tag{P9.12-10}$$

(b) Armed with this information, retain two modes for the beam and derive the equations of motion for the combined vehicle–beam system.

One more piece of information that might prove useful in the future is that to obtain accuracy in the slope calculation for a distributed system, more modes are needed than might be suggested by other considerations. Thus, if a slope calculation is required in a future model, it is suggested that a few additional modes be retained over the number chosen based on frequency considerations alone.

APPENDIX A

THREE-DIMENSIONAL RIGID-BODY MOTION IN A ROTATING COORDINATE SYSTEM

In many cases the description of rigid-body dynamics is best done in a frame moving with the body itself. An example is the study of vehicle dynamics, in which it is common to refer all forces, moments, and acceleration to a coordinate frame attached to the vehicle. This is a rotating and thus noninertial frame, so the equations of motion take on different forms from those usually presented in introductory dynamics textbooks. In their most general form, even the equations for a single rigid body are quite complex, as will be seen below, but often one does not need all possible degrees of freedom to study a particular phenomenon, and vehicles typically have symmetries that reduce the complexities of the equations considerably. In this appendix, the basic dynamic principles presented in Chapters 1 and 2 will be converted to a body-centered coordinate system and developed for the general three-dimensional case.

Figure 1 shows a coordinate frame attached to the center of mass and moving with the body. The body in question is represented as a vehicle since a body-centered coordinate system is often used in vehicle dynamics, but the equations apply to any rigid body. A standard notation is used for the velocity of the center of mass and angular velocity components, as indicated in Figure A.1. The x-axis points in the forward direction and the y-axis point to the right from the point of view of the pilot or driver in Figure A.1. For a right-handed system, the z-axis then points down. Some authors of vehicle dynamics papers prefer the z-axis to point up rather than down. In this case, the y-axis points to the left. It is important to use a right-handed coordinate system because the dynamic equations involve cross products, and if a left-handed coordinate system is used inadvertently, a number of terms will have incorrect signs.

The velocity of the center of mass, $\mathbf{v}$, and the angular velocity of the body, $\boldsymbol{\omega}$, can be expressed as follows:

$$\mathbf{v} = U\mathbf{i} + V\mathbf{j} + W\mathbf{k} \tag{A.1}$$

$$\boldsymbol{\omega} = p\mathbf{i} + q\mathbf{j} + r\mathbf{k} \tag{A.2}$$

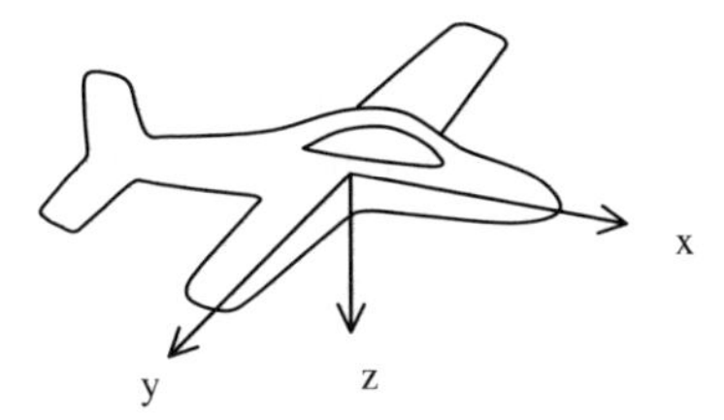

Figure A.1 Standard body-fixed coordinate system.

Alternatively, these quantities can be expressed as column vectors:

$$[v] = \begin{bmatrix} U \\ V \\ W \end{bmatrix} \tag{A.3}$$

$$[\omega] = \begin{bmatrix} p \\ q \\ r \end{bmatrix} \tag{A.4}$$

It may be worthwhile to note here that the notation for angular velocity components is not as easy to remember as one might think at first. Referring to Figure A.1 , it is clear that p, q, and r are angular rates around the x, y, and z axes, respectively. But the names given to these quantities do not correspond very well with their symbols: p is called the roll rate, q is called the pitch rate, and r is called the yaw rate. It might be nicer if the first letters of the names "pitch" and "roll" corresponded to p and r, but this is unfortunately not the case.

Another notable feature of the body-fixed coordinate system is that while velocities and accelerations can be expressed in the coordinate system, the position of the body is not represented at all. In contrast, when inertial coordinates are used, it is common to describe the position of a body first and then proceed to find expressions for velocities and accelerations. There are cases in which the position or motion path is not important for certain purposes. An example is for stability analysis [1].

The net force, **F**, on the body and the net moment or torque about the center of mass acting on the body, $\boldsymbol{\tau}$, are defined as vectors using the symbols X, Y, Z and L, M, N for components:

$$\mathbf{F} \equiv X\mathbf{i} + Y\mathbf{j} + Z\mathbf{k} \tag{A.5}$$

$$\boldsymbol{\tau} \equiv L\mathbf{i} + M\mathbf{j} + N\mathbf{k} \tag{A.6}$$

These general force and moment components can always be defined, of course, no matter how the forces and moments are determined, but in certain cases, these definitions are particularly useful. This is the case when the force components remain in the x–y–z directions as a vehicle moves. For example, it may be that the thrust force from an airplane propeller can be assumed always to act in the x-direction or that the lateral force on a rear tire of a car always acts in the y-direction. On the other hand, forces or moments that depend on the linear or rotational position of the body are harder to represent in the forms of Eqs. (A.5) and (A.6), because the body-centered coordinate system itself is incapable of representing the position of the body in inertial space.

The linear momentum, **P** is a vector. It is proportional to the velocity of the center of mass, with the scalar mass m being the proportionality constant:

$$\mathbf{P} = m\mathbf{v} \tag{A.7}$$

The angular momentum evaluated about the center of mass is also a vector, but it is best represented as a column vector related to the angular velocity:

$$[H_c] = [I_c][\omega] \tag{A.8}$$

The relationship between the angular momentum and the angular velocity in Eq. (A.8) involves the *centroidal inertia matrix*, which has nine components:

$$[I_c] = \begin{bmatrix} I_{xx} & I_{xy} & I_{xz} \\ I_{yx} & I_{yy} & I_{yz} \\ I_{zx} & I_{zy} & I_{zz} \end{bmatrix} \tag{A.9}$$

The components are called *moments of inertia* (along the diagonal of the matrix) and *products of inertia* (for the off-diagonal terms) and are defined by the well-known relationships

$$I_{xx} = \int (y^2 + z^2)\, dm, \quad I_{xy} = -\int (xy)\, dm, \quad \text{etc.} \tag{A.10}$$

Refer to any standard dynamics text, such as ref. [2].

The defining relationships in Eq. (A.10) imply that the matrix is symmetrical, so that only six of the nine components are independent. Because most vehicles are nearly symmetrical about certain planes, at least some of the products of inertia are commonly assumed to be zero. If it happens that the x–y–z axes are the principal axes for the body, all the off-diagonal products of inertia in Eq. (A.9) vanish. All rigid bodies have three orthogonal directions in the body that are the principal directions. If the x–y–z axes are aligned with the principal directions, then the axes become principal axes. When it is reasonable to assume that the x–y–z axes are principal axes, the equations of motion take on a highly simplified form.

Basic Dynamic Principles The basic dynamic principles for a rigid body can be stated in a form that is valid for any coordinate frame:

$$\mathbf{F} = \frac{d\mathbf{P}}{dt} = m\mathbf{a} \tag{A.11}$$

$$\boldsymbol{\tau} = \frac{d\mathbf{H}_c}{dt}, \quad [H_c] = [I_c][\omega] \tag{A.12}$$

Note that because Eq. (A.11) involves the scalar mass, it can be written as a simple vector equation. On the other hand, Eq. (A.12) is complicated by the tensor or matrix nature of $[I_c]$ in Eq. (A.9). As will be seen when Eqs. (A.11) and (A.12) are written out in component form, the dynamic equations for rotation in three dimensions are considerably more complicated than those for linear motion.

General Kinematic Considerations When the velocity components are defined with respect to a coordinate frame that is itself rotating, a standard technique must be used to find the absolute acceleration in terms of the acceleration with respect to the frame plus the effect of the frame rotation. This relation is typically presented in basic dynamics texts but is rarely applied to generate equations of motion in body-centered frames, as done here. Let an arbitrary vector **A** be expressed in a coordinate frame rotating with respect to an inertial frame $\mathbf{A} = A_x\mathbf{i} + A_y\mathbf{j} + A_z\mathbf{k}$, and let $\boldsymbol{\omega}$ be the angular velocity of the rotating frame $0xyz$. Then the derivative of the vector as it would be seen in an inertial frame can be related to the derivative in the rotating frame by the relation

$$\frac{d\mathbf{A}}{dt} = \left(\frac{\partial \mathbf{A}}{dt}\right)_{\text{rel}} + \boldsymbol{\omega} \times \mathbf{A} \tag{A.13}$$

The first term on the right of Eq. (A.13) is what an observer in the moving frame would think the change in the vector **A** would be. Expressed in terms of column vectors, the relative change in the vector **A** as it would appear in the rotating coordinate frame is simply

$$\left[\frac{\partial \mathbf{A}}{\partial t_{\text{rel}}}\right] = \begin{bmatrix} \dot{A}_x \\ \dot{A}_y \\ \dot{A}_z \end{bmatrix} \tag{A.14}$$

The second term on the right of Eq. (A.13) can be evaluated using the standard determinate form for a cross product and the definition of $\vec{\omega}$ from Eq. (A.4):

$$[\omega \times \mathbf{A}] = \text{Det}\begin{vmatrix} \mathbf{i} & \mathbf{j} & \mathbf{k} \\ p & q & r \\ A_x & A_y & A_z \end{vmatrix} = \begin{matrix} \mathbf{i}(qA_z - rA_y) \\ -\mathbf{j}(pA_z - rA_x) \\ +\mathbf{k}(pA_y - qA_x) \end{matrix} \tag{A.15}$$

The final result is best expressed in a column vector form:

$$\left[\frac{d\mathbf{A}}{dt}\right] = \begin{bmatrix} \dot{A}_x + qA_z - rA_y \\ \dot{A}_y + rA_x - pA_z \\ \dot{A}_z + pA_y - qA_x \end{bmatrix} \tag{A.16}$$

This result will now be used in the dynamic equations. First, Eq. (A.16) will be applied to the velocity of the center of mass to find the acceleration of the center of mass. Equation (A.11) becomes the following in the body-fixed system:

$$[F] = \begin{bmatrix} X \\ Y \\ Z \end{bmatrix} = m \begin{bmatrix} \dot{U} + qW - rV \\ \dot{V} + rU - pW \\ \dot{W} + pV - qU \end{bmatrix} \tag{A.17}$$

When Eq. (A.16) is applied to the angular momentum law, Eq. (A.12), the result is quite complex in general. First the angular momentum is expressed in column vector form using Eqs. (A.8) and (A.9):

$$[H_c] = \begin{bmatrix} H_x \\ H_y \\ H_z \end{bmatrix} = [I_c][\omega] = \begin{bmatrix} I_{xx}p + I_{xy}q + I_{xz}r \\ I_{yx}p + I_{yy}q + I_{yz}r \\ I_{zx}p + I_{zy}q + I_{zz}r \end{bmatrix} \tag{A.18}$$

Then Eq. (A.16) can be applied to the angular momentum column vector. The result is

$$\begin{aligned} [\tau] = \begin{bmatrix} L \\ M \\ N \end{bmatrix} &= \begin{bmatrix} I_{xx}\dot{p} + I_{xy}\dot{q} + I_{xz}\dot{r} \\ I_{yx}\dot{p} + I_{yy}\dot{q} + I_{yz}\dot{r} \\ I_{zx}\dot{p} + I_{yz}\dot{q} + I_{zz}\dot{r} \end{bmatrix} \\ &+ \begin{bmatrix} q(I_{zx}p + I_{zy}q + I_{zz}r) - r(I_{yx}p + I_{yy}q + I_{yz}r) \\ r(I_{xx}p + I_{xy}q + I_{xz}r) - p(I_{zx}p + I_{zy}q + I_{zz}r) \\ p(I_{yx}p + I_{yy}q + I_{yz}r) - q(I_{xx}p + I_{xy}q + I_{xz}r) \end{bmatrix} \end{aligned} \tag{A.19}$$

This is the general dynamic equation relating the moments about the x, y, and z axes to the rates of change of the angular velocity components. The contrast between Eqs. (A.17) and (A.19) is striking and is due to the differences between the otherwise analogous Eqs. (A.11) and (A.12).

The situation is often not quite as complicated as these general equations might suggest. For many vehicles, the x–y–z axes are lined up at least partially with the principal inertial axes of the vehicle. Recall that this means that if this is the case, some (or perhaps even all) of the products of inertia vanish, which simplifies Eqs. (A.18) and (A.19). For example, if the x–z plane is a plane of symmetry, $I_{xy} = I_{yx} = I_{yz} = I_{zy} = 0$. Then Eq. (A.18) becomes

$$\begin{bmatrix} H_x \\ H_y \\ H_z \end{bmatrix} = \begin{bmatrix} I_{xx}p + I_{xz}r \\ I_{yy}q \\ I_{zz}r + I_{zx}p \end{bmatrix} \tag{A.20}$$

and Eq. (A.19) simplifies to

$$\begin{bmatrix} L \\ M \\ N \end{bmatrix} = \begin{bmatrix} I_{xx}\dot{p} + I_{xz}\dot{r} \\ I_{yy}\dot{q} \\ I_{zx}\dot{p} + I_{zz}\dot{r} \end{bmatrix} + \begin{bmatrix} q(I_{xz}p + I_{zz}r) - r(I_{yy}q) \\ r(I_{xx}p + I_{zx}r) - p(I_{zx}p + I_{zz}r) \\ p(I_{yy}q) - q(I_{xx}p + I_{xz}r) \end{bmatrix} \tag{A.21}$$

It is also common that one does not always need to study complete three-dimensional motion. For land vehicles, for example, the heave velocity W and the pitch angular velocity q are sometimes ignored. Assuming the symmetry discussed above leading to Eqs. (A.20) and (A.21), and assuming further W and q are zero, a smaller number of fairly simple dynamic equations result when compared with the general case:

$$X = m(\dot{U} - rV) \tag{A.22}$$

$$Y = m(\dot{V} + rU) \tag{A.23}$$

$$L = I_{xx}\dot{p} + I_{xz}\dot{r} \tag{A.24}$$

$$N = I_{zz}\dot{r} + I_{zx}\dot{p} \tag{A.25}$$

Equations about as complex as Eqs. (A.22)–(A.25) are often used to study the stability of automobiles (by neglecting roll and heave motions) and the longitudinal stability of airplanes. Both cases involve plane motion rather than general three-dimensional motion.

REFERENCES

[1] D. Karnopp, *Vehicle Stability*, Marcel Dekker, New York, 2004.

[2] S. H. Crandall, D. C. Karnopp, E. F. Kurtz, Jr., and D. C. Pridmore-Brown, *Dynamics of Mechanical and Electromechanical Systems*, McGraw-Hill, New York, 1968.(Reprint edition, Krieger Publishing, Melbourne, FL.)

APPENDIX B

MOMENTS OF INERTIA FOR SOME COMMON SOLID BODY SHAPES

Figure B.1 shows a number of common solid body shapes and their moments of inertia assuming that they are made of a material with constant mass density. The notes for each case indicate how the results may be specialized for related shapes. For all the examples shown, the coordinate system is aligned with the principal axes of the bodies. The parallel axis theorem can be used to find the inertia properties for the bodies relative to a coordinate system centered at a point other than the point shown in the figure. In each of the formulas below, m is the mass of the body.

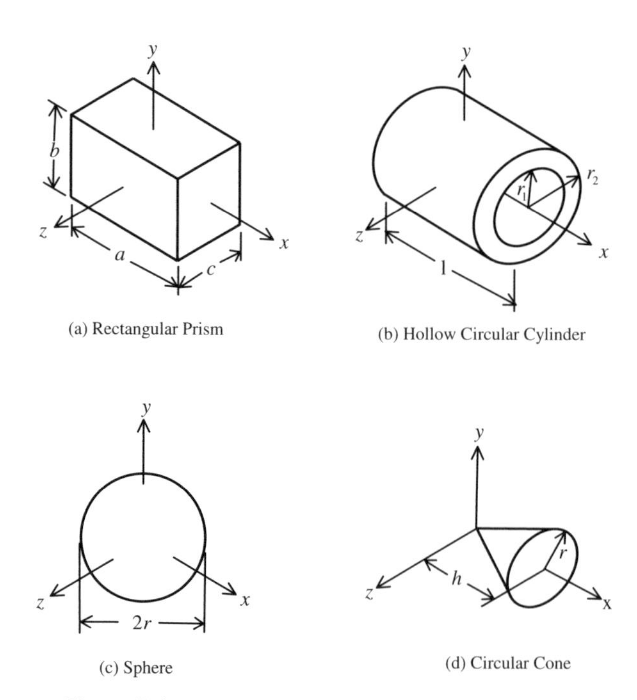

Figure B.1 Moments of inertia for some solid bodies.

Rectangular Prism

$$
\begin{aligned}
I_x &= \tfrac{1}{12}m(b^2 + c^2)\\
I_y &= \tfrac{1}{12}m(a^2 + b^2)\\
I_z &= \tfrac{1}{12}m(c^2 + a^2)
\end{aligned}
$$

Special Cases

- Thin rectangular plate in the x–y plane: set $c = 0$.
- Slender rod, length a in the x-direction: set $b = c = 0$.

Hollow Circular Cylinder

$$
\begin{aligned}
I_x &= \tfrac{1}{2}m(r_1^2 + r_2^2)\\
I_y = I_z &= \tfrac{1}{12}ml^2 + \tfrac{1}{4}m(r_1^2 + r_2^2)
\end{aligned}
$$

Special Cases

- Solid circular cylinder: set $r_1 = 0$.
- Slender rod: set $r_1 = r_2 = 0$.
- Thin circular disk, hollow: set $l = 0$.
- Thin circular disk, no hole: set $l = r_1 = 0$.
- Thin circular ring: set $l = 0,\ r_1 = r_2$.

Sphere

$$I_x = I_y = I_z = \tfrac{2}{5}mr^2$$

Circular Cone

$$
\begin{aligned}
I_x &= \tfrac{3}{10}mr^2\\
I_y = I_z &= \tfrac{3}{20}mr^2 + \tfrac{3}{5}mh^2
\end{aligned}
$$

Note: The center of mass of the cone is at $x = \frac{3}{4}h$. The moments of inertia of the cone evaluated for a coordinate system centered at the center of mass are

$$
\begin{aligned}
I_x &= \tfrac{3}{10}mr^2\\
I_y = I_z &= \tfrac{3}{20}mr^2 + \tfrac{3}{80}mh^2
\end{aligned}
$$

APPENDIX C

THE PARALLEL AXIS THEOREM

The parallel axis theorem tells how the inertia matrix evaluated at one point in a rigid body changes when the matrix is evaluated at another point with the coordinate axes parallel to the original axes. Although the theorem can be stated for two arbitrary points, it has a simple form if the first point is the center of mass. Figure C.1 shows how the $x'y'z'$ coordinate system centered at point B is shifted by amounts a, b, c from the coordinate system xyz with origin at the center of mass C. The relations between a point described in the two coordinate systems is

$$\begin{aligned} x' &= x + a \\ y' &= y + b \\ z' &= z + c \end{aligned}$$

Equivalently, one can say that the coordinates of the center of mass in the xyz system are $x = y = z = 0$ and in the $x'y'z'$ system are $x' = a,\ y' = b,\ z' = c$. The parallel

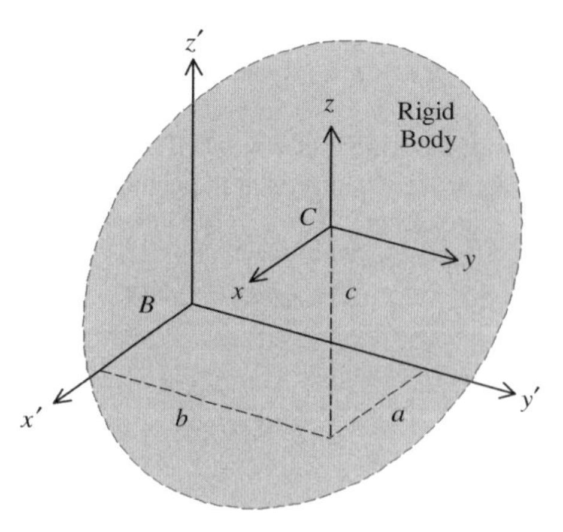

Figure C.1 A Coordinate system with origin at the center of mass of a rigid body C, and another parallel coordinate system with origin at an arbitrary point B.

axis theorem is then

$$[I_B] = [I_C] + M \begin{bmatrix} b^2 + c^2 & -ab & -ac \\ -ab & c^2 + a^2 & -bc \\ -ac & -bc & a^2 + b^2 \end{bmatrix}$$

where M is the mass of the body.

INDEX